INTRODUCTION TO POLYMER SCIENCE AND TECHNOLOGY: AN SPE TEXTBOOK

SPE MONOGRAPHS

Introduction to Polymer Science and Technology: An SPE Textbook

HERMAN S. KAUFMAN, Editor

Ramapo College of New Jersey
Mahwah, New Jersey

JOSEPH J. FALCETTA, Associate Editor

Bausch and Lomb Inc.
Rochester, New York

A Wiley-Interscience Publication

JOHN WILEY & SONS, New York · London · Sydney · Toronto

QD
381
I57

Library of Congress Cataloging in Publication Data

Main entry under title:

Introduction to polymer science and technology.

(SPE monographs; no. 2)
"A Wiley-Interscience publication."
Includes bibliographical references and index.
1. Polymers and polymerization. 2. Plastics.
I. Kaufman, Herman S., 1922– II. Falcetta, Joseph J. III. Series: Society of Plastics Engineers. SPE monographs; no. 2.
QD381.I57   547'.84   76-16838
ISBN 0–471–01493–1

Printed in the United States of America

10 9 8 7 6 5 4 3 2 1

SERIES PREFACE _____

The Society of Plastics Engineers is dedicated to the promotion of scientific and engineering knowledge of plastics and to the initiation and continuation of educational activities for the plastics industry.

An example of this dedication is the sponsorship of this and other technical books about plastics. These books are commissioned, directed, and reviewed by the Society's Technical Volumes Committee. Members of this committee are selected for their outstanding technical competence; among them are prominent authors, educators, and scientists in the plastics field.

In addition, the Society publishes *Plastics Engineering, Polymer Engineering and Science (PE & S)*, proceedings of its Annual, National and Regional Technical Conferences (*ANTEC, NATEC, RETEC*) and other selected publications. Additional information can be obtained by writing to the Society of Plastics Engineers, Inc., 656 West Putnam Avenue, Greenwich, Connecticut 06830.

William Frizelle,

Chairman, Technical Volumes Committee
Society of Plastics Engineers
St. Louis, Missouri

PREFACE ───────────────────────────

Several years ago one of us (HSK) began to give short courses entitled "Introduction to Polymer Science and Technology." These were usually offered under the auspices of the Society of Plastics Engineers at Annual Technical Conferences or at local section meetings. The courses were designed to provide the fundamentals of the preparation, characterization, and utilization of plastics to technical, production, sales, and management personnel who had limited scientific background but who had an interest in learning about polymers. None of the books available seemed suitable for this introductory type of course. Discussions with others in the field confirmed the need for such a book. This led to the development of an outline for a text. We were fortunate to obtain the cooperation of several outstanding teachers and participants in the plastics field in preparing this book, which is designed to fill the needs of short courses as well as of a regular undergraduate upper-level elective course. The material is presented in such a fashion that it can be useful to the person with relatively poor technical background as well as the educated scientist or engineer. Several of the chapters go into some technical detail for those who are interested, but skipping it will certainly not destroy the overall understanding of the basic principles, which is the primary objective of the book.

Because the plastics industry is such a large part of the technological economy, either directly or indirectly (packaging, textiles, etc.), it seems eminently desirable to encourage undergraduates to become familiar with the field.

The book is designed so that a teacher of chemistry or engineering can use it as a text for a one-semester course meant for junior or senior students. By the same token it should be useful as a basis for self-study for those interested in or involved with plastics. The references and the discussion questions and problems help to provide additional information and insight to the basic material.

Mahwah, New Jersey
Rochester, New York
July 1977

HERMAN S. KAUFMAN
JOSEPH J. FALCETTA

vii

CONTENTS —————————————————————

ix

CHAPTER 1

Introduction to Polymer Science

HERMAN F. MARK
Polytechnic Institute of New York
Brooklyn, New York

SHELDON ATLAS
Bronx Community College
Bronx, New York

A. HISTORICAL BACKGROUND

Natural polymers such as wood, cotton, wool, silk, lacquers, rubber, and many types of gums have been used for centuries in all kinds of practical applications. Their chemical composition and structure were unknown; improvements over the course of time were made mainly through breeding,

1

selection of the best raw materials, and advances in the mechanics of the manufacturing processes. A few very important though probably accidental chemical discoveries were made which affected the chemical composition and structure of certain natural polymers, such as the vulcanization of rubber, the mercerization of cotton, hemp, and flax, the tanning of leather, and the loading of silk. They led to significant technical results and were constantly improved in an empirical fashion.

Scientific work began (mostly in Europe) around 1880 on the chemical composition, structure, and morphology of cellulose, wool, silk, starch, and rubber; it was largely descriptive and qualitative but provided a very useful background for further quantitative approaches. During this period the first synthetic (or artificial) polymers were made, particularly condensation products of formaldehyde with phenol (bakelite), urea with proteins and derivatives of cellulose (nitrate and acetate); they acquired considerable commercial importance but their development was purely empirical and their properties were in most cases inferior to those of the corresponding natural products. Hence they were generally considered as substitutes or ersatz and were given such names as Kunstseide (artificial silk), Kunstleder (artificial leather), or Kunststoffe (artificial materials). Most of these developments took place in Europe, particularly in Germany, France, and England.

The groundwork for the organic chemistry of polymers or macromolecules was laid around 1905 in the Institute of Emil Fischer in Berlin. His work on sugars and amino acids clarified in a complete manner the composition, structure, and stereochemistry of these substances and opened the way to a stepwise synthesis of progressively larger and larger molecular species. Emil Fischer himself remained strictly in the domain of classical organic chemistry, of which he was the unsurpassed master, and reached only the lower limits of polymer chemistry (molecular weights between 1000 and 1500), but his co-workers pioneered in all fields of true polymer research: Freudenberg, Helferich, and Hess on polysaccharides; Leuchs and Bergmann on polypeptides; and Harries and Pammerer on polyhydrocarbons, particularly on rubber. At the same time Ostwald and Svedberg developed physicochemical methods for the investigation of colloidal systems by the measurement of diffusion, sedimentation, viscosity, and turbidity and laid the fundamentals for the quantitative study of polymer solutions. Finally, after the basic discovery of X-ray diffraction by von Laue, Bragg, Debye, and Scherrer showed the way to apply this method to the elucidation of the fine structure of the solid state even in microcrystalline materials; this eventually led to the X-ray investigation of the structure of such polymeric materials as cellulose, lignin, proteins, and rubber.

At the beginning of polymer science, which commenced with the systematic study of these natural organic compounds, the principal objective was to determine the chemical composition of these materials, to present structural formulas for them, and to understand their behavior on the basis of composition and structure.

The establishment of the chemical composition encountered difficulties because the above-named substances could not be purified according to the accepted methods of organic chemistry—crystallization and distillation—without losing some of their essential properties. New methods—adsorption and fractionation—had to be developed and less stringent definitions for purity has to be accepted. The presentation of a structural formula in the sense of classical organic chemistry was inhibited by the insolubility and infusibility of these natural products, which prevented the application of well established laws and methods of physical chemistry. It became necessary to amend the laws and to improve the methods in order to arrive at enough data to formulate intelligent conclusions. A correlation between structure and properties was not possible without knowing the structure and was made even more difficult because many important properties were found to change drastically without any apparent change of the basic structure. Again new research had to be initiated and pursued in order to arrive at a consistent picture.

In the mid-1920s (1925–1928) all these efforts led to the concept that the organic products that serve in nature as building materials all consist of very large molecules, which in most cases have the character of long flexible chains: they were called polymers or macromolecules. As soon as this principle was established, systematic efforts were directed at the synthesis of similar materials from well-known small and simple organic substances available in large quantities and at low cost from the industries of coal, oil, and natural gas and as by-products of agriculture and forestry. In rapid succession a considerable number of such synthetic polymers were prepared, which had very interesting applications as fibers, plastics, rubbers, coatings, or adhesives. Table 1 lists important polymers and the year of their commercial introduction. Their existence broadened the spectrum of materials available for basic structure and property research and greatly accelerated the accumulation of quantitative data on different types of macromolecules.

The combined application of all available methods, organic, physico-chemical, and physical, was systematically promoted and led to the synthesis of many new polymers, determination of their average molecular weight, and clarification of a number of basic facts such as the behavior of their solutions and the details of their structure in the solid state. Most results were still qualitative or semiquantitative but a wide field was opened for further research and numerous problems posed themselves on various topics. As the industrial application of different polymers began to develop, it became evident that many valuable properties could be obtained in great

Table 1 Some commercial polymers and approximate year of introduction

Date	Material	Typical applications
1868	Cellulose nitrate	Mirror frames
1909	Phenol–formaldehyde	Electrical insulators
1926	Alkyds	Electrical insulators
1927	Cellulose acetate	Packaging films
1927	Polyvinyl chloride	Flooring
1929	Urea–formaldehyde	Electrical switches and parts
1935	Ethyl cellulose	Moldings
1936	Polymethyl methacrylate	Display signs
1936	Polyvinyl acetate	Adhesives
1938	Cellulose acetate Butyrate	Sheets
1938	Polystyrene	Kitchenware, toys
1938	Polyamides (nylons)	Fibers, films
1939	Melamine–formaldelhyde	Tableware
1939	Polyvinylidene chloride	Films, paper coatings
1942	Polyesters (cross-linkable)	Boat hulls
1942	Polyethylene (low density)	Squeeze bottles, films
1943	Silicone	Rubber goods
1943	Fluoropolymers	Industrial gaskets, slip coatings
1943	Polyurethane	Foam goods
1947	Epoxies	Molds
1948	Acrylonitrile–butadiene–styrene copolymer	Radio cabinets, luggage
1955	Linear high-density polyethylene	Detergent bottles
1956	Acetal resin	Auto parts
1957	Polypropylene	Carpet fiber, moldings
1957	Polycarbonate	Appliance parts
1962	Phenoxy resin	Adhesives, coatings
1964	Ionomer resins	Moldings
1964	Polyphenylene oxide	High-temperature moldings
1965	Polyimides	High-temperature films and wire coatings
1965	Polybutene	Films
1965	Polysulfone	High-temperature thermoplastic
1965	Poly(4-methyl-1-pentene)	Clear moldings
1968	Phenylene ether sulfone	High-temperature films and moldings
1970	Ethylene–tetrafluoroethylene copolymer	Wire insulation
1970	Ethylene–chlorotrifluoroethylene copolymer	Wire insulation
1970	Moldable elastomers	Molded rubber goods
1971	Hydrogels (hydroxy acrylates)	Contact lenses
1972	Acrylonitrile copolymers	Soft drink bottles
1972	Moldable polyesters	Engineering thermoplastic
1974	Aromatic polyamides	High-strength tire cord

variety and at low cost. A few prominent names of this period are W. A. Carothers, K. H. Meyer, E. K. Rideal, H. Staudinger and H. Mark.

During the next decade the systematic synthesis of many new polymers continued in the field of fibers, plastics, and rubbers, particularly through the preparation of new monomers, the discovery of new catalytic systems, and the principle of copolymerization. There also were several developments which materially aided the maturation of the field; these include quantitative formulation of the kinetics of polymerization processes and understanding of their individual steps; development of the statistical thermodynamics and hydrodynamics of polymer solutions leading to a quantitative understanding of osmotic pressure, diffusion, sedimentation, and viscosity; and clarification of the solid-state structure of polymers, that is, theory of rubber elasticity, transition phenomena, crystalline–amorphous system, and relaxation behavior. Some prominent names of this stage are P. J. Flory, E. Guth, L. M. Huggins, W. Kuhn, W. H. Melville, and G. V. Schulz.

After 1940 successful extension, amplification, and refinement occurred in all directions. Specifically many new monomers, many new catalytic systems, and many new polymerization techniques were developed. Refinement of the knowledge of structure and properties came as a result of improved statistical treatments of macromolecules in solution and in the bulk state, development of new methods for the characterization of polymers such as light scattering, small-angle X-ray diffraction, polarized infrared adsorption spectroscopy, rotatory dispersion, nuclear magnetic resonance, differential thermal analysis, and sedimentation and diffusion in a density gradient cell. Clarification of the mechanism of polymerization under various conditions, such as in solution, suspension, or emulsion, at high pressures, at high or low temperatures, and in the form of living polymers, materially aided the manufacturing processes.

In general, polymeric materials consist of the elements, C, H, N, and O and are therefore conventionally classified as organic polymers. In some instances, however, other elements such as B, Si, P, S, F, and Cl are present in certain proportions and have a more or less important influence on the ultimate properties of the products. Together with the large family of metallic compounds and of ceramics the organic polymers represent the essential engineering materials in the construction of buildings, vehicles, engines, appliances, textiles, paper, rubber goods, and household articles of all kinds.

The rapid growth of these relatively new engineering materials in the recent past is due to several factors:

1. The basic raw materials for their production are readily available in large quantities and are, in general, inexpensive even though petroleum products have increased in price. Natural organic polymers such as cellulose

(paper, textiles), proteins (wool, silk, leather), starch (food, adhesives), and rubber (surgical gloves) are mainly available as products of farming and forestry activities, whereas the raw materials for the production of synthetic polymers come essentially from the industries of oil and coal. The simplest building units are called *monomers*, some of which, for example, ethylene, propylene, isobutylene, butadiene, and styrene, are by-products of the manufacture of gasoline and lube oils and are of relatively low cost (pennies per pound) and available in very large quantities (ethylene plants of 1 billion lb/yr capacity exist). Many others are simple derivatives of ethylene, benzene, formaldehyde, phenol, urea, and other basic organic chemicals; they range in cost up to about 50¢/lb and are also large-scale industrial products. Thus the basic building units for organic polymers, as defined above represent a large variety of compounds, readily available and of low or moderate cost.

2. Intensive research activities in many laboratories during the last 30 years have succeeded in elucidating the mechanism of the reactions with the aid of which long-chain molecules are formed from the above-mentioned basic units. They are called *polymerization* reactions and represent either typical *chain reactions* of highly exothermic character or *step reactions*, in the course of which chains of systematically repeating units are formed. During the same period systematic engineering efforts developed a number of relatively simple processes that permit the translation of polymerization and polycondensation reactions into large-scale industrial operations. Several types, such as polymerization in the gas phase at high pressures, in solution, suspension, emulsion, and even in the solid state are today well-developed unit processes that permit conversion of monomers into polymers rapidly, conveniently, and at low cost. In fact, the actual conversion cost, which for certain polymerization reactions can be as low as $0.05 to $0.10/lb, has greatly contributed to the rapid expansion of this field.

3. Thirty years ago there existed only few well-developed processes and machines to convert organic polymers from the manufactured state of a latex or a molding powder into the ultimate commercial products such as fibers, bristles, films, plates, rods, tubes, bottles, cups, combs, and other salable commodities. Today there exist many continuous automatic, rapid, and inexpensive methods for spinning, casting, blowing, injection and compression molding, stamping, and vacuum forming that make it possible for each polymer with attractive properties to be converted into useful consumer goods. Obviously the existence of many applications and the availability of standardized and automated methods to lead a new polymer into many different channels stimulates the synthesis and development of such new members of the organic polymer family.

4.   The large number of available monomers and the even larger number of polymers and copolymers made from them have provided us with an almost continuous spectrum of composition and structure of organic macromolecules. On the other hand, the systematic exploration of their mechanical, optical, electrical, and thermal behavior has provided an equally dense spectrum of characteristic practical properties and the study of their correspondence has led to a relatively profound and dependable understanding of structure–property relationships. This has the great advantage that new polymers or copolymers with desired and prescribed properties need not be developed by an empirical, more or less random system of synthetic efforts but can be designed on paper, and a successful elimination of many possibilities can be effected before laboratory work is actually started. This approach has been so successful in many instances that one can speak of a molecular engineering or "tailor-making" approach in the synthesis and development of new polymeric materials.

All fundamental and applied efforts of monomer synthesis, polymerization techniques, and manufacturing processes can eventually be condensed in a few guiding principles which represent, so to speak, the essence of our present understanding and knowhow. Such principles are, of course, only qualitative generalizations and must be supplemented in each individual case by quantitative considerations and numerical refinements, but they give a convenient and clarifying "helicopter view" of the present state of our knowledge and its practical application. As a consequence they are good working hypotheses or guideposts if they are used with caution and with the realization of their character as approximation and illustration.

In the following sections the most important principles of this type are enumerated and discussed briefly to provide a base for the detailed development that follows in later chapters.

B.   POLYMERIZATION

The word polymer is of Greek origin and means literally a molecule that consists of many (*poly*) parts (*meros*); the units that build up a polymer are called monomers (*monos* = one; *meros* = part). The process itself is known as polymerization; it leads from a small molecule in the molecular weight range between about 30 and 150 to a large molecule (giant molecule or macromolecule) in the range between 10,000 and 10 million. It has been found that in many cases the polymer molecules have the form of linear or slightly branched chains.

Although there are many different types of polymerization known, two of them are prevalent because of their simplicity and efficiency: polymerization by *addition* and polymerization by *condensation*.*

In the first case the individual monomers add to each other directly without any change in composition and form a long chain. For instance, ethylene, C_2H_4, is converted into polyethylene, $(C_2H_4)_n$, according to the scheme

$$CH_2{=}CH_2 \longrightarrow -CH_2-CH_2-CH_2-CH_2-CH_2-CH_2-$$

$$\text{(generally } n \geq 1000) \quad (1)$$

or ethylene oxide, C_2H_4O, is converted into polyethylene oxide, $(C_2H_4O)_n$ according to

$$H_2C\overbrace{}^{}CH_2 \longrightarrow -O-CH_2-CH_2-O-CH_2-CH_2-$$
$$\underset{O}{\diagdown\diagup}$$

$$\text{(where } n \geq 1000) \quad (2)$$

Processes of this type initiated by certain catalysts—free radical, cationic, and anionic—have the character of an exothermic chain reaction and are usually operated in the temperature range from -80 to $120°C$ to substantially quantitative yields of the polymer.

In the second case two or more monomers react with each other with elimination of a small part of them—usually water—to form a linear chain molecule. An important condition is that all reacting monomers carry two functional groups (bifunctionality) which can react with each other continuously. For instance, hexamethylenediamine, $H_2N{+}CH_2{\rightarrow}_6NH_2$, reacts with adipic acid, $HOOC{+}CH_2{\rightarrow}_4COOH$,

$$H_2N{+}CH_2{\rightarrow}_6NH-CO{+}CH_2{\rightarrow}_4COOH + H_2O \quad (3)$$

and this reaction continues, because the resulting intermediates are all bifunctional and keep reacting with more monomers or with each other. The final result are long linear chains consisting of building or repeat units of ${+}CO{+}CH_2{\rightarrow}_4CO-NH{+}CH_2{\rightarrow}_6NH{+}$. Polymers of this type are called polyamides because they contain amide groups, $-CO-NH-$; they are better known as nylons, which is a generic term for polyamides.

Another large family of polycondensation products are the polyesters; a particularly important member is the polymer obtained by the reaction of

---

* During recent years the terms chain reaction and step reaction have been frequently used to differentiate the two types of polymerizations. The use of these terms allows a more complete classification, especially of the more recently developed polymerization systems. Although it is not exactly correct to do so, throughout this book the terms chain reaction and addition, as well as step reaction and condensation, are used interchangeably.

terephthalic acid, $HOOC-C_6H_4-COOH$, and ethylene glycol, $HO-CH_2-CH_2-OH$, which consists of the repeating unit of

$$\left[CO-\bigcirc\!\!\!\bigcirc-CO-O-CH_2-CH_2-O \right]$$

All linear, or slightly branched, polymers—addition and condensation alike—are *thermoplastic.* This means that they can be repeatedly softened at elevated temperatures and solidified again by cooling; no chemical changes occur during this essentially physical process of softening and it is quite similar to the repeated melting and crystallization of metals.

If, in the course of polymerization process (either addition or condensation) nonlinear products are produced that do not form linear chains but rather three-dimensional networks, the materials are usually hard, infusible bodies that cannot be softened again without chemical decomposition. Materials of this kind are called *thermosetting*; the irreversible hardening is brought about by a chemical reaction called cross-linking and is not a purely physical process.

An important addition reaction of this type is the polymerization of divinylbenzene,

$$CH_2=CH-\bigcirc\!\!\!\bigcirc-CH=CH_2$$

where two available double bonds lead to the formation of a three-dimensional network that cannot be softened by a physical process but, eventually, only by a chemical decomposition. Three-dimensional network polymers (such as phenol–formaldehyde) can also be prepared by condensation polymerization (Chapter 2).

This leads to the simple and logical classification that most linear or branched polymers are thermoplastic, whereas highly cross-linked networks are thermosetting. The structure of the individual units and their functionality permit a ready prediction as to which type of plastic will be obtained upon polymerization of a given monomeric composition.

Before leaving the principle of polymerization, we must draw attention to the concept of copolymerization. In the previous examples given (e.g., ethylene or ethylene oxide polymerization), it was assumed that the polymer was built from a single monomeric species. The product is called a *homopolymer* in contradistinction to the polymer made by chemical combination of *two* or *more* different monomeric species, which is called a *copolymer*.

By suitable polymerization techniques several different types of co-polymers may be prepared:

Random Copolymer

$$-A-A-B-A-B-B-B-A-A-B-A-A-A-$$

This system is characterized by random placement of the respective monomer units along the polymer chain.

Alternating Copolymer

$$-A-B-A-B-A-B-A-B-A-B-$$

As the name implies, the monomer units react with a high degree of order, forming a polymer chain having a high degree of alternation.

Block Copolymer

$$-A-A-A-A-A-A-A-B-B-B-B-B-B-$$

This type of copolymer involves extended lengths or blocks of a given species, chemically bonded to extended lengths of another species.

Graft Copolymers

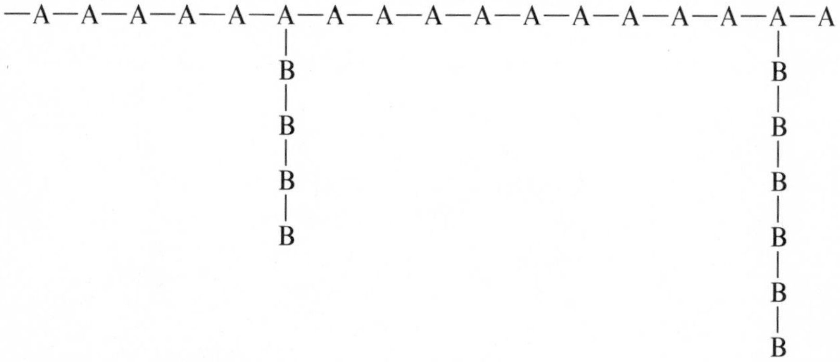

$$-A-A-A-A-A-A-A-A-A-A-A-A-A-A-A-A-A$$

$$
\begin{array}{ccc}
& | & | \\
& B & B \\
& | & | \\
& B & B \\
& | & | \\
& B & B \\
& | & | \\
& B & B \\
& & | \\
& & B \\
& & | \\
& & B
\end{array}
$$

These copolymers involve a primary backbone chain that has attached to it at several places other, different polymeric chains.

In all the above cases the properties of the copolymers are considerably different from those of either of the pure homopolymers. The reasons for this are discussed later. In any case this principle of copolymerization greatly broadens the possibility of "tailor-making" polymers with specific properties.

C. MOLECULAR WEIGHT

A factor of great importance in the synthesis and application of organic polymers is their *molecular weight*. The terms macromolecule, giant molecule, high polymer, and polymer already indicate that the molecules of this class of compounds are large and hence consist of many parts. In fact, all existing experience indicates that valuable and interesting properties of natural and synthetic polymers can be obtained only if their molecular weight (MW) is high, which in effect means long chains and greater opportunity for molecular entanglements and interactions. Since many important materials consist of chain molecules with repeating units, the concept of the degree of polymerization (DP) can be introduced; it represents the number of basic units in a give macromolecule. In general, polymer molecular weight (MW) and degree of polymerization are used interchangeably in the sense that they are related by the equation:

$$MW = DP \times (MW)_u$$

where $(MW)_u$ is the molecular weight of the repeating unit or monomer. For example, a polymer of polyethylene consisting of 1000 ethylene units has a DP of 1000 and a molecular weight of 28,000; that is,

$$1000 \ (DP) \times 28 \ (MW \ of \ CH_2{=}CH_2)$$

It has also been established by many contributors throughout the last 30 years that polymeric materials do not consist of strictly identical molecules but always represent a mixture of many species, each of which has a different molecular weight or DP. The characterization of these products is given by a molecular weight distribution function, which can be virtually expressed by a curve in which the frequency or percentage of each species is plotted against the molecular weight or the DP or this particular species. The narrower the distribution curve of a given polymer, the more homogeneous is the material. As a consequence of this molecular weight distribution of all polymeric compounds, we can not simply speak of a molecular weight or of a degree of polymerization, but must operate with an *average* molecular weight or degree of polymerization. (Molecular weight and molecular weight distributions are discussed more fully in Chapter 4.)

Many tests have established the fact that all important mechanical properties, particularly tensile strength, elongation to break, impact strength, and reversible elasticity of polymers, depend in a very definite way on the average molecular weight or DP in the sense that up to a certain, relatively low DP value no strength at all is developed. From then on, there is a steep rise of mechanical performance with DP until, at still larger molecular weights, the curve flattens out and there begins a domain of diminishing return of strength of further DP increments. Figure 1 shows the characteristic shape of this curve, which is typical for all polymers and differs for each individual material only in the numerical details. Each polymer has a critical DP value, DP_c, above which it develops toughness, but the numerical values are different; polyamides start to develop strength at DP's around 40, whereas cellulose requires 60 and many vinyl polymers need 100. The break

K of the curve is at a DP of 150 for polyamides, 250 for cellulose, and 400 for many vinyls. However, all polymers exhibit no strength below DP's of 30 and approach limiting strength above DP's of 600. Even if there are still small gains in the high molecular weight range of the curve they are difficult to attain practically because very long chains have high viscosities in the dissolved and molten state and become increasingly difficult and impractical to process. The characteristic shape of the curve in Fig. 1 has the consequence that a large number of practically useful polymers fall in the DP range of 200 to 2000, which in general corresponds to molecular weights from 20,000 to 200,000.

Evidently the curve in Fig. 1 is of great importance for anybody who wants to prepare a new and useful polymer. If at a certain point of his research he finds that the molecular weight of his present samples is around 5000, his attention and efforts will be mainly concentrated on changing the polymerization conditions in order to penetrate into a higher molecular weight range. If, on the other hand, he establishes that his present samples have molecular weight around 60,000 he will focus his interests on other factors such as molecular weight distribution, branching, influence of reactive groups, or stereoregulation (structure).

Flory and others have shown that a critical DP can be rationalized by considering that the bonds along each individual chain are much stronger

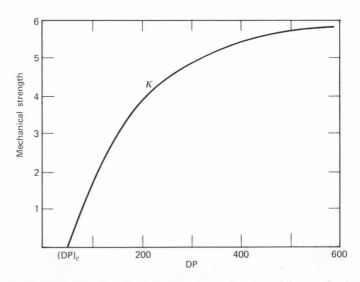

Figure 1   Idealized plot of mechanical strength as a function of degree of polymerization (DP).

(chemical bonds) than those between chains (van der Waals bonds), with the consequence that short chains slip very easily along each other and offer little or no stress-transferring strength.

On the other hand, as the chains become very long molecular entanglements increase and the accumulated resistance of many van der Waals bonds against slippage becomes great. It is clear that from this point on a further lengthening of the chains has little influence on physical properties.

Thus far we have discussed the influence of only the length of the chain molecules; let us now pass to the consideration of other chain properties which are also important for the mechanical and thermal behavior of polymeric systems.

D. CRYSTALLIZATION TENDENCY

One of the first important notions about the basic properties of chain molecules was that some of them exhibit a tendency to form crystalline materials, characterized by three-dimensional order. In using the concept of crystallinity in connection with polymers, it must be understood that in a mass of twisted and entangled polymer chains only certain volume elements comprised of chain segments of the system have reached a state of three-dimensional order that approaches in certain respects the order of a conventional crystal of materials such as sugar, naphthalene, or stearic acid. The three-dimensional order may be either the fringed micelle type (chains packed as a sheaf of wheat) or the chain folded type. The latter is most common in high polymers in the form of plastics. The crystalline domains in a polymeric material do not have the regular shape of normal crystals; they are much smaller in size, contain many more imperfections, and are connected with the disordered, amorphous regions by through-going polymer chains (which are responsible for polymer toughness), so that there are by no means any sharp boundaries between the ordered (crystalline) and disordered (amorphous) parts of the system (Fig. 2). On the other hand, crystalline polymers have high melting points, high densities, and high moduli of rigidity (they are stiff), resist dissolution and swelling, and are virtually impenetrable to diffusion of small molecules. They give sharp X-ray and electron diffraction patterns and show characteristic absorption peak splitting in the infrared. In some instances (cellulose, polyvinyl alcohol, and certain proteins) there is a reasonably sharp dividing line between a crystalline and amorphous portion of the system; in other cases (linear polyethylene, polyesters, and polyacetals) it appears to be more appropriate to consider the entire system as crystalline with the understanding that flaws and imperfections (twisted or folded chain segments) are more or less uniformly spread out over its

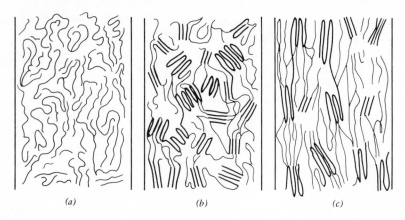

(a)   (b)   (c)

Figure 2 Schematic representation of (*a*) amorphous, (*b*) crystalline (showing chain folding), and (*c*) oriented crystalline polymers.

length and width and are responsible for the deviation of its behavior from that of a perfect crystal (Chapter 5).

Whatever the best mental picture for incomplete three-dimensional order in any special system may be, there can be no doubt that the tendency to crystallize plays a very important role in the thermal and mechanical behavior of polymers, and it will be useful to connect this tendency with the chemical composition and the structural details of the individual polymers. There are, of course, many contributing factors to such a complicated process as three-dimensional order; in this chapter it is appropriate to discuss only the most important of them, as follows:

1. Structural regularity of the chains, which can readily lead to the establishment of identity periods.

2. Free vibrational and rotational motions in the chains so that different conformations can be assumed.

3. Presence of specific groups which produce strong lateral intermolecular bonds (van der Waals or hydrogen bonds) and regular, periodic arrangement of such groups.

4. Absence of bulky, irregularly spaced substituents which inhibit the chain segments from fitting into a crystal lattice and prevent the laterally bonding groups from approaching each other to the distance of best interaction.

Before we undertake further discussion of these factors let us briefly discuss two important parameters which are very much involved with interchain effects. These are the glass-transition temperature, T_g, and the melt temperature, T_m.

Since the chains, on the average, contain several hundred monomeric units, their molecular motion is located in certain segments which vibrate, rotate, and wiggle independently of each other. At low temperatures (below the T_g) this segment motion is virtually absent or so slow that a noncrystalline polymer often has the character of a glass, which means that it is amorphous, hard, and brittle. If the material contains a certain percentage of crystalline domains, they act as reinforcing elements embedded in an amorphous matrix, and the sample is hard, tough, high melting, and difficultly soluble. If a polymer in this state is heated, the motion of the segments is accelerated and at a certain temperature becomes so vigorous that the material is transformed from a hard and brittle glass into a softer, more plastic state. This temperature is called the brittle point or glass point, T_g. In the case of a partially crystalline material this softening occurs only in the amorphous domains; the crystalline areas still remain unaffected and such a sample becomes more flexible and supple above T_g but still remains reinforced by its crystalline domains. If heating is continued eventually a temperature is reached at which the crystalline areas melt; it is known as T_m and represents the melting point of the crystalline phase. Above it the system behaves like a melt which has the character of a very viscous liquid.

All thermal and mechanical properties of polymers are therefore influenced by two temperatures, the glass point, T_g, of the amorphous phase and the melting point, T_m, of the crystalline phase. Table 2 contains these two crucial temperatures for a number of commercially important plastics. Let us now analyze these factors and estimate how much each of them contributes to the crystallizability of a given material.

1. Structural Regularity

The simplest polymer molecules are linear polyethylene and polyformaldehyde; their chains can readily assume a planar zigzag conformation characterized by a sequence of *trans* bonds and can therefore produce a very short identity period along the length of the chain (Chapter 5, Fig. 3). This evidently favors the establishment of lateral order, particularly if the macromolecules are oriented by stress or shear; in fact, both polymers are easily orientable and can attain very high degrees of crystallinity. The polyethylene chains are nonpolar and all intermolecular attraction is due to dispersion forces; the rotation around the C–C bond is inhibited by an energy barrier of about 2.7 k-cal/mole of bonds. This limited flexibility and the dispersion forces between adjacent chains are responsible for the high melting point, high rigidity, and low solubility of this material. In the

Table 2   Glass point and melting point for several important plastics materials[a]

Polymer	$T_g(°C)$	$T_m(°C)$
Polyethylene (high density, high crystallinity)	−125	135
Polyethylene (low density, low crystallinity)	−125	110
Polypropylene	−10	176
Polystyrene	100	240
Polyvinyl chloride	81	273
Polyvinylidene chloride	−20	190
Polymethyl methacrylate	105	200
Polyacrylonitrile	105	317
Nylon-6	50	223
Nylon-6,6	57	265
Polyethylene terephthalate	69	265
Polyformaldehyde	−85	181
Polyethylene oxide	−67	70
Cellulose triacetate	127	306
Polytetrafluoroethylene	−113	327

[a] In general it may be noted that the T_g increases with chain stiffness and T_m increases with intermolecular forces and degree of perfection of the crystallization. (The thermal transitions of polymers are discussed in detail in Chapter 6.)

case of polyformaldehyde the rotation about the C–O bond is less inhibited than that about the C–C bond, but the dipole character of the

$$\underset{C \qquad C}{\overset{O}{\diagup \diagdown}}$$

group produces polar forces between adjacent chains, which act over a longer range and are stronger than the dispersion forces. As a consequence polyformaldehyde has a higher rigidity and a higher melting point than polyethylene; it is also more difficultly soluble.

If one substituent is attached to a vinyl-type chain, there exist two simple ways to establish regular geometric placement of the substituents along the length of the chain, the isotactic placement in which all substituents (or at least a long row of them) have the same configurational position (either D or L) and the syndiotactic placement in which there exists a regular alternation between D and L over the entire molecule or at least over longer stretches of it. Any deviation from these two cases or any mixture of them is called atactic and refers to more or less random geometric positions of the substituents along the length of the chain (Chapter 2, Fig. 8). Natta, who first succeeded in preparing pure (or almost pure) representatives of these cases with polypropylene and other ethylenic polymers, has demonstrated that the stereoregular or stereospecific species are rigid, crystallizable, high melting, and difficultly soluble materials, whereas the atactic or irregular species are comparatively soft, low melting, and easily soluble polymers which do not crystallize under any conditions. The spectacular influence of this structural regularity on crystallization and on most mechanical and thermal properties has been established by many investigators for numerous vinyl, acrylic, and allylic polymers and has become an important principle in designing and "tailoring" new polymers, particularly since it is possible to aim at a definite prevalence of one of the structures by the appropriate choice of experimental conditions such as catalyst, solvent, temperature, and additives. (Stereochemistry of polymers is discussed further in Chapter 2.)

Another equally important influence of structural regularity on ultimate properties has been found to exist in the polymerization of conjugated dienes. If a butadiene molecule becomes the unit of a long chain the following three structures are possible:

$$1,4\text{-}cis \qquad \diagdown \diagup \overset{\displaystyle HC = CH}{} \diagdown \diagup \qquad \qquad (I)$$
$$CH_2 \qquad\quad CH_2$$

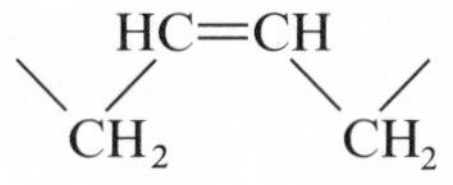

$$1,4\text{-}trans \qquad \overset{\displaystyle CH_2}{\diagup \diagdown} \qquad (II)$$
$$HC = CH$$
$$\diagdown \diagup$$
$$CH^2$$

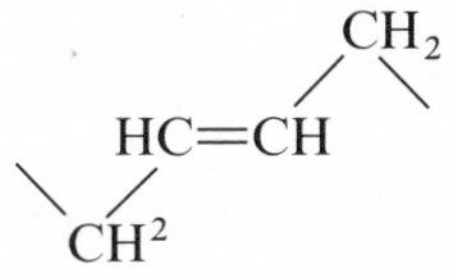

$$1,2 \qquad \overset{\diagdown}{\underset{|}{HC} - CH} \diagdown \qquad (III)$$
$$CH$$
$$\diagup\!\diagup$$
$$H_2C$$

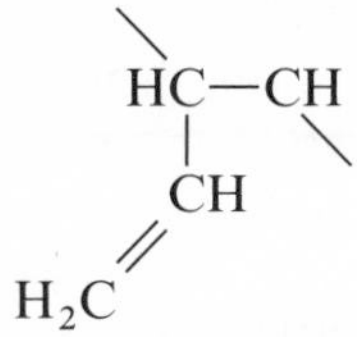

Through the use of appropriate initiators, solvents, temperatures, and additives it has been possible to synthesize each of these structures in almost pure form. Even though the same chemical monomer is used the three different structural arrangements shown above produce noticeably different behavior.

Form I is a soft, easily soluble elastomer with a T_g of $-108°C$ and a high retractive force; it crystallizes on stretching more than 200% and the crystalline phase has a melting point around 3°C (hevea rubber belongs to this species if isoprene is the monomer instead of butadiene).

Species II is a hard, difficultly soluble polymer which crystallizes readily without elongation and has a melting point around 70°C and a T_g of -83°C (balata and gutta-percha belong to this species if isoprene is the monomer instead of butadiene).

Species III exists in stereoregular forms (isotactic, syndiotactic) and in an irregular atactic form, all of which have been prepared by Natta and his collaborators with the aid of Ziegler catalysts (Chapter 2). The stereo-regular forms are rigid, crystalline, and difficultly soluble materials; the atactic species is a soft elastomer with slow and sluggish recovery character-istics. Many other vinyl and acrylic polymers have been obtained in atactic and stereoregular forms, and in all cases there is a pronounced influence of the structural character on the ultimate properties in the sense that regularity favors crystallizability and with it rigidity, high melting points, and resistance against dissolution.

Extensive work on copolymers of all kinds fully confirms the importance of structural regularity on crystallization tendency and consequently on properties. Those copolymers that are built by a regular alternation of the two components A and B show a distinct tendency to crystallize, whereas the corresponding one-to-one copolymers with random geometric distri-bution of the two components are intrinsically amorphous and represent nonrigid, soluble, and low softening resinous materials.

2. Chain Flexibility

The term flexibility in the context of linear macromolecules refers to the activation energies required to initiate vibrational and rotational motions as a consequence of which different conformations (or spatial arrangements) of the chain can be assumed at moderate temperatures in relatively short times. The energy barriers that separate different individual conformations in organic molecules have been assiduously studied with many ordinary, low molecular weight organic compounds with the aid of the specific heat, the infrared absorption spectrum, and magnetic resonance methods. The results have led to a rather complete knowledge of the stability of different conforma-tional isomers and the rate with which the equilibrium between them is established. (Polymer conformations are discussed in more detail in Chapters 4 and 5.) This information has been applied, with the necessary caution, to macromolecules and has led to the following general conclusions.

1. Linear polymers that contain only or mainly single bonds between C and C, C and O, or C and N allow rapid conformational changes (owing to ease of rotation around the single bonds); if they are regular and/or if there exist considerable intermolecular forces the materials are crystallizable,

relatively high melting, rigid, and difficultly soluble; however, if they are irregularly built they are amorphous, soft, and rubbery materials.

2. Ether and imine bonds or double bonds in the *cis* form reduce the energy barrier for rotation of the adjacent bonds and "soften" the chain in the sense that the polymers are less rigid, more rubbery, and more readily soluble than the corresponding chain of consecutive carbon–carbon bonds. This is particularly true if the "plasticizing" bonds are irregularly distributed along the length of the chains so that they do not favor but rather inhibit crystallization.

3. Cyclic structures in the backbone of the chain can very drastically inhibit conformational changes. *para*-Polyphenylene, for instance,

is a chain that cannot fold even at rather high temperatures because rotation about the carbon–carbon single bond between the *para*-combined phenylene rings can lead only to different angles between the planes of consecutive rings but not to a kink or bend in the main chain. In fact, representatives of this species are very rigid, very high melting, with a pronounced tendency to crystallize, and highly insoluble. This combination of valuable properties has not yet been fully brought to fruition because the presently known polyphenyls are in the relatively low molecular weight range.

para-Polyphenylene oxide and polyphenylene sulfide are other cases showing the chain-stiffening effect of a *para*-phenylene unit. Rotation

about the bonds between an ether oxygen atom and adjacent carbon atoms normally leads to bends and kinks in the chain, but this rotation is noticeably inhibited by the presence of the aromatic rings on each side of the oxygen or sulfur atom. As a consequence it has been found that chains of this type also represent high-melting, rigid, and difficultly soluble materials with high T_g values.

A well-known and famous example of the beneficial effect of *para*-phenylene units in a linear macromolecule is polyethylene glycol terephthalate, which has unusually high rigidity and melting point if compared with polyesters consisting of aliphatic acids. Chain stiffening can even be more pronounced if condensed ring systems are used such as in the polyimides

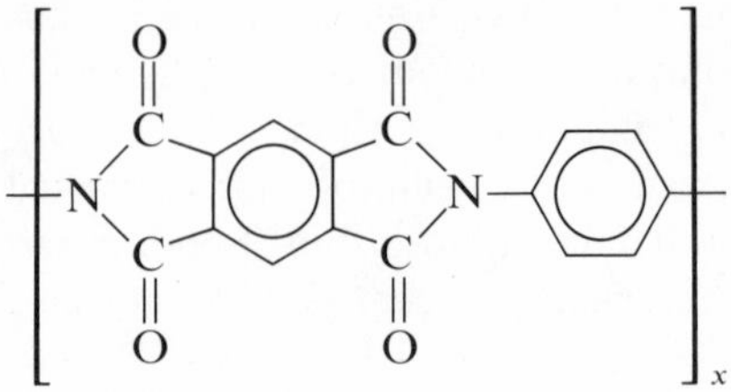

These materials are characterized by their hardness, high melting point, very high T_g, insolubility, and heat resistance because the chains are strongly stabilized by resonance between the aromatic ring systems.

Another interesting way to arrive at chains made up of condensed rings is the synthesis of so-called ladder polymers. The first case of such a structure was prepared by exposing polyacrylonitrile to elevated temperatures which causes the formation of rows of six rings by an electron pair displacement:

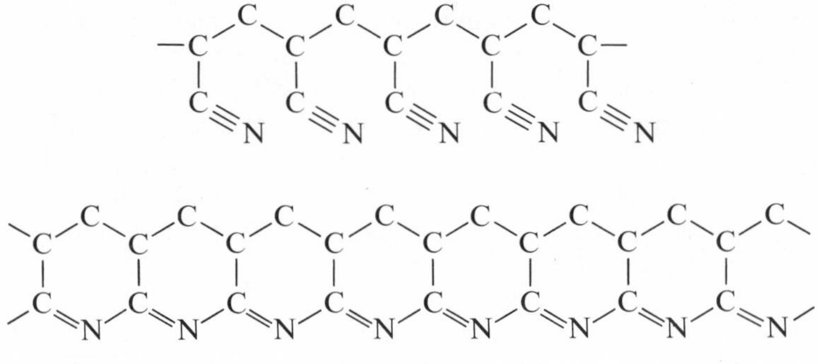

which involves stiffening, insolubility, and discoloration. Further heating leads to evolution of H_2 and to aromatization

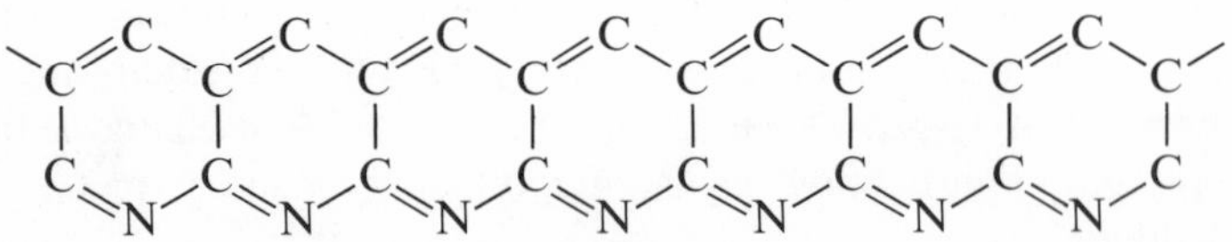

whereby a black, completely infusible and insoluble material is obtained that corresponds in its structure to a linear graphite in which one carbon atom of every ring has been replaced by nitrogen.

These examples clearly show that there are many possibilities for the formation of long stiff chains and that, in all cases, the properties of the resulting materials confirm the expectations.

3. Inter- and Intramolecular Forces

Since rigidity, high melting, and difficult solubility all depend on a firm cohesion between neighboring chains it is obvious that specific groups, which establish strong intermolecular bonds between such chains will favorably affect these properties. This will be particularly true if these groups are arranged along the macromolecules in regular distances so that they can get into each other's neighborhood without causing any valence strain in the chains themselves. In fact, it has been found that all groups which carry dipoles or are highly polarizable or permit the development of interchain hydrogen bonds favor crystallinity and with it all the other valuable properties already mentioned.

Examples of this are polyvinylchloride and polyvinylidene chloride where the C—Cl dipoles increase the lateral cohesive energy density of the system and with it rigidity, softening temperature, and resistance against dissolution and swelling. An example for the beneficial influence of polarizable groups can be found in polyethylene terephthalate where the phenylene rings are polarized by the C=O dipoles resulting in a firm and rigid lattice structure. Interchain hydrogen bonding is important for crystallinity, rigidity, high melting, and difficult solubility. A well-known example is polyvinyl alcohol with an —OH group at every other carbon atom. Even atactic chains can establish frequent hydrogen bonds if they are first parallelized by stretch or sheer, and the polymer is, therefore, a high melting, rigid fiber former. The syndiotactic species is even more rigid, higher softening and less soluble in water. Another case of strong lateral hydrogen bonding exists in cellulose, where its effect is enhanced by the rigidity of the glucopyranose rings; as a consequence cellulose is highly crystalline, infusible, and insoluble (or very difficultly soluble) and has an unusually high modulus of rigidity for an organic polymer. Perhaps the most striking effect of lateral hydrogen bonding between carefully spaced groups is demonstrated by the linear polyamides, where the —CO—NH— group is responsible for the establishment of these bonds. Since one can space these groups apart at different distances by introducing between them paraffinic —CH$_2$— chains of different length it was possible to establish the fact that small and regular distances between successive amide group along the chains leads to rigid, high melting, and difficultly soluble types, whereas reduction of the lateral hydrogen bonding by large and/or irregular distances between the amide groups produces low melting and even rubbery types, which are easily soluble in many organic liquids.

A combination of pronounced chain rigidity with lateral hydrogen bonding would lead to polymers of the structure:

$$-\overset{\overset{\displaystyle O}{\|}}{C}-\langle\bigcirc\rangle-\overset{\overset{\displaystyle O}{\|}}{C}-\underset{\underset{\displaystyle H}{|}}{N}-\langle\bigcirc\rangle-\underset{\underset{\displaystyle H}{|}}{N}-\overset{\overset{\displaystyle O}{\|}}{C}-\langle\bigcirc\rangle-\overset{\overset{\displaystyle O}{\|}}{C}-\underset{\underset{\displaystyle H}{|}}{N}-\langle\bigcirc\rangle-\underset{\underset{\displaystyle H}{|}}{N}-$$

In fact it was found that polyamides of this general type are extremely rigid, high melting, and resistant against heat, dissolution, and swelling.

4. Bulky substituents

The vibrational and rotational mobility of an intrinsically flexible chain can also be inhibited by bulky substituents; the degree of stiffening depends on the size, shape, and mutual interaction of the substituents. Methyl- and phenyl groups have, in general, a strong inhibiting influence on the segmental mobility of linear macromolecules as shown by the high glass transition and heat distortion point of polymethylmethacrylate compared with polymethylacrylate of polystyrene as compared with polyethylene. Larger aromatic systems have an even stronger influence and polyvinyl naphthalene, -anthracene and -carbazol are amorphous polymers of remarkable high heat distortion points and of unusual rigidity; they are also very difficultly soluble. Substituents from ethyl- to hexyl in general exhibit a softening influence because they increase the average distance between the main chains and prevent their dipole groups to approach each other to distances which are close enough for favorable interaction. The substituents, in these cases are open chains from two to six carbon atoms and have, themselves, a considerable internal mobility—they act as plasticizers which are chemically attached. Striking examples for this behavior are the polyacrylic- and methacrylic esters from ethyl to hexyl and the polyvinylesters from the propionate to hexoate, when rigidity, softening temperature, and resistance against dissolution and swelling decrease sharply with increasing number of carbon atoms in the chain of the substituent. If these chains become still longer—from twelve to eighteen carbon atoms, and remain unbranched (normal paraffinic alcohols or acids), a new phenomenon occurs—the tendency of the side chains to form crystalline domains of their own in which the side chains of neighboring macromolecules arrange themselves in bundles of laterally ordered units with each other. The geometric arrangement of the monomeric units in the chains is of importance because of the fact that polymeric materials possess the capacity to crystallize if the chains are regularly built and if they are brought to a laterally ordered state by orientation or annealing. High degrees of crystallinity produce rigidity, high softening ranges, and resistance against swelling and dissolution, whereas low degrees of crystallinity lead to softer products with lower heat distortion points and easier solubility.

Thus we have seen in this introductory chapter some of the basic reasons for the unusual behavior associated with polymeric materials and we have defined some of the important terms. Subsequent chapters enlarge on several of these important aspects.

DISCUSSION QUESTIONS AND PROBLEMS

1. The primary reason for the unusual properties of high polymers as compared to more conventional organic compounds is the long chain length or high molecular weight. From the material presented in this chapter, select the specific factors or characteristics which make polymers so useful in so many different applications.

2. From Table 1 it may be noted that the number of *new* commercially introduced polymers has declined considerably during the 1960s and 70s. Can you suggest reasons for this? Remember that the market usually requires maximum properties at the most economic cost.

3. Polymers can be made to serve as arctic rubbers, as fibers, as rigid engineering plastics, as high temperature insulators, and so forth; "tailor-making" of polymers is indeed an achievable goal. Consider and list the factors that make this possible. (These will be amplified in following chapters).

CHAPTER 2

Polymerization

M. GOODMAN
Chemistry Department
University of California,
San Diego
La Jolla, California

J. FALCETTA
Bausch and Lomb Inc.
Rochester, New York

This chapter describes the organic aspects of polymer chemistry. It deals consecutively with polymerizations which proceed (a) stepwise, (b) by ring opening, and (c) by "true addition." This systematic approach is possible only because polymer chemistry has now developed sufficiently, kinetically and mechanistically, to allow reaction classification.

In general, polymerization requires a higher degree of purity and control over reaction conditions (such as reactant ratio, solvent, temperature, and catalyst) than does a "normal" organic reaction. The requirements for making nylon, polypropylene, and other materials illustrate this. Impurities as low as a few parts per million can destroy polymerization control. Small defects in the placement of monomer residues in the chain severely alter the properties of the polymer, as do even traces of side reactions.

We believe that new and fascinating polymerization reactions remain to be discovered. In this chapter, we hope to show that polymer chemistry is alive and growing, playing an ever more dominant role in the forefront of modern chemistry.

A. STEPWISE POLYMERIZATION

1. Condensation Polymerization

A condensation reaction that proceeds stepwise to produce a large molecule is called a condensation polymerization. Such a reaction is chemically the same as a condensation reaction that produces a small, or "normal," organic molecule.

Fundamentals of Condensation Polymerization (1, 20)

Whether a condensation reaction produces a polymer or a small molecule depends upon the *functionality* of the reactants. Functionality is the average number of reacting groups per reacting molecule. A functionality less than two produces a low molecular weight condensation product. A functionality of two ordinarily produces a linear polymer, and a functionality greater than two allows branching and cross-linking (see Table 1). However, when one reactant is monofunctional, a low molecular weight compound invariably results regardless of the overall functionality of the system.

A condensation polymerization starts by producing mainly bimolecular products:

$$
\underset{\displaystyle HO-\overset{\displaystyle O}{\overset{\|}{C}}-R-\overset{\displaystyle O}{\overset{\|}{C}}-OH}{} + H_2N-R'-NH_2
$$

$$
\downarrow
$$

$$
HO-\overset{O}{\overset{\|}{C}}-R-\overset{O}{\overset{\|}{C}}-\overset{H}{\overset{|}{N}}-R'-NH_2 + H_2O \tag{1}
$$

28

Table 1 Relation of functionality and polymerization product

Compound A	f_A	Compound B	f_B	Functionality $f = \dfrac{f_A + f_B}{2}$	Product
$CH_3CH_2CH_2-\overset{O}{\underset{\|}{C}}-OH$	1	CH_3CH_2OH	1	1	$CH_3CH_2CH_2-\overset{O}{\underset{\|}{C}}-OCH_2CH_3$
$HO-\overset{O}{\underset{\|}{C}}-CH_2CH_2-\overset{O}{\underset{\|}{C}}-OH$	2	CH_3CH_2OH	1	1.5	$CH_2CH_3O-\overset{O}{\underset{\|}{C}}-CH_2CH_2-\overset{O}{\underset{\|}{C}}-OCH_2CH_3$
$HO-\overset{O}{\underset{\|}{C}}-CH_2CH_2-\overset{O}{\underset{\|}{C}}-OH$	2	$HOCH_2CH_2OH$	2	2	$\left(-\overset{O}{\underset{\|}{C}}-CH_2CH_2-\overset{O}{\underset{\|}{C}}-OCH_2CH_2O-\right)_n$ linear polymer
$HO-\overset{O}{\underset{\|}{C}}-CHCH_2-\overset{O}{\underset{\|}{C}}-OH$, with $COOH$	3	$HOCH_2CH_2OH$	2	2.5	Branched polymer, possible gel formation

As the concentration of bimolecular product increases, further reactions, forming tri- and tetramolecular products, become important. This stepwise mechanism causes nearly complete conversion of original reactants to low molecular weight products early in the reaction. High polymer forms later and more slowly, because the concentration of reactive groups decreases as the length of the chain increases.

A complete kinetic analysis of condensation polymerization would be extremely difficult, if not impossible, because of the large number of different molecules present. Flory (1, 20) simplified the analysis by assuming that the intrinsic chemical reactivity of a reactive group is unaffected by the size of the entire molecule. This proposal has been validated on mechanistic and experimental grounds.

Another fundamental assumption is that the reaction rate is independent of gross factors such as the diffusion rate and the viscosity of the medium. Although the mobility of the entire chain decreases with size, the intrinsic reactivity of the end group does not change. A long chain, then, may be considered simply as a diluent to the reactive end group. Experiment supports this concept.

Kinetics

The above assumptions amount kinetically to the proposition that all steps in a condensation polymerization have equal rate constants. The following treatment is based on this proposition.

Adipic acid and diethylene glycol when heated form polyester through a typical condensation polymerization:

$$
\underset{\substack{\| \\ O}}{HO-C}+CH_2\!\!\to_4\!\!\underset{\substack{\| \\ O}}{C}-OH + HO-CH_2CH_2-O-CH_2CH_2-OH
$$

$$
\left[\underset{\substack{\| \\ O}}{-C}+CH_2\!\!\to_4\!\!\underset{\substack{\| \\ O}}{C}-O-CH_2CH_2-O-CH_2CH_2-O\right]_n + H_2O \quad (2)
$$

polyester

This and other polyesterifications are acid catalyzed. The catalyst may be either the carboxylic acid reactant or an added strong acid.

Polymerization without Added Strong Acid. Here the rate of the polyesterification is proportional to the concentration of the alcohol group [OH] and to the square of the concentration of the carboxylic acid group [COOH], since the carboxylic acid is both reactant and catalyst.

$$-\frac{dc}{dt} = k\,[OH][COOH]^2 \qquad (3)$$

When the functional group concentrations are equal,

$$-\frac{dc}{dt} = kc^3 \qquad (4)$$

On integration, this becomes

$$2kt = \frac{1}{c^2} - \text{constant} \qquad (5)$$

If the reaction were uniformly third order throughout, the constant of integration would be $1/c_0^2$, where c_0 is the initial concentration of one of the functional groups. The equation then becomes

$$2kt = \frac{1}{c^2} - \frac{1}{c_0^2} \qquad (6)$$

For convenience, let p be the extent of reaction, that is, the fraction of c_0 that has reacted at time t. Then the concentration of *unreacted* functional group at time t is $c = (1 - p)c_0$. On substitution, then,

$$2c_0^2 kt = \frac{1}{(1 - p)^2} - 1 \qquad (7)$$

Figure 1 shows $1/(1 - p)^2$ plotted against time. When $1/(1 - p)^2$ is less than about 25, the plot is not linear. The quantity $1/(1 - p)$ is called the *degree of polymerization*, denoted by $\overline{DP}_n$. Its value corresponds to the number of monomer molecules in the average chain. The linear relationship, then, begins at a degree of polymerization of approximately five. After initially complex kinetics, the polymerization of diethylene glycol (DE) and adipic acid (A) proceeds in good agreement with third-order kinetics up to $\overline{DP}_n = 14$. When ω-hydroxyundecanoic acid, $HO-(CH_2)_{10}-COOH$, forms polyester, this agreement holds up to $\overline{DP}_n = 70$.

Polymerization with Added Strong Acid. The kinetic equation is simpler when diethylene glycol and adipic acid condense in the presence of a strong acid, such as toluenesulfonic acid, because the acid catalyst concentration is constant. This concentration, then, is incorporated into the rate constant.

$$k' = k\,[\text{catalyst}] \qquad (8)$$

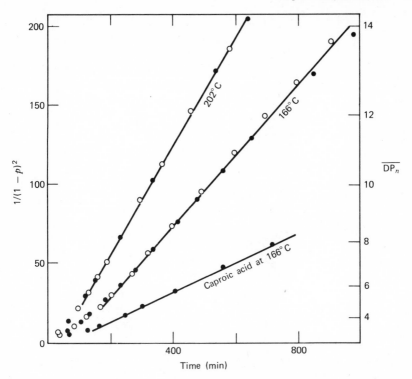

Figure 1   Reactions as a function of temperature of diethylene glycol (DE) with adipic acid (A) and of diethylene glycol (DE) with caproic acid (C) as a function of temperature. Time values at 202°C have been multiplied by two (reference 21).

If the alcohol and carboxylic acid concentrations are equal,

$$-\frac{dc}{dt} = k'c^2 \qquad (9)$$

This is treated as above so that finally

$$c_0 k't = \frac{1}{1-p} - 1 \qquad (10)$$

Figure 2 shows a straight-line dependence up to $\overline{DP}_n = 90$ from $\overline{DP}_n = 10$. In this case, a degree of polymerization of 90 corresponds to a molecular weight of 10,000.

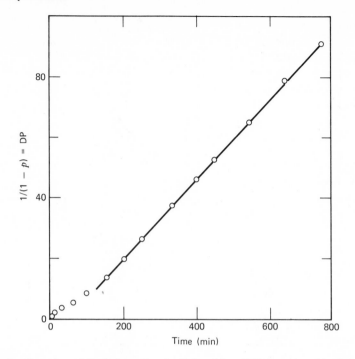

Figure 2   Reaction of diethylene glycol-adipic acid at 109°C catalyzed by 0.4 mole %
toluene sulfonic acid (reference 21).

Molecular Weights of Linear Polymers

A condensation reaction produces a linear polymer when the functionality
of the system is two. The system may contain either a single bifunctional
material, A–B, or it may contain two bifunctional compounds, A–A and
B–B.

In the first case, the starting material may be either an amino acid, a
hydroxy acid, or an alkylene chlorohydrin. The polymerization can be
represented as

$$A-B \longrightarrow \quad (A-B)(A-B)(A-B) + \text{coproduct} \quad (11)$$

Regardless of the degree of polymerization, each molecule in the system
must have one unreacted end group of type A and one unreacted end group
of type B. Therefore, the number of molecules in the system can be determined by measuring the number of unreacted end groups. This quantity
in moles per gram is the reciprocal of molecular weight (grams per mole).

$$\overline{M}_n = \frac{1}{\text{moles of unreacted A per gram}} \quad (12)$$

$\overline{M}_n$ is called the *number* average molecular weight, since it is based on the *number* of end groups per gram.

In the second case, the starting material is a pair of bifunctional compounds such as a dicarboxylic acid and a dialcohol, a dicarboxylic acid and a diamine, or a dialcohol and a dihalide. The polymerization can be represented as

$$A{-}A + B{-}B \longrightarrow {-}A{-}A{+}B{-}B{-}A{-}A{-}_x B{-}B \qquad (13)$$

As in the first case, the number of unreacted end groups provides a measure of the number average molecular weight.

In both cases, the degree of polymerization is simply the ratio of the number of starting molecules to the number of product molecules.

$$\overline{DP}_n = \frac{\text{initial number of molecules}}{\text{final number of molecules}} = \frac{c_0}{c_0(1-p)} = \frac{1}{1-p} \qquad (14)$$

As before, c_0 is the initial concentration of one of the functional groups and p is that fraction of c_0 which has reacted at a particular time (t).

Because degree of polymerization is the number of monomer molecules in the average polymer molecule, the average molecular weight of the product molecules is simply $\overline{DP}_n$ times the molecular weight of the starting material, M_0.

$$\overline{M}_n = M_0(\overline{DP}_n) = \frac{M_0}{1-p} \qquad (15)$$

With two monomers, A—A and B—B, M_0 is the average molecular weight.

In condensation polymerization, high molecular weight material forms only when nearly 100% of the reactive groups have reacted. Table 2, obtained from Eq. 14, shows this.

Table 2

% Reaction	$\overline{DP}_n$
50	2
90	10
95	20
99	100
99.9	1,000
99.99	10,000

A nearly quantitative reaction, then, is essential for high polymer since even a 99% conversion of starting material with a molecular weight of 100 results in an average molecular weight of only 10,000. This is insufficient for attaining many important physical properties.

Also extremely important in the A–A, B–B condensations is the accurate measurement of equimolar quantities of the two reactants. An excess of diacid, say, in a polyesterification terminates all chain ends with carboxylic acid groups, thereby making the formation of high polymer impossible. Thus, the degree of polymerization depends upon the reactant ratio (r).

$$r = \frac{N_A}{N_B} \qquad (16)$$

N_A is the number of A groups in the starting material mixture and N_B is the number of B groups. (By convention, r is less than 1, to avoid a negative $\overline{DP}_n$ value.)

$$\overline{DP}_n = \frac{1 + r}{2r(1 - p) + 1 - r} \qquad (17)$$

where p is the fraction of N_A, the minor component, already reacted. When all of N_A has reacted, $p = 1$ and the degree of polymerization is at its maximum.

$$\overline{DP}_n = \frac{1 + r}{1 - r} \qquad (18)$$

Thus a 2% deficiency of one component provides an r value of 0.98 and a maximum degree of polymerization of 99. Equation 18 can be used to evaluate the degree of polymerization in a system containing one bifunctional species and one monofunctional species, if the reactant ratio is defined as

$$r = \frac{N_A}{N_A + 2N_M} \qquad (19)$$

where N_M is the number of monofunctional molecules present.

Molecular Weight Distribution

During a condensation polymerization, each unreacted functional group is equally likely to react, regardless of the size of its appended chain. Therefore, at any instant a *range* of molecular sizes exists.

What is the probability that an A–B–A–B type of polyester will contain exactly x monomeric units, that is, be an x-mer? The probability that a given carboxyl group has reacted is equal to the extent of reaction (p). A molecule containing x units has undergone $x - 1$ reactions. The probability of this occurring is the product of the individual reaction probabilities. It is, therefore, p^{x-1}. The probability that the end carboxyl group has not reacted is $1 - p$. The total probability (P_x) that a given molecule contains x units, then, is

$$P_x = p^{x-1}(1 - p) \tag{20}$$

This must also be equivalent to the mole fraction of x-mers.

$$P_x = \frac{N_x}{N} \tag{21}$$

N_x is the number of x-mers and N is the total number of molecules in the system. The value of N at any instant is $N_0(1 - p)$, where N_0 is the initial number of monomer molecules. On substitution,

$$N_x = N_0 p^{x-1}(1 - p)^2 \tag{22}$$

This equation provides the number of x-mers in terms of the initial amount of monomer and the extent of reaction.

Distribution curves (Fig. 3) from this equation show wide molecular weight ranges at various extents of reaction. Even at high p values, low molecular weight chains are abundant. However, on a weight basis (Fig. 4) they are less significant. The weight fraction, W_x, is the weight of x-mers divided by the weight of all the molecules.

$$W_x = \frac{xN_x}{N_0} \tag{23}$$

Substitution produces an equation for calculating weight fraction from extent of reaction. This equation is the basis of Fig. 4.

$$W_x = xp^{x-1}(1 - p)^2 \tag{24}$$

The maxima in these weight fraction curves occur near the number average value of x.

Relationship of Molecular Weight Averages

The number average molecular weight $(\overline{M}_n)$ (see Eqs. 12 and 15) is defined as follows:

$$\overline{M}_n = \frac{\sum_{x=1}^{x=\infty} M_x N_x}{\sum_{x=1}^{x=\infty} N_x} = \frac{\sum_{x=1}^{x=\infty} (M_0 x)N_x}{\sum_{x=1}^{x=\infty} N_x} \tag{25}$$

where x, N_x, and M_0 represent previously defined quantities and M_x is the molecular weight of an x-mer. Substituting for N_x from Eq. 23, etc., we obtain

$$\overline{M}_n = M_0 \sum x(1 - p)p^{x-1} \tag{26}$$

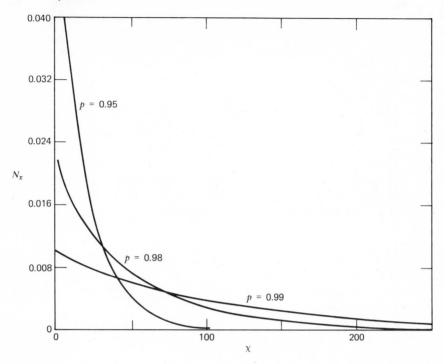

Figure 3 Mole fraction distribution of chain molecules in a linear condensation polymer for several extents of reaction (reference 20).

The summation $\sum x(1-p)p^{x-1}$ may be shown to equal $1/(1-p)$ and therefore

$$\overline{M}_n = \frac{M_0}{1-p} \tag{15}$$

which is the relationship previously obtained.

The weight average molecular weight is defined by

$$\overline{M}_w = \frac{\sum_{x=1}^{x=\infty} M_x W_x}{\sum_{x=1}^{x=\infty} W_x} = \frac{\sum_{x=1}^{x=\infty} M_0(x)^2 N_x}{\sum_{x=1}^{x=\infty} M_0(x)N_x} \tag{27}$$

In a manner similar to the above, an expression for $\overline{M}_w$ may be obtained for a condensation polymerization.

$$\overline{M}_w = M_0 \frac{1+p}{1-p} \tag{28}$$

The ratio of $\overline{M}_w/\overline{M}_n$ is two for values of p near 1.

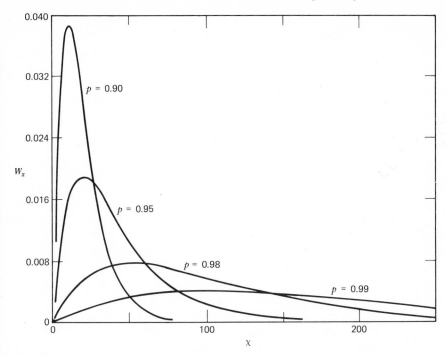

Figure 4 Weight fraction distribution of chain molecules in linear condensation polymers for several extents of reaction (reference 20).

This distribution is referred to as the most probable distribution. It is valid for any polymerization following the statistics developed in Section A.1.

Ring Formation versus Chain Polymerization

Implicit in our discussion of the limitation of $\overline{DP}$ by stoichiometric imbalance or incomplete reaction is the assumption that no side reactions occur during polymerization. For example, a side reaction that affects one functional group more than the other in an A–A, B–B polymerization would result in a stoichiometric imbalance and consequently limit molecular weight.

In many condensation polymerizations a significant side reaction is the formation of cyclic products through an intramolecular reaction. This has been observed in both AA–BB and AB polymerizations. For example, in the polymerization of lactic acid, there is obtained both the linear polymer and the dimeric cyclic ester, lactide. Cyclic products (in small amounts) are also isolated in the preparation of polyethylene terephthalate.

The determining factor in whether or not ring formation occurs is the size of the ring (or rings) that would be obtained upon cyclization. When the ring size is less than five or greater than seven, polymer formation is favored. However, when a five-membered ring can be formed this is invariably the exclusive product. For example, when γ-hydroxybutyric acid or γ-aminobutyric acid is subjected to condensation the exclusive product is the lactone or lactam, respectively.

Three-Dimensional Network Polymers

We previously stated (Table 1) that when the average functionality in a condensation polymerization exceeds two, there exists the potential for forming a branched polymer. An example of this would be the condensation between glycerol and a dibasic acid. At an early stage of the polymerization the structure of the polymer becomes complex. For every trifunctional glycerol molecule that reacts, the number of reactive functional groups on the polymer chain is increased by one. Therefore, as the polymerization proceeds, the number of possible ways in which the monomers may add on to the growing polymer rapidly increases. Many branching sites are thus formed. When large polymeric molecules condense, the number of reactive groups per polymer chain is further increased. The size of the polymer molecules rapidly increases, resulting eventually in the formation of a three-dimensional network polymer of infinite molecular weight. At this point (the gel point) the polymer becomes gel. At the gel point polymer molecules of finite molecular weight remain (often referred to as the sol).

The extent of reaction at which the gel point occurs is related to the functionality of the monomers and the proportions in which they are used. In the condensation of glycerol and equivalent amounts of various dibasic acids the gel point is found to occur at $p = 0.765$.

Methods of Condensation Polymerization (5, 7, 22)

One method of making a condensation polymer is to heat the monomer above the melting point of the polymer (with a catalyst, if required). Byproducts are removed from the reaction mixture. This method, called melt polymerization, fits the general pattern of condensation polymerizations described above. In addition, the monomer and the polymer must be stable at the reaction temperature, and the polymer must be fusible. Melt polymerization is used widely, both commercially and in the laboratory.

Another method, called interfacial polymerization, involves reaction at the interface between two immiscible liquids (one organic, the other aqueous). The preparation of polyhexamethylenesebacamide is typical. The starting materials are hexamethylenediamine, $H_2N\!\!-\!\!(CH_2)_6\!\!-\!\!NH_2$, and sebacyl chloride, $Cl-\overset{\overset{O}{\|}}{C}+CH_2\!\!\!+_8\!\!\overset{\overset{O}{\|}}{C}-Cl$. The upper layer is an aqueous solution of the diamine and sodium carbonate. The lower layer is a perchloroethylene solution of the diacid chloride. The polyamide forms rapidly at the interface and may be continuously removed through the aqueous phase as a collapsed film. (The sodium carbonate neutralizes the by-product hydrochloric acid.)

Since the reaction is diffusion controlled, the starting materials need not be present in exactly equivalent amounts. The diamine, soluble in both solvents, diffuses into the organic layer, where it reacts with the acid chloride. As the chains lengthen, the molecular weight reaches the limit of solubility and the polymer precipitates at the interface. The method is useful for making other polyamides, polysulfonamides, polyurethanes, and polyphenyl esters.

A third method, solution polymerization, uses an inert solvent as the reaction medium. A base is often used to neutralize acidic by-products. The starting material must react in high yield, but at a moderate rate. The solvent must be inert, must dissolve the starting material, and must either dissolve or swell the polymer.

Most solution polymerizations are done at temperatures below 40°C. Some, however, are run at high temperature as melt polymerizations, where the solvent acts merely as diluent. In true (low-temperature) solution polymerization, the solvent (or a base) plays a more active role. For condensation polymerizations, the solution method is restricted to specialty cases, such as polysulfones and high-temperature polymers.

Examples of Condensation Polymerization

Polyamides (23–25). Polyamides are important commercially and also from a theoretical viewpoint. Nylon and many other synthetic fibers are polyamides. Condensation polyamides are usually prepared from a diacid and a diamine, although polyamides are also prepared from lactams by ring opening, as discussed later.

$$HO-\overset{\overset{O}{\|}}{C}+CH_2\!\!\!+_n\!\!\overset{\overset{O}{\|}}{C}-OH + H_2N+CH_2\!\!\!+_m\!\!NH_2$$

$$\downarrow \text{ heat}$$

$$\left[\overset{\overset{O}{\|}}{{-C}}+CH_2\!\!\!+_n\!\!\overset{\overset{O}{\|}}{C}-NH+CH_2\!\!\!+_m\!\!NH\!-\right]_n + H_2O \qquad (29)$$

When made from diacids and diamines, nylons are usually named as nylon-M,N, where M and N represent the number of carbon atoms in the diamine and diacid, respectively. The most important example, nylon-6,6, is made from adipic acid and hexamethylenediamine. An exact one-to-one ratio of starting materials is obtained by preparing and recrystallizing a nylon salt before polymerization. When heated to about 260°C, this one-to-one salt eliminates water to form a high molecular weight polymer.

$$H_2N\!\!-\!\!(CH_2)_6\!\!-\!\!NH_2 \;+\; HO\!\!-\!\!\overset{\displaystyle O}{\overset{\displaystyle \|}{C}}\!\!-\!\!(CH_2)_4\!\!-\!\!\overset{\displaystyle O}{\overset{\displaystyle \|}{C}}\!\!-\!\!OH$$

$$\downarrow$$

$$H_3\overset{\oplus}{N}\!\!-\!\!(CH_2)_6\!\!-\!\!\overset{\oplus}{N}H_3 \quad \overset{\ominus}{O}\!\!-\!\!\overset{\displaystyle O}{\overset{\displaystyle \|}{C}}\!\!-\!\!(CH_2)_4\!\!-\!\!\overset{\displaystyle O}{\overset{\displaystyle \|}{C}}\!\!-\!\!\overset{\ominus}{O}$$

nylon salt

$$260°C \downarrow$$

$$\left[\!-\!NH\!\!-\!\!(CH_2)_6\!\!-\!\!NH\!\!-\!\!\overset{\displaystyle O}{\overset{\displaystyle \|}{C}}\!\!-\!\!(CH_2)_4\!\!-\!\!\overset{\displaystyle O}{\overset{\displaystyle \|}{C}}\!\!-\!\right]_n \qquad (30)$$

nylon-6,6

The molecular weight of the polymer may be controlled by using a slight excess of one of the starting materials. This is called *molecular weight stabilization.* A 2% excess of diacid results in polymer with a molecular weight of 12,000, a value suitable for fiber preparation. Of all the nylons, nylon-6,6 has the best physical properties and is the most economical.

In general, a polyamide with an aromatic ring in the chain melts higher and is less soluble than an aliphatic polyamide. This is especially true for *para*-bifunctional aromatics. For example, using terephthalic acid,

$$HO\!\!-\!\!\overset{\displaystyle O}{\overset{\displaystyle \|}{C}}\!\!-\!\!\langle\!\bigcirc\!\rangle\!\!-\!\!\overset{\displaystyle O}{\overset{\displaystyle \|}{C}}\!\!-\!\!OH$$

instead of adipic acid increases polymer melting points by 100–170°C. This may be due to increased stiffness in the chain.

Fully aromatic polyamides can also be prepared (26, 27), even though aromatic diamines are much less reactive than aliphatic diamines. Because these polymers melt at very high temperatures, melt polymerization is unsuitable and low-temperature techniques are used. For example, 1,3-phenylenediamine and terephthalyl chloride react either interfacially or in solution to form poly(m-phenyleneterephthalamide) (mp 365°C).

Polyesters (28–30). Polyesters are also commercially important. They are prepared either from a hydroxy acid or from a dialcohol and a diacid.

The lower diacids, however, produce side reactions. Malonic acid,

$$\underset{\displaystyle HO-\overset{\displaystyle O}{\overset{\|}{C}}-CH_2-\overset{\displaystyle O}{\overset{\|}{C}}-OH,}{}$$

decarboxylates. Succinic acid,

$$HO-\overset{\displaystyle O}{\overset{\|}{C}}\!\!-\!\!(CH_2)_2\overset{\displaystyle O}{\overset{\|}{C}}\!\!-\!\!OH,$$

and glutaric acid,

$$HO-\overset{\displaystyle O}{\overset{\|}{C}}\!\!-\!\!(CH_2)_3\overset{\displaystyle O}{\overset{\|}{C}}\!\!-\!\!OH,$$

form cyclic anhydrides. The best linear aliphatic polyester is obtained from adipic acid and ethylene glycol.

$$HO-\overset{\displaystyle O}{\overset{\|}{C}}\!\!-\!\!(CH_2)_4\overset{\displaystyle O}{\overset{\|}{C}}\!\!-\!\!OH + HO-CH_2CH_2-OH$$

$$\downarrow$$

$$\left[-\overset{\displaystyle O}{\overset{\|}{C}}\!\!-\!\!(CH_2)_4\overset{\displaystyle O}{\overset{\|}{C}}\!\!-\!\!O-CH_2CH_2-O-\right]_n \tag{31}$$

Unlike aliphatic polyamides, aliphatic polyesters melt too low to be commercially important. But like the polyamides, an aromatic group in the chain raises the melting point. Hence polyesters based on terephthalic acid have found commercial success.

As the number of methylene groups in the diol increases, the flexibility of the polymer chain increases, and the melting point decreases. Also, polyester based on isophthalic acid has more chain flexibility and therefore melts lower than terephthalic analogues. For example, polyethylene isophthalate melts at about 108°C, compared with 280°C for the terephthalate.

Polyethylene terephthalate is available commercially as a fiber and as a film. The high polymer is made from low polymer (oligomer) containing glycol end groups by catalytic ester interchange at high temperature. The oligomer is also made by ester interchange, starting with dimethyl terephthalate and excess ethylene glycol.

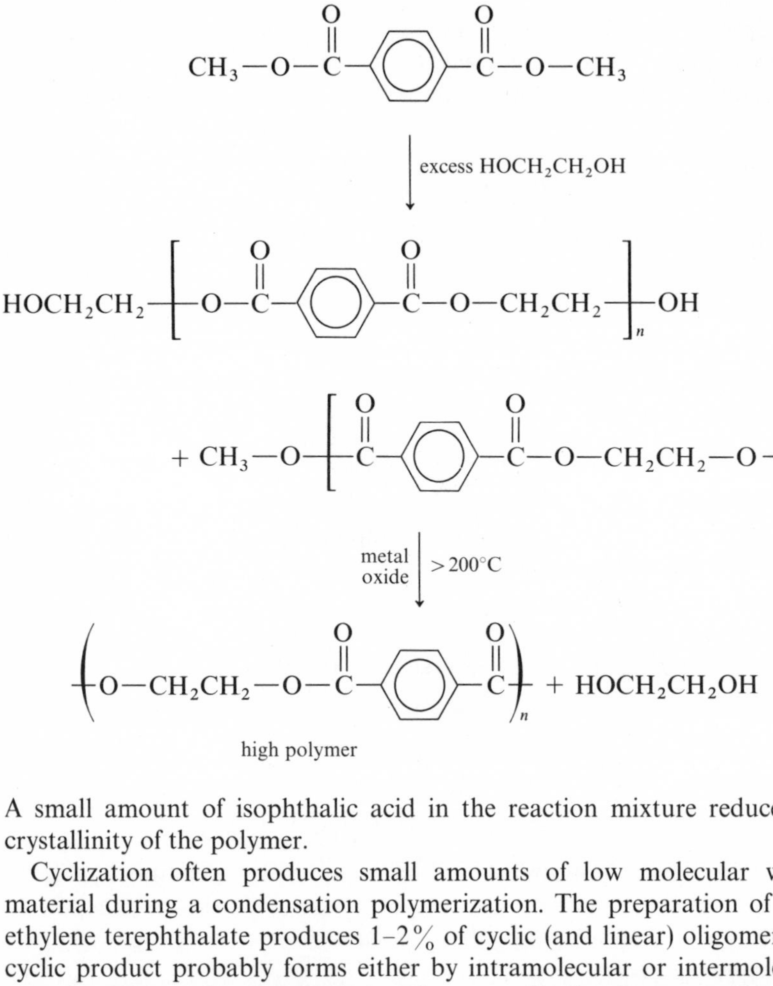

$$CH_3-O-\overset{\overset{\displaystyle O}{\|}}{C}-\underset{\bigcirc}{}-\overset{\overset{\displaystyle O}{\|}}{C}-O-CH_3$$

$$\downarrow \text{ excess } HOCH_2CH_2OH$$

$$HOCH_2CH_2\left[O-\overset{\overset{\displaystyle O}{\|}}{C}-\underset{\bigcirc}{}-\overset{\overset{\displaystyle O}{\|}}{C}-O-CH_2CH_2\right]_n OH$$

$$+ CH_3-O\left[C-\overset{\overset{\displaystyle O}{\|}}{}-\underset{\bigcirc}{}-\overset{\overset{\displaystyle O}{\|}}{C}-O-CH_2CH_2-O\right]_m H$$

$$\overset{\text{metal}}{\underset{\text{oxide}}{\Big\downarrow}} > 200°C$$

$$\left(O-CH_2CH_2-O-\overset{\overset{\displaystyle O}{\|}}{C}-\underset{\bigcirc}{}-\overset{\overset{\displaystyle O}{\|}}{C}\right)_n + HOCH_2CH_2OH \qquad (32)$$

high polymer

A small amount of isophthalic acid in the reaction mixture reduces the crystallinity of the polymer.

Cyclization often produces small amounts of low molecular weight material during a condensation polymerization. The preparation of polyethylene terephthalate produces 1–2% of cyclic (and linear) oligomer. The cyclic product probably forms either by intramolecular or intermolecular ester interchange by an alcohol end group. Cyclic oligomers also form during melt extrusion of the polymer.

Because alcohol is usually easier to remove from the reaction than water, ester interchange is usually preferred over direct esterification. Other ester-forming materials include diacid chlorides and diols, and anhydrides and diols.

Ester interchange occurs between polymer molecules during polymerization. Since an interchange reaction does not change the number of molecules

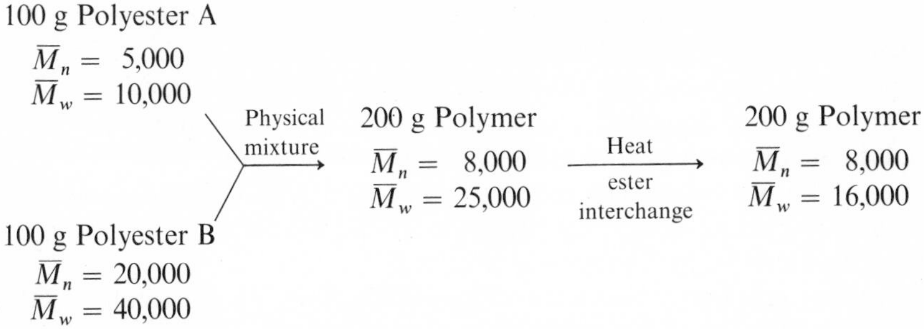

100 g Polyester A

$\overline{M}_n = 5,000$
$\overline{M}_w = 10,000$

Physical mixture → 200 g Polymer
$\overline{M}_n = 8,000$
$\overline{M}_w = 25,000$

— Heat ester interchange → 200 g Polymer
$\overline{M}_n = 8,000$
$\overline{M}_w = 16,000$

100 g Polyester B

$\overline{M}_n = 20,000$
$\overline{M}_w = 40,000$

Figure 5 Illustration of the molecular weight changes that take place upon ester interchange.

in the system, it does not change the number average molecular weight. But it may drastically change the weight average molecular weight.

If two samples of polyester having different molecular weights are mixed and heated, ester interchange occurs. At equilibrium, the polymer will have reattained the most probable $\overline{M}_w/\overline{M}_n$ value of two. For example, if 100 g of a polyester with $\overline{M}_n = 5000$ and $\overline{M}_w = 10,000$ is mixed with 100 g of polyester with $\overline{M}_n = 20,000$ and $\overline{M}_w = 40,000$, the physical mixture has $\overline{M}_n = 8000$ and $\overline{M}_w = 25,000$. Heating then causes ester interchange and a redistribution of molecular weight. The $\overline{M}_n$ remains 8000 because the number of molecules does not change. The $\overline{M}_w$ value becomes 16,000 to reestablish the ratio $\overline{M}_w/\overline{M}_n = 2$ (Fig. 5).

Polyurethanes (31, 32). A urethane can be prepared by an addition rather than a condensation reaction.

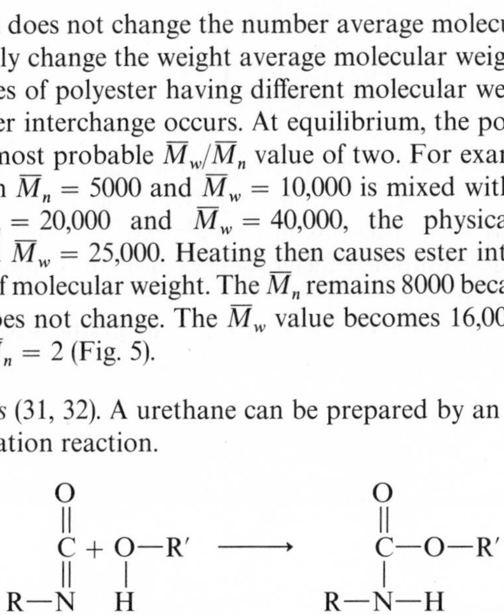

$$\begin{array}{c} O \\ \| \\ C \\ \| \\ R-N \end{array} + \begin{array}{c} O-R' \\ | \\ H \end{array} \longrightarrow \begin{array}{c} O \\ \| \\ C-O-R' \\ | \\ R-N-H \end{array} \qquad (33)$$

The reaction is catalyzed by a variety of acids, bases, and metal salts. Tertiary amines are common catalysts.

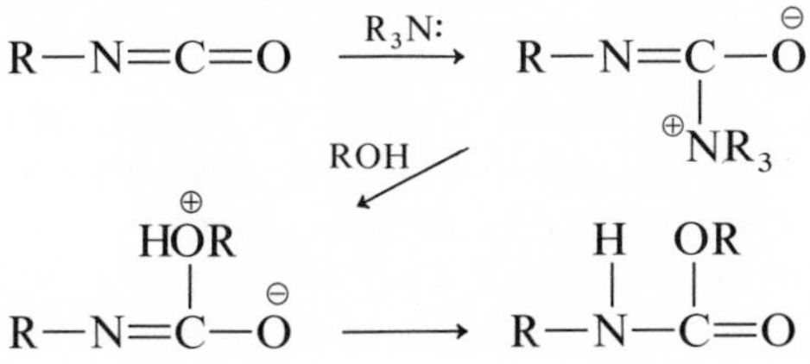

$$R-N=C=O \xrightarrow{R_3N:} R-N=C-\overset{\ominus}{O} \\ \overset{\oplus}{N}R_3$$

$$\text{ROH} \swarrow \qquad (34)$$

$$\begin{array}{c} \overset{\oplus}{HOR} \\ | \\ R-N=C-\overset{\ominus}{O} \end{array} \longrightarrow \begin{array}{c} H \quad OR \\ | \quad | \\ R-N-C=O \end{array}$$

The properties of polyurethanes vary with varying starting materials, functionality, and polymerization method. Highly elastic fibers, such as Spandex, are made from diisocyanates and special oligomeric glycols. These hydroxy-terminated oligomers are flexible, linear molecules with molecular weights of 800 to 3000. One such is made from adipic acid and excess 1,2-propanediol. The oligomer then reacts with excess diisocyanate to form isocyanate-terminated prepolymer. The final step is conversion to polymer of desired high molecular weight.

Polyurethane foams, important commercially, form when carbon dioxide evolves from the reaction between isocyanate and water.

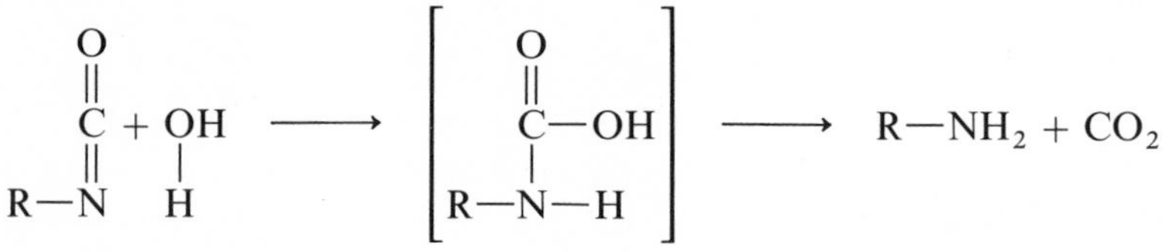

$$
\begin{matrix} O \\ \| \\ C \\ \| \\ R-N \end{matrix} + \begin{matrix} OH \\ | \\ H \end{matrix} \longrightarrow \left[\begin{matrix} O \\ \| \\ C-OH \\ | \\ R-N-H \end{matrix} \right] \longrightarrow R-NH_2 + CO_2 \quad (35)
$$

Choice of starting materials determines the rigidity of the foam. A prepolymer containing excess isocyanate groups reacts with water or a dicarboxylic acid to cause cross-linking or chain extension. This is accompanied by carbon dioxide evolution, which produces the foam. Catalysts and wetting agents are also used.

An example of a condensation reaction that produces a urethane group is that between a chloroformate and an amine.

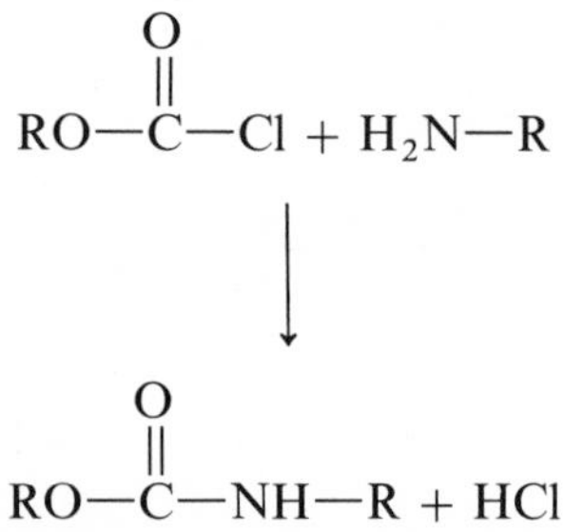

$$
\begin{matrix} O \\ \| \\ RO-C-Cl \end{matrix} + H_2N-R
$$

$$
\downarrow
$$

$$
\begin{matrix} O \\ \| \\ RO-C-NH-R \end{matrix} + HCl \quad (36)
$$

Bifunctional molecules produce polyurethanes.

Heteroaromatic Condensation Polymers. Recent interest in making polymers with high-temperature stability has prompted research into materials containing heterocyclic or heteroaromatic rings in the polymer chain. A leading example of this kind of polymer is polybenzimidazoles (33–35).

Benzimidazole derivatives have long been known as high-melting, chemical-resistant compounds. They are prepared from o-phenylenediamine and a carboxylic acid (or derivative).

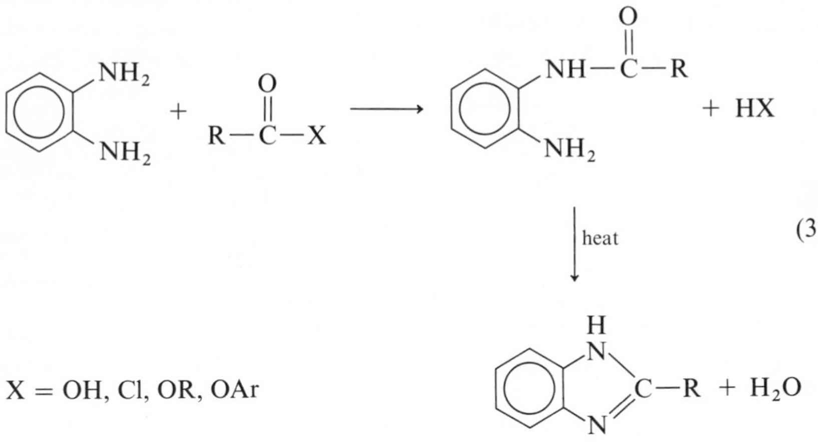

$$X = OH, Cl, OR, OAr \tag{37}$$

Polybenzimidazoles are made from an aromatic tetramine and a dicarboxylic acid. They are used as high-temperature resistant laminates and adhesives, and are being developed as fibers.

The first polybenzimidazoles were made from aliphatic diacids (36). Fully aromatic polybenzimidazoles, however, have better thermal stability. For example, the fully aromatic polymer made from 3,3'-diaminobenzidine and diphenyl isophthalate is stable in nitrogen at 600°C (37).

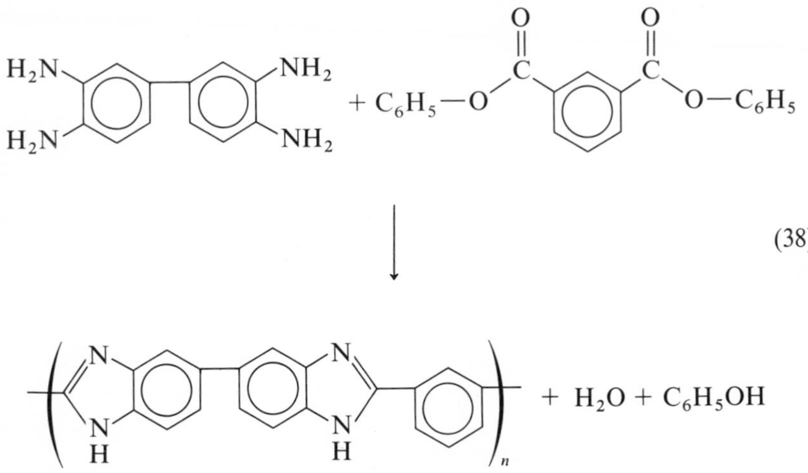

$$\tag{38}$$

The diphenyl ester of isophthalic acid gives optimum molecular weights. Alkyl esters and the free diacid react more slowly, and the latter tends to decarboxylate. The diacid chloride reacts too fast.

The low solubility and high melting points of these polymers make their processing difficult. Normally, a prepolymer is applied either as a hot melt or as a solution. Raising the temperature to 400°C then forms high polymer.

Phenol–Formaldehyde Resins (38). The first commercially significant synthetic polymers, the bakelite resins (phenoplasts), are included in this type of polymer. They are three-dimensional, cross-linked condensation polymers of phenol and formaldehyde. The polymerization, either basic or acidic, is done in stages. First, a linear soluble polymer is prepared. This is then converted into a cross-linked infusible and insoluble polymer.

Phenoplasts include two basic types, called resols and novolaks. Resols form by base-catalyzed condensation, usually using the hydroxides of sodium, potassium, or calcium. The novolaks form under acidic conditions. Resols contain aromatic methylol groups, and novolaks contain aromatic methanes. Resols tend to cross-link, whereas novolaks tend to be linear.

The base-catalyzed reaction includes three stages. Stage A resins, *resols* (Fig. 6), have low molecular weights and are soluble in base. Stage B resins, *resitols*, are slightly cross-linked, insoluble in base, but soluble in acetone. Stage C polymer, *resit*, is the final product, insoluble and infusible. Stage A molecules contain six or seven phenolic and methylol groups.

The formation of infusible product depends upon whether the catalyst is acidic or basic, the mole ratio of formaldehyde to phenol, and the temperature. With a basic catalyst and a mole ratio greater than one, the resol is convertible to insoluble, infusible product. An acid catalyst and the above mole ratio produces a linear polymer (a novolak).

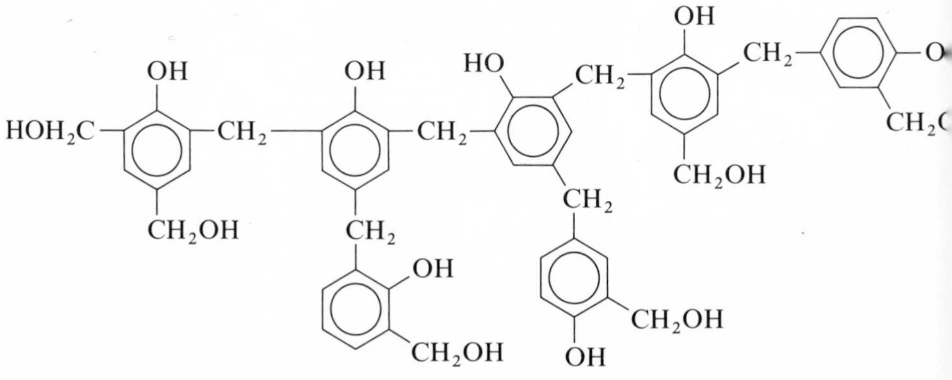

Figure 6   Structure of a resol.

Polymerization requires addition and condensation. Addition occurs only at the *ortho* and *para* positions of the phenol.

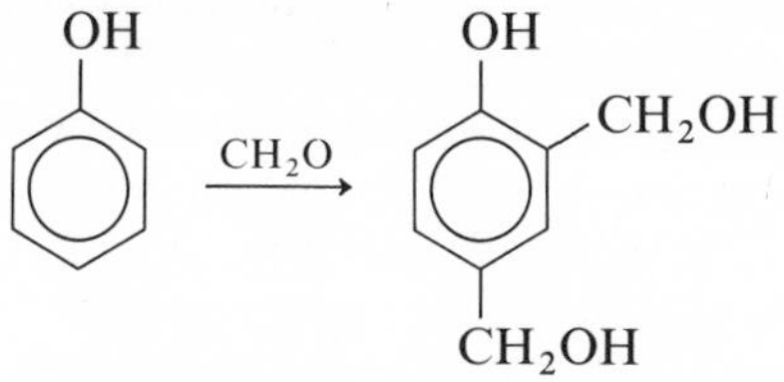

$$(39)$$

Therefore, up to three formaldehydes can add to one phenol.
Condensation also occurs only at the *ortho* and *para* positions.

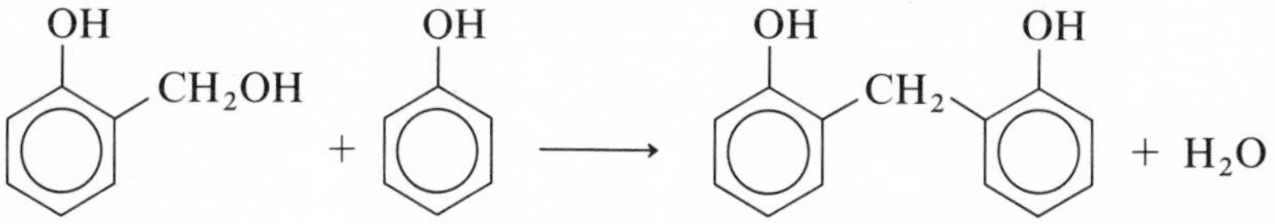

$$(40)$$

Since each phenol contains three reactive positions, cross-linking and gel formation are possible. Gelation should occur when the extent of reaction is 0.667.

Polysulfones (39–41). Aromatic polysulfones provide a class of thermoplastic polymer with exceptional physical and electrical properties at elevated temperatures. The polymers are prepared either by nucleophilic displacement reactions of 4,4'-dihalodiphenylsulfones or a Friedel–Crafts catalyzed polysulfonylation reaction.

The nucleophilic displacement of aryl halogens proceeds only with difficulty. However, when the compound contains a strong electron-withdrawing group such as a sulfone, the halogen is more prone to nucleophilic displacement. This is the principle used by the Union Carbide Corporation (39, 40) to prepare polysulfones based on bisphenol A:

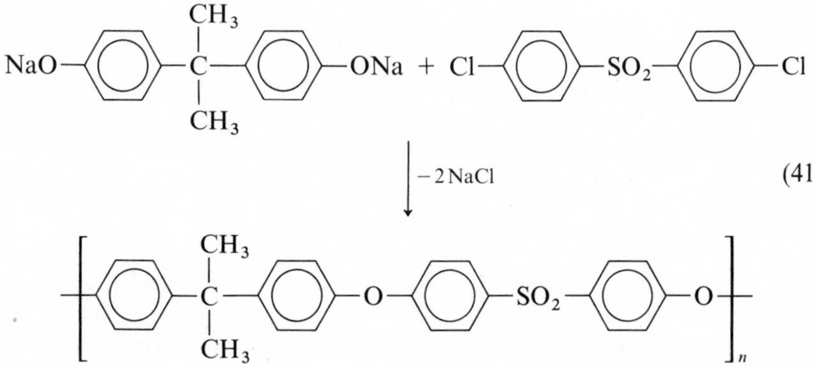

$$(41)$$

The choice of a solvent for the polymerization is quite limited. Dimethyl sulfoxide has been shown to be the best solvent. It is a solvent for the monomers and polymer and also increases the reaction rate. Water must be excluded from the reaction because it hydrolyzes the bisphenate salt. The molecular weight of the polymer sharply decreases in the presence of water.

The polymerization mechanism is that of an AA–BB polymerization previously discussed. Exact stoichiometric balance is required for the attainment of high molecular weight. Spectroscopic evidence indicates that the substitution is exclusively *para*.

Polyarylsulfones are also prepared by a Friedel–Crafts catalyzed polysulfonylation reaction (41–43). This is based on the classical Friedel–Crafts catalyzed reaction of a sulfonyl chloride and an aromatic ring to yield an arylsulfone:

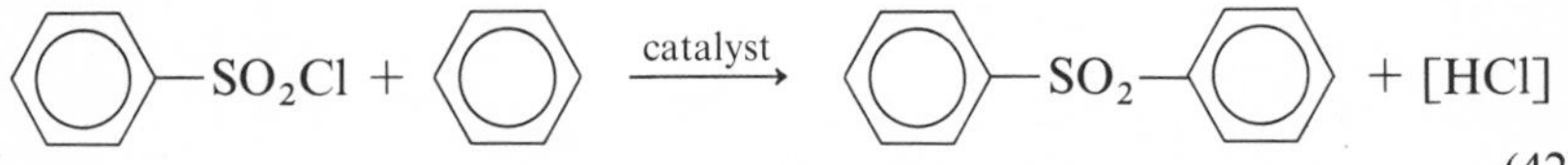

$$
\langle \text{O} \rangle\!-\!SO_2Cl + \langle \text{O} \rangle \xrightarrow{\text{catalyst}} \langle \text{O} \rangle\!-\!SO_2\!-\!\langle \text{O} \rangle + [HCl]
$$

(42)

The polymer-forming reactions may be depicted in the following manner:

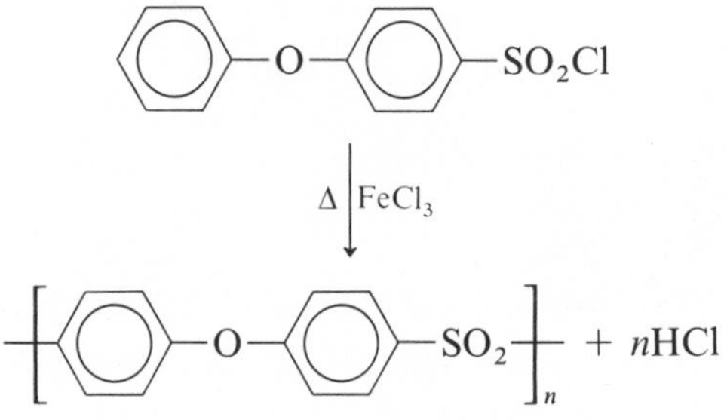

$$
\langle \text{O} \rangle\!-\!O\!-\!\langle \text{O} \rangle\!-\!SO_2Cl
$$

$$
\Delta \Big| FeCl_3
$$

(43)

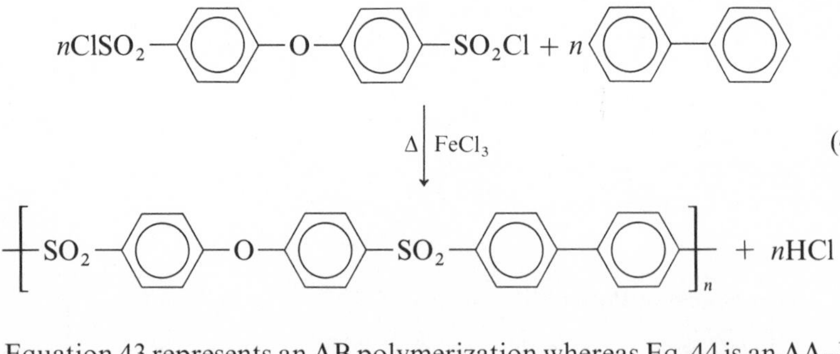

$$
\Big[\!\langle \text{O} \rangle\!-\!O\!-\!\langle \text{O} \rangle\!-\!SO_2\!\Big]_n + nHCl
$$

$$
nClSO_2\!-\!\langle \text{O} \rangle\!-\!O\!-\!\langle \text{O} \rangle\!-\!SO_2Cl + n\langle \text{O} \rangle\!-\!\langle \text{O} \rangle
$$

$$
\Delta \Big| FeCl_3
$$

(44)

$$
\Big[\!SO_2\!-\!\langle \text{O} \rangle\!-\!O\!-\!\langle \text{O} \rangle\!-\!SO_2\!-\!\langle \text{O} \rangle\!-\!\langle \text{O} \rangle\!\Big]_n + nHCl
$$

Equation 43 represents an AB polymerization whereas Eq. 44 is an AA–BB reaction. The sulfone group deactivates the aromatic ring to which it is attached. Therefore, with the proper choice of reaction conditions chain branching or cross-linking can be avoided or minimized. This is a very important concept. The functionality of the monomers without the deactivation is greater than two. As we have discussed previously, this leads to chain branching and/or cross-linking. This can be seen more clearly by looking at the structure of the product from the first step of Eq. 43 (the same principle holds for Eq. 44):

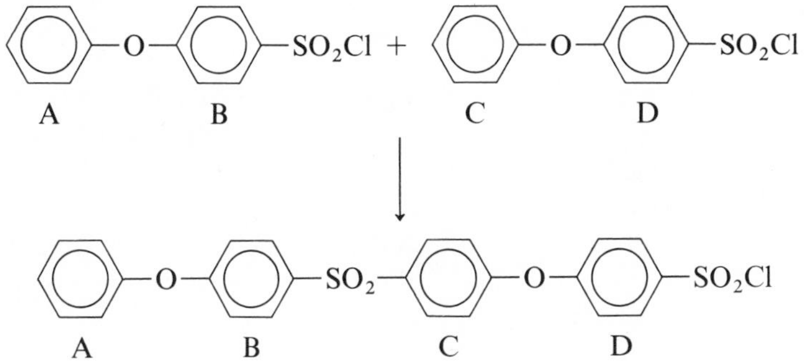

$$(45)$$

Spectroscopic evidence indicates that attack at the *para* position of the aromatic is almost exclusively favored. In the starting compounds the aromatic nuclei A and C are favored for attack. In the dimer, rings B, C and D are now deactivated to further sulfonylation because of the powerful electron-withdrawing effect of the sulfone group. Ring A, however, is still active and further sulfonylation takes place on its *para* position. We can also see now why only dinuclear aromatics are used for this polymerization. This is to provide active terminal aromatic nuclei which can participate in the propagation steps (ring A).

The most popular catalyst for the polymerization is anhydrous ferric chloride. Certain other Lewis acid catalysts may also be used. At the reaction temperatures normally employed (80–250°C) 0.1–1.0 mole % of catalyst is used. The polymerization can be carried out using melt techniques. However, it is preferable to employ an inert diluent as the reaction medium because of the lower temperatures required. The solvent must be inert toward electrophilic attack by the sulfonyl chloride and compatible with the Lewis acid catalysts. Nitro compounds such as nitrobenzene, sulfones, and a number of chlorinated hydrocarbons have been used as solvents.

2. Other Stepwise Polymerizations

This section presents selected examples of stepwise polymerizations which are not condensation reactions. These include 1,3-dipolar polymerization, Diels–Alder polymerization, and polymerization by oxidative coupling.

1,3-*Dipolar Polymerization*

The 1,3-dipolar addition reaction is a well-known organic reaction that has only recently been adapted to polymer synthesis. In general this reaction may be depicted as follows:

$$\begin{array}{ccc} \overset{\oplus}{a}\diagup^{\displaystyle b}\diagdown \underset{:}{c}^{\ominus} & \longrightarrow & a\diagup^{\displaystyle b}\diagdown c \\ \Vert & & | \qquad | \\ d\!=\!=\!e & & d\!-\!\!-\!\!-\!e \end{array} \tag{46}$$

Compound a–b–c is the 1,3-dipole and may be a nitrile ylide, a nitrilimine, or similar compound. Compound d–e is a dipolarophile and may be almost any double or triple bond.

One of the attractive features of adapting this reaction to polymer synthesis is that with the proper choice of reactants it is possible to prepare polymers with recurring five-membered aromatic heterocyclic rings. As described in an earlier section, these materials are being extensively studied as thermally stable polymers.

There are numerous examples of 1,3-dipolar polymerization. A representative one is the preparation of a polypyrazole from a diacetylene compound and a dinitrilimine:

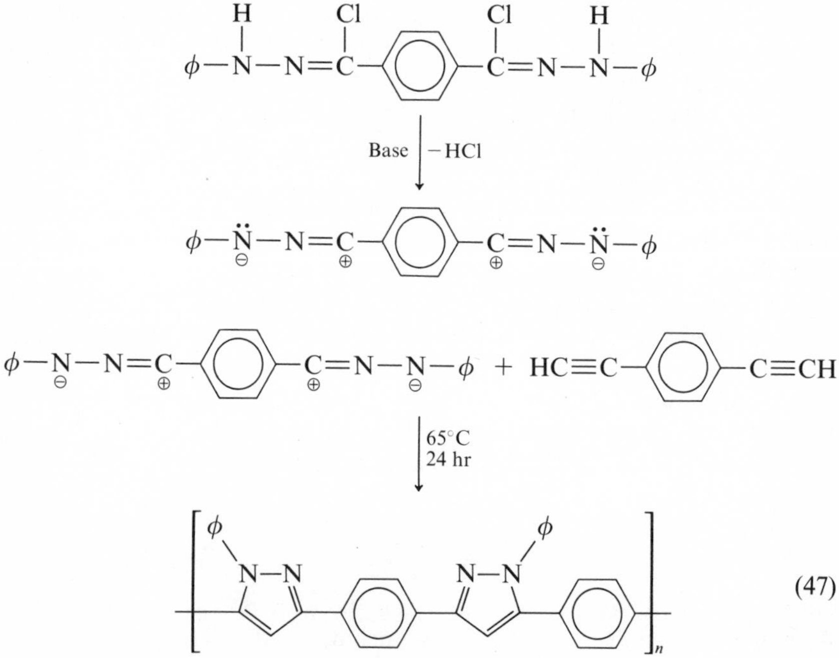

The polymer was obtained in 80–90% yield ($[\eta] = 0.32$).

The polymerization proceeds through a step reaction process similar to that discussed under condensation polymerization for the reaction of an A–A compound with a B–B compound. This also applies to the polymerization of an A–B compound, one which has both a dipolar group and a dipolarophile group. For example, when a benzene solution of p-azidophenylacetylene is heated at 60°C for 16 hr a polymer is obtained having the repeat unit of 1,4- and 1,5-substituted triazole rings in a random distribution (46):

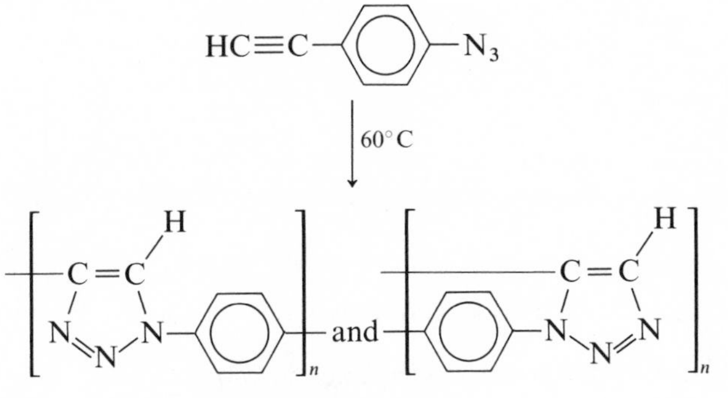

$$HC\equiv C - \underset{}{\bigcirc} - N_3$$

$$\Big\downarrow 60°C \tag{48}$$

Diels–Alder Polymerization (47, 48)

Diels–Alder polymerization occurs when a bifunctional diene reacts with a bifunctional dienophile, or when a monomer contains both types of reactive groups. Many such reactions have been carried out. However, high polymer does not usually form in good yield because of side reactions and because the Diels–Alder reaction itself is reversible. The reaction can sometimes be made irreversible by choosing a system in which the adduct eliminates a molecule such as carbon dioxide, carbon monoxide, or sulfur dioxide.

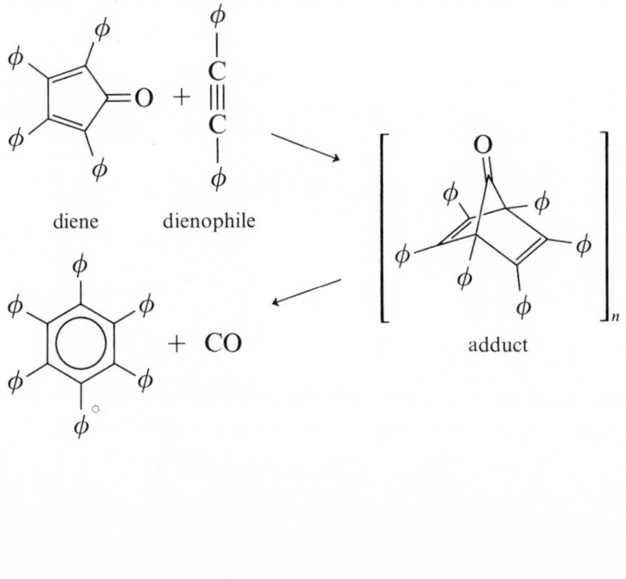

diene   dienophile   adduct   (49)

$+ CO$

Phenylated polyphenylenes can be prepared with $\overline{M}_n$ of about 40,000 in good yield (90%) using this technique (49).

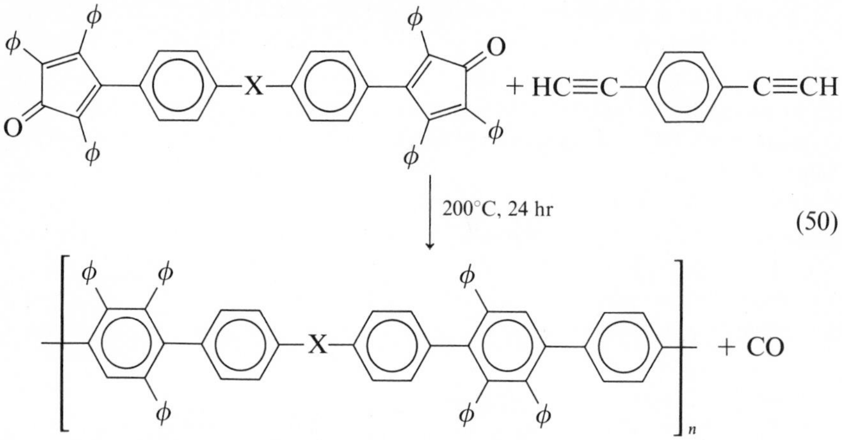

$$200°C, 24 \text{ hr} \tag{50}$$

$+ \text{CO}$

X = O, S, or nothing. This kind of polymer has good thermal stability.

Polymerization by Oxidative Coupling (50–53)

Oxidative coupling involves intermolecular dehydrogenation.

$$X{-}H + H{-}X \xrightarrow{\;[O]\;} X{-}X + H_2O \tag{51}$$

X represents a group that activates the hydrogen atom being removed. If this group contains two active hydrogen atoms, the reaction can lead to polymer.

$$H{-}X{-}H \xrightarrow{\;[O]\;} {+}X{+} + H_2O \tag{52}$$

Aromatic polyethers have been prepared this way by General Electric Co. (50, 52).

Oxygen passing into a solution of a 2,6-disubstituted phenol, a cuprous salt, and an amine may cause two reactions. One is a carbon–oxygen coupling leading to polymer. The other is a carbon–carbon coupling producing quinone dimer.

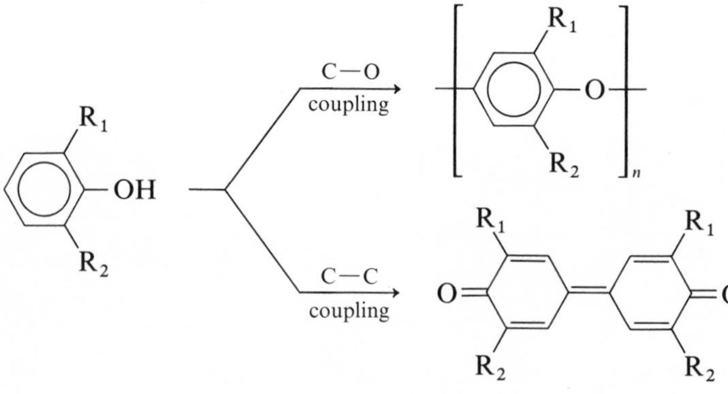

(53)

The nature of the product depends upon the structure of the phenol and the reaction conditions. If either *ortho* substituent is a bulky group, such as *tert*-butyl, the diphenoquinone results.

The reaction mechanism has received considerable study, but the exact nature of the active species responsible for polymerization is not known. It is probably a phenoxy–cuprous complex containing two amines. The oxygen raises copper(I) to copper(II), the oxidation state necessary for coupling. With an equimolar amount of copper(II), the oxygen is unnecessary.

One would probably expect that the propagation reaction of this polymerization is a chain process, with a monomeric species adding to the end of a growing chain. However, the evidence points to a step reaction process. For example, one can isolate dimers, trimers, and higher oligomers during the polymerization. It is also possible to conduct a "normal" polymerization starting with the dimer.

Oxidative coupling also produces polymer from other materials containing two active hydrogen atoms, for example, diethynyl compounds. In the presence of cuprous chloride, an amine, and oxygen; *m*-diethynylbenzene forms moderately high molecular weight polymer (54).

$$HC\equiv C \underset{}{\overset{}{\longmapsto}} C\equiv CH \xrightarrow{[O]} \left[-C\equiv C \underset{}{\overset{}{\longmapsto}} C\equiv C - \right]_n + H_2O \tag{54}$$

B. RING-OPENING POLYMERIZATION (55, 56)

A cyclic compound may polymerize by a ring-opening reaction. Sometimes this is the only way to obtain high polymer containing a particular repeating unit. Cyclic amides, esters, ethers, siloxanes, and *N*-carboxyanhydrides polymerize by ring opening. Although the polymer may contain a condensation-type repeating unit, the kinetics of the reaction indicates a chain mechanism similar to vinyl polymerization.

Ring stability is important in this reaction. Some five- or six-membered rings are so stable they do not polymerize. Of the many important examples of ring-opening polymerization, only lactam (cyclic amide) polymerization is discussed here (57).

$$\begin{array}{c} (CH_2)_x \quad C=O \\ \Big| \\ NH \end{array} \longrightarrow \left[NH\!\!+\!\!CH_2\!\!\xrightarrow{}_x\!\!\overset{O}{\overset{\|}{C}} \right]_n \tag{55}$$

Lactams polymerize either by melt polymerization or by anionic polymerization. Melt polymerization requires a catalytic amount of water. The water opens the lactam to form an amino acid

$$\underset{\underset{NH}{|}}{(\overset{\frown}{CH_2})_x}\ C{=}O + H_2O \longrightarrow H_2N{+}CH_2{\xrightarrow{}_x}CO_2H \qquad (56)$$

The amino acid then initiates polymerization through amide interchange.

$$\underset{\underset{NH}{|}}{(\overset{\frown}{CH_2})_x}\ C{=}O + H_2N{+}CH_2{\xrightarrow{}_x}CO_2H$$

$$\big\downarrow \qquad\qquad (57)$$

$$H_2N{+}CH_2{\xrightarrow{}_x}\overset{\overset{O}{\|}}{C}{-}NH{+}CH_2{\xrightarrow{}_x}CO_2H$$

The water is necessary, since pure dry caprolactam does not polymerize on heating to 250°C. Alternatively, an amino acid may be added to initiate polymerization. Besides the normal ring-opening propagation reaction, some *stepwise* propagation may occur between chains.

$$\sim{+}CH_2{\xrightarrow{}_x}NH_2 + HO{-}\overset{\overset{O}{\|}}{C}{+}CH_2{\xrightarrow{}_x}\sim$$

$$\big\downarrow \qquad\qquad (58)$$

$$\sim{+}CH_2{\xrightarrow{}_x}NH{-}\overset{\overset{O}{\|}}{C}{+}CH_2{\xrightarrow{}_x}\sim$$

The melt technique does not usually provide complete reaction. Besides polymer, the product mixture contains lactam, monomeric amino acid, and cyclic oligomers.

Anionic polymerization is extremely effective. Many lactams, unpolymerizable in the melt, form high polymer when initiated anionically. The six-membered lactam α-piperidone is one such.

Anionic polymerization of caprolactam normally involves adding about 1% each of sodium caprolactam and N-acetylcaprolactam to the pure dry monomer. The mixture is heated above the melting point of the monomer. To initiate, caprolactam anion opens the activated ring of N-acetylcaprolactam.

$$(\text{CH}_2)_5 \quad \text{C}=\text{O} \qquad (\text{CH}_2)_5 \quad \text{C}=\text{O}$$
$$\text{N}^{\ominus} \longrightarrow \qquad \text{N}-\overset{\displaystyle }{\underset{\displaystyle \underset{\parallel}{\text{O}}}{\text{C}}}-\text{CH}_3$$

$$\Big\downarrow \tag{59}$$

$$(\text{CH}_2)_5 \quad \text{C}=\text{O}$$
$$\text{N}-\underset{\underset{\text{O}}{\parallel}}{\text{C}}\!+\!\text{CH}_2\!\!+_{\!\!5}\!\overset{\ominus}{\text{N}}-\underset{\underset{\text{O}}{\parallel}}{\text{C}}-\text{CH}_3$$

The new anion takes a proton from a monomer, forming another caprolactam anion (activated monomer):

$$(\text{CH}_2)_5 \quad \text{C}=\text{O}$$
$$\text{N}-\underset{\underset{\text{O}}{\parallel}}{\text{C}}\!+\!\text{CH}_2\!+_{\!\!5}\!\overset{\ominus}{\text{N}}-\underset{\underset{\text{O}}{\parallel}}{\text{C}}-\text{CH}_3 \quad + \quad (\text{CH}_2)_5 \quad \text{C}=\text{O}$$
$$\text{NH}$$

$$\Big\downarrow \tag{60}$$

$$(\text{CH}_2)_5 \quad \text{C}=\text{O} \quad + \quad (\text{CH}_2)_5 \quad \text{C}=\text{O}$$
$$\text{N}^{\ominus} \qquad \qquad \text{N}-\underset{\underset{\text{O}}{\parallel}}{\text{C}}\!+\!\text{CH}_2\!+_{\!\!5}\!\text{NH}-\underset{\underset{\text{O}}{\parallel}}{\text{C}}-\text{CH}_3$$

Propagation occurs by repetition of these steps. Caprolactam anion opens the activated ring.

$$(\text{CH}_2)_5 \quad \text{C}=\text{O} \qquad (\text{CH}_2)_5 \quad \text{C}=\text{O}$$
$$\text{N}^{\ominus} \longrightarrow \qquad \text{N}-\underset{\underset{\text{O}}{\parallel}}{\text{C}}\!+\!\text{CH}_2\!+_{\!\!5}\!\text{NH}-\underset{\underset{\text{O}}{\parallel}}{\text{C}}-\text{CH}_3$$

$$\Big\downarrow \tag{61}$$

$$(\text{CH}_2)_5 \quad \text{C}=\text{O}$$
$$\text{N}-\underset{\underset{\text{O}}{\parallel}}{\text{C}}\!+\!\text{CH}_2\!+_{\!\!5}\!\overset{\ominus}{\text{N}}-\underset{\underset{\text{O}}{\parallel}}{\text{C}}\!+\!\text{CH}_2\!+_{\!\!5}\!\text{NH}-\underset{\underset{\text{O}}{\parallel}}{\text{C}}-\text{CH}_3$$

Then the new anion takes a proton from monomer, forming activated monomer, and so on.

The proton-transfer step is extremely fast. Nuclear magnetic resonance kinetic studies indicate a rate constant of 10^5 l/mole-sec (58). Thus the rate-determining step is probably the anion attack to open the activated ring. This mechanism presumably operates in the anionic polymerizations of other lactams.

Some lactams polymerize readily even without the N-acetyl activator. Here anion attack on the lactam itself initiates polymerization. As before, however, propagation is by anion attack on an activated ring. Lactam reactivity varies with the amount of ring strain. The highly strained β-lactams (four-membered rings) polymerize even in solution at room temperature. Activating compounds, though not necessary here, accelerate the reaction.

In general, polymerizability depends upon ring size and substituents. Very few five- or six-membered-ring lactams polymerize. About half of the seven-membered ring lactams do, and all four-, eight-, and nine-membered-ring lactams do. Polymerizability decreases when the ring contains an alkyl or aryl substituent. A substituent on the lactam nitrogen atom usually prevents polymerization. Another hetero atom in the ring either has no effect or reduces polymerizability.

C. ADDITION POLYMERIZATION

1. Introduction

Most of the remainder of this chapter deals with polymerizations that involve addition reactions, but that are neither stepwise nor ring-opening polymerizations; these are mainly vinyl-monomer polymerizations. The addition mechanism may be free radical, ionic, or coordination.

Table 3 lists some common vinyl monomers and the types of systems in which they polymerize at least moderately well. The nature of the system is crucial to the polymerizability of vinyl monomers. Succeeding sections discuss the choice of a suitable system for a given monomer.

2. Stereochemistry of Vinyl Polymers (60, 61)

The fundamental properties of a vinyl polymer depend heavily on the stereochemistry of the polymer. A vinyl monomer can polymerize either in a head-to-tail fashion, or in a head-to-head fashion (Fig. 7). In practice, all but a few percent (maximum) of the units are head-to-tail.

Another stereochemical factor is the relative configurations of neighboring tertiary carbon atoms. This also affects the polymer properties. If the relative

$$-CHR-CH_2-CHR-CH_2-CHR-CH_2-CHR-CH_2-CHR-CH_2-$$
(a)

$$-CH_2-CHR-CHR-CH_2-CH_2-CHR-CHR-CH_2-CH_2-CHR-$$
(b)

Figure 7   Structure of a head-to-tail and head-to-head arrangement of repeat units.

configurations fall into a regular pattern, the polymer is called *stereoregular*. Natta and co-workers prepared and characterized the first stereoregular polymers. According to Natta's terminology, identical relative configurations constitute an *isotactic* stereochemistry. Alternating opposite configurations constitute a *syndiotactic* stereochemistry. If no regular pattern exists, the polymer is called *atactic* (Fig. 8).

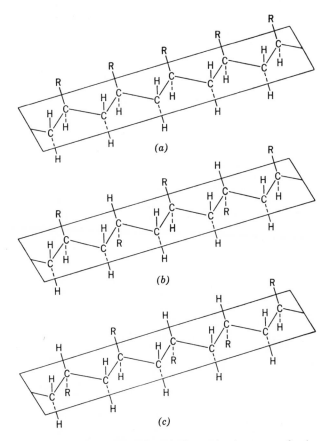

Figure 8   (a) Isotactic; (b) syndiotactic and (c) atactic placement of polymeric units.

Table 3   Addition mechanisms for polymerization of vinyl compounds (59)

Cationic only		Free Radical only		Anionic only	

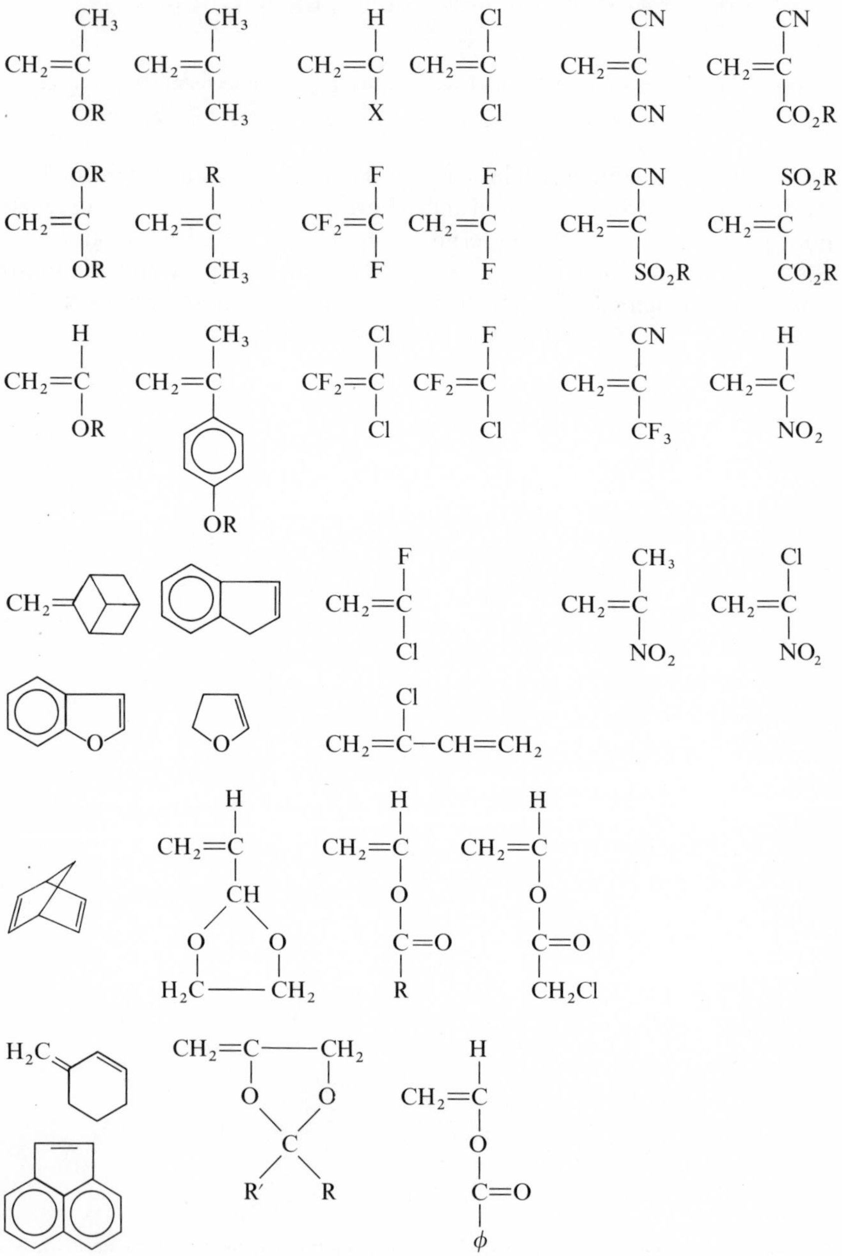

CH$_2$=C(CH$_3$)(OR)   CH$_2$=C(CH$_3$)(CH$_3$)   CH$_2$=C(H)(X)   CH$_2$=C(Cl)(Cl)   CH$_2$=C(CN)(CN)   CH$_2$=C(CN)(CO$_2$R)

CH$_2$=C(OR)(OR)   CH$_2$=C(R)(CH$_3$)   CF$_2$=C(F)(F)   CH$_2$=C(F)(F)   CH$_2$=C(CN)(SO$_2$R)   CH$_2$=C(SO$_2$R)(CO$_2$R)

CH$_2$=C(H)(OR)   CH$_2$=C(CH$_3$)(—C$_6$H$_4$—OR)   CF$_2$=C(Cl)(Cl)   CF$_2$=C(F)(Cl)   CH$_2$=C(CN)(CF$_3$)   CH$_2$=C(H)(NO$_2$)

CH$_2$=C(F)(Cl)   CH$_2$=C(CH$_3$)(NO$_2$)   CH$_2$=C(Cl)(NO$_2$)

CH$_2$=C(Cl)—CH=CH$_2$

CH$_2$=C(H)(CH⟨O—CH$_2$/O—CH$_2$⟩)   CH$_2$=C(H)(O—C(=O)—R)   CH$_2$=C(H)(O—C(=O)—CH$_2$Cl)

CH$_2$=C—CH$_2$ with O—C(R')(R)—O bridge   CH$_2$=C(H)(O—C(=O)—φ)

58

Table 3 (*Continued*)

Cationic or Free Radical	Anionic or Free Radical

Cationic or Free Radical structures:
- Carbazole with N–CH=CH₂
- Pyrrolidinone (N–CH=CH₂)
- Dibenzofuran with –C(H)=CH₂
- $CH_2=CH_2$
- benzene with CH₂=CH₂ and –CH=CH₂ substituents

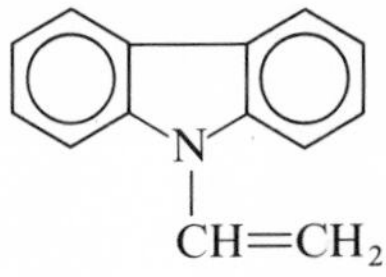

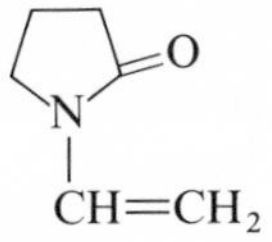

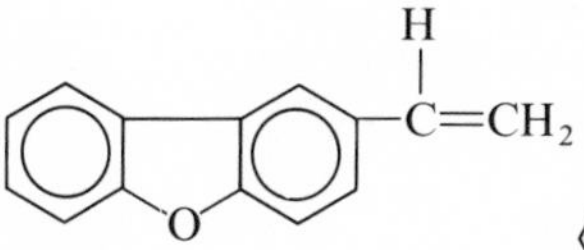

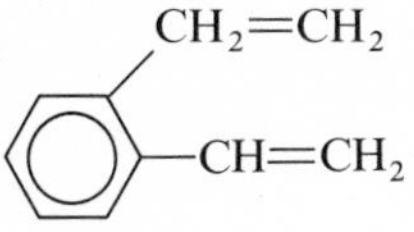

Anionic or Free Radical structures:

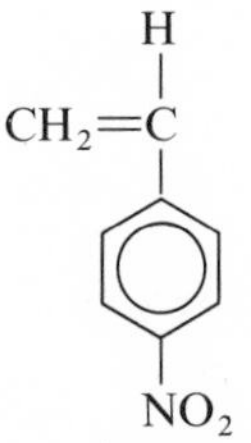

$$CH_2=\overset{\overset{\displaystyle H}{|}}{C}\!-\!CO_2R \qquad CH_2=\overset{\overset{\displaystyle H}{|}}{C}\!-\!O=C-NR_2$$

$$CH_2=\overset{\overset{\displaystyle CH_3}{|}}{C}\!-\!CO_2R \qquad CH_2=\overset{\overset{\displaystyle CH_3}{|}}{C}\!-\!\overset{\underset{\displaystyle O}{}}{C}NR_2$$

$$CH_2=C{\overset{\displaystyle CO_2R}{\underset{\displaystyle CO_2R}{}}} \qquad CH_2=\overset{\overset{\displaystyle H}{|}}{C}\!-\!CN$$

$$CH_2=\overset{\overset{\displaystyle H}{|}}{\underset{\underset{\displaystyle NO_2}{\bigcirc}}{C}} \qquad CH_2=\overset{\overset{\displaystyle CH_3}{|}}{C}\!-\!CN$$

$$CH_2=\overset{\overset{\displaystyle Cl}{|}}{\underset{\underset{\displaystyle OR}{C=O}}{C}} \qquad CH_2=\overset{\overset{\displaystyle Cl}{|}}{\underset{\underset{\displaystyle SR}{C=O}}{C}}$$

Free Radical, Cationic, or Anionic

$$CH_2=CH_2 \qquad CH_2=\overset{\overset{\displaystyle CH_3}{|}}{\underset{\underset{\displaystyle \phi}{}}{C}} \qquad CH_2=\overset{\overset{\displaystyle CH_3}{|}}{C}\!-\!CH=CH_2 \qquad CH_2=\overset{\overset{\displaystyle H}{|}}{C}\!-\!CH_3$$

$$CH_2=\overset{\overset{\displaystyle H}{|}}{\underset{\underset{\displaystyle \phi}{}}{C}} \qquad \text{(phenanthrene)} \qquad CH_2=CH-CH=CH_2 \qquad CH_2=\overset{\overset{\displaystyle H}{|}}{\underset{\underset{\displaystyle CH_3}{C=O}}{C}}$$

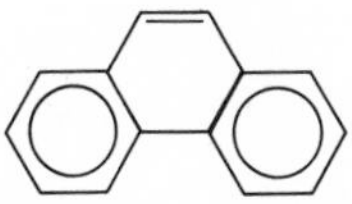

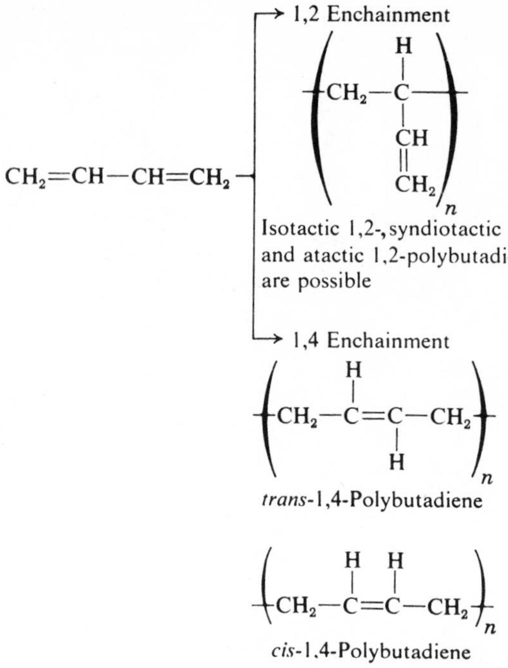

Figure 9   Possible structures for polybutadiene.

If 1,3-butadiene polymerizes by 1,2 addition, it may form isotactic, syndiotactic, or atactic polymer. However, if it polymerizes by 1,4 addition, it can only form *trans* or *cis* polymer (Fig. 9). The *trans* polymer is brittle or powdery, and the *cis* polymer is an excellent elastomer.

D. FREE-RADICAL POLYMERIZATION

Many common synthetic materials, such as polyethylene, polystyrene, and polymethyl methacrylate, are made by free-radical polymerization. All these polymerizations involve the same mechanistic sequence. First a free-radical initiator adds to a monomer:

$$R\cdot + CH_2{=}CHX \longrightarrow R{-}CH_2{-}\dot{C}H{-}X \tag{62}$$

and produces a new free radical. The new radical starts a chain by adding to another monomer.

$$R{-}CH_2{-}\dot{C}HX + CH_2{=}CHX \longrightarrow R{-}CH_2{-}CHX{-}CH_2{-}\dot{C}HX \tag{63}$$

This step repeats, increasing the chain by one monomer unit at a time. The chain terminates by coupling with another free radical, or by disproportionation.

A free radical is usually obtained by homolysis of an ordinary covalent bond. Examples include peroxide dissociation, halogen photolysis, and azo-compound decomposition.

$$R-O-O-R \xrightarrow{\Delta} 2R-O\cdot$$

$$Cl-Cl \xrightarrow{hv} 2Cl\cdot \tag{64}$$

$$CH_3-N{=}N-CH_3 \longrightarrow 2H_3C\cdot + N{\equiv}N$$

A free radical can also form in a one-electron oxidation–reduction reaction. Ferrous ion and cumene hydroperoxide produce oxy radicals even at 0°C.

$$Fe^{2+} + \phi-\underset{\underset{CH_3}{|}}{\overset{\overset{CH_3}{|}}{C}}-O-OH \longrightarrow Fe^{3+} + \phi-\underset{\underset{CH_3}{|}}{\overset{\overset{CH_3}{|}}{C}}-O\cdot + OH^{\ominus} \tag{65}$$

Alkyl radicals, the type responsible for propagation can be destroyed in two ways, by combination and by disproportionation.

$$R-CH_2-CHX\cdot + \cdot XHC-CH_2-R$$
$$\downarrow \tag{66}$$
$$R-CH_2-\underset{\underset{X}{|}}{CH}-\underset{\underset{X}{|}}{CH}-CH_2-R$$

$$R-CH_2-CHX\cdot + R-CH_2-CHX\cdot$$
$$\downarrow \tag{67}$$
$$R-CH_2-CH_2X + R-CH{=}CHX$$

In the first case, the product contains two initiator units per molecule. In the second case, half the product molecules are terminal olefins.

Radicals react mainly in two ways, by atom transfer and by addition. Equation 67 is an example where a hydrogen atom is transferred from one molecule to the other. In addition, a radical adds to an unsaturated center and forms a new radical.

$$R-\dot{C}H_2 + CH_2{=}\underset{\underset{X}{|}}{CH} \longrightarrow R-CH_2-CH_2-\underset{\underset{X}{|}}{\dot{C}H} \tag{68}$$

The latter is the basis for free-radical polymerization. The characteristic of this kind of propagation is that the chain grows by one monomer unit at a time.

Besides terminating by combination or by disproportionation, a chain may abstract an atom from another molecule. This terminates the original chain's growth, but it creates another free radical.

$$R-CH_2-\dot{C}H_2 + R-H \longrightarrow R-CH-CH_3 + R\cdot \qquad (69)$$

This reaction is called *chain transfer* and may occur with initiator monomer, solvent, polymer, or an impurity. A material deliberately added to a system to control molecular weight by chain transfer is called a *chain-transfer* agent. The new radical derived from the chain-transfer agent may initiate a new chain in the usual way.

$$R\cdot + CH_2 = \underset{\underset{X}{|}}{CH} \longrightarrow R-CH_2-\underset{\underset{X}{|}}{\dot{C}H} \qquad (70)$$

Because the process does not decrease the concentration of radicals in the system, it does not affect the overall rate of polymerization. It does, however, reduce the molecular weight of the product. If a material reduces both molecular weight and rate, it is called a *retarder*.

A material that actually prevents polymerization is called an *inhibitor*. The period of prevention is called the induction period. Most vinyl monomers contain inhibitors to prevent polymerization during storage.

1.   Kinetics of Free-Radical Polymerization (1, 6, 11, 12)

The overall rate of polymerization as well as the length of the polymeric chains formed in addition polymerization are determined by the rates of the individual processes of initiation, propagation, and termination. The overall mechanism for the conversion of monomer to polymer by use of a typical free-radical initiator, I, can be described by the following set of rate equations:
Initiation:

(i) $$I \xrightarrow{k_d} 2R\cdot$$

(ii) $$R\cdot + M \xrightarrow{k_i} RM\cdot$$

Chain propagation:

$$RM\cdot + M \xrightarrow{k_p} RMM\cdot$$

$$RMM\cdot + M \xrightarrow{k_p} RMMM\cdot$$

In general,

(iii) $$RM_{n-1}{}^{\textbf{·}} + M \xrightarrow{\ k_p\ } RM_n{}^{\textbf{·}}$$

Chain termination:

(*a*) Termination by combination

$$M_n{}^{\textbf{·}} + M_m{}^{\textbf{·}} \xrightarrow{\ k_{t,c}\ } M_{n+m}$$

(*b*) Termination by disproportionation

$$M_n{}^{\textbf{·}} + M_m{}^{\textbf{·}} \xrightarrow{\ k_{t,d}\ } M_n + M_m$$

If we assume that both termination reactions are kinetically equivalent, we may then write a generalized rate equation for the termination process as

(iv) $$M_n{}^{\textbf{·}} + M_m{}^{\textbf{·}} \xrightarrow{\ k_t\ } P \qquad \text{where} \quad k_t = k_{t,c} + k_{t,d}$$

In the above equations, I is an initiator such as an organic peroxide or an azo compound; $R{\textbf{·}}$ is a low molecular weight radical (also known as a primary radical); M is a molecule of monomer; and $M_n{\textbf{·}}$ is a polymeric radical with n monomer units in the chain. k_d, k_i, k_p, and k_t are the individual rate constants for the four reactions.

By making several assumptions, relatively simple expressions for the rate of polymerization (R_p) and degree of polymerization ($\overline{DP}$) may be obtained:

1. The rate of formation of free radicals ($R{\textbf{·}}$) is equal to the rate of consumption of free radicals (Eq. ii).

2. In the propagation step, radical activity is independent of chain length (i.e., only one k_p).

3. The rate of production of chain radicals is equal to the rate of termination of chain radicals:

$$\frac{d[RM_n{\textbf{·}}]}{dt} = 0 \tag{71}$$

4. The rate of polymerization is equal to the rate of propagation (i.e., the monomer consumed in Eq. ii is insignificant compared to that consumed in Eq. iii).

The rates of the four steps can be obtained from Eqs. i–iv as follows:

$$R_d = 2k_d[I] \tag{72}$$

$$R_i = k_i[R{\textbf{·}}][M] \tag{73}$$

$$R_p = k_p C^*[M] \tag{74}$$

$$R_i = 2k_t C^{*2} \tag{75}$$

The quantity $C*$ is defined as

$$C* \equiv \sum_{i=1}^{\infty} [RM_i\cdot] \tag{76}$$

Taking into account assumption 1, R_d may be set equal to R_i so that

$$R_i = 2k_d[I] \tag{77}$$

From the second steady-state assumption 3 it is found that

$$R_i = R_t$$

$$2k_d[I] = 2k_t C*^2$$

or

$$C* = \left(\frac{k_d}{k_t}[I]\right)^{1/2} \tag{78}$$

The rate of polymerization (R_p) (assumptions 2 and 4) may now be written as

$$R_p = v_p k_p C*[M] \tag{79}$$

Combining Eqs. (78) and (79) gives

$$R_p = \frac{k_p}{k_t^{1/2}}[M](k_d[I])^{1/2} \tag{80}$$

Equation 80 predicts that *the rate of formation of polymer should be proportional to the square root of the initiator concentration and the first power of the monomer concentration.*

2.   Kinetic Chain Length

The kinetic chain length, v, is the number of monomer molecules consumed by each primary radical. This is given by

$$v = \frac{\text{rate of propagation}}{\text{rate of initiation}} = \frac{R_p}{R_i}$$

$$= \frac{k_p C*[M]}{2k_t C*^2} = \frac{k_p[M]}{2k_t C*} \tag{81}$$

Substituting for $C*$ from Eq. 78

$$v = \frac{k_p[M]}{2(k_d k_t)^{1/2}[I]^{1/2}} \tag{82}$$

It can be seen from Eq. 81 that the kinetic chain length is inversely proportional to the rate of initiation (or rate of initiator decomposition). This has been proved experimentally in many vinyl polymerization reactions. It is also seen from Eq. 81 that the chain length obtained for a monomer at a given rate of reaction is a function of the nature of the monomer (as shown by the value of $k_p/k_t^{1/2}$) and not of the initiator.

The actual size of the polymer molecule depends on the mechanism of termination, whereas the kinetic chain length is independent of the termination mechanism. Thus if termination occurs by coupling, the number of monomer units in the average polymer molecule ($\overline{DP}$) is twice the kinetic chain length ($\overline{DP} = 2v$). If termination occurs by disproportionation, the degree of polymerization is equal to the kinetic chain length. In the presence of a solvent or other chain-transfer agent, the degree of polymerization can be arrived at by the following kinetic treatment:

Let XY represent the chain-transfer agent.

$$M_n\cdot + XY \xrightarrow{k_{tr}} M_n X + Y\cdot$$

$$Y\cdot + M \xrightarrow{k'_p} YM\cdot, \text{ etc.} \tag{83}$$

Unless $k'_p \ll k_p$ (in which case the agent would also act as an inhibitor), the chain-transfer agent has negligible effect on the rate of polymerization. The expression for the kinetic chain chain length can then be written as

$$v = \frac{R_p}{R_i + k_{tr}[XY]C^*} \tag{84}$$

Substituting for C^* from Eq. 78 into Eq. 84 and inverting, we obtain

$$\frac{1}{\overline{DP}} = \frac{1}{\overline{DP}_0} + \frac{k_{tr}}{k_p}\frac{[XY]}{[M]} \tag{85}$$

where $\overline{DP}$ and $\overline{DP}_0$ are the degree of polymerization of the polymer obtained with and without the transfer agent. The ratio k_{tr}/k_p is known as the chain-transfer coefficient. It is seen from Eq. 85 that by plotting $1/\overline{DP}$ against $[XY]/[M]$, a straight line with intercept $1/\overline{DP}_0$ and with slope k_{tr}/k_p is obtained.

The significance of the chain-transfer constant, k_{tr}/k_p (commonly designated as C_s) lies in the fact that the successful control of molecular weights in many commercially important vinyl polymerizations by use of these agents depends on the numerical values of these constants for a given monomer–transfer agent system. Thus the use of agents with transfer constants near unity ensures that the modifier is consumed at about the same rate as the monomer so that the ratio $[XY]/[M]$ remains constant during the entire reaction.

3.   Techniques of Radical Polymerization

The physical method by which a monomer is converted to a high polymer depends on the nature of the monomer, the end use of the polymer, and the economics of the various techniques. Other factors to be taken into consideration in choosing a technique and reaction conditions are conversion, molecular weight, molecular weight distribution, control of side reactions, and control of the rate of polymerization. The four common techniques of radical polymerization are bulk, solution, suspension, and emulsion. It is also possible to conduct polymerizations in the solid or gaseous state, but these techniques are not generally useful.

Bulk or Mass Polymerization (62, 63)

Bulk polymerizations are carried out by adding a soluble initiator (and possibly modifiers) to pure monomer in the liquid state. The temperature of the polymerization depends upon the monomer–initiator system used. These systems may be homogeneous, in which case the polymer is soluble in the monomer. In a heterogeneous system, the polymer precipitates from the molten monomer.

An excellent example of a homogeneous polymerization is the bulk polymerization of methyl methacrylate. The viscosity of the liquid reaction medium gradually increases until the mass solidifies to a clear homogeneous plastic.

Heterogeneous systems are obtained from the polymerization of monomers such as acrylonitrile and vinyl chloride. In these systems, two phases are present in the earlier stages of polymerization. The presence of two phases sometimes leads to undesirable side effects. The kinetics of bulk polymerization follow those outlined in Section D.1 at low conversions.

Bulk polymerization has several advantages over other techniques. The system is simple and requires no elaborate isolation or purification steps. The polymer is obtained pure (contaminated only by initiator fragments). Large castings may be prepared directly. Very high molecular weights are obtainable.

There are, however, several severe limitations in this technique. Mixing and heat transfer become difficult as the viscosity of the reaction mass increases. The problem of heat transfer is compounded by the highly exothermic nature of most free-radical polymerization reactions.

Several problems arise from the viscosity of the polymerization mass and the lack of good heat transfer. One is that the polymers are obtained with a broad molecular weight distribution. This may result in undesirable mechanical properties. Another serious problem is that for many bulk systems an autoacceleration occurs in the rate of polymerization. The phenomenon also occurs in concentrated solutions. This autoacceleration (also referred to as the Trommsdorff effect) takes place during the bulk polymerization of methyl methacrylate and methyl acrylate and to a lesser extent with styrene and vinyl acetate. It is attributed to a decrease in the termination rate of the polymerization. As the medium becomes more viscous the polymer chains lose mobility. The rate of bimolecular termination therefore decreases and the rate of polymerization increases, since the concentration of active radicals increases. If not properly controlled, the system could explode.

Solution Polymerization (64)

Solution polymerization has found limited commercial applicability in free-radical polymerizations. There are several major drawbacks to the technique. One is the need for an inert solvent to eliminate chain transfer. Problems are also encountered in the removal of solvent and isolation of pure polymer. Furthermore, in concentrated solutions, Trommsdorff effects are often observed.

However, the technique is used for the preparation of two industrially important polymers, polyacrylonitrile and poly-N-vinylpyrrolidone. Acrylonitrile is generally polymerized either by precipitation polymerization in water (the monomer is soluble whereas the polymer is not) or in dimethylformamide, dimethylacetamide, or dimethyl sulfoxide solution. Although some of the difficulties noted above do occur, the other methods are even more disadvantageous. Poly-N-vinylpyrrolidone is most often prepared in aqueous solution using hydrogen peroxide as the initiator.

Solution polymerization is also useful when the end use of the polymer requires a solution (certain adhesive and coating processes). It is also widely used for making polymers by ionic or coordination initiation. Polymers such as high-density polyethylene, butyl rubber, and ethylene–propylene copolymers are prepared this way. In these and other cases, the other techniques outlined in this section cannot be used. Reasons for this will be seen when these types of polymerizations are discussed.

Suspension Polymerization (65–67)

Suspension polymerization (also known as bead or pearl polymerization) is a widely used technique. It is essentially a modified bulk system in which the monomer is suspended as small droplets in an inert aqueous matrix. The water acts as a heat exchanger to remove the heat of polymerization from each of the monomer droplets. The suspension is maintained by vigorous stirring and the use of suspending agents and protective colloids. The polymer is obtained in the form of beads. After washing and drying it may be used directly in molding operations. Because the bead size obtained is a function of the stabilizing agent used, of the rate and type of agitation, and of the ratio of monomer to water, the particle size of the beads may be controlled by experimental conditions. Typically, a relatively narrow distribution of sizes is obtained.

Suspension polymerization has many attractive features. It is economically advantageous in that water is used as the heat exchange medium, compared to more expensive organic solvents for most solution polymerizations. Also, since the polymer is obtained directly in bead form, there are minimal problems in isolating the product, compared to emulsion or solution polymerization. Another advantage is in the purity of the polymer product. Although there are more additives present than either bulk or solution polymerization, the product is far easier to purify than in emulsion systems.

A key factor in carrying out a successful suspension polymerization is the composition of the aqueous phase. In it are dissolved organic polymers which act as protective coatings to prevent the agglomeration of the droplets. Polyvinyl alcohol, gelatin, and methyl cellulose are among the polymers used for this purpose.

Surface-active agents are also used to balance surface tension and interfacial tension of the droplets. This is important in allowing the initiator molecules to diffuse in and out of the droplets. Water-insoluble inorganic powders (uniform in size) are often added to ensure that monomer droplets are never larger than the inorganic powder. Such materials as talc, $BaSO_4$, $MgCO_3$, and $Ca_3(PO_4)_2$ are used.

The monomers must, of course, be water insoluble. When partial or slight solubility occurs, salts are added to force the monomer out of solution. The initiators are also water insoluble. Initiation therefore takes place in the dispersed phase. Peroxides are commonly used.

From a kinetic point of view, suspension polymerization is identical to bulk polymerization. Also, the technique is similar to emulsion polymerization in that both are heterogeneous systems involving a dispersed phase and a dispersion medium.

Emulsion Polymerization (68–73)

Emulsion polymerization is widely used in the commercial preparation of polymers. It has several distinct advantages over other techniques of polymerization. A unique characteristic of emulsion polymerization is that it attains high rates and high molecular weights.

In general, an emulsion polymerization would contain some or all of the following ingredients: dispersion medium, monomer, emulsifying agents, initiator, modifier, surface tension regulators, buffers, and protective colloids.

The dispersion medium, invariably water, serves not only as the suspending medium but also as a heat transfer medium. Most vinyl monomers used in emulsion polymerization are slightly water soluble. Monomers such as styrene, acrylates and methacrylates, vinyl chloride, butadiene, chloroprene, and many others have been used. Copolymers may also be prepared by this technique.

In a typical emulsion system the initiator must be water soluble (or have a water-soluble component) and be able to form free radicals below 100°C. Persulfate salts are commonly used as initiators. They may be decomposed either thermally (Eq. 86) or by a redox system, using the thiosulfate ion as the reducing agent (Eq. 87). A possible side reaction when using persulfate is also shown (Eq. 88):

$$K_2S_2O_8 \xrightarrow{\text{heat}} 2K^+ + 2SO_4^- \cdot \tag{86}$$

$$S_2O_8^{2-} + S_2O_3^{2-} \longrightarrow SO_4^- \cdot + SO_4^= + S_2O_3^- \tag{87}$$

$$SO_4^- \cdot + H_2O \longrightarrow HO \cdot + HSO_4^- \tag{88}$$

From this brief description of the components of emulsion systems it can be seen that a large number of ingredients are often necessary. Thus the main disadvantage of this method is often the difficulty of removing all impurities from the polymers. Another disadvantage is that the polymer is produced as a latex and cannot be agglomerated simply by dilution. Most often the latex is used directly.

It would be impossible to give a quantitative mechanistic and kinetic treatment of all types of emulsion polymerization. However, a quantitative treatment may be given for the case of a monomer slightly soluble in water, disregarding any role played by added ingredients such as modifiers or buffers.

The first step in an emulsion polymerization is to add the emulsifier to water. Part of the emulsifier dissolves in water but the greater part forms micelles. As the concentration of the emulsifier increases, a point called the "critical micelle concentration" (CMC) is reached. Below this point, no micelles exist and a polymerization in such a system would tend to be of the suspension type. Above the CMC, however, stable micelles exist. The micelles contain a relatively small number of molecules (roughly 100) and, if spherical, would be about 50 Å in diameter. There would be approximately 10^{18} micelles/ml of water. The hydrophobic portion of the emulsifier molecule forms the interior of the micelle, and the hydrophilic portion forms the surface of the micelle.

When a slightly water-soluble monomer is introduced, it distributes itself in three ways. A small portion forms solute molecules and another portion dissolves in the micelles. The micelles now swell to roughly double their original size. The greatest portion of the monomer is dispersed in small droplets with diameters usually larger than 1 μ. Part of the emulsifier has been transferred from the micelles to the monomer droplets, where it acts as a stabilizing agent.

When the water-soluble initiator is decomposed to form free radicals, they react with the small amount of monomer dissolved in the aqueous phase. The radical thus formed diffuses into the micelles where monomer is highly concentrated. No polymer is formed in the monomer droplets, because the micelles have a much higher surface/volume ratio and thereby provide a more favorable environment for the radicals. The polymer chain inside the micelle grows by the diffusion of monomer into the micelles from the monomer droplets.

As the micelle enlarges, it absorbs emulsifier molecules from the aqueous phase to stabilize the new polymer–water interface. The nonactivated micelles disintegrate and the emulsifier molecules dissolve. Only a small portion of the original micelles form polymer particles and at about 10–20% conversion, all the nonactivated micelles have disappeared. The polymerizing chain inside a micelle or polymer particle does not deplete the particle of monomer, for fresh monomer is continuously diffusing from the monomer droplets. Because of osmotic forces, the amount of monomer diffusing from the monomer droplets to the particles exceeds the amount consumed by polymerization. A particle may contain up to 50% monomer. Owing to this phenomenon, the monomer droplets disappear at about 60–80% conversion.

From the above description of an emulsion polymerization, the reason for the high molecular weights and high rates obtained should be obvious. As the emulsion particles separate from the aqueous phase, termination by combination is greatly reduced. Another point is that it can be shown that two free radicals in the same polymer particle would terminate in a time period very small compared to what would be required for further polymerization. Therefore each particle must contain either one growing chain or none. On the average the number of radicals per particle would be $\frac{1}{2}$.

The kinetics of emulsion polymerization are therefore different from either mass or solution polymerization. Smith and Ewart (74) have derived kinetic equations for the idealized emulsion polymerization described above. The rate of polymerization per cubic centimeter of water is given as follows:

$$R_p = \frac{N k_p [\text{M}]}{2 N_A} \tag{89}$$

where [M] is the monomer concentration *in the polymer particles*, N is *the number of polymer particles in a cubic centimeter of the aqueous phase*, and N_A is Avogadro's number. Since the monomer concentration varies little as long as excess monomer exists in the droplets, the rate should depend *principally on the number of particles*. The rate should also be independent of the rate of generation of free radicals, and should increase with increasing soap concentration. The degree of polymerization is given by

$$\overline{DP}_N = \frac{k_p N[M]}{\rho} \tag{90}$$

where ρ is the rate of generation of radicals from the initiator. Unlike the rate, the $\overline{DP}$ depends on the rate of generation of radicals. However, the $\overline{DP}$ also increases with increasing soap concentration.

The validity of Eqs. 89 and 90 has been experimentally confirmed in the emulsion polymerization of styrene, butadiene, and isoprene.

E.   COPOLYMERIZATION (9, 75, 76)

The simultaneous polymerization of two or more monomers is called copolymerization. A copolymer is thus defined as a polymer having at least two different monomers incorporated into one polymeric chain, whereas the term homopolymer describes a polymeric chain derived from a single monomer.

Interest has been centered on copolymerization, because it is often found that the copolymer has more desirable mechanical properties than the respective homopolymers.

One may also introduce a comonomer into a polymer to render it receptive to dyes or provide sites for cross-linking the polymer chains. Many commercial polymers are copolymers.

1.   The Copolymer Equation

To be able to predict the composition of a copolymer before the polymerization is extremely useful. This can be accomplished by the copolymer equation, a derivation of which follows.

The four possible propagation steps in a copolymerization are the following:

	Growing Chain	Monomer Added		Product	Rate
A	$\sim\!\!\sim\!\!-m_1\cdot$	M_1	$\xrightarrow{k_{11}}$	$\sim\!\!\sim\!\!-m_1\cdot$	$k_{11}[m_1\cdot][M_1]$
B	$\sim\!\!\sim\!\!-m_1\cdot$	M_2	$\xrightarrow{k_{12}}$	$\sim\!\!\sim\!\!-m_2\cdot$	$k_{12}[m_1\cdot][M_2]$
C	$\sim\!\!\sim\!\!-m_2\cdot$	M_1	$\xrightarrow{k_{21}}$	$\sim\!\!\sim\!\!-m_1\cdot$	$k_{21}[m_2\cdot][M_1]$
D	$\sim\!\!\sim\!\!-m_2\cdot$	M_2	$\xrightarrow{k_{22}}$	$\sim\!\!\sim\!\!-m_2\cdot$	$k_{22}[m_2\cdot][M_2]$

where M_1 and M_2 are monomer one and two, respectively, $m_1\cdot$ denotes the polymer chain with a terminal monomer one, and $m_2\cdot$ is the polymer chain with a terminal monomer two.*

The above assumes that the reactivity of a growing chain is governed only by the last unit (i.e., there is no penultimate unit effect).

The analysis is further simplified by a steady-state assumption: *the rate of disappearance of a chain end equals the rate of its appearance*; that is,

$$\frac{d[m_1\cdot]}{dt} = 0 = \frac{d[m_2\cdot]}{dt} \tag{91}$$

Therefore the rate of reaction B must equal the rate of reaction C:

$$k_{12}[m_1\cdot][M_2] = k_{21}[m_2\cdot][M_1] \tag{92}$$

The rates of disappearance of monomers one and two are

$$\frac{-d[M_1]}{dt} = k_{11}[m_1\cdot][M_1] + k_{21}[m_2\cdot][M_1] \tag{93}$$

$$\frac{-d[M_2]}{dt} = k_{22}[m_2\cdot][M_2] + k_{12}[m_1\cdot][M_2] \tag{94}$$

After combining Eqs. 93 and 94 and rearranging, Eq. 95 is obtained:

$$\frac{d[M_1]}{d[M_2]} = \frac{[M_1]}{[M_2]} \frac{k_{11}\dfrac{[m_1\cdot]}{[m_2\cdot]} + k_{21}}{k_{22} + k_{12}\dfrac{[m_1\cdot]}{[m_2\cdot]}} \tag{95}$$

Since the concentrations of the growing chains ($[m_1\cdot]$ and $[m_2\cdot]$) are extremely difficult to obtain, these quantities are eliminated by employing the steady-state assumption, Eq. 92. In rearranged form this becomes

$$\frac{[m_1\cdot]}{[m_2\cdot]} = \frac{k_{21}[M_1]}{k_{12}[M_2]} \tag{96}$$

Substituting Eq. 96 into Eq. 95 and rearranging gives:

$$\frac{d[M_1]}{d[M_2]} = \frac{[M_1]}{[M_2]} \frac{(k_{11}/k_{12})[M_1] + [M_2]}{(k_{22}/k_{21})[M_2] + [M_1]} \tag{97}$$

From Eq. 97 we can now define the important quantities termed the reactivity ratios (r_1 and r_2):

$$r_1 = \frac{k_{11}}{k_{12}}; \qquad r_2 = \frac{k_{22}}{k_{21}} \tag{98}$$

* Bracketed quantities indicate the concentration of the specific species.

On substituting the quantities defined in Eq. 98 into Eq. 97 we obtain:

$$\frac{d[M_1]}{d[M_2]} = \frac{[M_1]}{[M_2]} \frac{r_1[M_1] + [M_2]}{r_2[M_2] + [M_1]} \tag{99}$$

Equation 99 when integrated is valid for all degrees of conversion (if the previously discussed assumptions are valid). However, the integrated form is difficult to apply. Therefore a further assumption is made to cover the range of low percent conversion (below 5%). For the initial copolymer composition (i.e., $[M_1]$ and $[M_2]$ do not change) the disappearance of the monomer equals its incorporation into the polymer. That is,

$$\frac{d[M_1]}{d[M_2]} = \frac{[m_1]}{[m_2]} \tag{100}$$

On substituting Eq. 100 into Eq. 99 the final form of the *copolymer equation* is obtained:

$$\frac{[m_1]}{[m_2]} = \frac{[M_1]}{[M_2]} \frac{r_1[M_1] + [M_2]}{r_2[M_2] + [M_1]} \tag{101}$$

This equation would now yield the composition of the copolymer $(m_1)/(m_2)$ knowing only the monomer feed $[M_1]/[M_2]$ and the reactivity ratios r_1 and r_2. The equation may also be derived without making the steady state assumption, by a statistical treatment.

2. Reactivity Ratios

From the definition of reactivity ratios in Eq. 98 we can see that this ratio is an indication of the preference of a growing chain for adding a monomer identical to the terminal unit or for adding the other monomer. Therefore if $r < 1$ the growing chain prefers to add the comonomer rather than its own kind, and if $r > 1$ the growing chain prefers to add its own kind rather than the other comonomer. There are a number of different combinations of reactivity ratio values that are of special interest:

(a) $\qquad\qquad\qquad r_1 \gg 1, \qquad r_2 \gg 1$

In this case, two homopolymers would be formed.

(b) $\qquad\qquad\qquad r_1 r_2 = 1$

A copolymer that is described by the above reactivity ratios is termed an *ideal copolymer*, in which the monomers are randomly distributed throughout the chain. The amount of any monomer incorporated into a polymer

chain is simply dependent on the concentration of that monomer in the feed
and the relative reactivities of the two monomers.

$$(c) \qquad\qquad r_1 = r_2 = 0$$

An alternating copolymer results from a system having the above reactivity
ratios. The copolymer has alternating units of monomers one and two,
because each monomer reacts exclusively with the other monomer. There
are many monomers, such as maleic anhydride, which do not homopolymerize except under special conditions (therefore $r_1 = 0$) but readily form
copolymers. There are also monomer pairs, such as stilbene–maleic anhydride, neither of which homopolymerize (except under special conditions)
but readily form alternating copolymers.

3. Dependency of Copolymer Composition on Monomer Feed

In the preceding section it was pointed out that copolymer composition is a
function of both the monomer feed and the reactivity ratios (Eq. 101). This
dependency is usually depicted as shown in Fig. 10, where the mole fraction
of the monomers in the copolymer is plotted as a function of the monomer
feed for various reactivity ratios. It can be seen that for a given monomer feed,

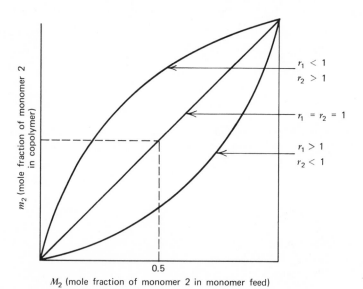

Figure 10   The dependency of copolymer composition as a function of monomer feed
and reactivity ratios.

the composition of the resulting copolymer varies widely, depending on the reactivity ratios.

To apply Eq. 101 in calculating the predicted copolymer composition, the reactivity ratios for the system must be known. [There exist several extensive complications of reactivity ratios (77, 78).] If they are not, they can be found by preparing a number of copolymers of varying feed, finding the copolymer composition, then solving Eq. 101. Many methods are available for determining copolymer composition, including elemental analysis and various spectroscopic methods.

4. Determination of Reactivity Ratios (79)

Three methods may be used to obtain reactivity ratios from a series of m_1/m_2 and M_1/M_2 data, other than by solving a series of simultaneous equations.

Direct Curve Fitting

Reactivity ratios may be obtained from direct curve fitting of polymer–monomer composition curves such as Fig. 10. This is not a favored method, however, because the composition curve is insensitive to small changes in the reactivity ratios.

Mayo and Lewis Method (80)

The most common method used is that of Mayo and Lewis. Rearranging the copolymer equation (Eq. 101) into the form of the equation for a straight line ($y = mx + b$),

$$r_2 = \left(\frac{[M_1]}{[M_2]}\right)^2 \frac{[m_2]}{[m_1]} r_1 + \frac{[M_1]}{[M_2]} \frac{[m_2]}{[m_1]} - 1 \qquad (102)$$

Every set of M_1, M_2 and m_1, m_2 values produces a straight line. By setting r_1 equal to arbitrary values, corresponding values of r_2 can be found and a straight line drawn for a specific set of M_1, M_2 and m_1, m_2 values. The probable values of r_1 and r_2 lie in the area cut out by the intersecting lines as in Fig. 11.

Method of Fineman and Ross (81)

In this method Eq. 102 is used as a basis. The equation is put into different form by setting $F = [M_1]/[M_2]$ and $f = [m_2]/[m_1]$. Therefore,

$$r_2 = F^2 f r_1 + F(f - 1)$$

or, rearranging,

$$F(1 - f) = -r_2 + r_1 F^2 f \qquad (103)$$

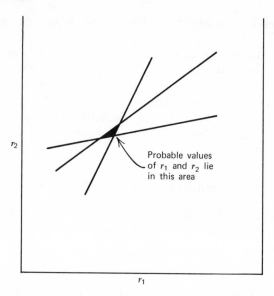

Figure 11   Calculation of reactivity ratios by the Mayo–Lewis method.

This equation is of the form $y = mx + b$, where

$$y = F(1 - f)$$
$$x = F^2 f$$
$$m = r_1$$
$$b = -r_2$$

Therefore each experiment defines one pair of x and y, and determines one point on a straight line. The slope of this line equals r_1 and the intercept $-r_2$.

5. Q–e Scheme (82)

Alfrey and Price (82) have developed a semiquantitative method for estimating the reactivity ratios of any pair of monomers, without the above procedures. This is the Q–e scheme. It attempts to combine the recognized effects of resonance stabilization and polarity on the relative reactivities of various monomers with various free radicals. This is done by defining the rate constant in the following manner:

$$k_{ij} = P_i Q_j \exp(-e_i e_j) \tag{104}$$

where P_i is characteristic of radical i, Q_j is the mean reactivity of monomer j (resonance stabilization of the double bond), and e_i, e_j refers to the polarity of the monomers. (It is assumed that e is the same for the monomer and the radical derived from it.) Going back to the definition of reactivity ratios (Eq. 98), a new relationship for reactivity ratios can be defined.

$$r_1 = \frac{k_{11}}{k_{12}} = \frac{Q_1}{Q_2} \exp\{-[e_1(e_1 - e_2)]\}$$

$$r_2 = \frac{k_{22}}{k_{21}} = \frac{Q_2}{Q_1} \exp\{-[e_2(e_2 - e_1)]\}$$

(105)

If the Q and e values are known for a pair of monomers, the two reactivity ratios may be calculated. A table of Q and e values has been built up using styrene as a standard with values of $Q = 1.0$, $e = -0.8$. Knowing the Q and e values for styrene, the values can be calculated for any monomer copolymerized with styrene (from a knowledge of the reactivity ratios). Working from there, reactivity ratios may be calculated for any pair of monomers. It would then be possible to calculate the reactivity ratios for a pair of monomers that have not yet been copolymerized. There are many compilations of Q and e values (83, 84).

The Q–e scheme gives reasonable values for most monomer pairs. It usually breaks down for systems with a large difference in Q, and specifically for sterically hindered monomers such as 1,2-disubstituted monomers. It should also be pointed out that Eqs. 105 predict that the product of two reactivity ratios cannot be larger than one, since

$$r_1 r_2 = \exp[-(e_1 - e_2)^2]$$

(106)

However, it has been shown that there are a small number of systems that have $r_1 r_2 > 1$. This cannot be explained on the basis of the simple Q–e treatment.

As mentioned earlier, this treatment takes into account both resonance and polarity effects. The resonance factor affects the chemical reactivity of both the monomer and the radical formed from it. Conjugation in a monomer (styrene, for example) increases its reactivity but decreases the reactivity of the adduct radical, because of resonance stabilization. In constrast, an unconjugated monomer (i.e., vinyl acetate) is relatively unreactive with free radicals, but yields an adduct radical of high reactivity.

The polarity factor (e value) attempts to take into account the observation that there is a strong alternating tendency when a monomer with a strong electron-withdrawing substituent (acrylonitrile) is copolymerized with a monomer with a strong electron-donating substituent (p-methoxystyrene).

6. Ionic Copolymerization (85–87)

The foregoing sections discuss only free-radical copolymerizations, but the basic concepts are valid for ionic copolymerizations. Both the copolymer equation and the concept of reactivity ratios apply to ionic copolymerizations. There is, however, no simple method, such as the $Q-e$ scheme, for estimating reactivity ratios. In a cationic or anionic copolymerization, the reactivity ratios depend on the system (initiator and solvent and temperature). The reason for this will be seen in the following discussions of propagation in anionic and cationic systems. Here the rate and direction of propagation of ionic species are profoundly influenced by the system employed, and so the reactivity of an ion in both homopolymerization and copolymerization often changes with different polymerization conditions.

Since the reactivity of an anionic or cationic chain end is very much different from that of a free-radical chain end, one would expect copolymer composition to depend heavily on the mode of initiation. The classic example of this is the styrene–methyl methacrylate pair, the composition curve of which is shown in Fig. 12 (88). Often with a new initiator system, copolymerizations are run to elucidate the mode of propagation (anionic, cationic, or free radical).

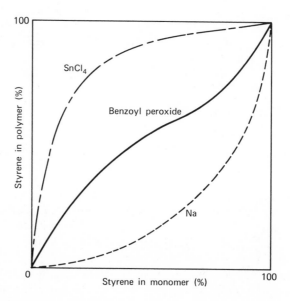

Figure 12   Copolymer composition of styrene-methyl methacrylate as a function of different catalysts (reference 88).

Ziegler–Natta catalysts may also be used to prepare copolymers (89, 90). Several commercially important materials are prepared in this manner, including rubber based on ethylene and propylene.

7. Block and Graft Copolymers (91–94)

There are two types of copolymers of special interest, block and graft copolymers. A block copolymer is made up of long blocks of one monomer followed by long blocks of the second monomer (Fig. 13).

$$-AA\text{---}\sim\sim\text{---}AA-BB\text{---}\sim\sim\text{---}BB-AA\text{---}\sim\sim\text{---}AA-$$

Figure 13   Schematic representation of a block copolymer.

In a graft copolymer, sequences of one monomer are "grafted" onto a backbone of the other monomer (Fig. 14).

```
—AAAAAAAAAA—
  |    |    |
  B    B    B
  B    B    B
  B    B    B
  B    B    B
  B    B    B
  B    B    B
  |    |    |
```

Figure 14   Schematic representation of a graft copolymer.

The properties of a block or graft copolymer differ from those of either the "random" copolymer or a physical mixture of the homopolymers. Conditions are usually designed to retain the desirable properties of the individual block or graft components and to eliminate their less desirable properties.

One of the most common methods of preparing a block copolymer is through the use of living polymers. This is discussed in Section F.5.

There are numerous ways to form graft copolymers. One method generates a free radical on the chain backbone. The graft then grows from these new active centers. This can be done with a copolymer of styrene and about 5% vinyl bromide; a schematic representation of this would be as follows:

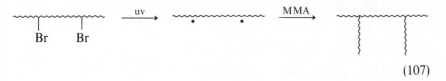

(107)

F. ANIONIC POLYMERIZATION (15, 95–99)

Anionic polymerizations are polymerizations that proceed by the addition of monomer to a reactive anionic chain end. For a vinyl monomer,

$$CH_2{=}C\overset{\displaystyle R_1}{\underset{\displaystyle R_2}{\big<}}$$

, to polymerize anionically, R_1 and/or R_2 must be electron-withdrawing groups or unsaturated functions which stabilize a negative charge by resonance or inductive effects. A number of vinyl monomers satisfy this requirement (Table 3). Other monomers also polymerize anionically. These include lactams, N-carboxyanhydrides, epoxides, aldehydes, lactones, isocyanates, and others.

This section describes the basic principles of anionic polymerization, using vinyl monomers as examples. The same basic principles apply to non-vinyl monomers.

1. Initiation

An active center for an anionic polymerization may be generated in two ways.

$$\overset{\delta^+ \ \ \delta^-}{M{-}B} + CH_2{=}C\overset{\displaystyle R_1}{\underset{\displaystyle R_2}{\big<}} \longrightarrow B{-}CH_2{-}\overset{\ominus}{C}\overset{\displaystyle R_1}{\underset{\displaystyle R_2}{\big<}} \quad M^{\oplus} \qquad (108)$$

M–B is a covalent or ionic metal amide, alkoxide, hydroxide, alkyl, or aryl.

$$M^{\circ} + CH_2{=}C\overset{\displaystyle R_1}{\underset{\displaystyle R_2}{\big<}}$$

$$\downarrow$$

$$M^{\oplus} \left[{}^{\ominus}CH_2{-}\overset{\cdot}{C}\overset{\displaystyle R_1}{\underset{\displaystyle R_2}{\big<}} \longrightarrow {\cdot}CH_2{-}\overset{\ominus}{C}\overset{\displaystyle R_1}{\underset{\displaystyle R_2}{\big<}} \right] \qquad (109)$$

M° is usually an alkali metal.

Typical initiators include potassium amide in liquid ammonia, organo-alkali compounds such as butyl lithium, alkali metals such as sodium or lithium, and Grignard reagents such as phenyl magnesium bromide. The initiators are usually basic. Often the more basic an ion is, the more effective it is as an initiator.

One of the factors determining whether or not a particular ion will initiate polymerization is the nature of the monomer. As a rough correlation, we conclude that the more susceptible a monomer is to initiation the less basic the initiator must be. Szwarc (100) points out that although the $Ph_3CO^{\ominus}$ ion readily initiates the polymerization of vinylidene dicyanide, it is less efficient in polymerizing methyl methacrylate, and incapable of initiating the polymerization of styrene. The natures of the counterions and solvents represent additional factors that contribute to the rate and direction of anionic initiation.

2. Propagation

$$B-CH_2-C\overset{R_1}{\underset{R_2}{\diagup}}\ M^{\oplus} \ \xrightarrow{\ CH_2=C\overset{R_1}{\underset{R_2}{\diagup}}\ }\ B-CH_2-C\overset{R_1}{\underset{R_2}{\diagup}}-CH_2-C\overset{R_1}{\underset{R_2}{\diagup}}\ M^{\oplus}\ etc.$$

$$(110)$$

The above representation shows that an ion pair or an ion aggregate must interact with the π-electron system of the monomer to form an active ionic species. The exact nature of the transition state involved in the propagation is exceedingly difficult to describe.

Reactivity varies with the solvation of the ion pair and the nature of the counterions. Current research aims to explain the effect of solvation of monomer, the reactivity of the growing ion pair, and whether propagation proceeds through free ions, ion pairs, or ion aggregates.

Recent studies have provided several general principles concerning the effect of an alkali metal counterion and solvent on the rate constant for propagation (k_p). It has been shown that the rate constant increases for a given ion pair, as the dielectric constant of the solvent increases. This effect becomes less important as the size of the cation increases. In a solvent of high dielectric constant such as tetrahydrofuran, the rate constant for propagation of styrene decreases on going from lithium to cesium among the alkali metals. However, in dioxane the order of reactivity is reversed. This reversal of reactivity may be ascribed to the propagation occurring mainly through ion pairs in dioxane and through free ions in tetrahydrofuran.

3. Termination (101)

The termination process in an anionic polymerization is more complex than that observed for free-radical polymerization, where termination invariably proceeds via disproportionation or coupling. A number of anionic systems do not terminate, if impurities (acid, alcohols, carbon dioxide, oxygen, water, etc.) are excluded.

The termination process for the system potassium amide–liquid ammonia–styrene has been studied. Termination appears to proceed by transfer to solvent:

$$\sim\!\!-\overset{|}{\underset{|}{C}}{}^{\ominus} + NH_3 \longrightarrow \sim\!\!-\overset{|}{\underset{|}{C}}-H + NH_2^{\ominus} \tag{111}$$

In the presence of water, alcohol, acids, or any other species capable of donating a proton, the chain is terminated by proton abstraction:

$$\sim\!\!-\overset{|}{\underset{|}{C}}-\overset{|}{\underset{|}{C}}{}^{\ominus} + HA \longrightarrow \sim\!\!-\overset{|}{\underset{|}{C}}-\overset{|}{\underset{|}{C}}-H + A^{\ominus} \tag{112}$$

In the presence of carbon dioxide or oxygen, the propagating chain retains its anionic character. However, it is incapable of further propagation:

$$\sim\!\!-\overset{|}{\underset{|}{C}}-\overset{|}{\underset{|}{C}}{}^{\ominus} + CO_2 \longrightarrow \sim\!\!-\overset{|}{\underset{|}{C}}-\overset{|}{\underset{|}{C}}-C\!\!\overset{O^{\ominus}}{\diagdown_{O}} \tag{113}$$

$$\sim\!\!-\overset{|}{\underset{|}{C}}-\overset{|}{\underset{|}{C}}{}^{\ominus} + O_2 \longrightarrow \sim\!\!-\overset{|}{\underset{|}{C}}-\overset{|}{\underset{|}{C}}-O-O^{\ominus} \tag{114}$$

A growing chain may also terminate by chain transfer to monomer or by the elimination of a hydride ion from the carbon adjacent to the growing end, resulting in a terminal double bond.

4. Polymerization of Dienes (102)

The anionic preparation of polymers and copolymers from dienes is receiving great attention academically and industrially. Several commercial polymers are now prepared this way, and the list should grow during the next few years.

An anionic system has advantages over the more conventional free-radical system. It provides better control over stereochemistry and allows a given molecular weight to be attained with a narrow distribution.

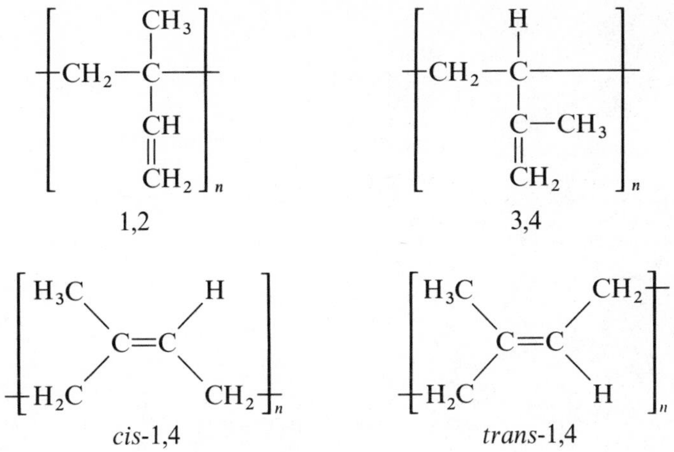

Figure 15   Possible structures for polyisoprene.

Diene polymers may contain a variety of structural units, since several modes of addition are possible (Fig. 9). Isoprene, for example, may undergo 1,2 addition, 3,4 addition, *cis*-1,4 addition, or *trans*-1,4 addition (Fig. 15).

The 1,2 and 3,4 additions may also vary in stereoregularity. Changing the initiator system often changes the structure of the polymer. Lithium metal produces an isoprene polymer different from the other alkali metals (Fig. 16). The lithium-initiated polymer is very like natural rubber (*cis*-1,4-polyisoprene) except that it contains about 6% of the 3,4-addition units. The other alkali metals (through cesium) cause 5–10% 1,2 addition, 35–50% 3,4 addition and 40–50% *trans* addition structure with only about 7% of the *cis* addition structure. In tetrahydrofuran, *no cis* structure is produced. The polyisoprene forms by 16% 1,2 addition, 51% 3,4 addition, and 33% *trans*-1,4 addition.

The mechanism of the reaction accounts for the specificity of lithium metal and for the polar-solvent effect. Structure II indicates that only a small unsolvated cation, such as $Li^{\oplus}$, could fit into the position necessary to promote *cis*-1,4 addition.

Compound I is called a Schlenk adduct. Indirect evidence for its existence lies in Ziegler's showing the existence of analogous compounds using 2,3-dimethylbutadiene and piperylene (103).

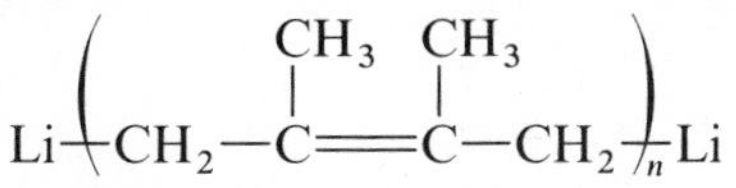

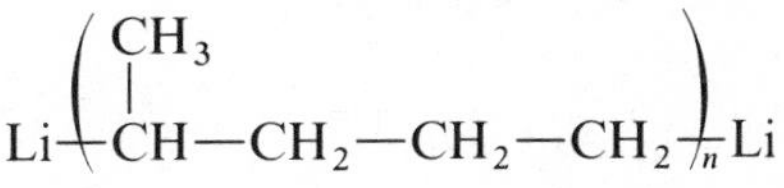

$$Li \Big(CH_2 - \overset{\overset{\displaystyle CH_3}{|}}{C} = \overset{\overset{\displaystyle CH_3}{|}}{C} - CH_2 \Big)_{\!\!n} Li \qquad Li \Big(\overset{\overset{\displaystyle CH_3}{|}}{CH} - CH_2 - CH_2 - CH_2 \Big)_{\!\!n} Li$$

where $n = 1, 2, 3, 4, 5, 6$.

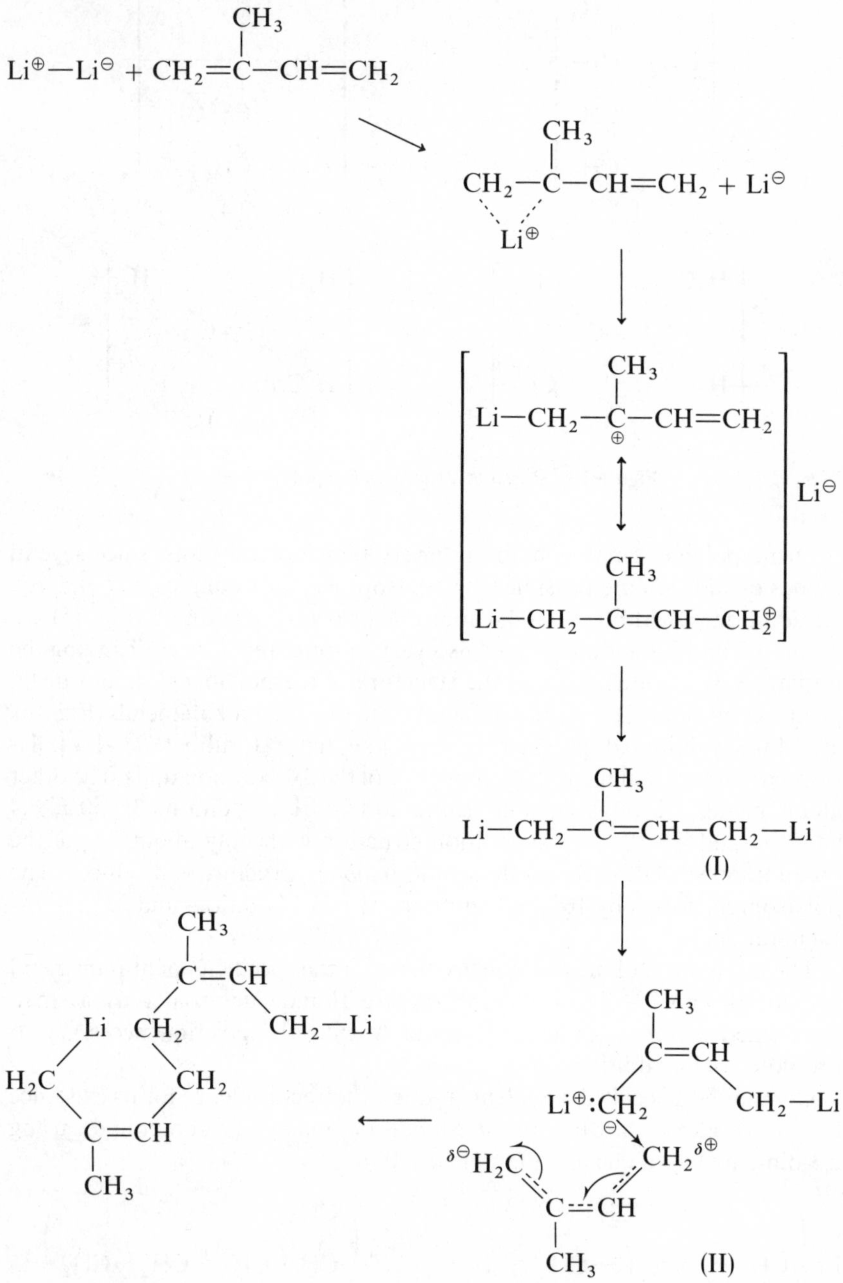

$$Li^{\oplus}-Li^{\ominus} + CH_2=\underset{\underset{\displaystyle CH_3}{|}}{C}-CH=CH_2$$

$$\underset{\underset{\displaystyle Li^{\oplus}}{\vdots}}{CH_2}-\underset{\underset{\displaystyle CH_3}{|}}{C}-CH=CH_2 + Li^{\ominus}$$

$$\left[\begin{array}{c} Li-CH_2-\underset{\underset{\displaystyle \oplus}{|}}{\overset{\overset{\displaystyle CH_3}{|}}{C}}-CH=CH_2 \\ \updownarrow \\ Li-CH_2-\overset{\overset{\displaystyle CH_3}{|}}{C}=CH-CH_2^{\oplus} \end{array} \right] Li^{\ominus}$$

$$Li-CH_2-\overset{\overset{\displaystyle CH_3}{|}}{C}=CH-CH_2-Li$$
(I)

(II)

Figure 16 Mechanism of stereoregular polymerization of isoprene by lithium.

84

The organolithium mechanism is similar. Here, too, the lithium directs polymerization at the chain end. To initiate, however, the isoprene first forms a π complex with the lithium cation. The Schlenck adduct is replaced by

$$Bu-CH_2-\underset{\underset{CH_3}{|}}{C}=CH-CH_2-Li$$

and propagation proceeds as with lithium metal.

Polar solvent destroys the stereospecificity by solvating the lithium cation. This disrupts the geometry of the transition state necessary for *cis*-1,4 addition.

5. Living Polymers (104, 105)

As stated earlier, many anionic polymerizations do not terminate unless an impurity is present. Such a polymer is called a *living* polymer. Szwarc (106) provided the first example, polystyrene initiated by sodium naphthalene in tetrahydrofuran at 0 to $-78°C$. Interestingly, the naphthalene–sodium complex is dark green. Addition of styrene causes a color change to red, the color of styryl anions. When polymerization is complete, addition of more styrene causes more polymerization. At 100% conversion, the $\overline{DP}$ is proportional to $2[M]/[C]$ where $[M]$ and $[C]$ are the concentrations of monomer and catalyst. Another feature here is that molecular weight distribution is very narrow. $\overline{M}_w/\overline{M}_n = 1.06\text{–}1.12$.

Szwarc proposed the following mechanism for his system:

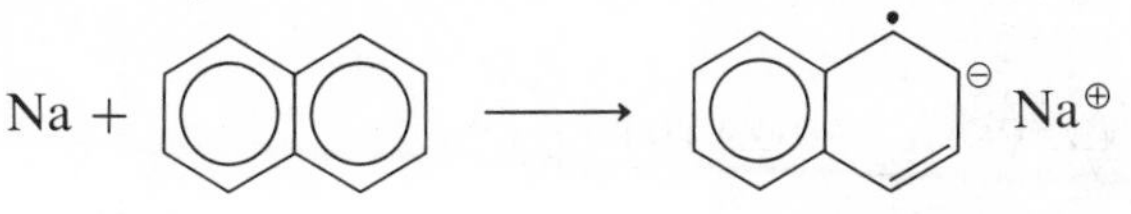

$$\text{(115)}$$

sodium complex
of naphthalene

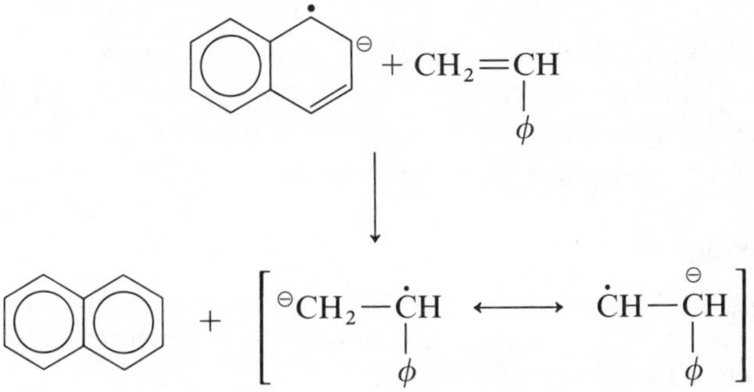

$$\text{(116)}$$

$$2(\overset{\cdot}{C}H_2-CH^{\ominus}) \rightleftharpoons {}^{\ominus}CH-CH_2-CH_2-\overset{\ominus}{C}H$$

$$\underset{\phi}{|} \qquad \underset{\phi}{|} \qquad \underset{\phi}{|}$$

$$\begin{array}{c} H \\ | \\ CH_2=C \\ | \\ \phi \end{array} \qquad (117)$$

$$\left(\overset{\ominus}{\underset{\phi}{|}}CH-CH_2\right)_n\overset{}{\underset{\phi}{|}}CH-CH_2-CH_2-\overset{}{\underset{\phi}{|}}CH\left(CH_2-\overset{}{\underset{\phi}{|}}CH\right)_n^{\ominus}$$

This mechanism holds for other monomers with this catalyst.

Block copolymers can be prepared using living polymers (99, 107–110). For example, addition of methyl methacrylate to a solution of living polystyrene causes polymerization of methyl methacrylate at the living ends of the polystyrene chains (111). The reverse is not true, however. Living polymethyl methacrylate cannot initiate styrene polymerization because the methacrylate anion is not basic enough. Polymers containing up to nine blocks have been prepared from living polymers. Some of the block pairs include styrene–isoprene, styrene–butadiene, styrene–ethylene oxide, ethylene oxide–n-butyl isocyanate, and methyl methacrylate–acrylonitrile.

Living polymers are also noteworthy in that they may be selectively terminated to form a polymer with predetermined end groups (99, 105). For example, a polystyrene or a polybutadiene may be prepared with carboxylate end groups by termination with carbon dioxide.

G. CATIONIC POLYMERIZATION (112–115)

A cationic polymerization is one that proceeds by addition of monomer to a cationic chain end.

$$R^{\oplus} + CH_2{=}CR_2 \longrightarrow R-CH_2-\overset{\oplus}{C}R_2, \text{ etc.} \qquad (118)$$

Catalysts are usually electron deficient, and so electron-rich double bonds make good monomers (Table 3). These include strained cyclic compounds containing a Lewis-base function, epoxides, ethylenimines, and aldehydes. This section emphasizes vinyl monomers.

1. Initiation

In general, the catalyst creates an electron-deficient site. The catalyst may be a simple cation, or it may require a co-catalyst to generate a more complex species before inducing polymerization.

The following represent types of materials which initiate cationic polymerization:

1. Protonic acids: strong mineral acids such as H_2SO_4 and H_3PO_4 (strong organic acids such as CF_3COOH are poor catalysts and generally lead to low molecular weight polymers).
2. Lewis acids with and without co-catalysts: $AlCl_3$, BF_3, ZnO, $FeCl_3$, R_3Al, R_2Zn, PF_5.
3. Carbonium ion sources: $\phi_3C{-}X$, $\phi_3\overset{\oplus}{C}HgX_3^{\ominus}$, $R\overset{\oplus}{C}O\ Y^{\ominus}$ where $Y^{\ominus}$ is $BF_4^{\ominus}$ or $ClO_4^{\ominus}$.
4. Halogens: I_2.
5. Coordination complexes: $TiCl_4 + ClR \rightarrow [TiCl_5^{\ominus}\ R^{\oplus}]$.

When iodine is used, the initiating species is $I^{\oplus}$:

$$2I_2 \longrightarrow I^{\oplus} + I_3^{\ominus} \tag{119}$$

The protonic acids and the carbonium ion sources presumably initiate by the addition of a cation to the double bond:

$$M^{\oplus}X^{\ominus} + CH_2{=}C\overset{\displaystyle R'}{\underset{\displaystyle R''}{\diagup\diagdown}} \longrightarrow M{-}CH_2{-}C\overset{\displaystyle R'}{\underset{\displaystyle R''}{\diagup\diagdown}} \oplus\ X^{\ominus} \tag{120}$$

Determining the necessity of a co-catalyst is difficult, primarily because it requires an impurity-free experiment. Carefully purified isobutylene and boron trifluoride do not react, but addition of water causes polymerization. A co-catalyst is also required to polymerize isobutylene by $SnCl_4$ or $TiCl_4$, styrene by $AlCl_3$, $SnCl_4$, or $TiCl_4$, and for a few other cases.

Co-catalysts appear to be necessary for all Friedel–Crafts halides. If so, initiation is similar to that by strong acids, by transfer of a cation (proton) to monomer:

$$\underset{\text{(co-catalyst)}}{MX_n + BA} \longrightarrow MX_NBA \longrightarrow [MX_NB]^{\ominus}A^{\oplus} \tag{121}$$

$$A^{\oplus} + CH_2{=}C\overset{\displaystyle R'}{\underset{\displaystyle R''}{\diagup\diagdown}} \longrightarrow A{-}CH_2{-}C\overset{\displaystyle R'}{\underset{\displaystyle R''}{\diagup\diagdown}} \oplus \tag{122}$$

It has been shown, for example, that boron trifluoride and water react to give a highly acidic monohydrate; the initiating species would be a proton:

$$BF_3 + H_2O \longrightarrow HBF_3OH \longrightarrow H^{\oplus}[BF_3OH]^{\ominus} \tag{123}$$

2. Propagation

In our discussion of anionic polymerization it was emphasized that it is often difficult to determine the exact nature of the propagation species (i.e., free ions, ion pairs, or higher aggregates). The same is true for cationic polymerization. Except perhaps for reactions in nonpolar solvents it would be an oversimplification to visualize a single species as the propagation mechanism. For polar solvents the hypothesis that the chain carriers are ions implies the existence of free cations and cations forming part of the ion pairs and possibly higher aggregates.

It is possible to view most cationic polymerizations as propagating via numerous active species. For example, free ions and ion aggregates may simultaneously be active. Situations where the rate-determining step involves association produce different propagating species as a function of the concentration of monomer, catalyst, and polymeric growing chain.

3. Termination

In cationic polymerizations there are two main classes of termination reactions:

1. Charge neutralization, which can arise from dissociation of an ion pair into free ions, which then recombine:

$$
\underset{\overset{|}{R}}{\overset{\overset{R}{|}}{\sim\!\!\sim\!\!-C^{\oplus}}}\ominus BF_3OH \longrightarrow \underset{\overset{|}{R}}{\overset{\overset{R}{|}}{\sim\!\!\sim\!\!-C^{\oplus}}} + BF_3 + OH^{\ominus}
$$

$$
\underset{\overset{|}{R}}{\overset{\overset{R}{|}}{\sim\!\!\sim\!\!-C}}-OH + BF_3 \quad (124)
$$

A variant of this type of reaction can occur by reaction of the growing chain with impurities such as alcohols and water.

2. Formation of a cation too stable to propagate. The inhibiting or poisoning effect of such impurities as oxygen, amines, and sulfur compounds can lead to stable cations:

$$
\underset{\overset{|}{R}}{\overset{\overset{R}{|}}{\sim\!\!\sim\!\!-C^{\oplus}}} + R_3'N \longrightarrow \underset{\overset{|}{R}}{\overset{\overset{R}{|}}{\sim\!\!\sim\!\!-C}}-\overset{\oplus}{N}R_3' \quad (125)
$$

Another possible termination mode is hydride transfer coupled with allylic stabilization. This has been proposed for the polymerization of isobutylene (116), and may be applicable to other systems.

$$
\begin{array}{c}
\text{H} \quad\quad \text{H} \ \ \text{H} \ \ \text{H} \\
| \quad\quad\quad | \ \ \ | \ \ \ | \\
\sim\!\!\!\sim\!\!-\!C^{\oplus} + \text{C}\!=\!\text{C}\!-\!\text{C}\!-\!\text{H} \\
| \quad\quad\quad | \ \ \ \ \ \ \ | \\
\text{H} \quad\quad \text{H} \quad\ \ \text{H}
\end{array}
\longrightarrow
\begin{array}{c}
\text{H} \quad\quad \text{H} \quad\quad \text{H} \quad\quad \text{H} \\
| \quad\quad\quad\quad \diagdown \quad\ | \quad\ \diagup \\
\sim\!\!\!\sim\!\!-\!\text{CH} + \delta^{\oplus}\ \ \text{C}\!\!=\!\!\!=\!\!\text{C}\!\!=\!\!\!=\!\!\text{C}\ \ \delta^{\oplus} \\
| \quad\quad\quad\quad \diagup \quad\quad\quad \diagdown \\
\text{H} \quad\quad \text{H} \quad\quad\quad\quad\quad \text{H}
\end{array}
\quad (126)
$$

The above reactions would result in the irreversible destruction of the growing ion.

Most of the other terminating reactions involve transfer reactions. Polymer chains, monomer, and solvent can participate in such reactions. Thus in order to reduce the contribution of these side reactions, cationic polymerizations are carried out using extremely pure components and low temperature.

Because of the ease with which carbonium ions terminate or engage in transfer reactions, living cationic polymers are rare and require special conditions. For example, the polymerization of tetrahydrofuran by benzene-diazonium hexafluorophosphate leads to a living polymer of the type discussed under anionic polymerization (117).

4. Kinetics of Polymerization

Most cationic polymerizations can be described by the following kinetic expression:

$$
\frac{-dM}{dt} = k[M]^x[\text{Cat}]^y \tag{127}
$$

where M denotes monomer and Cat denotes catalyst or catalyst–co-catalyst complex.

For isobutylene, x and $y = 1$. This is an extremely rare finding. For most other cationic polymerizations, the rate depends on higher orders of monomer and catalyst. This is not surprising in the light of the above discussion on initiation and propagation. However, the $\overline{\text{DP}}$ for most cationic systems appears to be independent of concentration of catalyst and co-catalyst. Generally, the following dependence of $\overline{\text{DP}}$ on monomer concentration is observed:

$$
\overline{\text{DP}} = \frac{k_1[\text{M}]}{k_2[\text{M}] + k_3} \tag{128}
$$

where k_1 is the propagation rate constant, k_2 is the termination rate constant, and k_3 is the transfer rate constant.

5. Polymerization of Isobutylene (118)

Isobutylene can be polymerized at extremely low temperatures ($-100°C$) to produce high molecular weight polymers. Liquid ethylene is usually used as a diluent. Typical catalysts include BF_3 and $AlCl_3$.

$$CH_2{=}C\overset{\displaystyle CH_3}{\underset{\displaystyle CH_3}{\bigg\langle}} \;+\; BF_3 \text{ or } AlCl_3 \quad \xrightarrow[\substack{\text{solvent for}\\ \text{catalyst,}\\ \text{ethylene as}\\ \text{diluent}}]{-100°C} \quad \substack{\text{high polymer produced}\\ \text{very rapidly}}$$

$$(129)$$

This type of polymerization has been called flash polymerization, since the heat of polymerization causes vaporization of the ethylene. If no heat-transfer agent is used, a large induction period is observed and low polymer is obtained. In a typical reaction at $-100°C$, 10 g each of isobutylene and ethylene and 4 g of 5% $AlCl_3$ in ethyl chloride produce polymer with a molecular weight of 110,000 almost instantaneously.

Isobutylene is used commercially to prepare butyl rubber, a copolymer of isobutylene and 1–2 mole % isoprene. The diene provides the unsaturation necessary for vulcanization.

6. Isomerization Polymerization (119, 120)

Isomerization polymerization is a polyaddition reaction in which the propagating species rearranges prior to propagation. Thus the repeat unit of the polymer does not possess the structure of the starting monomer.

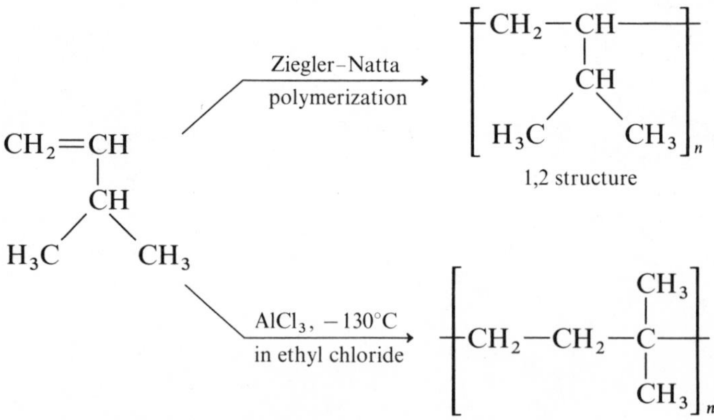

Figure 17   Polymerization of 3-methyl-1-butene.

This occurs in certain anionic, cationic, and free-radical polymerizations. However, it is most prevalent in cationic systems since carbonium or oxonium ions rearrange easily. Isomerization polymerization may involve isomerization by material transport or isomerization by bond (electron) migration.

The classic example of isomerization polymerization by material transport is the polymerization of 3-methyl-1-butene. A Ziegler–Natta catalyst produces an unrearranged polymer. However, under cationic conditions at low temperatures, the polymer has an entirely different structure (Fig. 17).

The mechanism proposed for this cationic polymerization involves the rearrangement of the propagating species through a 1,2-hydride shift.

$$CH_2{=}CH \overset{A^\oplus}{\longrightarrow} A{-}CH_2{-}\overset{H}{\underset{}{C}}{}^\oplus \xrightarrow[\text{shift}]{\text{1,2-hydride}} A{-}CH_2{-}CH_2$$

(with pendant $\overset{\diagup\diagdown}{H_3C \quad CH_3}$ CH groups and terminal $C^\oplus$)

$$\xrightarrow{\text{monomer}}$$

$$A{-}CH_2{-}CH_2{-}\underset{CH_3}{\overset{CH_3}{C}}{-}CH_2{-}\underset{\underset{\diagup\diagdown}{H_3C \quad CH_3}}{\overset{}{C}}{}^\oplus \xrightarrow[\text{shift}]{\text{1,2-hydride}} \text{etc.} \quad (130)$$

The driving force for this rearrangement is the formation of the more stable (~ 11 kcal/mole) tertiary carbonium ion. Propagation occurs through this species. Polymerization at $-130°C$ produces a crystalline polymer with almost exclusively the 1,3 structure. However, at temperatures above $-100°C$, the polymer is rubbery and contains both 1,2 and 1,3 structures.

This type of polymerization can also involve shifts of other groups (119, 120). For example, 3,3-dimethyl-1-butene polymerizes under similar conditions to a polymer with a 1,3 structure:

$$CH_2{=}CH,\ CH_3{-}\underset{CH_3}{\overset{}{C}}{-}CH_3 \longrightarrow \left[{-}CH_2{-}\underset{H}{\overset{CH_3}{C}}{-}\underset{CH_3}{\overset{CH_3}{C}}{-}\right]_n \quad (131)$$

The mechanism here is the same as the one above, except that a methyl group shifts instead of a hydride ion.

Isomerization polymerizations also proceed through bond or electron migration. In many polymerizations involving bond migration, one can polymerize a linear monomer to a polymer containing cyclic repeat units. This type of polymerization is often called cyclopolymerization (121, 122). It is a fairly general phenomenon with certain 1,5- and 1,6-dienes.

For example, the cationic polymerization of 2,6-diphenyl-1,6-hexadiene produces a polymer with a cyclic repeating unit (123).

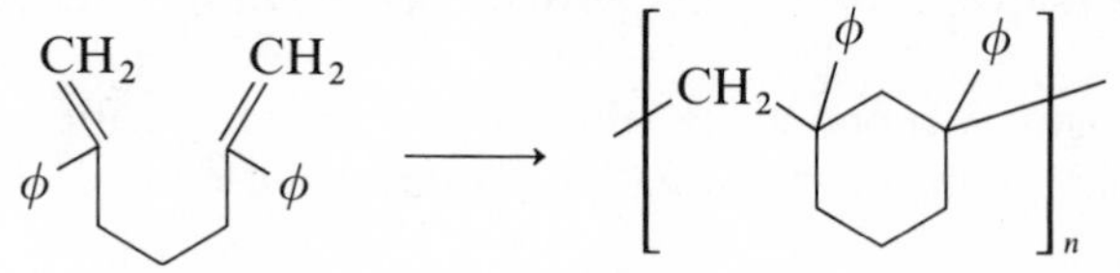

$$\tag{132}$$

Propagation probably involves the following bond migration:

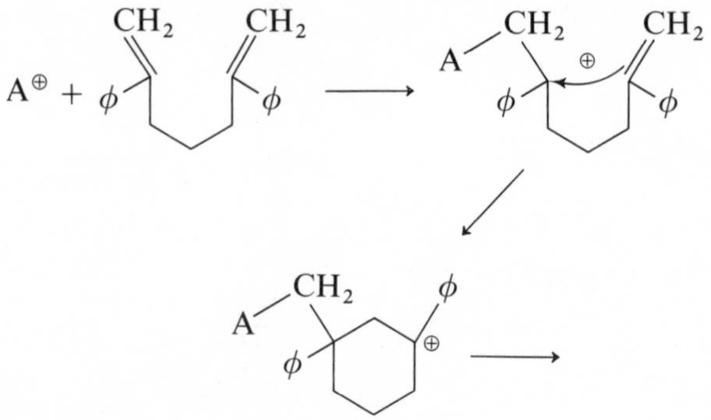

$$\tag{133}$$

H.   ZIEGLER–NATTA POLYMERIZATION (124–130)

1.   Introduction

Coordination-catalyzed polymerization was announced in 1955 by Ziegler (131). He reported the low-pressure polymerization of ethylene, using a catalyst formed by mixing solutions of triethylaluminum and titanium tetrachloride. Prior to this, ethylene was polymerized under extreme conditions of temperature and pressure, yielding a polymer containing many branches, unsaturation, and ketonic groups. Polyethylene prepared using the Ziegler system has a higher melting point (124–134°C versus 111°C), higher density (0.95–0.97 versus 0.92), a higher crystallinity ($\sim$95 versus 60%) and is almost completely linear.

Natta and his co-workers (126) used modified Ziegler-type catalysts to polymerize vinyl monomers such as propylene, styrene, and 1-butene to highly crystalline linear products. These polymers have lower solubilities, higher densities, and higher melting points than the corresponding amorphous polymers. Previously, crystalline synthetic polymers were rare. These types of structures were limited to polyethylene and polymers derived from symmetrical vinylidene monomers.

Natta showed that the crystallinity of poly-α-olefins prepared with the new coordination catalysts is due to the stereoregular structure of the polymers. As noted in Section C.2 two stereoregular structures were possible for a head-to-tail vinyl polymer (namely, isotactic and syndiotactic).

A Ziegler–Natta catalyst is prepared by combining a transition metal halide (e.g., titanium chloride) with a reducing agent (e.g., aluminum alkyls) in an inert atmosphere. In addition, modified Ziegler–Natta catalysts substitute vanadium halides for titanium halides or use titanium alkyl halides and/or alkyl aluminum halides. Sometimes hydride reducing agents are used, or complexing agents such as acetylacetonates are added. In this section we consider only standard Ziegler–Natta systems (using $TiCl_3$ and AIR_3) and deduce general mechanistic considerations from them.

A wide variety of catalyst systems and monomers have been studied, but most of the effort has been on olefin monomers. Special conditions must be employed when using a Ziegler–Natta type of catalyst to polymerize a monomer containing a heteroatom such as oxygen or nitrogen. Under normal conditions, these monomers react with the catalyst through their unshared electron pair. To obtain a stereoregular polymer of 5-hydroxy-1-pentene, the monomer is first converted to a silicon derivative. This decreases the nucleophilicity of the oxygen. A branched alkyl group used in the alkyl-aluminum further decreases interaction between the catalyst and monomer.

Often a catalyst is active with one monomer (or group of monomers) but not with others. For example, catalysts based on group VIII transition metals ($AlEt_2Cl + CoCl_2$ or $NiCl_2$) polymerize dienes, but not ethylene or α-olefins.

In the following sections, several proposals for the structure of the Ziegler–Natta catalysts are discussed, followed by a brief outline on the mechanism of polymerization and the forces of stereoregulation.

2. Structure of the Catalyst (60, 124)

The exact nature of the insoluble materials that comprise the typical Ziegler–Natta catalyst cannot be determined precisely from average composition or by identification of the constituents. On mixing the insoluble titanium halide with the soluble alkyl aluminum, a precipitate forms. The reaction presumably includes reduction of the titanium halide to a lower oxidation state, and formation of organoaluminum compounds. Most of the proposed structures

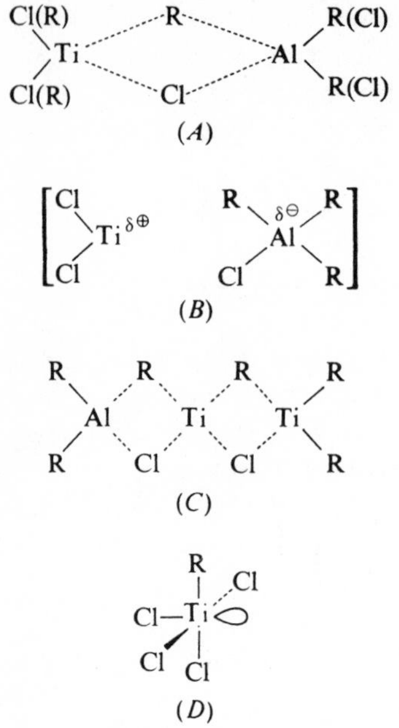

$$Cl(R) \diagdown \quad \cdots R \cdots \quad R(Cl)$$
$$\diagup Ti \diagdown \quad \cdots \quad Al \diagdown$$
$$Cl(R) \quad \cdots Cl \cdots \quad R(Cl)$$

(*A*)

$$\left[\begin{matrix} Cl \diagdown \\ \diagup Ti^{\delta \oplus} \\ Cl \end{matrix} \qquad \begin{matrix} R \diagdown \quad \delta \ominus \diagup R \\ Al \\ Cl \quad R \end{matrix} \right]$$

(*B*)

$$R \diagdown \quad R \quad R \quad R$$
$$Al \quad Ti \quad Ti$$
$$R \quad Cl \quad Cl \quad R$$

(*C*)

$$\begin{matrix} R \quad Cl \\ | \quad \diagup \\ Cl-Ti \bigcirc \\ \diagup | \\ Cl \quad Cl \end{matrix}$$

(*D*)

Figure 18 Suggested structures for active Ziegler–Natta catalysts (reference 60).

(some of which are shown in Fig. 18) are based on this and on the assumption that, since aluminum halides and alkyls are associated molecules, the same type of bonding is probably present in the complexes.

The bridged bonds in structures *A* and *C* explain the facile bond making and breaking in the region of complexation with monomer. Ionic and polarization effects produce the activity in structure *B*, whereas *D* uses a vacant orbital near an organometallic bond. Catalyst *D* probably forms by reaction between $AlEt_3$ and $TiCl_3$ in the manner (132) indicated in Fig. 19.

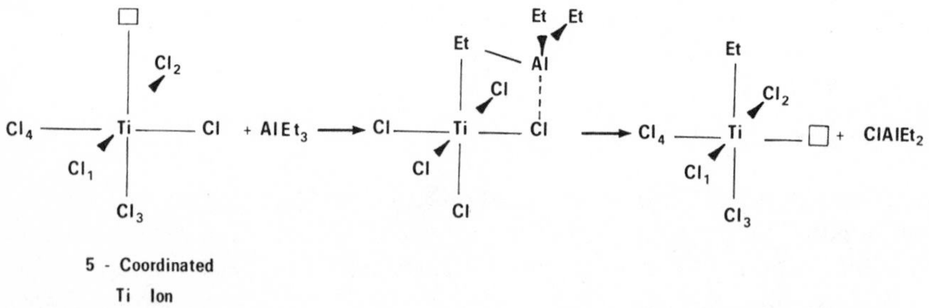

5 - Coordinated
Ti Ion

Figure 19 Proposed formation of the active catalyst from $AlEt_3$ and $TiCl_3$ (reference 132).

The ions Cl_1 and Cl_4 are attached to a second titanium in the crystal lattice of $TiCl_3$. Cossee (132). who proposed this structure, believes that these are nonexchangeable, whereas the fifth chloride ion is replaced by an ethyl group. The octahedral structure is thus provided by the four chloride ions anchored in the interior of the solid lattice. The ethyl group is attached by a σ bond to the titanium and the sixth position is a vacant orbital.

To obtain a stereoregulated polymerization, the monomer must be preoriented in some manner. This may be done in several ways. It has been found that α-olefins adsorb to the surface of nickel metal or a titanium salt, forming a complex between the π electrons of the double bond and the vacant d orbitals of the metal by opening the double bond (Fig. 20) (133). This adsorption and complexing can be used to explain stereoregular polymerization.

Many factors influence the activity of the catalyst. Understanding the exact nature of active sites is thus a difficult task. The crystal structure of both the catalyst components and the complex affect the activity. For example, titanium trichloride may exist in several crystalline modifications (134), which have different activities when combined with an alkylaluminum. Other factors that influence catalyst activity are the molar ratio of components, aging of the complex, temperature of preparation, and impurities. The effect of additives (or impurities) on catalyst activity is a complex topic. Hydrogen is often used as an additive to decrease polymer molecular weight, and oxygen in catalytic amounts often acts as a promoter of catalyst activity.

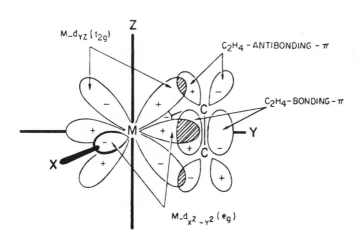

Figure 20 Schematic representation of the π-bond between a transition metal and ethylene (reference 133).

In some systems, water decreases polymer molecular weight and in others it increases stereoregularity.

Understanding the exact nature of the active sites in Ziegler–Natta type of catalysts is further complicated by the fact that in several systems active sites of differing activity are found. This difference in activity may arise either from differences in the oxidation state of the transition metal or from differing environments of the active site. The presence of active sites of differing activity can affect both the molecular weight distribution and the stereoregularity of the polymer. This has been shown to be true for several systems.

Thus Ziegler–Natta systems are quite complex. It would be impossible to present a mechanistic scheme to cover even a small portion of the many catalytic systems. Therefore, only general statements concerning the most widely used systems are made in the following sections.

3.   Polymerization Mechanisms (135)

The general kinetic scheme for polymerization is similar to the other types of addition polymerization in that initiation, propagation, and termination steps are involved.

Initiation

$$Cat-R' + CH_2=CHR \longrightarrow Cat-CH_2\underset{\displaystyle R}{CH}-R' \qquad (134)$$

Propagation

$$Cat-CH_2\underset{\displaystyle R}{CH}-R' + nCH_2=CHR \longrightarrow Cat\left[CH_2\underset{\displaystyle R}{CH}\right]_{n+1}R' \qquad (135)$$

Termination
By an active hydrogen compound:

$$Cat-CH_2\underset{\displaystyle R}{CH}\left[CH_2\underset{\displaystyle R}{CH}\right]_n R' + HY$$

$$\downarrow$$

$$Cat-Y + CH_3\underset{\displaystyle R}{CH}\left[CH_2\underset{\displaystyle R}{CH}\right]_n R' \qquad (136)$$

By transfer with monomer:

$$\text{Cat} - \text{CH}_2\text{CH} \left[\text{CH}_2\text{CH} \right]_n R' + \text{CH}_2 = \text{CHR}$$
$$\quad\quad\;\; | \qquad\quad |$$
$$\quad\quad\;\; R \qquad\quad R$$

$$\downarrow$$

$$\text{Cat} - \text{CH}_2 - \text{CH}_2 + \text{CH}_2 = \text{C} \left[\text{CH}_2\text{CH} \right]_n R' \qquad (137)$$
$$\quad\quad\quad\quad\quad | \qquad\quad\quad | \qquad\;\; |$$
$$\quad\quad\quad\quad\quad R \qquad\quad\quad R \qquad R$$

By spontaneous internal hydride transfer:

$$\text{Cat} - \text{CH}_2 - \text{CH} \left[\text{CH}_2\text{CH} \right]_n R'$$
$$\quad\quad\quad\quad | \qquad\quad |$$
$$\quad\quad\quad\quad R \qquad\quad R$$

$$\downarrow$$

$$\text{Cat} - \text{H} + \text{CH}_2 = \text{C} \left[\text{CH}_2\text{CH} \right]_n R' \qquad (138)$$
$$\quad\quad\quad\quad\quad | \qquad\quad | \qquad\;\; |$$
$$\quad\quad\quad\quad\quad R \qquad\quad R \qquad R$$

Experimental evidence (124) indicates that initiation (Eq. 134) takes place at a transition metal–carbon bond with an alkyl group becoming attached to the monomer. Several mechanistic models for this are discussed below.

The propagation step (Eq. 135) consists of repeated insertion of the olefin into the transition metal–carbon bond. The stereoregularity of the resulting polymer is controlled by the propagation step. Models for this steric control are presented in the following section.

The termination step in a Ziegler–Natta polymerization may be quite complex. Three possibilities are presented above (Eqs. 136–138) and other termination processes are also possible (136).

Several variations of a bimetallic mechanism have been proposed for the polymerization mechanism. In 1960, Natta (137) proposed that the bridged-bond structure of the catalyst (cf. Fig. 18) can be applied mechanistically. Although Natta did not specifically state it, the monomer may be regarded as being coordinated in the manner depicted in Fig. 20. According to Natta, the olefin is simply inserted between the existing titanium–carbon bond through the polarization mechanism in Fig. 21. The bridged bond opens, producing an electron-deficient metal which attracts the π electrons from the monomer into a σ-type bond. At the same time, the resulting carbanion attacks the

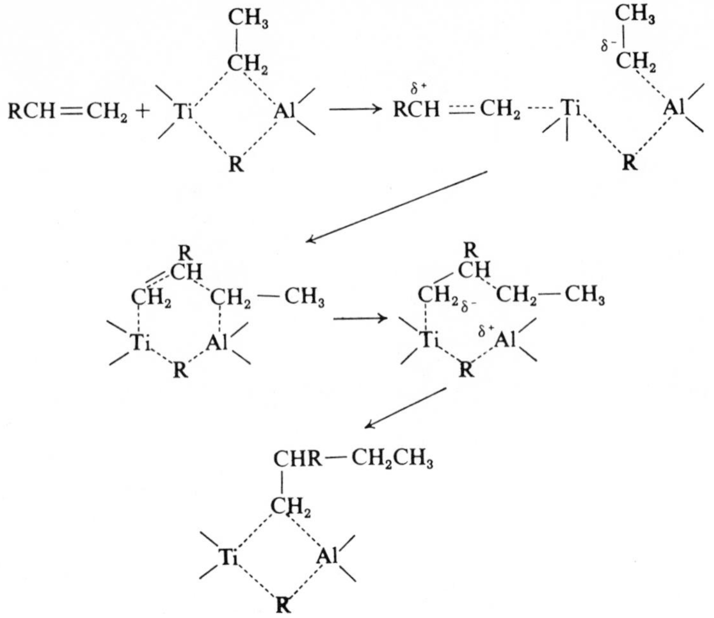

Figure 21 Bimetallic mechanism proposed by Patat and Sinn (reference 138).

electron-deficient center in the monomer. When this occurs, a new bridged-bond structure is generated, which includes the formerly complexed monomer. A related bimetallic mechanism had been proposed previously by Patat and Sinn (138).

These bimetallic mechanism are attractive in that they provide for an active site which can complex an approaching monomer and maintain the terminal group of the polymer chain in a position suitable for reaction. Naturally the bimetallic structure need not be made up of titanium and aluminum. It is possible to devise a mechanism similar to those proposed by Natta (137) and Patat and Sinn (138) by utilizing two titanium atoms in the active site (cf. Fig. 18).

As previously stated, the preponderance of evidence indicates that growth of the polymer chain in Ziegler–Natta catalysis involves participation of the transition-metal centers. Why, therefore, is it necessary to propose a mechanism using two metal sites? As a result of this question, several monometallic mechanisms have been proposed.

Cossee (132, 133, 139) has proposed a monometallic mechanism involving an active site in which the transition metal has an octahedral configuration

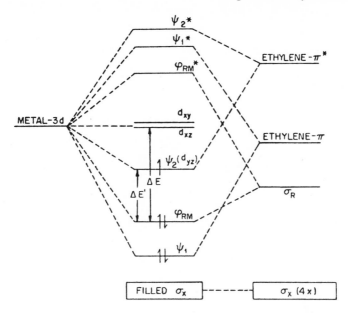

Figure 22 Proposed molecular orbital energy diagram for the octahedral complex RTiCl$_4$ (C$_2$H$_4$). For reasons of simplicity, 4s and 4p orbitals are not taken into account and the Ti–Cl bond is supposed to be 100% ionic (reference 133).

(Fig. 18). The driving force for the polymerization can be explained on the basis of a molecular orbital treatment (Fig. 22).

When an olefin such as ethylene is coordinated to the complex (Fig. 20), the metal dyz orbital becomes combined with the π^* orbital of ethylene, forming the Ψ_2 orbital. This molecular orbital is considerably lower in energy than the uncomplexed antibonding orbital of ethylene (π^*). The energy gap between the ϕ_2 orbital and the filled orbital ϕ_{RM} (representing the bond between the alkyl group and titanium) is reduced below the ΔE between ϕ_{RM} and the empty d orbitals ($\Delta E'$). This reduction in the energy gap more readily allows the promotion of an electron from the ϕ_{RM} orbital. If $\Delta E'$ is smaller than the critical energy gap related to the stability of metal-to-alkyl bonds, the alkyl group is expelled and attaches itself in a concerted process to the nearest carbon atom of the olefin. At the same time the other side of the olefin connects itself to the metal (Fig. 23). This leads to the re-generation of a vacant orbital with a different orientation.

This molecular orbital treatment is not confined to the Cossee proposal of an octahedral complex as the active site. This treatment can be generalized to show the vital role played by the empty (or half-filled) 3d orbitals of the

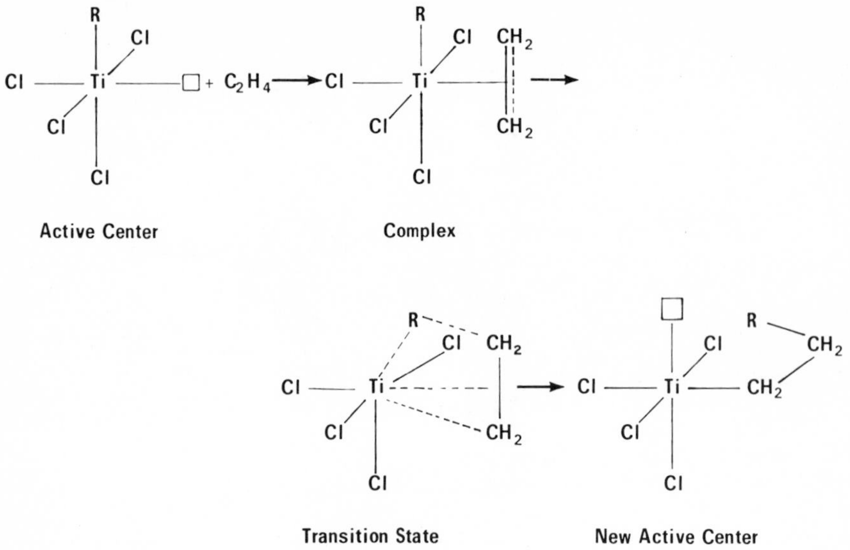

Figure 23   Mechanism for the polymerization of ethylene proposed by Cossee (reference 132).

transition metal in forming a π complex with the olefin to lower the energy barrier for the insertion of the olefin between the transition metal–alkyl bond.

At present it is not possible to distinguish between the mono- and bimetallic mechanisms. The bimetallic mechanisms appear attractive because they do not involve growing-chain migration for isotactic propagations. They also provide a method for introducing an optically active monomer with the same approach and monomer conformation at all times. On the other hand, the monometallic suggestions are significantly simpler than the bimetallic mechanisms. It is difficult to require two complicated metal sites to participate in the myriad catalytic situations which exist under Ziegler–Natta conditions. In addition, Boor (124) believes that the geometry of the monometallic active site does not present a serious problem for the migration. He also suggests that insertion of the monomer with complexation could lead to propagation without a change of orbitals even with a monometallic mechanism. Lastly, Boor claims that the Cossee octahedral structure for the active site leads to diastereomeric centers after reaction with monomer. Thus any given site preferentially complexes with one antipode rather than with the other. This can explain experiments conducted by Pino (140) on α-olefins, in which racemic polymers were resolved into optically active fractions and also were polymerized asymmetrically by optically active catalysts. Although these ideas are attractive, it has been suggested (60) that it is inconsistent to do away with complexation in one case (i.e., to avoid migration) and to use it for diastereomeric selectivity in another (i.e., optically active polymers).

4. Forces of Stereoregulation

The weight of experimental evidence (124) indicates that the driving force of isotactic propagation at the active center of the heterogeneous catalyst arises from the steric interaction between the α-olefin and the ligands which form the environment of the growth center.

Cossee (132) has proposed a model for the isotactic propagation of α-olefins based on his propagation scheme (Fig. 23). The stereospecificity results from the placement of the octahedral active site in an asymmetric environment. From an investigation of a model for the active site in a solid lattice, two main conclusions are drawn: (1) a plane drawn through the alkyl group, titanium, Cl_3, or Cl_4 (cf. Figs. 19 and 24) is not a plane of symmetry. Therefore, Cl_1 and Cl_2 are nonequivalent. (2) The positions of the alkyl group and the vacant orbital are nonequivalent. With these observations in mind, the stereospecific propagation using propylene as an example can be depicted as in Fig. 24. In the activated complex between propylene and the active center (Fig. 24b), the propylene molecule is oriented with its methylene end downward. There are now two possible orientations for the methyl group. It could be placed on either side of the plane going through the alkyl group, titanium, and Cl_3, Cl_4. These orientations are not equivalent (see above comment on the nonequivalency of Cl_1 and Cl_2). Therefore, it may be reasoned that there will be a preference for one orientation. This is shown in Fig. 24 where the methyl is oriented on the side of Cl_1. The configuration of carbon atom 2 is now fixed (Figs. 24c,d).

When the insertion is complete (Figure 24d), the vacant orbital and the alkyl substituent have exchanged positions. Since we have previously observed that these positions are not equivalent (compare Figs. 24a and d) two possibilities now arise for the insertion of the next monomer unit. The alkyl group can remain as in Fig. 24d with monomer complexing in this position, or the positions can invert before the second monomer complexes (24e). Since a syndiotactic polymer would be obtained in the former case, inversion must occur before monomer complexation.

The monomer and complex are now oriented in the same manner as in structure 24b. Complexation and insertion would proceed as in 24b–d and an isotactic polymer would be obtained. (The first two units are shown in 24g.)

In the case of the preparation of syndiotactic polypropylene using soluble catalysts, it has been suggested (141) that propagation at the active center of the catalyst arises from methyl–methyl steric interactions which force the incoming monomer to have a configuration opposite to that of the last added monomer unit.

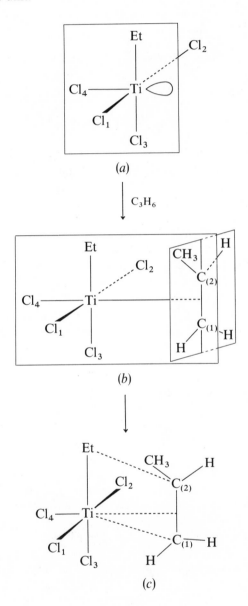

(a)

C_3H_6

(b)

(c)

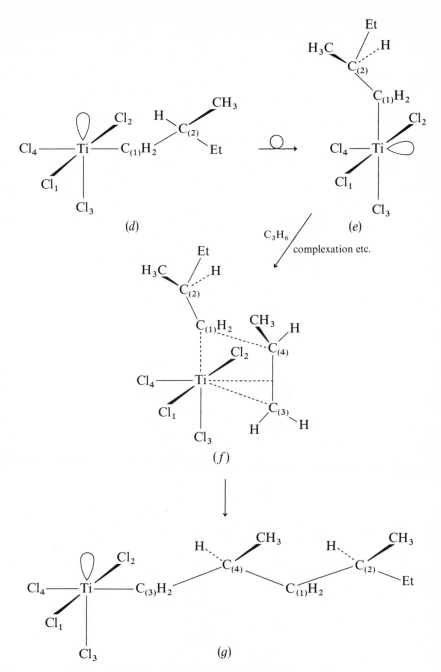

Figure 24 Stereospecific propagation of propylene.

GENERAL REFERENCES

1. P. J. Flory, *Principles of Polymer Chemistry*, Cornell University Press, Ithaca, N.Y. 1953.
2. F. W. Billmeyer, Jr., *Textbook of Polymer Science*, 2nd ed., Wiley-Interscience, New York, 1971.
3. R. W. Lenz, *Organic Chemistry of Synthetic High Polymers*, Wiley-Interscience, New York, 1967.
4. G. E. Ham, (Ed.), *Vinyl Polymerization*, Dekker, New York, 1967, 1969.
5. W. R. Sorenson and T. W. Campbell, *Preparative Methods of Polymer Chemistry*, 2nd ed., Wiley-Interscience, New York, 1968.
6. G. M. Burnett, *Mechanism of Polymer Reactions*, Interscience, New York, 1954.
7. P. W. Morgan, *Condensation Polymers: by Interfacial and Solution Methods*, Wiley-Interscience, New York, 1965.
8. L. S. Reich and A. Schindler, *Polymerization by Organometallic Compounds*, Wiley-Interscience, New York, 1966.
9. G. E. Ham, (Ed.), *Copolymerization*, Wiley-Interscience, New York, 1964.
10. C. E. Schildknecht, (Ed.), *Polymer Processes*, Interscience, New York, 1956.
11. C. Walling, *Free Radicals in Solution*, Wiley, New York, 1957.
12. C. H. Bamford, W. G. Barb, A. D. Jenkins, and P. F. Onyon, *The Kinetics of Vinyl Polymerization by Radical Mechanisms*, Academic, New York, 1958.
13. A. H. Frazer, *High Temperature Resistant Polymers*, Wiley-Interscience, New York, 1968.
14. A. D. Ketley, (Ed.), *The Stereochemistry of Macromolecules*, Vols. I–III, Dekker, New York, 1967, 1968.
15. M. Szwarc, *Carbanions, Living Polymers and Electron-Transfer Processes*, Wiley-Interscience, New York, 1968.
16. H. Lee, D. Stoffey, and K. Neville, *New Linear Polymers*, McGraw-Hill, New York, 1967.
17. H. F. Mark, N. G. Gaylord, and N. M. Bikales (Eds.), *Encyclopedia of Polymer Science and Technology*, Wiley-Interscience, New York, 1964–1971.
18. N. A. J. Platzer, (Ed.), *Addition and Condensation Polymerization Processes*, Advances in Chemistry Series, Number 91, American Chemical Society, Washington, D.C., 1969.
19. J. P. Kennedy and E. G. M. Tornquivst, (Eds.), *Polymer Chemistry of Synthetic Elastomers*, Wiley-Interscience, New York, 1968, 1969.

SPECIFIC REFERENCES

20. P. J. Flory, *Chem. Revs.*, **39**, 137 (1946).
21. P. J. Flory, *J. Am. Chem. Soc.*, **61**, 3334 (1939); **62**, 2261 (1940).
22. W. R. Sorenson, *Polym. Prepr.*, **10**, 84 (1969).
23. M. I. Kohan (Ed.), *Nylon Plastics*, Wiley-Interscience, New York, 1973.
24. P. W. Morgan, Ref. 7, Chap. 5.
25. W. Sweeny and J. Zimmerman, in Ref. 17, Vol. 10, 1969, p. 483, and related chapters.
26. A. H. Frazer, Ref. 13, Section III-A.
27. W. B. Black and J. Preston, in Man-Made Fibers, Vol. 2, H. F. Mark, S. M. Atlas, and E. Cernia (Eds.), Wiley-Interscience, New York, 1968, p. 297.
28. N. G. Gaylord (Ed.), *Polyesters*, Wiley-Interscience, New York, 1964.

29. V. V. Korshak and S. V. Vinogradova, *Polyesters*, Pergamon Press, London, 1965.

30. I. Goodman, in Ref. 17, Vol. 11, 1969, p. 62, related chapters.

31. K. A. Pigott, in Ref. 17, Vol. 11, 1969, p. 506.

32. J. H. Saunders and K. C. Frisch, *Polyurethanes: Chemistry and Technology*, Wiley-Interscience, New York, 1962, 1964.

33. H. Lee, D. Stoffey, and K. Neville, Ref. 16, Chap. 9.

34. A. H. Frazer, Ref. 13, Section IV-A.

35. H. H. Levine, in Ref. 17, Vol. 11, 1969, p. 188.

36. K. C. Brinker and I. M. Robinson, U.S. Pat. 2,895,948 (June 1959).

37. H. A. Vogel and C. S. Marvel, *J. Polym. Sci.*, **50**, 511 (1961).

38. W. A. Keutgen, in Ref. 17, Vol. 10, 1969, p. 1.

39. R. N. Johnson, in Ref. 17, Vol. 11, 1969, p. 447.

40. H. Lee, D. Stoffey, and K. Neville, Ref. 16, Chap. 5.

41. K. J. Ivin and J. B. Rose, *Advances in Macromolecular Chemistry*, Vol. I, W. M. Pasika (Ed.), Academic, New York, 1968, p. 336.

42. H. A. Vogel, *J. Polym. Sci.*, *A1*, **8**, 2035 (1970).

43. M. E. A. Cudby, R. G. Feasey, S. Gaskin, M. E. B. Jones, and J. B. Rose, *J. Polym. Sci.*, **C22**, 747 (1969).

44. J. K. Stille and F. W. Harris, *J. Polym. Sci.*, **B4**, 333 (1966).

45. J. K. Stille and F. W. Harris, *J. Polym. Sci.*, *A1*, **6**, 2317 (1968).

46. M. G. Baldwin, K. E. Johnson, J. A. Lovinger, and C. O. Parker, *J. Polym. Sci.*, **B5**, 803 (1967).

47. J. K. Stille, F. W. Harris, H. Mukamal, R. O. Rakutis, C. L. Schilling, G. K. Noren, and J. A. Reed, in Ref. 18, p. 628.

48. E. A. Kraiman, in Ref. 17, Vol. 5, 1966, p. 23.

49. H. Mukamal, F. W. Harris, and J. K. Stille, *J. Polym. Sci.*, *A1*, **5**, 2721 (1967).

50. H. Lee, D. Stoffey, and K. Neville, in Ref. 16, Chap. 3.

51. G. D. Cooper and A. Katchman, in Ref. 18, p. 660.

52. A. S. Hay, P. Shenian, A. C. Gowan, P. F. Erhardt, W. R. Haaf, and J. E. Theberge, Ref. 17, Vol. 10, 1969, p. 92.

53. H. C. Bach and W. B. Black, in Ref. 18, p. 679.

54. A. S. Hay, *J. Polym. Sci.*, *A1*, **7**, 1625 (1969).

55. K. C. Frisch and S. L. Reegen (Eds.), *Ring Opening Polymerization*, Dekker, New York, 1969.

56. O. Vogl and J. Furukawa, (Eds.), *Polymerization of Heterocyclics*, Dekker, New York, 1973.

57. J. Šebenda, in Ref. 56, p. 153.

58. S. Barzakay, M. Levy, and D. Vofsi, *J. Polym. Sci.*, **B3**, 601 (1965).

59. C. E. Schildknecht, *Ind. Eng. Chem.*, **50**, 107 (1958) (several changes have been made in this table including the addition of some recent material).

60. M. Goodman, in *Topics in Stereochemistry*, Vol. 2, N. L. Allinger and E. L. Eliel (Eds.), Wiley-Interscience, New York, 1967, p. 73.

61. G. Natta and G. Allegra, *Tetrahedron*, **30**, 1987 (1974).

62. C. E. Schildknecht, in Ref. 10, Chap. 2.

63. H. Ringdorf, in Ref. 17, Vol. 2, 1965, p. 642.

64. C. E. Schildknecht, in Ref. 10, Chap. 5.
65. E. Farber, in Ref. 17, Vol. 13, 1970, p. 552.
66. E. Trommsdorff and C. E. Schildknecht, in Ref. 10, Chap. 3.
67. G. M. Burnett, Ref. 6, pp. 301–303, 308–310.
68. F. A. Bovey, I. M. Kolthoff, A. L. Medalia, and E. J. Meehan, *Emulsion Polymerization*, Interscience, New York, 1955.
69. J. C. H. Hwa and J. W. Vanderhoff (Eds.), "New Concepts in Emulsion Polymerization, Symposium," *J. Polym. Sci.*, **C27**, (1969).
70. E. W. Duck, in Ref. 17, Vol. 5, 1966, p. 801.
71. B. M. C. van der Hoff, *Advan. Chem. Ser.*, **34**, 6 (1962).
72. "International Symposium on Emulsion Polymers," *Polym. Prepr.*, **16**, (1), 1975.
73. J. W. Vanderhoff, in Ref. 4, Part II, Chap. 1.
74. W. V. Smith and R. H. Ewart, *J. Chem. Phys.*, **16**, 592 (1948).
75. G. E. Ham, in Ref. 17, Vol. 4, 1966, p. 165.
76. T. Alfrey, Jr., J. Bohrer, and H. F. Mark, *Copolymerization*, Wiley-Interscience, New York, 1953.
77. H. F. Mark, B. Immergut, E. H. Immergut, L. J. Young, and K. I. Beynon, in Ref. 9, Appendix A.
78. J. Brandrup and E. H. Immergut (Eds.), *Polymer Handbook*, 2nd ed., Wiley-Interscience, New York, 1975, Chap. II-5.
79. R. M. Joshi, *J. Macromol. Sci.-Chem.*, **A7**, 1231 (1973).
80. F. R. Mayo and F. M. Lewis, *J. Am. Chem. Soc.*, **66**, 1574 (1944).
81. M. Fineman and S. D. Ross, *J. Polym. Sci.*, **5**, 259 (1950).
82. T. Alfrey, Jr., and L. J. Young, in Ref. 9, Chap. 2.
83. L. J. Young, in Ref. 9, Appendix B.
84. J. Brandrup and E. H. Immergut, in Ref. 133, Chap. II-6.
85. M. Morton, in Ref. 9, Chap. 7.
86. J. P. Kennedy, in Ref. 9, Chap. 5.
87. M. Szwarc, in Ref. 15, Chap. 9.
88. D. C. Pepper, *Quart. Rev.*, **8**, 88 (1954).
89. G. E. Ham (Ed.), in Ref. 9, Chap. 4.
90. I. Pasquon, A. Valvassori, and G. Sartori, in Ref. 14, Vol. 1, Chap. 4.
91. A. S. Hoffman and R. Bacskai, in Ref. 9, Chap. 6.
92. R. J. Ceresa, *Block and Graft Copolymerization*, Wiley-Interscience, New York, 1973.
93. D. C. Allport and W. H. James, *Block Copolymers*, Wiley-Interscience, New York, 1973.
94. H. A. J. Battaerd and G. W. Tregear, *Graft Copolymers*, Wiley-Interscience, New York, 1967.
95. C. G. Overberger, J. E. Mulvaney, and A. M. Schiller, in Ref. 17, Vol. 2, 1965, p. 95.
96. M. Morton, in Ref. 4, Part II, Chap. 5.
97. L. S. Reich and A. Schindler, in Ref. 8, Chap. 7.
98. H. L. Hsieh and W. H. Glaze, *Rubber Chem. Technol.*, **43**, 22 (1970).
99. M. Morton and L. J. Fetters, *Macromolecular Reviews*, Vol. 2, Wiley-Interscience, New York, 1967, p. 71.

100. M. Szwarc, *Makromol. Chem.*, **35**, 132 (1961).

101. M. Szwarc, in Ref. 15, Chap. 12.

102. L. E. Forman, in Ref. 19, Chap. 6.

103. K. Ziegler, L. Jakob, H. Wollthan, and A. Wenz, *Ann. Chem.*, **511**, 64 (1934).

104. M. Szwarc, in Ref. 15, Chap. 2.

105. J. F. Henderson and M. Szwarc, *Macromolecular Reviews*, Vol. 3, Wiley-Interscience, New York, 1968, p. 317.

106. M. Szwarc, M. Levy, and R. Milkovitch, *J. Am. Chem. Soc.*, **78**, 2656 (1956).

107. L. J. Fetters, *J. Res. Natl. Bur. Std.*, *A*, **70**, 421 (1966).

108. R. Zelinski and C. W. Childers, *Rubber Chem. Technol.*, **41**, 161 (1968).

109. M. Morton, *Polym. Prepr.*, **10**, 512 (1969).

110. J. Moacanin, G. Holden, and N. W. Tschoegl (Eds.), "Symposium on Block Polymers," *J. Polym. Sci.*, **C26** (1969).

111. R. K. Graham, D. L. Dunkelberger, and E. S. Cohn, *J. Polym. Sci.*, **22**, 189 (1956).

112. A. M. Eastman, in Ref. 17, Vol. 3, 1965, p. 25.

113. P. H. Plesch (Ed.), *Cationic Polymerization*, Pergamon Press, Oxford, 1963.

114. J. P. Kennedy and A. W. Langer, Jr., *Fortschr. Hochpolym.-Forsch.*, **3**, 508 (1964).

115. Z. Zlamal, in Ref. 4, Part II, Chap. 6.

116. J. P. Kennedy and R. G. Squires, *J. Macromol. Sci.—Chem.*, **A1**, 805 (1967).

117. M. P. Dreyfuss and P. Dreyfuss, *Polymer*, **6**, 93 (1965).

118. J. P. Kennedy, in Ref. 19, Chap. 5A.

119. J. P. Kennedy, in Ref. 17, Vol. 7, 1967, p. 754.

120. J. P. Kennedy, *Trans. N.Y. Acad. Sci.*, **28**, 1080 (1966).

121. G. B. Butler, in Ref. 17, Vol. 4, 1966, p. 568.

122. G. B. Butler et al., *J. Macromol. Sci.—Chem.*, **A8**, 1139, 1175, 1205, 1239 (1974).

123. C. S. Marvel and E. J. Gall, *J. Org. Chem.*, **25**, 1784 (1960).

124. J. Boor, Jr., *Macromolecular Reviews*, Vol. 2, 1967, p. 115.

125. L. S. Reich and A. Schindler, in Ref. 8, Chaps. 3 and 4.

126. G. Natta and F. Danusso (Eds.), *Stereoregular Polymers and Stereospecific Polymerizations*, Vols. 1 and 2, Pergamon Press, Oxford, 1967.

127. M. Goodman, J. Brandrup, and H. F. Mark, in *Crystalline Olefin Polymers*, R. A. V. Raff and K. W. Doak (Eds.), Wiley-Interscience, New York, 1965, Part I, Chap. 4.

128. A. D. Ketley (Ed.), in Ref. 14, Vol. 1 contains several chapters on different aspects of Ziegler–Natta polymerizations.

129. H.-O. Olivé and S. Olivé, *Fortschr. Hochpolym.-Forsch.*, **6**, 421 (1969).

130. J. Boor, *Polym. Prepr.*, **15**, (1), 359 (1974); and several other papers in this book.

131. K. Ziegler, E. Holzkamp, H. Breil, and H. Martin, *Angew. Chem.*, **67**, 541 (1955)

132. P. Cossee, in Ref. 14, Vol. 1, Chap. 3.

133. P. Crossee, *J. Catal.*, **3**, 80 (1964).

134. J. Boor, Jr., in Ref. 124, pp. 123–139.

135. T. Keii, *Kinetics of Ziegler–Natta Polymerization*, Halsted Press, New York, 1973.

136. R. W. Lenz, in Ref. 3, pp. 597–598.

137. G. Natta, *J. Polym. Sci.*, **48**, 219 (1960).

138. E. Patat and H. Sinn, *Angew. Chem.*, **70**, 496 (1958).
139. P. Cossee, *Rec. Trav. Chim.*, **85**, 1151 (1966).
140. P. Pino, F. Ciardelli, and G. Montagnoli, *J. Polym. Sci.*, **C16**, 3265 (1968) and references cited therein.
141. E. A. Youngman and J. Boor, Jr., *Macromolecular Reviews*, Vol. 2, 1967, p. 33.

DISCUSSION QUESTIONS AND PROBLEMS

1. Compare the basic characteristics of step or condensation polymerization with those of chain reaction polymerization with respect to the following:
   a. Methods of initiation.
   b. Time required for polymerization.
   c. Control of molecular weight.

2. Under cationic polymerization conditions, certain imines will polymerize via ring opening. What repeat unit would be expected from the ring opening of Conidine? Discuss why.

3. When butyl lithium is added to a toluene solution of butadiene and styrene the resulting polymer is essentially a block copolymer of styrene and butadiene. Explain why.

4. Vinyl ethyl sulfone can polymerize using either free radical or anionic initiators. Explain why.

5. Water acts as a chain terminator in an anionic polymerization. However, under very specific conditions, it is possible to prepare a living polymer in water. What are these conditions? Suggest at least one possible monomer.

6. Why would a polyester based on isophthalic acid have a lower melting point than an analogous polyester based on terephthalic acid?

7. If you prepared a previously unknown vinyl monomer, how would you decide what initiator system should be used first?

8. If your first attempt at preparing a polymer from an A–A, B–B polymerization resulted in very low molecular weight polymer, what steps would you take to try to increase the molecular weight?

9. What factors must be taken into consideration when choosing a solvent for the following types of polymerization: free radical, anionic, cationic, and condensation.

10. Natta and Ziegler were awarded a Nobel Prize for their discovery of stereo-regular polymerization systems. Review the basic concepts and indicate how they differ from those of the other polymerization systems (condensation, addition, etc.).

CHAPTER 3

Polymer
Modifications

NORMAN G. GAYLORD
DAVID S. HOFFENBERG
Gaylord Research Institute Inc.
Newark, New Jersey

Despite the very large number of monomeric compounds that have been synthesized and converted to corresponding homopolymers, the restless and probing minds of chemists and the needs and desires of an ever-expanding technology have continually sought to prepare novel compositions having unique and useful properties.

In this chapter we consider how such novel materials can and have been prepared by modifying existing polymers through copolymerization, polymer blending, post-reaction of polymers, or combinations of these techniques.

A. COPOLYMERIZATION

The history of polymer modification by copolymerization began around 1910 when it was discovered that copolymers of olefins and dienes yielded better elastomers than polydienes or polyolefins alone. However, it wasn't until the 1930s, when Staudinger made his pioneering contributions to the field of copolymerization, and the 1940s, when wartime shortages of natural rubber created the necessity for synthetic rubbers, that this field really came into its own.

Today, equipped with experience gained over the last 30 years, and armed with a fairly good understanding of the kinetics and mechanism of copolymerization, polymer architects who create new macromolecules by combining old monomers in multitudinous combinations are becoming so adept at this work that the day does not seem very far off when true custom synthesis of polymers with accurately predetermined properties will be a reality.

Like all chemical species, macromolecules have properties that, for convenience, may be divided into physical properties and chemical properties. Sometimes it is desirable to modify only physical properties, such as viscoelastic properties or solubility. At other times only modification of chemical properties is important and sometimes modification of both is required. In this connection, experience seems to indicate that considerably more of a modifying comonomer is necessary to achieve a significant change in physical properties than would be required to confer a useful degree of chemical reactivity.

1. Physical Properties of Random Copolymers from Vinyl Monomers

One of the major weapons in the arsenal of polymer architects is a good working knowledge of a rather loose classification of monomer types and specific monomers into categories. The categories in turn bear some relationship to ultimate polymer properties. For example, monomers may be classified into polar and nonpolar types, into high T_g (rigid) and low T_g (nonrigid) types, etc., depending on whether their homopolymers are classified as polar or nonpolar, rigid or nonrigid, etc., and whether those monomers have been demonstrated to confer such properties to copolymer compositions.

Before proceeding to specific illustrations of polymer modification through copolymerization, the reader should be apprised of several generalizations that, with some exceptions, apply to all random copolymer systems.

The most fundamental, perhaps, deal with crystallinity and crystalline melting point (T_m) (7). When copolymers are made from monomers that form crystalline homopolymers, the degree of crystallinity and the crystalline melting point decrease as the concentration of the second constituent is increased relative to either homopolymer.* This is true for the same reason that impurities depress the melting point of pure low molecular weight materials; they interfere with packing in the crystal lattice. The exception is when the comonomers are isomorphous, in which case they fit smoothly into the same lattice configuration. Examples of crystalline melting point variation as a function of copolymer composition are given in Fig. 1.

It should be clear from the foregoing that unless the comonomers are isomorphous (a rather rare occurrence in vinyl copolymers), efforts to raise the crystalline melting point of a polymeric species by copolymerizing with small amounts of a high melting comonomer are apt to be fruitless. In the example shown in Fig. 1, it would take more than 50% of comonomer B to give a material whose crystalline melting point is greater than homopolymer A. Under these circumstances, what we have really succeeded in doing is lowering the melting point of homopolymer B by inclusion of comonomer A.

A large number of polymer properties, including modulus, strength, orientability, and brittleness, are to a greater or lesser degree influenced by the degree of crystallinity of a polymer. It is therefore often desirable to modify the crystallinity of a polymer. Although it is relatively easy to reduce

---

* The melting point of the copolymer depends on the mole fraction n of one of the crystallizing constituents and is given by the equation:

$$\frac{1}{T_m} - \frac{1}{T_m^0} = - \frac{R}{\Delta H_m} (\ln n)$$

where T_m is the melting point of the copolymer, T_m^0 the melting point of the homopolymer, and ΔH_m the heat of fusion.

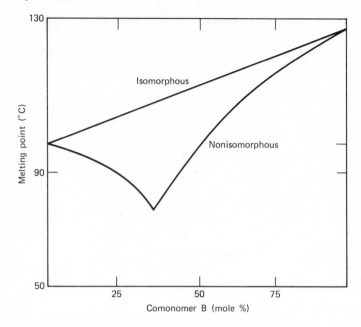

Figure 1   Crystalline melting point of an AB copolymer as a function of composition.

the crystallinity of a material, efforts to increase crystallinity by additions of small or modest amounts of comonomers known to form highly crystalline homopolymers suffer from the same problems as efforts to raise crystalline melting point by this method.

An outstanding example is the copolymer of ethylene with a small amount (1–3%) of butene, which lowers the crystallinity just enough to minimize stress cracking normally associated with the linear homopolymer, especially in blown bottle applications.

There is a second group of polymer properties wherein the copolymer property falls between those of the corresponding homopolymers. One such property is the glass-transition point, T_g,* which has been shown to be a weighted average (8, 9). The glass-transition temperature is that temperature at which chains begin to exhibit mobility and commence to flow more or less freely (Chapter 6).

---

* T_g can be calculated from the relationship

$$A_1 C_1 (T_g - T_{g1}) + A_2 C_2 (T_g - T_{g2}) = 0$$

where T_{g1} and T_{g2} are the glass transitions of the homopolymers, C_1 and C_2 are the weight fractions of each monomer in the homopolymer, and A_1 and A_2 are parameters which depend on monomer type.

Chain flexibility, a property important in determining elasticity, modulus, heat distortion, impact strength, and a number of other properties, is likewise predictable from a knowledge of the chain flexibility of the parent homopolymers. Chain flexibility arises from rotation about saturated chain bonds which are affected by steric and electrostatic interchain interactions, such as in the glass-transition temperature.

A third group of polymer properties includes those in which the overall effect of one monomer type with the other would appear to be synergistic; that is, the effect of adding reasonable amounts of a comonomer to a polymer produces an effect, usually in a desirable direction, which may be greater than would be expected by simply attributing a weighted portion to each homopolymer. These properties include the very important ones of melt viscosity, and properties such as solubility and its converse, solvent resistance. The latter, of course, applies more to crystalline polymers. Energies required to permeate and overcome crystal lattice forces in these materials are usually of a relatively high order of magnitude. Because incorporation of comonomers decreases crystallinity markedly, it also affects solvent resistance.

The dependence of many viscometric and solution properties in copolymer systems which do not crystallize results primarily from changes in inter- and intramolecular forces as measured by cohesive energy. Higher cohesive energy results in decreased solubility. It also usually results in higher stiffness and hardness and improved mechanical properties. Reducing solubility by increasing intermolecular cohesive energy is not unique to polymers, but is simply an extension of the same phenomenon in low molecular weight materials. However, because of the long chainlike nature of polymers, and the rather unique phenomenon of many intramolecular interactions, the intramolecular cohesive energy becomes significant. Thus a randomly coiled copolymer of high cohesive energy is more tightly coiled and more resistant to uncoiling than a parent homopolymer having less cohesive energy (Fig. 2).

(I)   (II)

Figure 2   (I) Lightly coiled homopolymer; (II) more tightly coiled copolymer.

The results of tighter coiling are improved viscometric properties (improved flow) and improved chemical and solvent resistance owing to resistance to penetration by solvent and reagent molecules of the tighter coil.

Thus the somewhat more than additive effect of copolymerization on solubility properties is a result of superposition of changes in intramolecular cohesive energy on intermolecular cohesive energy.

From the preceding discussion of what polymer modification by copolymerization can and cannot do, we see that those polymers that depend on crystalline forces for most of their outstanding properties are good candidates for modification by copolymerization. Those that depend on reasonably high glass transitions for their usefulness, that is, polystyrene, polymethyl methacrylate, polyvinyl chloride, and other amorphous thermoplastic materials, have been the subject of a reasonable amount of investigation, whereas materials that depend on low T_g's, rubbers and elastomers, have been the most fruitful areas for copolymer study and modification.

Perhaps the most important commercial copolymer is SBR (styrene-butadiene) rubber (10). The radical-initiated homopolymer of butadiene is a rubbery material but it has relatively poor properties in such areas as strength, abrasion resistance, flex fatigue resistance, solvent resistance, etc. By incorporating 28% of styrene (III) virtually all these properties are improved to a greater or lesser degree. Essentially all of the improvement is attributable to the greater rigidity of the styrene molecule and to a lesser degree to its slightly greater polarity (vis-à-vis butadiene). The microstructure of the butadiene units in the copolymer consists of about 70% trans-1,4, 15% cis-1,4, and 15% 1,2 units.

$$\left[CH_2-CH \atop \bigcirc \right]_x \left[CH_2-CH \atop CH=CH_2 \right]_y \left[CH_2-CH=CH-CH_2 \right]_z$$

(III)

The other side of the SBR coin is butadiene-modified styrene. Emulsion copolymers whose compositions are 75% styrene and 25% butadiene find wide application in paints and surface coatings, where the hardness of polystyrene is partially retained but its brittleness is modified so as to make the copolymer suitable as a paint.

Another important butadiene-modified elastomer is the butadiene-acrylonitrile copolymer (10). About the same weight percentage of modifier is used in nitrile rubber as in SBR rubber, that is, 25% for the most useful products. Nitrile rubbers are noted for their outstanding oil resistance. Table 1 shows the variation of properties with acrylonitrile content.

Table 1   Properties of butadiene–acrylonitrile rubber as a function of nitrile content

% Acrylonitrile in Copolymer	Properties of Vulcanizate
50–60	Tough leathery plastic with high resistance to aromatics
35–40	Leathery with high oil resistance
25	Rubbery with medium oil resistance
15	Rubbery with fair oil resistance
2–5	Rubbery with poor oil resistance

Diene-based elastomers generally suffer from poor oxygen and ozone resistance. This is due to unsaturation in the polymer. Saturated random copolymers of ethylene and propylene, both normally high melting crystalline homopolymers, yield noncrystalline elastomers with excellent oxygen and ozone resistance, but suffer from poor solvent resistance. Another example of crystallizable thermoplastic homopolymers that, when combined in a random copolymer relationship, form an extremely useful class of non-crystalline copolymers is the fluorinated rubbers, such as are sold under the trade names Fluorel (3M Co.) or Viton (Du Pont Co.), which contain hexa-fluoropropylene (50–60%) and tetrafluoroethylene (40–50%) (IV). The randomly spaced CF_3 group destroys crystallization.

$$\begin{array}{cccc} F & F & F & F \\ -C- & C- & C- & C- \\ F & F & F & \mid \end{array}$$
$$\begin{array}{c} FCF \\ F \end{array}$$

(IV)

Virtually all synthetic elastomers with the exception of certain stereo-specifically polymerized dienes are copolymeric in nature.

Copolymerization is also used to modify physical properties of amorphous thermoplastic materials. For the most part, the comonomer is added to improve flow and processability of a rather stiff homopolymer. Most often only relatively small amounts of comonomer (usually one possessing a lower T_g) are employed, since larger amounts tend to adversely affect other properties, the most important of which is often the maximum service temperature.

Thus the incorporation of up to 5–10% ethyl acrylate in polymethyl methacrylate results in a modest but significant reduction in melt viscosity and molding temperature, at the same time not causing an intolerable reduction in thermal properties. The same considerations govern vinylidene chloride–vinyl chloride copolymers (Saran–Dow Chemical Co.) and the multitudinous vinyl chloride copolymers, the most common of which is a vinyl chloride–vinyl acetate (3–5%) copolymer. Despite the improvement in processability of polyvinyl chloride by copolymerization, a large number of commercial grades require plasticization before use.

This section has dealt with modification of the physical properties of random copolymers from vinyl monomers. The title was chosen to distinguish these materials from block and graft copolymers, discussed in a later section.

2. Physical Properties of Heterochain Addition Copolymers

A heterochain addition polymer is one in which the backbone of the polymer contains atoms other than carbon, and whose mode of formation is via an addition type mechanism. Although there are numerous types actually and theoretically possible, including polycaprolactam, polyesters from lactones, and polyimines, perhaps the most important are those with either-linked oxygen in the backbone. In this class are polyethers made from cyclic ethers (ethylene oxide, propylene oxide, trimethylene oxide, tetrahydrofuran) and carbonyl compounds (formaldehyde, acetaldehyde), which may be considered the simplest form of cyclic ether (11). By starting with monomers of reasonably close reactivities, copolymers of all of these types may be prepared.

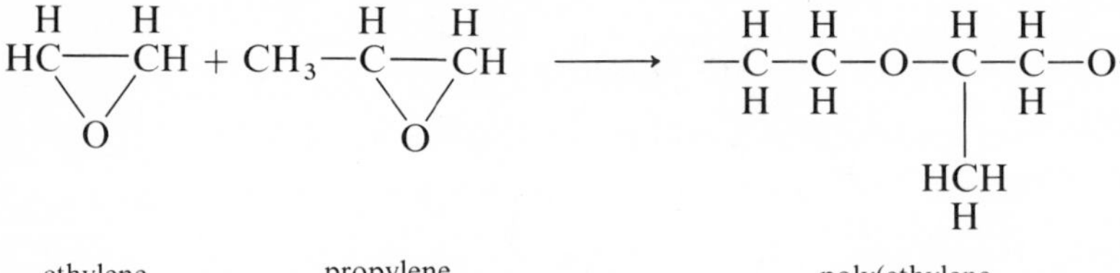

| ethylene oxide | propylene oxide | poly(ethylene oxide-co-propylene oxide) |

$$(1)$$

The physical properties of randomly copolymerized heterochain addition copolymers bear virtually the same relationship to corresponding homopolymers as do randomly copolymerized olefinic copolymers to their homopolymers.

The polycarbonyl compounds, of which polyformaldehyde (also known as acetal resin) is by far the most important, are considered engineering plastics which depend on crystalline forces for most of their desirable properties.

Therefore, it is not surprising that copolymerization of these materials is undertaken somewhat reluctantly. Of the several commercial polyformaldehyde resins competing for markets, one is virtually a homopolymer and most of the others contain the smallest possible amount of a comonomer, added for the purpose of improving thermal stability (see Section A.5).

Polyethers of the polyethylene oxide or polypropylene oxide type find utility in different application areas. Very low molecular weight polymers and copolymers are used as detergents and surfactants. In this end use solubility properties with respect to oils and water are most important, and the ratios of ethylene oxide (more hydrophilic) and propylene oxide (more hydrophobic) are adjusted so that the desired solubility characteristics are achieved in the final product.

3.   Physical Properties of Condensation Copolymers

Condensation polymers, as indicated in the preceding chapter, cover the broad area of those materials which are prepared by the co-elimination of usually small molecules from di- or polyfunctional reactants. Those condensation polymers which have achieved a large commercial position include some of the thermosetting materials such as phenolics and aminoplasts (urea or melamine–formaldehyde), 6,6-type nylons, polycarbonates, and polyesters, In phenolics and aminoplast resins, modification of physical properties by copolymerization is generally not important, and thus is not considered here. Nylons depend on crystallinity for their major properties, as do the fiber-forming polyesters, polyethylene terephthalate (Du Pont's Dacron) or poly(cyclohexanedimethanol terephthalate) (Eastman Kodak's Kodel); they are therefore usually used as homopolymers.

Other polyesters, however, find their way into a fairly wide spectrum of end use patterns, and in many of these copolymers are important. They find utility in surface coatings, resins, urethane elastomers, laminating resins, and thermosetting molding compounds. That one class of polymers can be so modified and adapted to meet the really diverse requirements of so many end products is a genuine tribute to the polymer architect, who in turn owes his success to the very large number of dibasic acids and dihydroxy compounds that are more or less readily available.

Of course, polyesters and copolyesters conform to the same general principles that govern the relationships between structure and physical properties, as discussed in the preceding sections. In condensation polymers, however, we have the additional advantage that it is fairly easy to modify the backbone structure of the polymer as well as the pendant structure. In olefinic addition polymers, virtually all backbones consist of a long string of carbon atoms, most of which are saturated. In polyethers, we get a linear carbon–oxygen backbone in which the quantity of oxygen may be varied. In polyesters, however, we may start with aromatic dibasic acids such as phthalic, isophthalic, or terephthalic, or aromatic diols (α,ω-dihydroxy-methyl-p-xylene), heterocyclic acids or diols, and unsaturated acids or diols. Also, by using ethylene glycol or 1,9-dihydroxynonane, for example, the repeat distance between ester groups can be controlled.

In general, aromatic and alicyclic rings in the backbone greatly increase chain stiffness. Pendant groups such as are found in 1,2-propylene glycol, 1,2- or 1,3-butylene glycol, and α-methylsuccinic acid, lower glass-transition temperatures and improve flexibility. Longer carbon–carbon sequences improve resistance to polar solvents whereas shorter segments improve resistance to hydrocarbon oils.

A typical surface coating resin (also known as an alkyd resin) is an interpolymer of phthalic anhydride, glycerol, and a naturally occurring, usually unsaturated fatty acid (linseed, cottonseed, etc.). Modifications of this would include replacing some of the phthalic with maleic acid or succinic acid (softer paint) or some of the glycerol with another triol or some diol to give products with modified properties (12).

Polyesters used in urethane elastomers are usually copolymers of ethylene glycol and propylene glycol with adipic acid, to give low crystalline melting and glass-transition point products with flexible backbones, comparable to the polyethers with which they compete.

Laminating and molding polyesters usually have in common the one feature that some of the dibasic acid is unsaturated (maleic, fumaric, itaconic) so as to provide sites for cross-linking with styrene. However, from that point the factors that determine the monomer mix are as complex and diverse as the people who do the formulating, the specific end use, and the economic factors. For example, a polyester compounded for high compressive strength comprises 1.5 moles of hydrogenated bisphenol A, 1.5 moles of 1,2-propylene glycol, 1.0 moles of phthalic anhydride, 1.0 moles of maleic anhydride, and 1.0 moles of fumaric acid.

Indeed, the almost infinite modifications that can be brought about relatively easily by copolymerization in polyester manufacture may very well have served as an impediment to commercial growth. The very multitude of choice may have served to confuse many users and required producers to maintain inventories of a large number of starting materials and products, thus keeping costs higher than necessary.

4.   Chemical Properties of Random Copolymers from Vinyl Monomers

For the most part, the ultimate utility of a polymeric material depends on how it functions in a given application. This function is most often dependent on the physical properties of a material, but quite often a certain amount and type of chemical reactivity is necessary for that material to perform as intended. For example, for an elastomer to develop maximum desirable properties, it must be cured or cross-linked. For a synthetic fiber such as an acrylic fiber or a polyolefinic fiber to be dyeable, it must have dye-reactive sites. Sometimes, with an eye toward certain post-polymerization reactions, a polymer may be prepared with certain reactive sites, or latently reactive sites.

In this section we give a few examples of how the chemical reactivity and usefulness of polymers are enhanced by copolymerization. It is tacitly assumed here that the physical properties of the polymer are satisfactory, and that only certain additional chemical reactivity is desired. It is therefore important to confer this chemical reactivity with as little comonomer as possible in order to maintain physical properties. Fortunately, for most purposes, only a small degree of reactivity is usually sufficient.

By far the most important chemical reactivity to be conferred on a polymer is cross-linkability. Such reactivity is important in rubbers, in surface coatings, and of course in thermosetting resins. Mention was made earlier of the fact that natural rubber and most of the early synthetic rubbers contained high amounts of unsaturation (from diene monomers) and were readily cured or vulcanized through these double bonds. The residual double bonds, however, made the rubber susceptible to oxygen and ozone degradation. Efforts to improve ozone and oxygen resistance led to the development of essentially saturated elastomers such as butyl rubber (polyisobutylene), EPR rubber (ethylene–propylene copolymer), and acrylic rubbers (polybutyl acrylate and other acrylic ester–acrylonitrile copolymers). However, these elastomers were not readily cured by conventional means so that small amounts (2–5 %) of unsaturated or other reactive groups had to be copolymerized into these polymers. With butyl rubber, the most common comonomer is isoprene; with EPR rubber, 1,4-hexadiene, dicyclopentadiene, and ethylidenenorbornene have been used. A number of different co-monomers have been used in the acrylic rubbers. In some cases, comonomers containing a pendant active chlorine group have allowed for cross-linking with metal oxide catalysts.

Another important area in which cross-linking is important is in surface coating materials. The incorporation of reactivity by copolymerization is nicely illustrated by a fairly new class of resins, the thermosetting acrylics (13). Ethyl acrylate is first copolymerized with enough styrene (mole ratio EA:S = 1.35:1) and acrylamide (ca. 10–15 mole %) so that reaction with formaldehyde leads to cross-linkable sites in the form of N-methylol derivatives. Adding other modifying resins or simply heating the base resin gives cured films having excellent surface coating characteristics. The reaction sequence is as follows:

$$
\begin{array}{c}
CH_2{=}CH \\
\bigcirc
\end{array}
\;+\;
\begin{array}{c}
CH_2{=}CH \\
|\\
COOC_2H_5
\end{array}
\;+\;
\begin{array}{c}
CH_2{=}CH \\
|\\
CONH_2
\end{array}
$$

$$\downarrow R^{\cdot}$$

$$
-CH_2-CH-CH_2-CH-\!\!-\!\!-\!\!-CH_2-CH-
$$

with substituents: phenyl ($\bigcirc$), $COOC_2H_5$, $CONH_2$

$$\downarrow \; ROH \mid HCHO$$

$$
-CH_2-CH-CH_2-CH-\!\!-\!\!-\!\!-CH_2-CH-
$$

with substituents: phenyl ($\bigcirc$), $COOC_2H_5$, $C{=}O$ / $NHCH_2OR$

(2)

$$\downarrow \Delta$$

$$
-CH_2-CH-CH_2-CH-\!\!-\!\!-\!\!-CH_2-CH-
$$

with substituents: phenyl, $COOC_2H_5$, and a chain:
$C{=}O$
NH
CH_2
NH
$C{=}O$

$$
-CH_2-CH-CH_2-CH-\!\!-\!\!-\!\!-CH_2-CH-
$$

with substituents: phenyl, $COOC_2H_5$

The chemistry of dyeing of fibrous materials is dependent on chemical reactions or extremely strong physical interactions between fiber molecules and dye molecules. It was found that polyacrylonitrile made an excellent fiber material, but dyes that had been developed to color natural fibers such as cotton and wool were not suitable for it. The approach to this problem taken by one producer was to develop special dyes that would be suitable. Another approach taken by other producers was to prepare copolymers in which the comonomer would provide enough dye-reactive sites. Thus an acrylonitrile copolymer containing a few percent of methylvinylpyridine comonomer has excellent dyeability. At the present time the dyeing of polypropylene fibers has raised similar problems.

5.   Chemical Properties of Heterochain Copolymers

Although there are really no fundamental differences between the types of chemical reactivity that can be conferred on heterochain addition polymers by copolymerization as opposed to olefinic copolymers, one very important type of chemical reactivity is nicely illustrated with this polymer type. It is conceivable that monomers that have specific antioxidant or antiozonant properties might be copolymerized to produce polymers having such characteristics. In actual practice, it is usually preferred to add such materials as separate entities. Aside from oxidative degradation, however, polymers may also undergo thermal degradation in inert atmospheres. Such degradations depend on the nature and the inherent thermal stability of the polymer backbone or, more specifically, the elements or links that form that backbone. Since most olefinic polymers having a simple carbon chain backbone thermally degrade by a random scission process (polyvinyl chloride, polymethyl methacrylate and polytetrafluoroethylene degrade by other mechanisms), improvement of thermal stability by copolymerization is difficult.

Many heterochain addition polymers, on the other hand, thermally degrade by a reaction which is the reverse reaction of polymerization, that is, unzipping, monomer unit by monomer unit, until the polymer has been converted back to monomer. If a comonomer can be incorporated into the chain backbone to interrupt the unzipping, it will improve the thermal stability of the polymer.

This is exactly what is done with some polyformaldehyde resins (11, 14). Thus a higher cyclic ether such as ethylene oxide or 1,3-dioxolane is copolymerized with formaldehyde. The unzipping of an oxyethylene unit from such a polymer is much more difficult than just removal of formaldehyde, so that thermal stability is imparted to the polymer by copolymerization:

$$\text{HCHO} \longrightarrow -\text{CH}_2-\text{O}-\text{CH}_2-\text{O}-\text{CH}_2-\text{O}- \qquad (3)$$

$$(\text{HCHO})_3 + \text{H}_2\text{C}\underset{\text{O}}{\overset{\diagup \diagdown}{-\!-\!-}}\text{CH}_2$$

$$-\text{CH}_2-\text{O}-\text{CH}_2-\text{O}-\text{CH}_2-\text{CH}_2-\text{O}- \qquad (4)$$

6. Chemical Properties of Condensation Copolymers

The inclusion of functionality in polymers by copolymerization is probably most easily accomplished in condensation copolymers, and indeed it is with this group of polymers that it all began. Of course the purposes here are much the same as with vinyl copolymers and heterochain copolymers, for example, the provision of cross-linking sites, grafting sites, or dyeing or finishing sites.

A good example may be found in the area of alkyd resin technology (12). An alkyd resin to be used as a paint or other surface coating material is usually a conglomeration of a number of monomeric species, each added to confer a special property to the product. Air dryability, of course, is one of these properties and it is conferred through the incorporation of unsaturated oils that cross-link on exposure to oxygen. Thus in a typical alkyd formulation we find a dibasic acid anhydride such as phthalic and/or maleic anhydride, a diol such as ethylene glycol or propylene glycol, a triol such as glycerin, and some long-chain unsaturated carboxylic acid. Although the actual structures present in alkyd resins are rather complex, one typical molecule containing all these monomeric species is shown in Eq. 5.

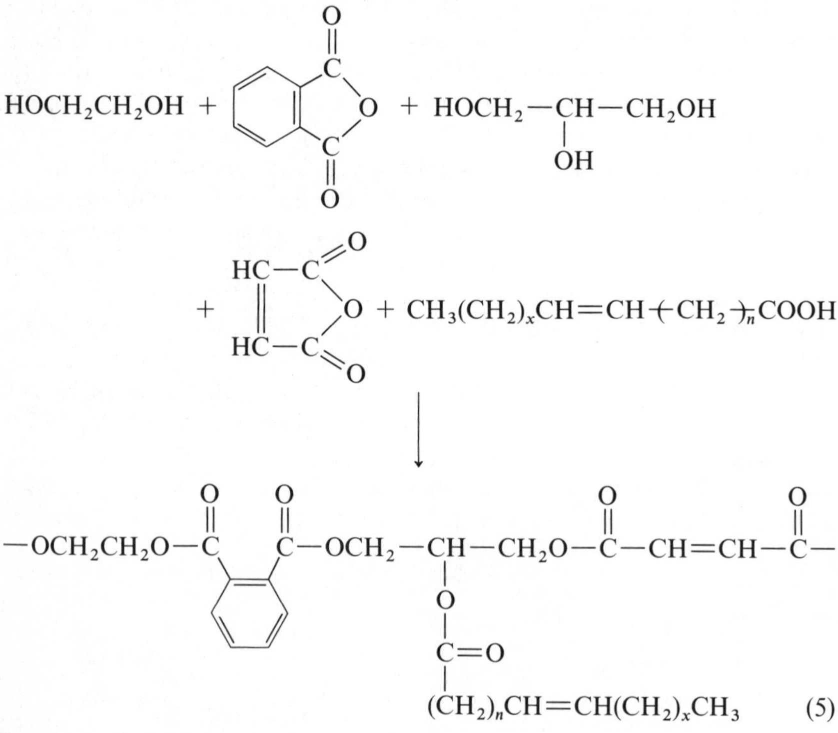

$$\text{(5)}$$

Cross-linking occurs via peroxides generated at the unsaturated carbon atoms of the oil (long-chain fatty acid). The degree of cross-linking and the speed of drying can be controlled by proper choice of the type and quantity of oily comonomers and of maleic anhydride included in the formulation. Of course, unsaturated polyesters such as those made from maleic and itaconic acids (used with fiber glass reinforcement for boat hulls, automobile bodies, and other structural applications) may also be made with various degrees of reactivity to cross-linking with a monomer by proper choice of monomer and maleic anhydride concentration.

Another important property of polymers that has been modified and improved by copolymerization is combustibility or flammability (15). Though it is true that some polymers have greater inherent flammability than others, virtually all hydrocarbon polymers are considered highly combustible. To the degree that a polymer contains other elements, its degree of combustibility is usually reduced, and no group of elements has found greater utility in making fire-retardant polymers than the halogens. Although most fire-retardant compositions usually depend on a weight percent of halogen, cases of synergism in fire retardance have been observed.

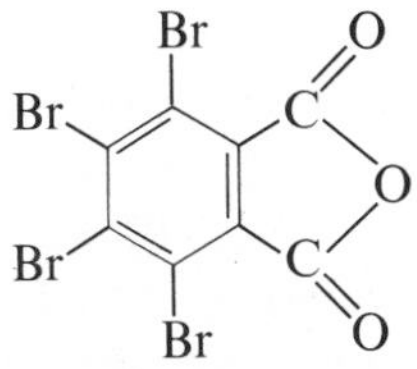

Tetrabromophthalic
anhydride

V

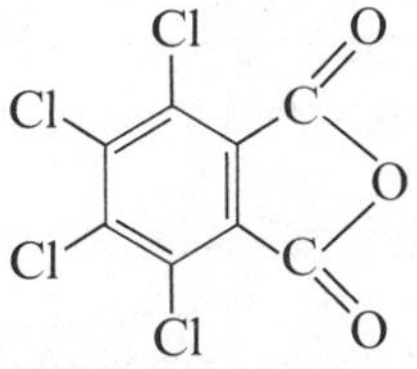

Tetrachlorophthalic
anhydride

VI

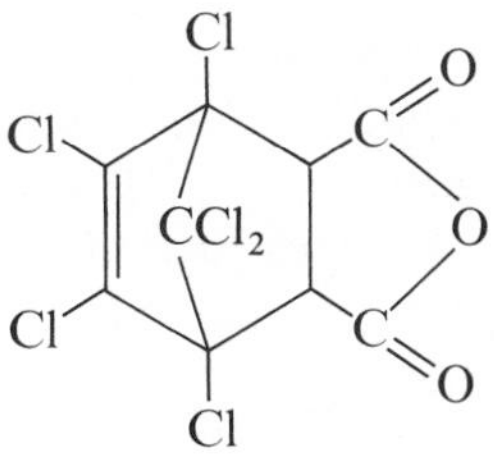

Chlorendic or HET anhydride

VII

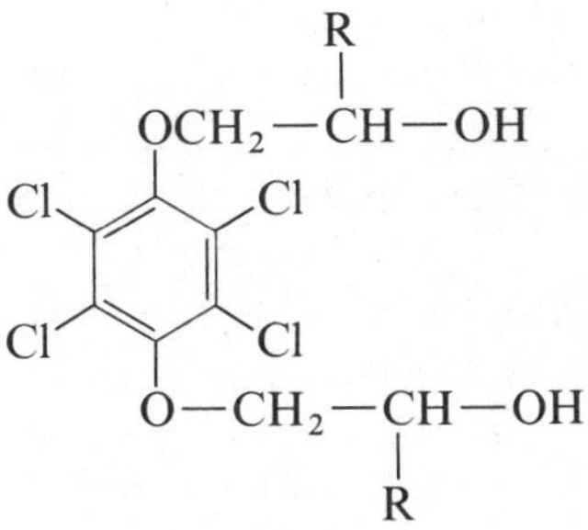

Oxyalkylated tetrachloro-
hydroquinone.

VIII

Thus certain combinations of halogen and nitrogen or halogen and phosphorus in a polymer impart a greater degree of fire resistance to the polymer than would be expected from equivalent amounts of each component.

The improvement in the fire-resistant properties of a polymer can thus be improved by copolymerization with monomers containing other elements. Obviously this procedure can and has been applied to vinyl and heterochain polymers. Many of the more notable successful applications have been in the field of condensation copolymers. For example, the substitution of brominated phenols for some of the phenol in phenolic resins has led to fire-resistant resins of this class. In polyesters, tetrachlorophthalic anhydride, tetrabromophthalic anhydride, chlorendic anhydride, and certain halogen-containing diols (V–VIII) have been used to prepare fire-resistant compositions.

Epichlorohydrinated polybromobisphenol A has been used as a comonomer in the preparation of fire-retardant epoxy formulations.

B. POLYMER BLENDING

1. High Polymer Blends

In the preceding section we have seen how the properties of a given polymer may be modified by copolymerizing with the initial monomer an amount of another monomer sufficient to bring about the desired modifications in final properties. A reasonable question then arises; "Cannot the same desired end result be achieved by simply mixing one homopolymer with the right amount of the other homopolymer?" The answer to this question, generally speaking, is "no." It has long been known, and has been shown thermodynamically, that because of unfavorable free energy considerations polymers prefer to intertwine among themselves rather than among other polymers. Thus solutions of two different but electronically similar polymers dissolved in a common solvent, when mixed, eventually separate into distinct phases, each phase containing virtually all of only one polymer, and practically none of the other polymer. Separations of this type are also observed in the absence of solvent. Materials that separate so are said to be incompatible. Not only do chemical differences between polymers result in incompatibility but even structural differences can cause this effect.

Thus it has been shown that high-density polyethylene is not infinitely compatible with low-density polyethylene.

Because of their large size and somewhat limited mobility at normal temperatures, physical mixtures of polymers may very well stay together for very long times, and the thermodynamic tendency to separate might not be noticeable during the time of the experiment. Nevertheless, measurement of certain telltale properties distinguishes polymer blends from copolymers. The most critical of these are transition temperature measurements, either glass transition or melting point. Homopolymers or random copolymers show single transitions unique for a single material, whereas blends or mixtures of polymers show a transition for each component present in the mixture.

With respect to properties such as tensile strength and other mechanical properties of blends, if the properties are determined by a single measurement during a short duration, the specimen appears to be a single material and the property is simply that of a weighted arithmetic average of the components of the blend. The above statement applies only to very well mixed and intimately blended polymer mixtures. Because amorphous polymers are most readily blended, ideal mixture behavior is most often found in mixtures of amorphous polymers. Deviations from mechanical behavior in which blend properties are an average of homopolymer properties reflect poor mixing and most likely occur in crystalline polymer–amorphous polymer blends and crystalline polymer–crystalline polymer blends. Indeed, in some cases, it may be virtually impossible to get satisfactorily intimate mixing between crystalline polymers except by the use of a compatibilizing agent as discussed below.

This is not to imply that polymer blending is not a useful and important technique for preparing modified polymers. In the first place, in certain cases where molecular weights are fairly low, compatibilization can be realized. In the second place, if mixtures can be co-reacted, such as co-curing of two unsaturated rubbers or of melamine–formaldehyde with urea–formaldehyde resins, very useful materials result. Modification of alkyd resins with other polymers is a favorite procedure for varying the properties of surface coating resins. In this case, too, the polymers are co-cured so that the final product is actually a copolymer (held together by chemical bonds) rather than a simple physical mixture. In the third place, if the expected useful life of a fabricated article is relatively short and the article is not to be subjected to elevated temperatures where phase separation might occur rapidly, a simple mixture of polymers might be perfectly suitable in achieving a certain specific property in the article

The fourth and the most common situation in which polymer blends are useful is when the blend is modified by an additional component called a compatibilizing agent (16, 17). Generally, these materials are to polymer blends what emulsifying agents are to oil-in-water or water-in-oil emulsions. An emulsifying agent is usually thought of in terms of having head and tail portions, the head being soluble in one of the phases and the tail in the other phase. Emulsifying agents are normally reasonably long molecules so that there is a reasonable distance between the head and the tail. If this analogy is carried through to polymer compatibilizers, they should likewise have a segment soluble in one polymer and a suitable distance away in the compatibilizer there should be a segment soluble in the other homopolymeric component. Of course, in polymer compatibilizers, the agent need not be composed of only two segments, but can be of many segments. It is not surprising, therefore, that the most successful compatibilizing agents are block and graft copolymers (see Section C.2), where one segment is the same as one of the homopolymers, and the other segment is the same as the second. An idealized version of the operation of this polymeric compatibilization is shown in Fig. 3. In this figure, it can be seen that the segments of A in the copolymers intertwine with A homopolymer and prevent its migration. Similarly, the B segments in the copolymer are intimately mixed with B homopolymer and prevent its migration. It is therefore seen that such systems are essentially stablized polymer blends. An example of this is the mixture of nylon-6 and polyethylene that can be stabilized in this way!

Although most properties of stabilized polymer blends are the same as unstabilized blends, certain differences exist. For one, there is the presence of a third component which affects properties. More important; however, is the fact that in the presence of compatibilizers, homopolymers cannot migrate, and in the environment in which they find themselves, they tend to occupy the least volume. Thus they curl up or form tighter coils within each molecule, and within the whole of the material, the number of polymer–polymer entanglements is less than it would be in a homopolymeric mass, where chains are extended to their most probable length and interchain entanglements are as frequent as intrachain entanglements. The most pronounced

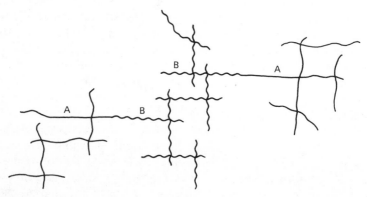

Figure 3 Compatibilization of homopolymer A (————) with homopolymer B (〜〜〜〜) by block copolymer A–B (——^{A}〜^{B}〜).

result of this state of affairs is that stabilized polymer blends show somewhat lower viscosities, that is, improved flow over what would be expected from the viscosities of each homopolymer. A compatibilized blend of two polymers having the same melt index number would be expected to have a higher melt index number (increased melt flow).

More often polymers are blended to improve properties other than melt flow properties. By far the most important and most common is the blending of rubbers with more brittle thermoplastics to improve toughness or impact strength.

The outstanding commercial products employing this technique are the impact polystyrenes, the ABS (acrylonitrile–butadiene–styrene) resins, and the impact methacrylate resins. Thus whereas polystyrene or polymethyl methacrylate homopolymers may have low impact strength, properly compatibilized blends of these materials with certain rubbers yield materials having impact strengths improved by as much as 10 to 20-fold.

Although there are several techniques employed in the manufacture of these high impact materials, they have in common that as part of the production procedure block and graft copolymers are created. Thus impact polystyrene may be prepared by a vigorous comastication of polystyrene and a rubber such as a butadiene–styrene (SBR) copolymer. It may also be prepared by the polymerization of styrene monomer containing dissolved SBR. The generation of block and graft copolymers under these conditions is discussed in Section C.2.

Physically, a compatibilized polymer blend can be thought of as islands of homopolymer A in a sea of homopolymer B on one side of an inversion point, and islands of homopolymer B in a sea of homopolymer A on the other side of this inversion point, the inversion point being a function of the nature of the materials. The compatibilizing agents would then surround the islands and extend into the sea, preventing the islands from migrating and coalescing, in short, stabilizing a basically unstable situation.

With this model, it can be seen how each homopolymer maintains its identity, and how properties such as transitions are observed for each component; it is in this very fundamental way that properties of polymer blends differ from random copolymers having the same ratios of the same monomers.

Although, as already indicated, most polymers are incompatible in the absence of a compatibilizing agent which contains segments structurally similar to and therefore capable of mixing with the otherwise incompatible polymers, polymers having different structures may be blended under certain circumstances.

The solubility of a simple compound may be calculated by determining the cohesive energy density (18). The latter, analogous to the molar refraction, is the sum of contributions from various structural elements. The solubility parameter, δ, is the square root of the cohesive energy density. The calculated δ are divided into highly hydrogen-bonded, weakly hydrogen-bonded, and non-hydrogen-bonded groups. Simple compounds having δ values that differ by as much as five units are mutually compatible. However, owing to the intermolecular (e.g., crystallization and hydrogen bonding) and intramolecular (e.g., entanglements) interactions in polymers, structural contributions are not sufficient to permit simple calculation of the δ value. The latter is therefore determined empirically by finding a suitable solvent for a given polymer and assigning the δ of the solvent to the polymer.

Polymers that have δ values differing by no more than 1 unit are generally compatible. Thus owing to the closeness of their solubility parameters,

	δ
Polyethyl acrylate	9.4
Polyvinyl chloride	9.5
Poly(butadiene-*co*-acrylonitrile) (70/30)	9.4

polyethyl acrylate as well as nitrile rubber may be added to polyvinyl chloride to impart impact resistance.

The application of solubility parameter concepts extends to the use of graft or block copolymers having segments with appropriate δ as compatibilizing agents for polymers with similar δ though different structures.

2. Plasticization

Plasticizers are added to plastics mainly to improve flow, and thereby processability, and to reduce brittleness (19, 20). The plasticizer added is usually the amount of that type required to lower the glass-transition temperature to below ambient temperature. Properties of thermoplastics then change from hard, brittle, glasslike materials to soft, flexible, tough materials. To be effective and useful, a plasticizer must first of all be compatible with the host thermoplastic material. Since the molecular weight and size of plasticizer molecules are relatively small, when compared with high polymer molecules, the thermodynamic difficulties in mixing are not nearly as great as with mixtures of high polymers. Nevertheless, since another requirement of a plasticizer is that it be nonvolatile and nonmigrating, substantial molecular weights are desirable and the choice of plasticizers for a given plastic is fairly narrow. Generally, plasticizers depend on polar intermolecular forces between the two components; this explains why plasticization is difficult to achieve in nonpolar polymers, such as the polyolefins. It is also important to note that plasticization is difficult to achieve with highly crystalline polymers.

Amorphous polymers and slightly crystalline materials are most usefully plasticized.

Commercially, the polymers that find the greatest utility in plasticized form are polyvinyl chloride and its copolymers, such as the vinylidene chloride copolymer and the vinyl acetate copolymer. Production of polyvinyl chloride resins exceeds 3 billion lb/year, and more than 90% of this is compounded with plasticizer for ultimate use. Polyvinyl chloride plasticizers are mainly esters of polybasic acids, such as phthalate esters, phosphate esters, adipates, and azelates.

3. Other Additives

Although polymer properties such as toughness and flexibility may be modified by the addition of plasticizers, and impact strength by the addition of rubbers (with compatibilization), a large number of other materials have been added to plastics and resins to modify and enhance specific properties which make the material more useful in any of a number of ways.

Fillers

Many plastics are useful only when combined with reinforcing materials, such as particulate or fibrous solids (21). Amino resins and phenolic resins are almost always used in conjunction with fillers, which include wood flour, cellulose, powdered mica, asbestos, and glass fibers. These fillers greatly enhance dimensional stability, strength, abrasion resistance, and heat resistance. Glass fibers and woven fiberglass are widely used to reinforce unsaturated polyester resins and epoxy resins.

Strength properties of many thermoplastics such as nylon, polystyrene, acrylic resins, polycarbonates, acetal resins, and polyethylene are also greatly enhanced by the incorporation of glass fibers. A final and extremely important example is the reinforcement of rubber with various types of carbon black.

Antioxidants

Antioxidants (22) and antiozonants (23), as their names imply, prevent or inhibit the oxidation of polymers, usually by becoming oxidized themselves, thereby enhancing the useful life of materials under conditions where they might become badly degraded by oxidation. Common antioxidants include phenols, aromatic amines, and certain condensation products of aldehydes. ketones, and sulfur-containing compounds.

Flame Retardants

Flame retardants usually function by making a material self-extinguishing; that is, they prevent the propagation of a flame front. The mechanism may be by action as a thermal barrier, as by charring, thus eliminating fuel by reducing heat transfer or by quenching the chain reactions in the flame in which the retardant acts as a radical scavenger. Materials employed as flame retardants include halogen compounds, phosphorus compounds, and antimony trioxide (15).

Stabilizers

Degradation of plastics by mechanisms other than oxidation and burning must also be considered (24). Thus in vinyl chloride resins, lead and tin compounds that scavenge hydrogen chloride act as stabilizers. Ultraviolet ray absorbers also function as stabilizers by preventing radiation-induced decomposition of materials. Polyolefins and unsaturated rubbers are particularly susceptible to this type of degradation.

Colorants and Pigments

Soluble colorants are termed dyes and are used to color transparent resins such as polyvinyl chloride, polystyrene, or polymethyl methacrylate. Insoluble colorants are called pigments and may be organic or inorganic. They yield opaque products. Generally, only very small amounts of colorants are necessary to impart the desired hue to a plastic. Thus the only property that is affected by dyes and pigments is color. However, enhanced susceptibility to degradation or, conversely, increased stability to environmental exposure may result from particular dye structures.

C. POST-REACTION OF POLYMERS

We have just seen how polymer properties may be modified by copolymerization and by blending or mixing with other materials which are effective in such a function. We now discuss yet another technique for changing properties of polymers. The technique involves chemical reactions of high molecular weight materials with reagents or with low molecular weight materials in order to effect certain desired changes. The reactions we consider here are essentially nondegradative, that is, the length of the backbone after reaction is not less than it was prior to reaction. In a number of cases, however, post-reaction involves increasing the length of chains even to the extent of infinite cross-linked networks.

The post-reactions to be discussed fall broadly into two classes. In one class are those reactions in which the length of chains is increased, and in the other are reactions on side chains, pendant groups, or on single atoms or groups in the backbone. Reactions of the first type include cross-linking, formation of block and graft copolymers, and chain extension reactions. In the second category are halogenation, epoxidation, sulfonation, chloro-sulfonation, hydrolysis, surface modification, and some miscellaneous chemical reactions.

Chemical reactions on polymers are influenced not only by the reactivity of a particular functional group but also by its accessibility (25). Although intrachain steric factors that result in neighboring group effects are well recognized, the interchain interactions that determine the accessibility of a potential reaction site must also be considered. The factors that influence polymer reactions include amorphous or crystalline content, orientation, solubility, and compatibility. The interchain interactions involved are not only those between chains of the initial polymer but, more significantly, those between polymer chains that have reacted or partially reacted and polymer chains that have not. Chemical reaction may change solvent–polymer inter-actions so that either chain extension or coiling occurs with resultant changes in the accessibility of reaction sites. Incompatibility between polymer that has partially reacted and polymer that has not reacted may result in phase separation. Increased compatibility may result from the generation of block copolymers, owing to localized or segmental reaction, which then act as compatibilizing agents and make the separation of reaction products, such as by extraction of unreacted material, extremely difficult.

1.   Cross-linking

In the section dealing with polymer modification by copolymerization we discussed copolymerization with certain functional or reactive monomers which were required to provide cross-linking sites during post-reaction of polymers. In that section we mentioned the curing of rubber and of certain surface coating resins. In this section we consider properties of cross-linked polymers broadly and indicate the chemical reactions involved in the for-mation of cross-linked networks for some of the more common materials which find their major usefulness as cured resins (26).

Cross-linked plastics are also known as thermoset resins, but are probably better called network polymers. Network polymers have many unique char-acteristics. Thus, depending on the degree of cross-linking, they tend to be virtually insoluble in all solvents, but tend to swell in good solvents. In-creasing amounts of crosslinking increase the modulus (stiffness and hardness) of polymers. Cross-linking increases the glass-transition temperature of polymers and rheological experiments show that network polymers exhibit no viscoelastic flow, even up to the point of decomposition.

The importance of cross-linking in the manufacture of rubber and of many surface coating materials has already been mentioned. A number of other useful polymers are normally employed in the form of networks. Among them are the amino resins, phenolic resins, unsaturated polyesters, epoxy resins, and polyurethanes. The major requirement for cross-linkability is as least two polymerizable groups per monomer unit and in some cases a larger number is desirable.

The chemistry of cross-linking of three of the various materials mentioned above is depicted in Eq. 6–8.

Amino resins (melamine)

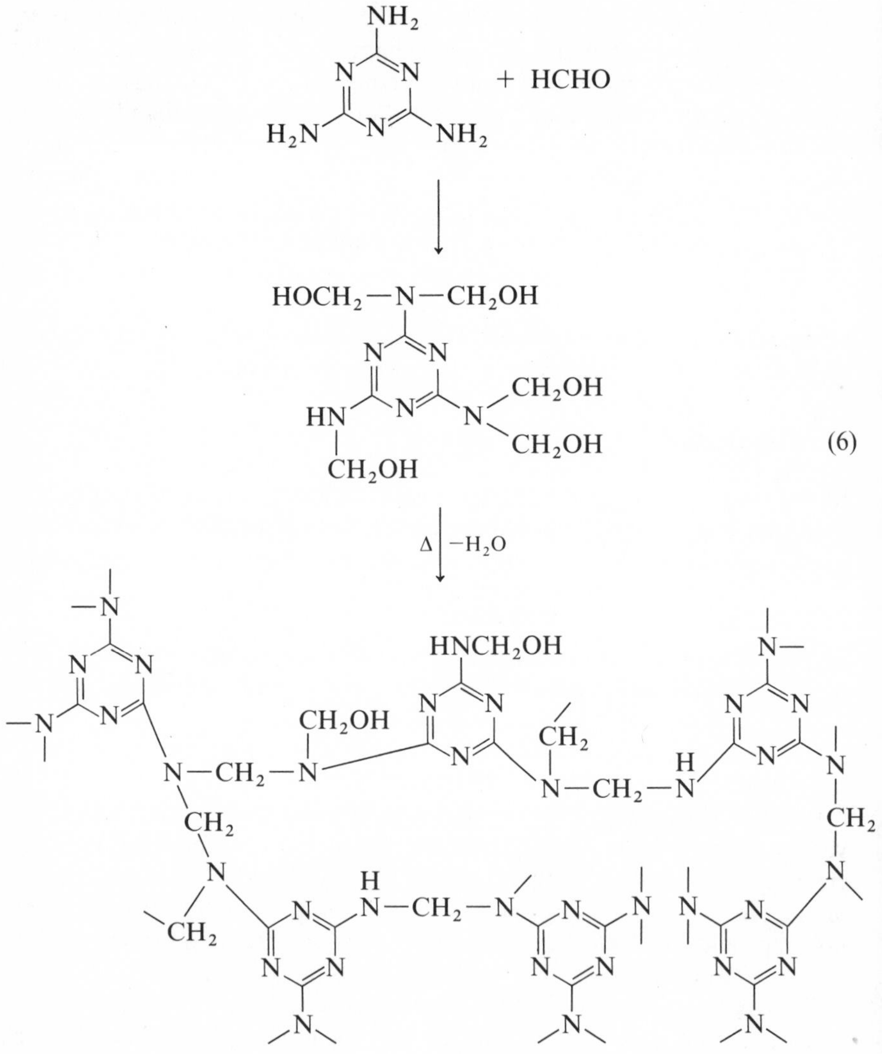

(6)

Unsaturated polyesters

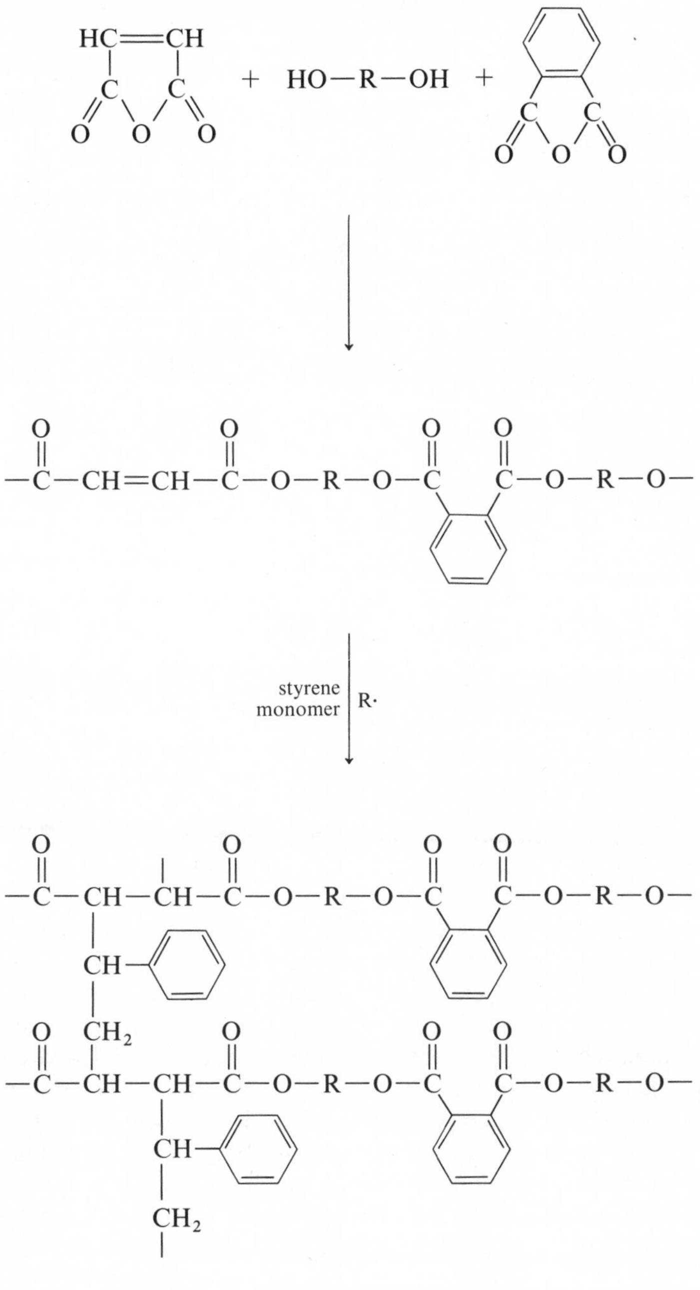

$$
\underset{\substack{\| \ \ \| \\ O \ \ O \ \ O}}{\overset{\displaystyle HC=CH}{\underset{C \diagdown \diagup C}{}}} \quad + \quad HO-R-OH \quad + \quad \underset{\substack{\| \ \ \| \\ O \ \ O \ \ O}}{\overset{\bigcirc}{\underset{C \diagdown \diagup C}{}}}
$$

$\downarrow$

$$
\overset{O}{\underset{\|}{-C}}-CH=CH-\overset{O}{\underset{\|}{C}}-O-R-O-\overset{O}{\underset{\|}{C}}\ \overset{O}{\underset{\|}{C}}-O-R-O-
$$

(7)

$$\text{styrene} \atop \text{monomer}\ \Big|\ R\cdot$$

$\downarrow$

$$
\overset{O}{\underset{\|}{-C}}-CH-\overset{|}{\underset{|}{CH}}-\overset{O}{\underset{\|}{C}}-O-R-O-\overset{O}{\underset{\|}{C}}\ \overset{O}{\underset{\|}{C}}-O-R-O-
$$

$$CH-\bigcirc$$

$$
\overset{O}{\underset{\|}{-C}}\ \overset{CH_2}{\underset{|}{-CH}}-CH-\overset{O}{\underset{\|}{C}}-O-R-O-\overset{O}{\underset{\|}{C}}\ \overset{O}{\underset{\|}{C}}-O-R-O-
$$

$$CH-\bigcirc$$

$$CH_2$$

Polyurethanes

$$O=C=N-R-N=C=O + HO-R'-OH$$

$$\downarrow$$

$$\overset{\displaystyle O}{\underset{\displaystyle \|}{}}\qquad\qquad\overset{\displaystyle O}{\underset{\displaystyle \|}{}}\qquad\qquad\overset{\displaystyle O}{\underset{\displaystyle \|}{}}\qquad\qquad\overset{\displaystyle O}{\underset{\displaystyle \|}{}}$$

$$-O-R'-O-\overset{\|}{C}-\overset{H}{N}-R-\overset{H}{N}-\overset{\|}{C}-O-R'-O-\overset{\|}{C}-\overset{H}{N}-R-\overset{H}{N}-\overset{\|}{C}-$$

$$\downarrow \quad O=C=N-R-N=C=O$$

$$-O-R'-O-\overset{O}{\overset{\|}{C}}-N-R-\overset{H}{N}-\overset{O}{\overset{\|}{C}}-O-R'-O-\overset{O}{\overset{\|}{C}}-N-R-\overset{H}{N}-\overset{O}{\overset{\|}{C}}-$$
$$C=OC=O$$
$$NHNH$$
$$RR$$
$$NHNH$$
$$C=OC=O$$
$$-O-R'-O-\overset{}{\underset{\underset{\displaystyle O}{\|}}{C}}-\overset{}{N}-R-\overset{H}{N}-\overset{}{\underset{\underset{\displaystyle O}{\|}}{C}}-O-R'-O-\overset{}{\underset{\underset{\displaystyle O}{\|}}{C}}-\overset{}{N}-R-\overset{H}{N}-\overset{}{\underset{\underset{\displaystyle O}{\|}}{C}}-$$

$$(8)$$

Another method for cross-linking polymers is by subjecting them to radiation. This procedure is used mainly with polyolefins such as polyethylene. Radiation induces carbon–hydrogen bond scission forming radicals, and when two radicals in adjacent chains combine to form a carbon–carbon bond, a cross-link is established (27).

Before going on to the next topic it is well to remind the reader that cross-linked network polymers may be prepared by copolymerization as well as by post-reaction. Thus a few percent of divinylbenzene copolymerized with styrene yields insoluble products. Similarly the addition of ethylene dimethacrylate to an acrylic system provides a network polymer. In condensation polymerization cross-linked materials can also be obtained by adding small amounts of trifunctional monomers (triols or tribasic acids in polyesters) and carrying the conversion to a high enough degree to form the network (see chapter 2).

2. Block and Graft Copolymer Formation

Block and graft copolymers were mentioned earlier and their usefulness in stabilizing polymer blends described (also see Chapter 2). Perhaps the most interesting characteristic of sequence copolymers, as both block and graft copolymers may be termed, is that they exhibit many properties that are characteristic of each homopolymer. Thus a sequence copolymer may show two glass-transition temperatures. If both polymer types are crystalline, two crystalline melting points are observed (although the overall degree of crystallinity in a block or graft copolymer is drastically reduced compared with the tendencies of the corresponding homopolymer to crystallize). Solubility of sequence copolymers is difficult to predict. However, if the sequences differ widely in polarity, good solvents for either homopolymer are ineffective in dissolving the copolymer. Although the solvent undoubtedly swells regions containing the polymer type for which it is a good solvent, the segments for which it is a poor solvent intertwine and act as cross-links. Plots of certain energy-absorbing phenomena such as dielectric constant as a function of frequency, or Youngs modulus as a function of temperature, definitely indicate the presence of two materials behaving virtually independently. Sequence polymers behave as polymer mixtures or blends, but are stable because unlike segments can not migrate away (17).

A number of techniques have been developed for the synthesis of block and graft copolymers (28, 29). We need mention only the most important and the most interesting here. The most industrially important procedure is to carry out the polymerization of one monomer in the presence of a polymer of the other material. Thus a rubber backbone–styrene graft copolymer results when styrene monomer containing dissolved rubber (SBR) is subject to polymerization conditions with radical initiators. A probable mechanism is shown in Eq. 9.

(9)

Of course, at the same time much styrene is homopolymerized and long segments of the rubber molecule remain unreacted. The result is a product that is very useful per se, exhibiting high impact strength.

Graft copolymerization may also be initiated by other mechanisms, such as by ionic initiation. Thus a styrene-p-chlorostyrene copolymer when treated with lithium metal provides a backbone with pendant lithium phenyl groups. These groups then initiate the polymerization of a number of monomers such as dienes and many acrylic monomers which polymerize well by anionic mechanisms. Pendant cations or potential cations may be employed to initiate cationic polymerization.

A great deal of work has been reported on the initiation of graft copolymerization by irradiation of a mixture, solution, or swollen mass of polymer and grafting monomer. Types of radiation found useful include gamma irradiation, X-rays, ultraviolet radiation, and electron bombardment.

3. Chain Extension

Although cross-linking and grafting may be considered chain extension reactions, for the narrower purposes of this section we limit the definition of chain extension reactions to those reactions that occur only at the ends of polymer chains and lead to longer backbones. Chain extension even in this narrow sense encompasses several catagories. Among these are the block copolymer type in which reasonably long segments of polymer A become connected through a long sequence of polymer B. Another type is one in which a number of long segments of polymer are "hooked up" through a simple monomeric reagent. Perhaps the best example of this is the preparation of polyurethanes, which are in reality polyesters or polyethers connected by a urethane or carbamate linkage. A third type is chain extension of condensation polymers which result from an elimination reaction; it is a normal consequence of the mechanism of condensation polymerization. These three types are shown in Eqs. 10–12 using a polyether, a polyester urethane, and a polyamide for illustration.

Type I. Poly(oxyethylene-block-co-oxypropylene)

$$H_2C{-}CH_2 \xrightarrow{\text{Base}} HO(CH_2CH_2O)_xH$$

$$\downarrow \quad CH_3CH{-}CH_2 \big/ O$$

$$\tag{10}$$

$$CH_3$$
$$|$$
$$HO(CH_2CH_2O)_x(CHCH_2O)_yH$$

Type II. Polyester urethane

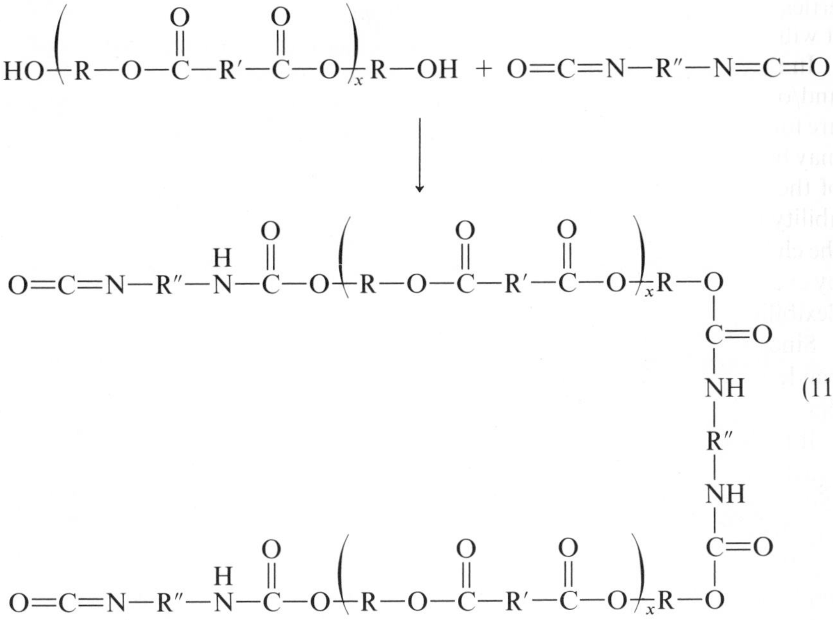

$$HO\left(R-O-\underset{\underset{O}{\parallel}}{C}-R'-\underset{\underset{O}{\parallel}}{C}-O\right)_x R-OH \;+\; O=C=N-R''-N=C=O$$

$$\downarrow$$

$$O=C=N-R''-\underset{H}{N}-\underset{\underset{O}{\parallel}}{C}-O\left(R-O-\underset{\underset{O}{\parallel}}{C}-R'-\underset{\underset{O}{\parallel}}{C}-O\right)_x R-O$$

$$C=O$$

$$NH \qquad (11)$$

$$R''$$

$$NH$$

$$O=C=N-R''-\underset{H}{N}-\underset{\underset{O}{\parallel}}{C}-O\left(R-O-\underset{\underset{O}{\parallel}}{C}-R'-\underset{\underset{O}{\parallel}}{C}-O\right)_x R-O \quad C=O$$

Type III. Polyamide

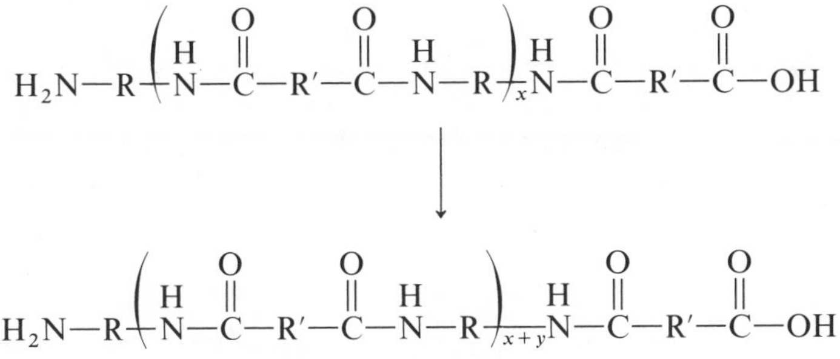

$$H_2N-R\left(\underset{H}{N}-\underset{\underset{O}{\parallel}}{C}-R'-\underset{\underset{O}{\parallel}}{C}-\underset{H}{N}-R\right)_x \underset{H}{N}-\underset{\underset{O}{\parallel}}{C}-R'-\underset{\underset{O}{\parallel}}{C}-OH$$

$$\downarrow \qquad (12)$$

$$H_2N-R\left(\underset{H}{N}-\underset{\underset{O}{\parallel}}{C}-R'-\underset{\underset{O}{\parallel}}{C}-\underset{H}{N}-R\right)_{x+y} \underset{H}{N}-\underset{\underset{O}{\parallel}}{C}-R'-\underset{\underset{O}{\parallel}}{C}-OH$$

Type I leads to a block copolymer. The characteristics and general properties of block copolymers have been discussed in the preceding section on block and graft copolymers. Type III leads to a higher molecular weight polymer having a recurring unit of essentially the same molecular weight, and as such is characterized simply as a higher molecular weight species of the unextended polymer. By increasing or doubling the molecular weight one usually increases the softening point and many strength properties. If, however, the unextended polymer already had the molecular weight at which those properties have leveled off, the effect of chain extension will be undesirable since it will probably make processing more difficult.

In the Type II polymers where the monomeric linking group is functional and/or provides a novel configuration or polarity to the chain, novel materials are formed. Thus in the example cited, where the polyester or polyether chain may be of molecular weight 2000 and represents more than 90% of the weight of the polymer, it is the unique properties of the urethane linkage, and its ability to hydrogen bond and provide cross-linking sites which really define the characteristics of the polymer. Polyurethanes, of course, are characterized by excellent toughness, abrasion resistance, low-temperature properties, and flexibility.

Since the properties of the Type III chain extended polymers are so dependent on whether or not the chain extending group is uniquely functional, one cannot discuss the properties of such materials in general terms.

It is obvious that a prerequisite to chain extending of polymers is to have starting materials with functional end groups, or at least with one reactive end. Perhaps the most interesting chemistry involves the synthesis of block copolymers (Type I). Techniques have been developed for carrying out such reactions on virtually all types of addition polymers, including radical and ionically initiated vinyl polymers (chapter 2).

If one initiates vinyl polymerization with hydrogen peroxide the end group is a hydroxyl group. If the termination mechanism is by coupling we have a diol that can be chain extended in any number of ways, for example, by reaction with a dibasic acid or with a diisocyanate, by conversion to alkoxide groups, by reaction with oxirane compounds, or even by ceric ion initiated vinyl polymerization. Polymers that terminate by coupling include styrene, butadiene, isoprene, chloroprene, and tetrafluoroethylene. If termination is by disproportionation the hydrogen peroxide initiated polymer will have a hydroxyl group at one end and the other end will be saturated or unsaturated. The hydroxyl or the unsaturated end can be used to add monomers using established procedures.

Polysulfide polymers also undergo chain extension reactions prior to cure (30, 31). These are brought about by reagents such as bis-2-chloroethylformal, alkylene dihalides, and others. These polysulfides are useful in adhesives, coatings, binders for rocket fuels, impregnants for leather and other porous materials, and as modifiers for other resin systems. The molecular weight of the mercaptan-terminated polymers may also be increased by reaction with metal oxides and a variety of oxidizing agents. Addition of the thiol group to difunctional materials susceptible to such addition reactions, for example, epoxy compounds or isocyanates, has also been exploited in making the polysulfides commercially useful materials.

4. Halogenation

With the topic of halogenation of polymers, we move into the area of modification of polymers by treatment with reagents, whose mode of action and utility has been long established in the classical organic chemistry of small molecules.

Halogenation of organic molecules has classically been considered to be of two types, addition of halogen to unsaturated molecules or substitution of halogen, usually for hydrogen, in saturated molecules, as follows:

$$-R-\underset{}{\overset{H}{C}}=\underset{}{\overset{H}{C}}-R- + X_2 \longrightarrow -R-\underset{X}{\overset{H}{\underset{|}{C}}}-\underset{X}{\overset{H}{\underset{|}{C}}}-R- \quad \text{(addition)} \quad (13)$$

$$-R-\underset{H}{\overset{H}{C}}-\underset{H}{\overset{H}{C}}-R- + X_2 \longrightarrow -R-\underset{X}{\overset{H}{\underset{|}{C}}}-\underset{H}{\overset{H}{\underset{|}{C}}}-R- + HX$$

$$\text{(substitution)} \quad (14)$$

$$X_2 = Cl_2 \text{ or } Br_2$$

It should be remembered that the halogenation of polymers is generally a second choice in preparing halogenated polymers. By far the vast majority of commercially important halogen-containing polymers are prepared by polymerizing the corresponding halogen-containing monomers.

Nevertheless, there are some important polymeric materials which are prepared by the halogenation, generally chlorination, of polymers, and both addition mechanisms and substitution mechanisms are employed (32, 33).

In the class of addition reactions is perhaps one of the oldest examples of polymer modification, the chlorination of natural rubber. Although a certain amount of substitution and cyclization of the rubber occur simultaneously with the addition, addition is the major reaction until the product contains 30–35 % of chlorine. After that point, substitution becomes the predominant reaction.

Commercially, the rubber is dissolved in carbon tetrachloride and chlorine added until the product contains 66 % or more chlorine and is still soluble. Chlorinated rubber is stable at ambient temperatures, but on moderate heating, it loses hydrogen chloride and undergoes decomposition. However, it has good stability to acid, alkali, and oxidizing agents. Further, it is non-flammable. It is hard, rigid, difficultly moldable, and is generally applied from solution as a coating material.

Unsaturated polymers may also be halogenated by the addition of hydrogen halide. Thus rubber hydrochloride resembles polyvinyl chloride in a number of ways, but because of its relative high cost, has not achieved much commercial acceptance, except in film form.

Most of the work on substitutive halogenation has been carried out with polyolefins, such as polyethylene and polypropylene. This substitutive halogenation is generally a radical chain process requiring some type of radical initiation. As expected, the properties vary as a function of chlorine content, as shown in Table 2.

Table 2   Chlorinated polyethylene

Chlorine (wt %)	$T_g(^\circ C)$	Description
0	-70	Hard plastic
30	-20	Rubbery
40	10	Soft, sluggish
50	20	Leathery
55	35	Rigid
60	75	Brittle
70	150	Brittle

Thus as chlorination increases, polyethylene changes from a tough thermoplastic to a rubbery material, then becomes more leathery, and finally becomes hard and brittle (much like chlorinated rubber). Brominated polyethylene is similar. Chlorination may be carried out at 65°C in aqueous suspension to about 40% chlorine content. However, as the chlorine content increases, crystallinity decreases and agglomeration becomes a problem, but techniques to overcome this have been developed. Chlorination of polypropylene generally results in polymer degradation as does chlorination of polyisobutylene. However, chlorination of butyl rubber to about 1–2% chlorine content results in rubbers which may be cured by metal oxide catalysts.

Related to chlorination of polyolefins is the chlorosulfonation of these materials, which is accompanied by chlorination (33):

$$R\!-\!H + Cl\cdot \xrightarrow{\ -HCl\ } R\cdot \xrightarrow{\ SO_2 + Cl_2\ } RSO_2Cl + Cl\cdot \qquad (15)$$

Hypalon, which is commercially available chlorosulfonated polyethylene, contains 22–26% chlorine and 1.3–1.7% sulfur. The polymer is elastomeric and may be cured with metal oxides or polyamines. If has improved solvent chemical and flame resistance over normal rubber, or even neoprene. Similar products can be made by chlorosulfonation of polypropylene, and chlorosulfonation has been employed to put cross-linkable groups into ethylene–propylene copolymer.

5.   Epoxidation

As with halogenation, the epoxidation of polymers is similar to the epoxidation of simple low molecular weight materials. However, in contrast to halogenation, epoxidation is applicable only to unsaturated polymers. Although simple olefins can be epoxidized with oxygen in the vapor phase, or via a hypochlorous acid reaction product intermediate, or with hydrogen peroxide, the preferred reagent for use with polymers are organic peracids:

$$
\begin{array}{c}
\quad\quad\quad\quad\quad\quad\quad O \\
\quad\ \ H\ \ H \quad\quad\quad \| \\
-R-C{=}C-R- + RC-O-OH
\end{array}
$$

$$
\begin{array}{c}
\searrow \\
\quad\quad\quad H\quad\quad\ H \\
\quad\quad -R-C\!\!-\!\!\!-\!\!\!-\!\!C-R- + RCOOH \quad\quad (16)\\
\quad\quad\quad\quad \diagdown\ \diagup \\
\quad\quad\quad\quad\ \ O
\end{array}
$$

In general, the unsaturated polymer to be epoxidized should be readily soluble in suitable solvents. Steric and electronic influences of neighboring groups are important in that the double bond must be accessible and receptive to the peracid. These groups have a great effect on the rate of epoxidation as well as the stability of the product to ring opening. Highly alkyl-substituted double bonds react much faster than isolated double bonds. However, it is just this type of epoxide that has a greater tendency to ring opening and isomerization.

The epoxidation of a wide variety of polymers has been described in the literature. These include homopolymers of dienes such as butadiene, isoprene, and cyclopentadiene (34, 35). A number of secondary reactions including ring opening to glycol, isomerization to carbonyl compounds, and polymerization may occur during epoxidation, and very often epoxidized polymers contain small amounts of the other functionalities.

The utility of epoxidized polymers resides in their reactivity and their ability to add a wide variety of nucleophilic and electrophilic reagents, as shown in the following equations:

$$
\begin{array}{c}
\quad H\quad\quad\ H \quad\quad\quad\quad\quad\quad\quad\quad\quad\quad B \\
\quad\quad\quad\quad\quad\quad\quad\quad\quad\quad\quad\quad\quad\quad\quad | \\
-R-C\!\!-\!\!\!-\!\!\!-\!\!C-R- + B{:}^{\ominus} \longrightarrow -R-C-C-R- \quad (17)\\
\ \ \diagdown\ \diagup \quad\quad\quad\quad\quad\quad\quad\quad\quad\quad | \ \ H \\
\quad\ O \quad\quad\quad\quad\quad\quad\quad\quad\quad\quad\quad\quad O^{\ominus}
\end{array}
$$

$$
\begin{array}{c}
\quad H\quad\quad\ H \quad\quad\quad\quad\quad\quad\quad\quad\quad\quad H\ \ \oplus \\
-R-C\!\!-\!\!\!-\!\!\!-\!\!C-R- + A^{\oplus} \longrightarrow -R-C-C-R- \quad (18)\\
\ \ \diagdown\ \diagup \quad\quad\quad\quad\quad\quad\quad\quad\quad\quad | \ \ H \\
\quad\ O \quad\quad\quad\quad\quad\quad\quad\quad\quad\quad\quad\quad OA
\end{array}
$$

Multiple epoxy groups can be built into the polymer chain and the polymer usually contains residual unsaturation. Further, any number of hydroxyl groups can be introduced. These resins are readily cross-linked with poly-functional nucleophiles or electrophiles, that is, polyamines or polybasic acids or by BF_3. Since the oxirane content is usually reasonably low, the physical properties of the epoxidized polymers are very much like their unsaturated polymer precursors, and the properties of the cross-linked products depend a great deal on the nature and functionality of the curing agent.

Generally, properties are typical of thermoset resins having good electrical properties. Compared to epoxy resins made from epichlorohydrin–bisphenol adducts, these resins have improved heat resistance. Epoxidized poly-butadienes are useful in potting, encapsulation, adhesives, surface coatings, tooling, flooring, road repair, molding, laminating, etc. In addition, these particular resins are compatible with rubber and improve a number of rubber properties.

6. Hydrolysis

If reactions that lead to severe backbone degradation, such as with polyesters, polyamides, and polycarbonates, are disregarded, hydrolysis is an important reaction of polymers. For some polymers, it is the only feasible method of preparation.

Polyvinyl alcohol, for example, is commercially prepared by hydrolysis of polyvinyl acetate (Eq. 19). Actually, a number of grades of polyvinyl alcohol are offered. They differ mainly in the degree of hydrolysis, for example, 80% hydrolyzed, 90% hydrolyzed, and 97% hydrolyzed (36).

$$
\begin{array}{c}
\mathrm{H\ \ H\ \ H\ \ H} \\
\mathrm{-C-C-C-C-} \\
\mathrm{H\ \ |\ \ H\ \ |} \\
\mathrm{O\ \ \ \ \ O} \\
\mathrm{|\ \ \ \ \ \ |} \\
\mathrm{C{=}O\ \ C{=}O} \\
\mathrm{|\ \ \ \ \ \ |} \\
\mathrm{CH_3\ \ CH_3}
\end{array}
\quad
\xrightarrow[\mathrm{H^{\oplus}\ or\ OH^{\ominus}}]{\mathrm{H_2O}}
\quad
\begin{array}{c}
\mathrm{H\ \ H\ \ H\ \ H} \\
\mathrm{-C-C-C-C-} \\
\mathrm{H\ \ |\ \ H\ \ |} \\
\mathrm{OH\ \ \ OH}
\end{array}
\qquad (19)
$$

The whole family of acrylic ester derivatives may also undergo hydrolysis. In this case, however, since the carboxylic acid group is pendant to the back-bone via a carbon–carbon bond, the products are polycarboxylic acids. Thus hydrolysis of polymethyl methacrylate leads to polymethacrylic acid, and from polymethyl acrylate the product is polyacrylic acid.

Conversely, polyacrylic acid and polymethacrylic acid may be poly-esterified to a wide variety of polyacrylic esters. Although a great many of these acrylic polymers are industrially valuable, it is preferred to initially prepare the desired monomer, for example, acrylic acid, acrylamide, or acrylic esters, and the polymer by direct polymerization.

7. Miscellaneous Chemical Reactions of Unsaturated Polymers

Aside from halogenation and epoxidation discussed above, the reactive nature of unsaturated, and particularly polyunsaturated, polymers (for the most part diene polymers) has been the subject of much investigation. The fact that these elastomeric materials are so industrially important has motivated much of this work and a number of reactions have served mainly to elucidate structure.

One of these is catalytic hydrogenation. Although hydrogenation of all of the double bonds of an unsaturated polymer is not easily accomplished, it is fairly easy to hydrogenate 85–95 % of the double bonds and, in certain cases, more than 99 % hydrogenation has been achieved. By and large, the same techniques, that is, high pressures and supported metal catalysts, used for hydrogenation of olefins and other low molecular weight unsaturated materials, are the ones that have found greatest utility in hydrogenation of polymers (37). It would be predicted that as 1,4-polybutadiene is hydrogenated, the product will begin to take on polyethylene characteristics. This is exactly what is observed. Similarly hydrogenated 1,2-polybutadiene is very similar to poly-1-butene. If these unsaturated elastomers are only partially hydrogenated, they will still be vulcanizable, but because the products have less unsaturation, they show improved ozone resistance.

Treating unsaturated polymers with acidic reagents, or heating them under nonvulcanizing conditions, leads to numerous molecular transformations, including cyclization, double bond shift, and *cis–trans* isomerization (38, 39). Cyclization of natural rubber is shown in Eq. 20.

(20)

The addition of a number of other materials to unsaturated polymers has received some attention. Included in these materials are thiols (Eq. 21) which are induced to add to the double bond in the presence of free-radical catalysts (40). The nature and extent of addition depend greatly on conditions and the specific thiol. Thus in the absence of oxygen at 50°C, the amount of ethanethiol reacting with SBR rubber appears to reach a limiting value of 25%, whereas in the presence of oxygen, 90% of the theoretical amount of thiol is added. However, as the degree of substitution approaches the theoretical amount, the molecular weight of the product decreases, indicating that degradative additions are occurring simultaneously. However, with methanethiol, nondegradative additions up to about 97–98% of theory have been achieved. These highly substituted polymers are very difficult to vulcanize using normal recipes, and are very resistant to attack by ozone and oxygen.

$$+CH_2-CH=CH-CH_2 \overset{}{)_n} + RSH$$

$$\text{catalyst} \downarrow$$

$$\left(CH_2-CH_2-\underset{\underset{SR}{|}}{CH}-CH_2 \right)_{\overline{n}} \tag{21}$$

The relative ease of addition of the SH group across polymer double bonds has been used to introduce reactive pendant groups into polymer backbones. Thus carboxylic acid groups may be introduced by the addition of thioglycolic acid to unsaturated polymers.

8. Surface Modification

Chemical reactions at the surface of polymers would have to include reactions of a polymer with its environment. This is usually referred to as aging or weathering, and is beyond the scope of this chapter. An effort is made to discuss here a number of deliberate modifications of surface properties (41).

Usually these reactions have been carried out on fibers and films rather than bulk polymers or molding beads or pellets, and the major objective has been to modify the surface without significantly altering the physical properties of the substrate. Commercially, such modifications are used to improve feel, washability, dye and ink retention, antistatic properties, and abrasion resistance of fibers and films. Permeability to liquids and vapors, solvent resistance, and adhesion may also be improved.

Of great technological importance is the deliberate and controlled treatment has also been employed to improve the adherability of poly-film. The initial products, formed on reaction of oxygen with methylene and methinyl groups, are hydroperoxides. These catalyze the formation of aldehydes and ketones and, finally, acids and esters.

Oxidation has been carried out by corona discharge and treatment with ozone, hydrogen peroxide, nitrous acid, or alkaline hypochlorite. Irradiation in air or oxygen also results in surface oxidation, as does treatment with an oxidizing flame. The oxidized surface is relatively polar, hydrophilic rather than hydrophobic, and as such is readily printable, adherable, and anti-static. Also surface oxidized polyethylene loses its "greasy feeling." Surface treatment has also been employed to improve the adherability of poly-tetrafluoroethylene.

Although polyolefins are generally unaffected by strong acids or bases, they may be sulfonated or chlorosulfonated under somewhat forcing conditions. Surface sulfonated polyethylene films, when first treated with ethylene-diamine and then a polar polymer, such as terpolymer of vinylidene chloride, acrylonitrile, and acrylic acid, yield films of improved clarity, scuff resistance, and reduced permeability.

Treating fluorosulfonated polyethylene with alkali gives a permanently amber colored surface suitable for storing light-sensitive materials.

Acid-treated polyethylene terephthalate has improved adherability, and exposure of polystyrene films to sulfonating conditions for a period of a few seconds to a few minutes provides surfaces with antifogging properties, improved adhesion, etc. Other polymers whose surfaces have been usefully modified with acids or alkali include polymethyl methacrylate, poly-acrylonitrile, wool, cotton, SBR, polyisobutylene, natural rubber, and nylon.

Surface halogenation of polyolefins has also been employed to yield a polar surface and the resulting improvements in properties of polar surfaces. The surface of nylon fibers has been modified by chlorination.

A variety of polymers and plastics have been surface treated with alkyl or alkenyl halosilanes. Generally, the materials become opaque, but improved heat, stain, and scratch resistance is imparted. The use of silanes with hydroxyl-containing polymers such as glass and cellulosics to improve adhesion and/or water repellency or reduce soiling is well known.

Surface treatment of such notoriously inert materials as polytetrafluoro-ethylene or polychlorotrifluoroethylene, by treatment with liquid ammonia solutions of active metals such as Li, Na, Cu, Ba, and Mg, leads to darkened but highly adherent surfaces. Presumably, this treatment results in the ammination of the surface through an unsaturated (dehalogenated) inter-mediate. Polychlorotrifluoroethylene may also be made adherable by treatment with organic amines. Similar results have been claimed on treat-ment of these fluoropolymers with BF_3.

Much more sophisticated surface modification of fibers, films, sheets, and webs involves graft copolymerization of monomers onto surfaces so treated as to provide initiation sites. Included in this category would be vinyl acetate or styrene grafted onto polytetrafluoroethylene, acrylonitrile grafted onto polyethylene sheet, and grafting of a wide variety of monomers to natural products, including paper, cotton, and wool. The grafted materials show much the same property modification as discussed above with the simple reagents. The techniques of graft polymerization have also been discussed above in about as much depth as is appropriate.

9. Modification of Cellulose

From historical, technological, and commercial viewpoints, no discussion of polymer modification could even hope to approach completeness without mentioning modification of cellulose (42).

In the cellulose structure each cyclic unit, or anhydroglucose unit, has three hydroxyl groups and an acetal linkage. As a result, cellulose is highly hydrogen bonded and highly crystalline. On heating, cellulose decomposes before it melts and is virtually insoluble in any inexpensive solvent that can be handled in normal fashion. Despite its many desirable properties and plentiful supply, this lack of processability is a serious handicap to the use of cellulose as a plastic material. Thus much effort has been expended in making it processable and the most successful efforts have involved its modification by permanent or temporary derivatization. Derivatization (in a sense, blocking of the hydroxyl groups) clearly eliminates hydrogen bonding, at the same time interfering with the crystallization of cellulose. Under these circumstances, the melting or flow point is decreased to below the decomposition temperature, and many cellulose derivatives can be molded, extruded, calendered, and spun into fibers. As derivatives, cellulosics are readily dissolved in commercially acceptable solvents and in this form find application in lacquers, coatings, and films.

The fine structure of cellulose is usually of great importance in determining the course and rate of reaction, and the properties of the products. In reacting with insoluble or swollen cellulose, reagents normally attack the noncrystalline or accessible regions first, so that if the reaction is terminated prior to complete hydroxyl group reaction, the reacted groups are not randomly distributed, but are concentrated in certain regions, whereas other regions are little affected.

The best known reaction of alcohols is esterification. It is therefore not surprising that cellulose esters have received a great deal of attention. The usual method of esterification is to treat the cellulose with carboxylic acids (with catalyst), acid anhydrides, or acid chlorides.

Cellulose esters of the simple lower aliphatic carboxylic acids have proved the most useful. Cellulose acetate has found large markets in plastics, sheeting, film, and fibers. It is characterized by good clarity, easy colorability, antistatic properties, and easy fabrication. Actually, cellulose monoacetate, diacetate, and triacetate are all important products, but the most common is that with about 65–75 % of the hydroxyl groups acetylated. In line with what was said earlier, it should not be surprising that 33 % acetylated cellulose, prepared by direct acetylation, is quite different in properties from the one prepared by hydrolysis of 66 % of the acetyl groups of cellulose triacetate. In the latter case, the acetyl groups are randomly distributed.

Somewhat more hydrophobic plastics, sheets, films, etc., are available through cellulose propionate, cellulose butyrate, or the mixed esters, that is, cellulose–acetate–butyrate or propionate–acetate.

Some other interesting esters of cellulose include those prepared from organic sulfonic acids, carbamate esters from cellulose and isocyanates, and those produced by reaction with ketene.

Inorganic esters, particularly the nitrate esters, go back to the earliest days of synthetic polymers and coatings. Cellulose nitrate, sometimes called nitrocellulose, is prepared by allowing cellulose to react with a nixture of nitric and sulfuric acids, for about $\frac{1}{2}$ hr. The reaction is very rapid and accessibility is not important in this case. Degradation occurs simultaneously and, depending on reaction and post-reaction conditions, a wide variety of cellulose nitrates, different in nitrate content, sulfate ester content, and viscosity are offered as lacquers.

Nitration may also be realized by allowing cellulose to react with nitrogen pentoxide, nitrogen trioxide, or dinitrogen tetroxide. At full substitution cellulose nitrate contains 14.14 % nitrogen. Useful products contain from about 11 to 13.5 % nitrogen. At higher nitrogen contents, the materials are employed as explosives. Even the lower nitrogen content materials which are used as lacquers are extremely flammable.

Sulfates, phosphates, phosphites, phosphinates, perchlorates, nitro-silicates, nitrites, and oxychlorides of cellulose are also known.

Next to cellulose esters in importance are cellulose ethers. They are most commonly prepared by reaction of sodium cellulose with the corresponding alkyl halide or sulfate.

Cellulose methyl ether may also be prepared by reaction with diazomethane. Because this synthesis is considerably more difficult and expensive than esterification, usually products having relatively low degrees of substitution are prepared. These have properties similar to cellulose esters of the same degree of substitution, but are much more resistant to hydrolysis.

Formation of cellulose ether linkages with other functionalities in the ether fragment is of considerable interest. Thus cellulosic polyethers (hydroxy terminated) may be prepared from sodium cellulose and ethylene or propylene oxide.

Pendant acid groups are introduced by reaction of sodium cellulose with chloroacetic acid or sodium chloroacetate. These lightly carboxyalkylated cellulosic materials are useful because of their affinity for certain acid reactive dyes and retard soiling.

Amine groups, reactive with basic reactive dyes, have been introduced by reaction of sodium cellulose with haloalkyl amines. A comparable product has been prepared by reaction with ethylenimine.

Alkali cellulose may add to a number of α–β unsaturated materials in Michael type condensations. Of most interest is the reaction with acrylonitrile to give cyanoethyl cellulose (Eq. 22). This material has shown outstanding rot resistance.

$$\text{Cell—OH} + \text{CH}_2\text{=CH—CN} \longrightarrow \text{Cell—O—CH}_2\text{—CH}_2\text{—CN}$$

$$(22)$$

Other unsaturated materials that have been similarly reacted include acrylamide, methyl acrylate, fumaronitrile, maleic acid, and acrylic acid. Reaction of cellulose with acetylene in alkali medium has yielded vinyl cellulose of a degree of substitution of about 0.5.

A very important modification of cellulose is the temporary modification alluded to earlier. This is the viscose process. It involves reaction of sodium cellulose with carbon disulfide to give sodium cellulose xanthate, which forms a thick dope in water. If this dope is extruded through spinnerettes into acid solution, the cellulose is regenerated and viscose rayon is formed. If the cellulose is regenerated in the form of sheets, the product is cellophane.

The sequence of transformations is as follows:

$$\text{Cell—OH} + \text{NaOH} \longrightarrow \text{Cell—O—Na}$$

$$\downarrow \text{CS}_2$$

$$\overset{\overset{\text{S}}{\underset{}{\|}}}{\text{Cell—O—C—SNa}} \xrightarrow{\text{HX}} \text{Cell—OH} + \text{CS}_2 + \text{NaX} \qquad (23)$$

Cellulose reacts with carbonyl compounds, particularly aldehydes, to form hemiacetals and acetals. The latter linkages are formed primarily, and this is a powerful method for cross-linking cellulose. Cross-linked cellulose has improved dimensional stability and, in textiles, is used for crease proofing. A similar mechanism accounts for the action of dimethylolurea as a cellulose cross-linker. Because of the commercial importance of cross-linked cellulose, almost all ether forming reactions have been employed with difunctional reagents, such as diepoxides and diisocyanates.

Metalation of cellulose, that is, the formation of sodium cellulose, has been indicated. Other metallocelluloses are obtained on treatment of cellulose with the corresponding metal hydroxide. The same products can be obtained by reaction of cellulose with metal amides in liquid ammonia. Cellulose forms chelate type compounds with cuprammonium salts or cupriethylenediamine, which are soluble in the medium. Similar complexes with cadmium, nickel, cobalt, and zinc are also known.

Silicone and silane derivatives of cellulose have been prepared. They are structurally analogous to ethers and, at low degrees of substitution in fibers and fabrics, provide water repellency and anti-soiling properties.

GENERAL REFERENCES

1. H. Mark, N. G. Gaylord, and N. M. Bikales (Eds.), *Encyclopedia of Polymer Science and Technology*, Wiley-Interscience, New York, 1964–1971.

2. P. J. Flory, *Principles of Polymer Chemistry*, Cornell Univerisity Press, Ithaca, N.Y., 1953.

3. R. W. Lenz, *Organic Chemistry of Synthetic High Polymers*, Wiley-Interscience, New York, 1967.

4. M. L. Miller, *The Structure of Polymers*, Reinhold, New York, 1966.

5. E. Baer (Ed.), *Engineering Design for Plastics*, Reinhold, New York, 1964.

6. E. M. Fettes (Ed.), *Chemical Reactions of Polymers*, Wiley-Interscience, New York, 1964.

SPECIFIC REFERENCES

7. R. L. Miller, "Crystallinity," in Ref. 1, Vol. 4, 1966, p. 449.

8. T. Alfrey, *Mechanical Behavior of High Polymers*, Interscience, New York, 1948.

9. V. A. Kargin and G. L. Slonimsky, "Mechanical Properties," in Ref. 1, Vol. 8, 1968, p. 445.

10. W. M. Saltman, "Butadiene Polymers," in Ref. 1, Vol. 2, 1965, p. 678.

11. N. G. Gaylord, (Ed.), *Polyethers. Part I. Polyalkylene Oxides and Other Polyethers*, Wiley-Interscience, New York, 1963.

12. R. G. Mraz and R. P. Silver, "Alkyd Resins," in Ref. 1, Vol. 1, 1964, p. 663.

13. "Symposium on Thermosetting Acrylic Resins," *Off. Dig. Fed. Soc. Paint Technol.* **33**, 679 (1961).

14. J. C. Bevington and H. May, "Aldehyde Polymers," in Ref. 1, Vol. 1, 1964, p. 609.

15. R. R. Hindersinn and G. M. Wagner, "Fire Retardancy," in Ref. 1, Vol. 7, 1967, p. 1.

16. R. F. Gould (Ed.), *Multicomponent Polymer Systems*, Advances in Chemistry Series No. 99, American Chemical Society, Washington, D.C., 1971.

17. G. E. Molau (Ed.), *Colloidal and Morphological Behavior of Block and Graft Copolymers*, Plenum, New York, 1971.

18. J. L. Gardon, "Cohesive Energy Density," in Ref. 1, Vol. 3, 1965, p. 833.

19. R. F. Gould (Ed.), *Plasticization and Plasticizer Processes*, Advances in Chemistry Series No. 48, American Chemical Society, Washington, D.C., 1965.

20. J. R. Darby and J. K. Sears, "Plasticizers," in Ref. 1. Vol. 10. 1969, p. 228.

21. W. J. Frissell, "Fillers," in Ref. 1, Vol. 6, 1967, p. 740.

22. G. C. Maassen, R. J. Fawcett, and W. R. Connell, "Antioxidants," in Ref. 1, Vol. 2, 1965, p. 171.

23. W. L. Cox, "Antiozonants," in Ref. 1, Vol. 2, 1965, p. 197.

24. L. I. Nass, "Stabilization," in Ref. 1, Vol. 12, 1970, p. 725.

25. N. G. Gaylord, *J. Polymer Sci.*, *C*, **24**, 1 (1968).

26. S C. Temin, "Chemical Crosslinking," in Ref. 1, Vol. 4, 1966, p. 331

27. A. R. Shultz, "Crosslinking with Radiation," in Ref. 1, Vol. 4. 1966, p. 398.

28. R. J. Ceresa, "Block and Graft Copolymers," in Ref. 1, Vol. 2, 1965, p. 485.

29. N. G. Gaylord and F. S. Ang, "Graft Copolymerization," in Ref. 6, p. 831.

30. M. B. Berenbaum, "Polysulfide Polymers. I. Chemistry," in *Polyethers. Part III. Polyalkylene Sulfides and Other Polythioethers*, N. G. Gaylord (Ed.), Wiley-Interscience, New York, 1962, p. 43.

31. J. R. Panek, "Polysulfide Polymers. II. Applications," in *Polyethers. Part III. Polyalkylene Sulfides and Other Polythioethers*, N. G. Gaylord (Ed.), Wiley-Interscience, New York, 1962, p. 115.

32. P. J. Canterino, "Halogenation of Unsaturated Polymeric Hydrocarbons," in Ref. 6 p. 142.

33. G. D. Jones, "Halogenation of Saturated Polymeric Hydrocarbons," in Ref. 6, p. 247.

34. D. Swern, "Epoxidation," in Ref. 1, Vol. 6, 1967, p. 83.

35. F. B. Greenspan, "Epoxidation," in Ref. 6, p. 152.

36. M. K. Lindemann, "Vinyl Alcohol Polymers," in Ref. 1, Vol. 14, 1971, p. 149.

37. J. Wicklatz, "Hydrogenation," in Ref. 6, p. 173.

38. M. A. Golub, "Isomerization of Unsaturated Polymeric Hydrocarbons," in Ref. 6, p. 103.

39. J. Scanlan, "Cyclization," in Ref. 6, p. 125.

40. G. E. Meyer, L. B. Tewksbury, and R. M. Pierson, "Addition of Thiols to Unsaturated Polymeric Hydrocarbons," in Ref. 6, p. 133.

41. D. J. Angier, "Surface Reactions," in Ref. 6, p. 1009.

42. K. Ward, Jr. and A. J. Morak, "Reactions of Cellulose," in Ref. 6, p. 321.

DISCUSSION QUESTIONS AND PROBLEMS

1. Major efforts in laboratories have shifted from the preparation of new polymers (whose cost is usually very high) to modification of existing relatively low-cost monomers and polymers. With the material presented in Chapters 2 and 3 indicate how you would "tailor-make" the following polymers:

    a. Stress and crack-resistant polyethylene suitable for detergent bottle applications.

    b. Impact-resistant polystyrene suitable for toys or radio cabinets.

    c. Flexible polyvinyl chloride.

    d. Moisture-resistant polyamides.

    e. Ozone-resistant rubber.

2. Cross-linking provides an important means of controlling properties especially those which depend on temperature. Review some of the systems that utilize this technique.

3. Discuss the use of compatibilizing agents in polymer blends and solutions.

4. Polyvinyl chloride is subject to degradation during processing and use. Discuss the role of stabilizers in polyvinyl chloride and give representative examples of the various types.

5. Why is it more difficult to impart flame retardance to nylon and acetal resins than to polyesters?

6. Discuss the critical factors in carrying out chemical reactions on polymers.

7. Discuss graft copolymerization as applied to the preparation of impact resistant polymers.

CHAPTER 4 _____

The Size and Weight
of Polymer Molecules

FRED W. BILLMEYER, Jr.
Department of Chemistry
Rensselaer Polytechnic Institute
Troy, New York

A. THE PHYSICAL DIMENSIONS OF LONG–CHAIN POLYMERS (9–11)

The picture of a polymer molecule that has been developed in the preceding chapters is largely based on chemistry. We have seen how macromolecules are made by polymerization, how they are cross-linked or plasticized, what kinds of chemical bonds they contain, and so on. At this point, as we begin the study of the properties of polymers, it is useful to develop a *physical* picture of what these long molecules are really like.

Polymer molecules are truly unique in their physical nature, as a result of their tremendous length compared to the dimensions of their cross section. This basic fact is responsible for almost all the interesting properties of plastics, fibers, rubbers, films, adhesives, and other materials for which long-chain molecules are essential.

To better appreciate this unique feature of polymers, let us consider a simple model of a typical polymer molecule. The model (Fig. 1) is made by snapping together 2000 plastic "pop-it" beads. If we let each bead represent a methylene ($-CH_2-$) group, the model represents a polyethylene chain of degree of polymerization 1000, or molecular weight 28,000. This is about right for the linear polyethylenes used commercially. The model is 83 ft long and about $\frac{3}{8}$ in. in diameter; both these dimensions are approximately 100,000,000 times those of the real polymer chain. Thus the model gives a vivid demonstration of the long, thin nature of a polymer molecule.

Another feature of this model which is approximately correct is its extreme flexibility. In fact, in a local sense, the bead model is not flexible enough, for it takes a dozen or so beads to bend the chain into a complete circle,

Figure 1   A model of a polymer molecule, consisting of 2000 plastic "pop-it" beads strung together. The shorter strands at the bottom right represent octane and butane.

whereas we know that six methylene groups can form a closed ring—cyclohexane—without strain. Let us now attempt to improve our physical picture of the real polymer chain by exploring the consequences of this flexibility.

To do this, we need to consider the properties of single polymer molecules, each one isolated from its neighbors. Because molecules this large do not exist in the vapor phase, we consider the chain suspended in a sea of small molecules, in other words, in dilute solution in a low molecular weight solvent.

Let us consider the arrangements in three dimensions, which we call the *conformations*, that a polymer chain can have in dilute solution. First, consider a short segment of the chain, just four methylene groups long. If we define a plane by three of the carbon atoms in this segment, as in Fig. 2, and allow free rotation about the carbon–carbon bond, the fourth atom can be anywhere on the circle indicated in the figure, although some positions are more likely than others, since steric hindrances prevent completely free rotation about the bonds. Because each successive carbon atom along the chain can, at random, take any of several positions on a similar circle based, again at random, on the position of the preceding atom, the total number of conformations possible in a chain of 2000 atoms must be formidably large indeed.

One of these conformations of particular interest is that in which each successive carbon atom lies in the same plane in the *trans* location with respect to the earlier atoms in the chain. This gives the fully-extended all-*trans* conformation shown in Fig. 3. In this conformation, the *fully extended length* of a 2000-atom chain would be 2540 Å, corresponding to the 83-ft length of our model. The *contour length* of the chain, following all the zigzag's from atom to atom, is $2000 \times 1.54 = 3080$ Å.

But this is only one out of a *very* large number of conformations, and we may expect that *on the average* the two ends of the chain are separated by a much smaller distance. As we shall see, this distance can be measured directly, and also calculated from theory with some simplifying assumptions.

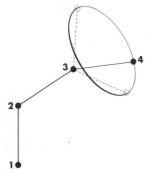

Figure 2   A segment of a polymer chain, showing four successive chain atoms. The first three of these define a plane, and the fourth can lie anywhere on the indicated circle which is perpendicular to and bisected by the plane.

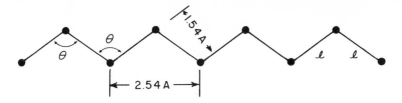

Figure 3   The fully extended all-*trans* conformation of a carbon–carbon chain.

1.   The Freely Jointed Chain

The simplest calculation of polymer chain dimensions is based on the assumption that the chain consists of x links of length l, with no restrictions at all on the angles between successive links. The calculation of the end-to-end distance of such a chain is the same *random-flight* problem that occurs in diffusion theory and elsewhere. Again, the results must be expressed in terms of average quantities for a large number of chains, or for many successive conformations taken up by one chain over a period of time.

If we ask, as in Fig. 4, what the chance W is of finding one end of the chain in a volume element $dx\,dy\,dz$ at a distance r from the other end, the solution of the random-flight problem is given in the curve of Fig. 5. This result shows

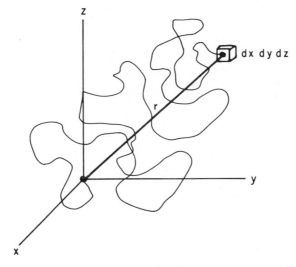

Figure 4   A highly schematic representation of a random-coil polymer chain, with one end at the origin of a coordinate system and the other in a volume element $dx\,dy\,dz$ at a distance $r - (x^2 + y^2 + z^2)^{1/2}$ from the origin.

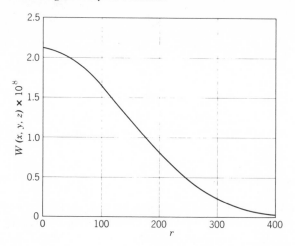

Figure 5   The probability $W(x, y, z)$ of finding the end of the chain of Fig. 4 in the volume element $dx\ dy\ dz$ as a function of r (in angstrom units). Calculated for a chain of 10^4 links, each 2.5 Å long.

that the chance is greatest of finding the two ends close together, and that the probability of finding the second end falls according to a Gaussian curve as the ends get farther apart. (For real chains, this cannot be quite correct, for the Gaussian function never goes to zero, whereas real chains have a finite contour length and can never get any longer. The difference is unimportant for the features we are interested in.) The curve of Fig. 5 also suggests that the density of chain segments is greatest at the center of the randomly coiling chain, and falls off toward the outside in the manner shown.

It is more instructive, in asking about the size of this random coil, to calculate the probability of finding one end of the chain in a spherical shell of thickness dr at a distance r from the other end. The answer to this problem is given in Fig. 6, where it is seen that there is a maximum probability, corresponding to a most probable dimension for the chain. If the root-mean-square end-to-end distance is selected as a measure of this dimension, the random-flight theory leads to a very simple result:

$$(\overline{r_f^2})^{1/2} = l(n)^{1/2} \tag{1}$$

where the symbol $\overline{r^2}$ designates the average of the square of r, the end-to-end distance, and the subscript f is added to signify that this is the result of the random-flight calculation.

The most significant feature of this result is that the dimension of the polymer chain is proportional to the *square root* of its number of links, and

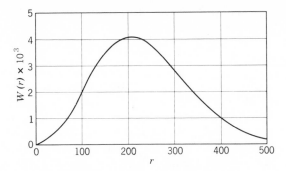

Figure 6   The probability of $W(r)$ of finding the end of the chain of Fig. 4 in a spherical shell of thickness dr at a distance r from the origin.

therefore to the square root of molecular weight. This proportionality is retained, for all practical purposes, even though the highly artificial model of the freely jointed chain is replaced by more realistic approximations.

Finally, calculation shows that $(\overline{r_f^2})^{1/2}$ is quite small compared to the contour length of the chain. For a 2000-atom chain with bonds 1.54 Å long, for which we calculated the fully extended length to be 2540 Å, $(\overline{r_f^2})^{1/2} = 69$ Å.

2.   Real Polymer Chains

Real polymer molecules differ from the freely jointed chain model in several respects, each of which leads to an increase in their physical dimensions.

Bond Angles

It is clear that restricting the angle between successive links in the chain to a fixed value will increase its dimensions. The increase in $\overline{r^2}$ is by a factor $(1 - \cos\theta)/(1 + \cos\theta)$, where θ is the bond angle. For carbon–carbon bonds, this is a factor of just 2.0.

Restricted Rotation

Similarly, restriction of free rotation about chain bonds, whether as a result of the steric hindrance of side groups or for other reasons, again leads to increases in the chain dimensions. The effect varies, of course, with the chemical nature of the polymer, and its recent calculation for a wide variety of polymer types has been a major contribution (12). For polyethylene, for example, the chain is further expanded by a factor of 1.8 owing to restricted rotation about the carbon–carbon chain bonds.

The dimension that results from including the effects of bond angle and restricted rotation with the random-flight calculation is known as the *unperturbed dimension* of the polymer chain, and is given the symbol $(\overline{r_0^2})^{1/2}$. For polyethylene, $(\overline{r_0^2})^{1/2} = 2.6(\overline{r_f^2})^{1/2}$.

The Excluded Volume

A more subtle difference between the freely jointed model and the real polymer chain results from the fact that, in the former, two joints or segments can occupy the same position in space at the same time, whereas for the real polymer chain any conformations for which this occurs at any point along the chain cannot exist. The number of such impossible conformations that must be excluded is greater for more compact arrangements, with smaller values of r. The distribution curve of Fig. 6 is distorted, therefore, in such a way that the root-mean-square end-to-end distance is increased.

The theoretical calculation of the excluded volume effect is extremely difficult, and remains a major unsolved problem of polymer science (13). A more fruitful approach has been to simulate chain conformations by computation on digital computers—the so-called "Monte Carlo" calculations (14). Still more practical at the present time is the experimental evaluation of the combined effects of all the above factors, as is now described.

3.   Polymer–Solvent Interactions and Unperturbed Chain Dimensions

Let us now consider the results of experimental measurements of the dimensions of polymer molecules. (The methods by which such measurements are made are discussed in Section F; we consider only the results here.) In general, we find that, as expected, the experimentally obtained dimensions are larger than the unperturbed dimension. For convenience, let us write the actual end-to-end distance as $(\overline{r^2})^{1/2}$ and define an *expansion coefficient* α to describe the increased size of the real chain:

$$(\overline{r^2})^{1/2} = \alpha(\overline{r_0^2})^{1/2} \tag{2}$$

We find that α depends on the solvent in which the polymer dimensions are measured. If the solvent is one in which strong interaction forces are developed between polymer and solvent molecules—for example, a polar polymer in a polar solvent—α is large and the chain is relatively expanded. Such a solvent is called a thermodynamically "good" solvent for that polymer.

Conversely, in a "poor" solvent, α is small, the chain having smaller dimensions. In fact, it is found that by proper choice of solvent and temperature, α can be made equal to one, but at values of α not much less than one, the polymer will not remain in solution.

It can be shown that α is an indirect measure of the polymer–solvent interaction forces, and that $\alpha = 1$ when these forces go to zero. The polymer molecules are said to be unperturbed by interactions with the solvent under these conditions, and by the definition of α the quantity $(\overline{r_0^2})^{1/2}$ is known as the *unperturbed dimension* of the chain.

The fact that a polymer chain has the same *average* dimension as the unperturbed end-to-end distance does not, of course, mean that it behaves in all respects as if the excluded-volume effect were absent. Real polymer chains still cannot assume impossible conformations, and the distribution of chain lengths about the average must be quite different from that of Fig. 6.

It is known that polymer–solvent interaction forces, and hence α, depend on temperature. For a given solvent, the temperature at which $\alpha = 1$ is called the Flory temperature Θ; a solvent used at $T = \Theta$ is called a *theta solvent*. In later sections we discuss ways of finding the theta temperature, and other consequences of working at $T = \Theta$.

Values of α (which varies slowly with molecular weight) for a typical polymer and solvent are given in Table 1.

Table 1   Values of the expansion coefficient for polystyrene in benzene at 20°C (15)

Molecular Weight	α
44,500	1.23
65,500	1.27
262,000	1.46
694,000	1.54
2,550,000	1.72
6,270,000	1.91

4.   Interpenetration of Polymer Molecules in Dilute Solution

In Table 2 are listed the dimensions discussed so far for a polyethylene chain of molecular weight 28,000, and for the "pop-it" bead model of such a chain. In addition to inquiring about the contour length and fully extended length of the chain, and its unperturbed and actual end-to-end distances, we may ask how much volume the chain occupies in solution.

The answer to this question must, of course, be an average figure, since we know that the chain takes up a variety of conformations. From the distribution curve of Fig. 6 it can be calculated that the diameter of a sphere which just encloses the polymer chain on the average is 2.36 times the chain's end-to-end distance, and the diameter of a sphere which just encloses the chain 95% of the time is about $5(\overline{r^2})^{1/2}$.

Table 2   Dimensions of a polyethylene chain of $M = 28,000$ and the corresponding "pop-it" bead model

Dimension	For the Molecule (Å)	For the Model (ft)
Contour length	3080	—
Fully extended length	2540	83
Random-flight end-to-end distance	69	2.25
Unperturbed end-to-end distance	178	6
Diameter of sphere just enclosing the chain on the average	420	14
Diameter of sphere just enclosing the chain 95% of the time	900	30

It is obvious that only a small fraction of the volume of such a sphere is occupied by the segments of the chain itself. The rest of the space must be taken up by solvent molecules, if the solution is sufficiently dilute, or by segments of other polymer chains in more concentrated solutions. A simple calculation shows that polymer segments occupy only about 0.02% of the volume of the sphere within which all the chain segments lie 95% of the time.

We can conclude, therefore, that polymer molecules must markedly interpenetrate each other's domains unless they are placed in extremely dilute solutions. Qualitatively, it is not surprising that intermolecular interactions exist among polymer chains at all concentrations at which measurements are normally made.

Finally, from what little we know of the amorphous state of polymers, as described further in the following chapters, there is no reason to doubt that the same sort of interpenetration of chains exists here also. This fact has important consequences in determining the rheological properties of polymers.

B.   COUNTING MOLECULES AND THE NUMBER-AVERAGE MOLECULAR WEIGHT

1.   The Existence of Molecular Weight Distributions

At this point we turn our consideration from the size of polymer molecules to the measurement of their mass. Here again, we find it necessary always to consider average quantities, for it is well known that, except for certain biological materials, all polymers are made up of molecules with a wide variety of molecular weights. As was pointed out in Chapter 2, the existence of a distribution of molecular weights in essentially all polymer samples results from the presence of a statistical element in all known polymerization processes (16). Consequently, the weights of the individual molecules are determined by the results of random processes. Although random, these processes are not uncontrolled, and the average properties of polymers, including their average molecular weights, are well defined.

Just as several types of polymer chain dimensions are of interest, so are several types of average molecular weights. The major ones are indicated in Fig. 7, which represents the molecular weight distribution of a typical polymer. The most important of these molecular weights, which we discuss in this and the following sections, are the number-average $\overline{M}_n$ and the weight-average $\overline{M}_w$.

2. The Number-Average Molecular Weight $\overline{M}_n$

The type of average molecular weight called the number average results from any process in which every molecule is counted in the same way, regardless of its mass. It is the simple average obtained by dividing the mass of the polymer sample by the number of molecules it contains.

If the sample contains N_i molecules of the ith kind, for a total number of molecules $\sum_{i=1}^{\infty} N_i$, and each of the ith kind of molecule has a mass m, then

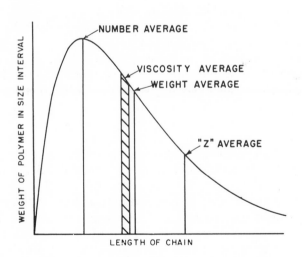

Figure 7   The distribution of molecular weights in a typical high polymer. The various average molecular weights located on the distribution curve are defined in this and the following sections.

the total mass of all the molecules is $\sum_{i=1}^{\infty} N_i m_i$. The number-average molecular mass is

$$\overline{m}_n = \frac{\sum_{i=1}^{\infty} m_i N_i}{\sum_{i=1}^{\infty} N_i} \tag{3}$$

and multiplication by Avogadro's number gives the number-average molecular weight (mole weight),

$$\overline{M}_n = \frac{\sum_{i=1}^{\infty} M_i N_i}{\sum_{i=1}^{\infty} N_i} \tag{4}$$

Number-average molecular weights of commercial polymers usually lie in the range 10,000–100,000, although some materials have values of $\overline{M}_n$ tenfold higher, and others tenfold lower. In most cases, however, the physical properties we associate with typical high polymers are not well developed if $\overline{M}_n$ is below about 10,000.

3.   Chemical Methods for Determining $\overline{M}_n$ (17, 18)

If it is known that some identifiable chemical group occurs the same number of times on each polymer molecule in a sample, then determination of the molar concentration of this group allows the evaluation of $\overline{M}_n$. A little consideration shows that, for all practical purposes, these groups must be on the ends of the polymer chains. (Further requirements are that the chains must be linear rather than branched and that it be known whether the group investigated is on one or both ends of the chain.) Hence the name *end-group analysis* is appropriate for the chemical determination of the number-average molecular weight.

Like all methods for determining $\overline{M}_n$, the techniques of end-group analysis become insensitive at high molecular weights, as the fraction of end groups becomes too small to be measured with precision. Although conventional chemical analyses often lose precision at molecular weights above 25,000, some methods based on physical techniques such as infrared spectroscopy can be used for values of $\overline{M}_n$ at least as high as 100,000.

Typical end-group analyses by chemical methods include the determination of carboxyl groups in polyesters and polyamides by direct titration with base in an alcoholic or phenolic solvent, and the titration of amino groups in polyamides with acid. Hydroxyl groups may be converted to acid functions by esterification, or acetylated and hydrolyzed with subsequent titration of the released acetic acid. Infrared analysis can also be used; this technique is also useful for determining unsaturated end groups in polyolefins, such as vinyl groups in certain types of polyethylene. Radioactive tagging of end groups has also been used.

C. PHYSICAL METHODS FOR DETERMINING $\overline{M}_n$ (3, 19–23)

1. The Colligative Properties

The thermodynamic methods by which, in effect, the number of molecules in a sample can be counted are known as the *colligative* methods. Measurement of the colligative properties, therefore, leads to the determination of the number-average molecular weight. The term "thermodynamic" is applied to these methods because their underlying theories are based on equilibrium considerations, as is all of thermodynamics. Properly carried out, a colligative property measurement can be relied upon to give an unambiguous and reliable result. This can not be said in general of methods based on other than thermodynamic considerations.

The colligative properties are as follows: (1) the lowering of the solvent vapor pressure in a polymer solution compared to that of the pure solvent, (2) the similar depression of the freezing point, (3) the elevation of the boiling point, and (4) the osmotic pressure developed between solvent and solution across a semipermeable membrane, through which solvent molecules can pass freely but polymer molecules can not.

The magnitude of the colligative properties for a typical polymer—polystyrene dissolved in benzene—has been calculated and is presented in Table 3. It is clear that the difference in vapor pressure between a polymer solution and the pure solvent is very small. In addition, equilibrium between the two is established very slowly—days to weeks—through the vapor phase, and during this time the temperature of the apparatus must be held very precisely constant. Although a few experiments have been reported (24, 25), the direct vapor-pressure lowering method is virtually never used for high polymers; we do not consider it further. The method of vapor-phase osmometry, discussed below, is closely related, however.

Table 3 **Values of the Colligative Properties of a Solution of Polystyrene, $\overline{M}_n = 20,000$, in Benzene at $c = 0.01$ g/cm^3**

Property	Value
Vapor pressure lowering	4×10^{-3} mm Hg
Freezing point depression	2.5×10^{-3} °C
Boiling point elevation	1.3×10^{-3} °C
Osmotic pressure	15 cm solvent

Cryoscopy

Despite the small temperature differences involved, the measurement of the freezing-point depression has been applied successfully for determining $\overline{M}_n$ for many years. Recently, in a refined application of the method to polyethylene (26), thermistors were used as temperature-sensing elements. The large temperature dependence of the electrical resistance of these solid-state devices allows them to measure temperature differences as small as $1\text{--}2 \times 10^{-5}\,°C$, when used in a Wheatstone bridge circuit with powerful electronic amplification.

To avoid complications from the precipitation of crystalline polyethylene, the freezing-point determinations were carried out with hexamethylbenzene, which melts at 165°C, as the solvent. The measurements were made in a simple cryoscopic cell maintained in a constant-temperature bath about 2° below the solvent freezing point. The cell was electrically heated to melt the solvent and magnetically stirred. After the solvent freezing point was determined, polymer was added and dissolved, and the freezing points of the solutions determined at several concentrations. Finally, a portion of a low molecular weight substance such as octacosane $(M = 395)$ or tristearin $(M = 892)$ was added and the resulting freezing point determined for calibration purposes.

To ensure reproducibility in the freezing point measurements, supercooling must be minimized and controlled. For this purpose, a nucleating agent for the crystallization of the solvent was used.

For reasons mentioned in Section H, all measurements of the properties of polymer solutions must be made at a series of concentrations and extrapolated to $c = 0$. The equation relating the freezing-point depression $-\Delta T_f$ to $\overline{M}_n$ and c is

$$-\frac{\Delta T_f}{c} = \frac{RT^2}{\rho \Delta H_f}\left(\frac{1}{\overline{M}_n} + A_2 c\right) \qquad (5)$$

where R is the gas constant and ρ and ΔH_f are the density and heat of fusion of the solvent. The constant A_2, known as the *second virial coefficient*, describes the polymer–solvent interaction forces. Its evaluation, and the concentration dependence of the colligative properties, are discussed below.

If ρ, ΔH_f, and the relation between the electrical resistance of the thermistors and temperature are known, Eq. 5 can be used to calculate $\overline{M}_n$ without the necessity for measuring any other material of known molecular weight. For this reason the cryoscopic method, like all the colligative techniques, is said to be an *absolute* method for determining $\overline{M}_n$. It is easier, however, to measure, in terms of electrical resistance, the freezing-point depression of a known substance such as one of the calibration standards mentioned, and calculate $\bar{M}_n$ for the polymer by a simple ratio.

Since ΔT_f is inversely proportional to $\bar{M}_n$, the cryoscopic method loses sensitivity as $\bar{M}_n$ increases. It is useful for molecular weights up to about 30,000.

Ebulliometry

In many respects, the measurement of the boiling-point elevation ΔT_b is similar to that of the freezing-point depression. Thermistors or multi-junction thermocouples, used in Wheatstone bridge circuits, are the usual temperature sensors. Calibration with known low molecular weight sub-stances is practiced for convenience. An equation of the type of Eq. 5 applies, with ΔT_b replacing $-\Delta T_f$ and the heat of vaporization of the solvent ΔH_v replacing ΔH_f.

Two types of ebulliometer are used for polymer solutions. In one, the polymer solution and the pure solvent are placed in similar vessels containing the thermistors, and rocked gently to ensure even boiling. In the other, the polymer solution is heated in a boiler and pumped over one temperature sensor, while the condensation temperature of the pure solvent is measured by the other sensor. The latter type of instrument is less satisfactory for polymer solutions which tend to foam on boiling, a common occurrence.

The ebulliometric method is usually considered applicable for molecular weights up to 30,000, but results stated to be reliable have been reported (27) for samples with $\bar{M}_n$ well in excess of 100,000.

Membrane Osmometry

If we judge solely from the magnitude of the colligative effect, the measure-ment of the osmotic pressure ought to be the most suitable thermodynamic method for determining $\bar{M}_n$. Indeed, the history of osmometry predates polymer science itself by several decades. Unfortunately, the properties of available semipermeable membrane materials severely limit the applicability of the method.

Conventional osmometers consist of two closed compartments separated by a semipermeable membrane and connected to capillary tubes. A typical instrument used some years ago is the block osmometer of Fuoss and Mead (28). The two compartments are filled with the solvent and the polymer solution. Thermodynamics requires that solvent molecules pass through the membrane into the solution, tending to dilute it and bring it closer to equi-librium with the solvent. Since the solution compartment is closed, the solution level in the capillary tube must rise to accommodate the solvent passing through the membrane. This creates a hydrostatic pressure tending to drive solvent back through the membrane to the solution side. When this

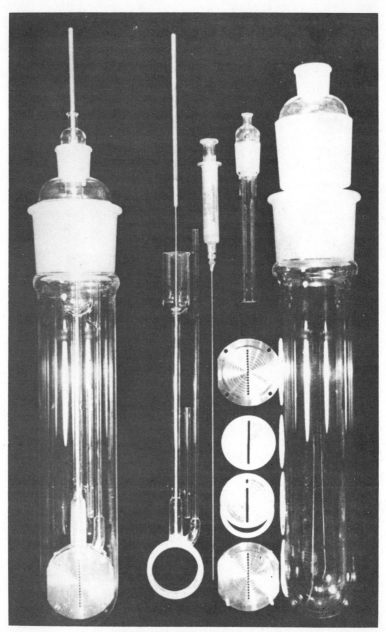

Figure 8 The Stabin osmometer (30). The membrane is placed between two metal plates; one of these assemblies is clamped on each side of the glass solution cell. From this cell rise the measuring capillary and a filling tube, closed with a metal rod. The entire assembly, including a short capillary section for the solvent reading, is immersed in a tube partially filled with solvent.

pressure just equals the osmotic pressure, the liquid level in the capillary stabilizes.

Although the large membrane area of the block osmometer favors rapid equilibration, it has largely been superseded by smaller and simpler instruments based on the designs of Zimm and Myerson (29) and Stabin and Immergut (30) (Fig. 8). These osmometers consist of a solution compartment and capillary, with a membrane on each side. The entire assembly is placed in a large tube containing the solvent, which can easily be thermostated.

In the last few years rapid automatic osmometers (31–33) (Fig. 9) have become available, in which one compartment is completely closed and fitted with a sensitive pressure-sensing device rather than a capillary. Some instruments are direct reading; in others, a servo system rapidly adjusts the liquid level in the solvent compartment to balance the osmotic pressure before an appreciable amount of solvent has to pass through the membrane. As a result of this rapid action, the osmometers come to equilibrium in 1–5 min instead of 10–20 hr as typically required in conventional instruments.

The osmotic membranes most commonly used are made from "gel cellophane," regenerated cellulose which has never been dried out since manufacture. Other membrane materials include films of several polymers swollen to some extent by the solvents used, and some porous inorganic substances such as special glasses. All these materials, however, are only approximately semipermeable; they allow some low molecular weight polymer species to diffuse through to the solution side of the osmometer. Any such molecules are effectively removed from the osmotic experiment; the resulting osmotic pressure is too low, and the molecular weight derived from it is too high.

This error, which can at times amount to a factor of two or more (34, 35), is the most serious limitation of the osmotic method. In severe cases, it can be detected by a continuing drop in the apparent osmotic pressure with time after the osmometer should have equilibrated (Fig. 10). Although it is tempting to extrapolate the data linearly back to zero time, the theories of the diffusion effect (36) clearly show that this is not appropriate. There is, in fact, no simple way to correct for the diffusion, since one cannot know what species went through the membrane before significant measurements of the osmotic pressure could be made.

It is difficult to state in advance at what levels of molecular weight the diffusion effect becomes serious, since it depends on the number of low molecular weight species in the polymer sample. It is thought to be conservative to state that osmometry is usually reliable for samples with $\overline{M}_n$ greater than 20,000 for fractionated samples, or greater than 50,000 for whole polymer samples. These limits vary widely with the type of membranes used. The use of rapid automatic osmometers does not reduce this lower limit

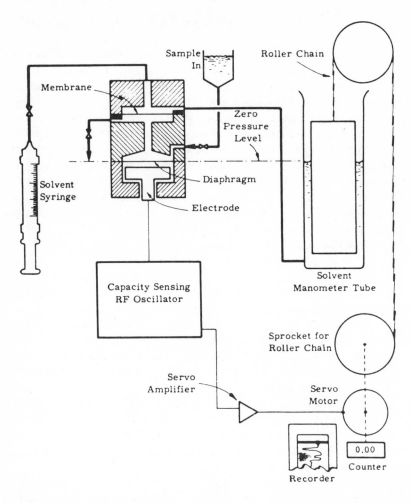

Figure 9 Diagram of the Stabin–Shell rapid automatic osmometer (31). The solvent compartment (above the membrane) is open to the atmosphere at the level-controlling manometer tube. The solution compartment (below) is completely closed. Passage of solvent through the membrane deflects the flexible metal diaphragm, causing a change in capacitance of a condenser of which the diaphragm is one electrode. Through a servo system, this change controls the solvent level to such a point that no net flow of solvent through the membrane occurs.

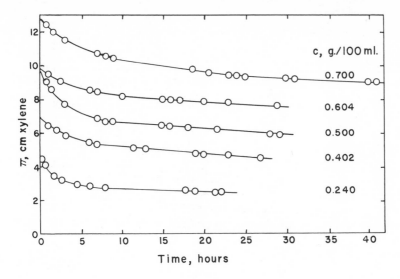

Figure 10 Plot of osmotic pressure as a function of time, showing the continuing slow drop after initial equilibration characteristic of the diffusion of low molecular weight polymer species through the membrane (35).

significantly for a given membrane type (34). The upper limit of applicability of osmometry, set at the point where the osmotic height becomes too small to measure conveniently, is at about $\overline{M}_n = 1,000,000$.

The equation relating to osmotic pressure π to $\overline{M}_n$ and c is

$$\frac{\pi}{RTc} = \frac{1}{\overline{M}_n} + A_2 c \tag{6}$$

As with the other colligative properties, osmotic data are plotted as a straight-line function of concentration whose intercept is inversely proportional to $\overline{M}_n$ and whose slope, the second virial coefficient A_2, measures the interaction forces between the polymer molecules and those of the solvent.

Typical osmotic data are shown in Figs. 11 and 12. The data (37) of Fig. 11 were obtained with samples of the same polymer type but different molecular weights, dissolved in the same solvent. Thus A_2 was constant (except for a slow decrease with increasing $\overline{M}_n$ not noticeable in the figure) and the data points lie on parallel lines. In Fig. 12 are shown the data (38) for a single polymer sample in three different solvents. Here the lines have the same intercept, as they must because $\overline{M}_n$ is constant, but different slopes representing different values of A_2. For nitrobenzene (line c), A_2 is very nearly zero. This solvent is therefore very nearly a theta solvent for nitrocellulose at the

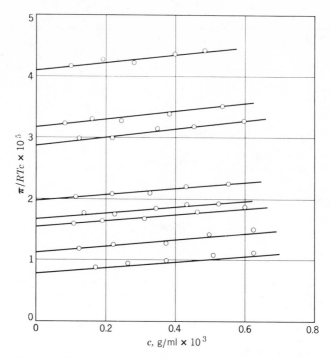

Figure 11 Plot of $\pi/RTc = 1/\overline{M}_n$ vs. c for fractions of cellulose acetate in acetone (37). As required by Eq. 6, the lines have the same slope but different intercepts.

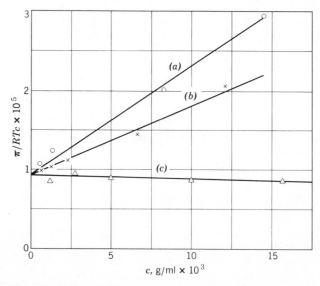

Figure 12 Plot of $\pi/RTc = 1/\overline{M}_n$ vs. c for nitrocellulose in three different solvents: (*a*) acetone, (*b*) methanol, and (*c*) nitrobenzene (38). As required by Eq. 6, the lines have the same intercept but different slopes.

172

temperature of these measurements. It follows that the theta temperature can be determined by finding that temperature at which the second virial coefficient is zero.

Vapor-Phase Osmometry

In this method (39, 40), the property measured, in a so-called "vapor pressure osmometer" (Fig. 13), is the small temperature difference resulting from different rates of solvent evaporation from droplets of pure solvent and polymer solution maintained in an atmosphere of solvent vapor. An important assumption of the method is that this temperature difference is proportional to the vapor pressure lowering of the polymer solution at equilibrium and thus to the number-average molecular weight: because of heat losses, the full temperature difference expected from theory is not attained. Measurements must be made at several concentrations and extrapolated to $c = 0$. Like ebulliometry and cryoscopy, the method is calibrated with low molecular weight standards and is useful for values of $\overline{M}_n$ at least up to 40,000 in favorable cases. It is rapid and has the additional

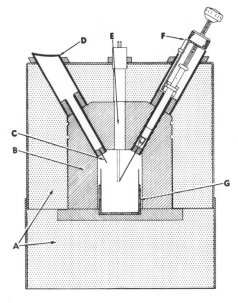

A—Foam
B—Aluminum block
C—Chamber
D—Syringe guide

E—Thermistor probe
F—Syringe in down
    (loading) position
G—Solvent cup and wick

Figure 13 The measuring chamber of the "vapor pressure osmometer" (39). A is insulating material surrounding a metal block B, within which is the solvent chamber C. Syringes introduced at D and F place drops of solvent and solution on the thermistors at the bottom of the probe E. Solvent vapor from a cup and wick G fills the chamber.

advantage that only a few milligrams of sample is required. A series of articles describing the technique and instrumentation has appeared (40). Successful measurement has been reported of number-average molecular weights as high as 160,000, but with commercial equipment the upper limit remains near 25,000. The lower limit is that at which the solute becomes appreciably volatile. Because the method does not measure equilibrium vapor pressure lowering but depends on the development of quasi-steady-state phenomena, care must be taken to standardize such variables as time of measurement and drop size between calibration and sample measurement. Significant nonlinearity has been observed in the extrapolation of data from relatively high concentrations to infinite dilution. The method is finding increasing use in the study of association and in the rapid estimation of osmotic coefficients of electrolytes.

2. Comparison of Data

Although it is conservative to state that most methods for measuring polymer molecular weights have an accuracy of only 5–10%, significantly closer agreement among methods can be achieved in favorable cases. An extensive intercomparison of methods for determining $\overline{M}_n$ (41) gave the data in Table 4. Here, despite additional difficulties imposed by high-temperature operation, chemical and several thermodynamic methods gave excellent agreement for samples of polyethylene. The osmotic molecular weights in parentheses show the effects of diffusion of low molecular weight species through the membranes. They were obtained with conventional osmometers; data on these samples obtained with rapid automatic osmometers are similar (34).

Table 4   Number-average molecular weight of polyethylene by various methods (41)

Method	Branched Polyethylene Samples			Linear Polyethylene Samples			
	75	76	77	99	99H	99L	101
End groups by infrared	—	—	—	8,100	37,400	4,800	11,700
Ebulliometry	11,150	—	18,400	—	—	—	—
Cryoscopy	10,700	13,300	19,100	8,100	—	3,320	11,800
Membrane osmometry	(30,300)	(26,600)	(31,400)	—	40,600	—	—
Vapor-phase osmometry	11,000	16,300	18,800	8,600	38,400	3,500	11,900

D. WEIGHT-AVERAGE MOLECULAR WEIGHT

After $\overline{M}_n$, the next higher average molecular weight which can be measured by absolute methods is the *weight-average molecular weight* $\overline{M}_w$. This quantity is defined as

$$\overline{M}_w = \frac{\sum_{i=1}^{\infty} N_i M_i^2}{\sum_{i=1}^{\infty} N_i M_i} \qquad (7)$$

It should be noted that each molecule contributes to $\overline{M}_w$ in proportion to the square of its mass: a quantity proportional to the first power of M measures only concentration, and not molecular weight. In terms of concentrations $c_i = N_i M_i$ and weight fractions $w_i = c_i/c$, where $c = \sum_{i=1}^{\infty} c_i$

$$\overline{M}_w = \frac{\sum_{i=1}^{\infty} c_i M_i}{c} = \sum_{i=1}^{\infty} w_i M_i \qquad (8)$$

Unfortunately, there appears to be no simple analogy for $\overline{M}_w$ akin to counting molecules to obtain $\overline{M}_n$.

Because heavier molecules contribute more to $\overline{M}_w$ than light ones, $\overline{M}_w$ is always greater than $\overline{M}_n$ except for a hypothetical monodisperse polymer. The value of $\overline{M}_w$ is greatly influenced by the presence of high molecular weight species, just as $\overline{M}_n$ is influenced by species at the low end of the molecular weight distribution curve.

The quantity $\overline{M}_w/\overline{M}_n$ is a useful measure of the breadth of the molecular weight distribution curve, and is the parameter most often quoted for describing this feature. The range of values of $\overline{M}_w/\overline{M}_n$ in synthetic polymers is quite large, as illustrated in Table 5. Among the narrowest distributions produced so far in such materials are those found in the so-called "living"

Table 5   Typical ranges of $\overline{M}_w/\overline{M}_n$ in synthetic polymers

Polymer	Range
Hypothetical monodisperse polymer	1.000
Actual "monodisperse" "living" polymers	1.01–1.05
Addition polymer, termination by coupling	1.5
Addition polymer, termination by disproportionation, or condensation polymer	2.0
High conversion vinyl polymers	2–5
Polymers made with autoacceleration	5–10
Addition polymers prepared by coordination polymerization	8–30
Branched polymers	20–50

polymers made by anionic polymerization (42), for which $\overline{M}_w/\overline{M}_n$ is well below 1.04. This figure is somewhat misleading, however, since it corresponds to a distribution with a half-width of roughly 30% of $\overline{M}_n$. Samples with about tenfold narrower distributions have been prepared, but significant "tails" to the distributions are still present.

Most polymers made with relatively simple polymerization kinetics (16) have $\overline{M}_w/\overline{M}_n$ in the range 1.5 to 5. Where the kinetics is complex, considerably higher values of $\overline{M}_w/\overline{M}_n$ can occur.

1.   Light Scattering (43–47)

The scattering of light is a very common phenomenon, occurring whenever a beam of light traverses matter. It accounts for such diverse phenomena as the blue color of the sky, the colors of sunsets and rainbows, and the colors of clouds, smokes, and most white objects.

In polymer solutions, light scattering can be used to measure $\overline{M}_w$ since the amplitude of the scattering is proportional to the mass of the scattering particle, but it is the intensity, the square of the amplitude, which is measured.

Light Scattering by Small Particles

Historically, the theory of light scattering in gases was developed by Lord Rayleigh (48) nearly 100 years ago, and that in liquids by Einstein and others (49) more than 50 years ago. Debye (50) extended these theories to solutions in 1944, deriving the *Debye equation*, on which the determination of $\overline{M}_w$ is based for scattering particles small compared to the wavelength of the light used,

$$K \frac{c}{R_{90}} = H \frac{c}{\tau} = \frac{1}{\overline{M}_w} + 2A_2 c \tag{9}$$

Here R_{90} is the *Rayleigh ratio* R_θ evaluated at the angle of observation (Fig. 14) $\theta = 90°$. The Rayleigh ratio relates the intensity of scattered light, i_θ, per unit volume V of scattering material, observed at a distance r, to the primary beam intensity I_0:

$$R_\theta = \frac{i_\theta r^2}{I_0 V} \tag{10}$$

Alternatively, scattering can be expressed in terms of the turbidity, τ, the total amount of light removed from the primary beam by scattering at all angles:

$$\tau = \frac{1}{I_0} \int_\theta i_\theta d_\theta \tag{11}$$

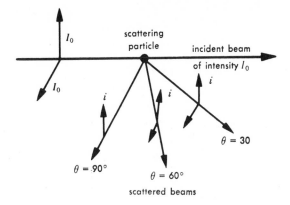

Figure 14.   Essential features of the light-scattering process.

If, over a length of sample l, the primary beam intensity is reduced from I_0 to I,

$$\frac{I}{I_0} = e^{-\tau l} \qquad (12)$$

The analogy between Eq. 9 and Eqs. 5 and 6 for the determination of $\overline{M}_n$ is clear. Note, however, that in the case of light scattering the measured quantity is directly rather than inversely proportional to M, both of these quantities appearing in the denominator when the equation is written as a linear function of c. This means that there is a lower, rather than an upper, limit to values of $\overline{M}_w$ which can be measured, reached as τ or R_θ becomes too small to be measured. In most cases this limit is well below the high polymer range.

The second virial coefficient A_2 can be determined from light scattering as well as from the colligative properties, although the slope term in Eq. 9 is twice A_2.

The light-scattering calibration constants K and H are given by

$$K = \frac{2\pi^2 n^2}{N_0 \lambda^4}\left(\frac{dn}{dc}\right)^2 \qquad \text{and} \qquad H = \frac{32\pi^3 n^2}{3N_0 \lambda^4}\left(\frac{dn}{dc}\right)^2 \qquad (13)$$

where N_0 is Avogadro's number, n is the refractive index of the solution, and dn/dc is the *specific refractive increment*, measuring the change in refractive index of the solution with polymer concentration. This quantity is independent of concentration, and can be written as $(n - n_0)/c$, where n_0 is the refractive index of the solvent. The specific refractive increment is also independent of the molecular weight of the polymer. It is usually too small (ca. 0.1–0.2) to be measured on an ordinary refractometer, but *differential refractometers* are widely used for its determination.

The appearance of λ^4 in the denominator of Eq. 13 is characteristic for all scattering processes involving particles small compared to λ, including gases and liquids. It means that blue light is scattered more strongly than red light, and accounts for the blue color of the sky (48) and of many large bodies of water.

As written, Eq. 9 is correct for the usual experimental arrangement of vertically polarized incident light and a detector viewing only vertically polarized scattered light. Minor correction terms can be added if other arrangements are used.

Light Scattering by Larger Particles

The requirement that the scattering particles must be small compared to (0.05–0.10 of) the wavelength of the light is very restrictive. As we have seen in Section A, the real end-to-end distance of even relatively low molecular weight polymers approaches 0.05λ (visible light has $\lambda = 4000$–7000 Å); most polymers have $\overline{M}_w$ much larger than that assumed in our model. When the size of a scattering particle exceeds this lower limit, different parts of it are exposed to incident light of different phase (Fig. 15). The light waves scattered from these different parts interfere with one another, with the net result that R_θ depends on the angle of observation θ. For monodisperse particles this angular dependence can be derived explicitly for particles of known shape, such as spheres and rods. For random-coil particles, for example, it is given by the function

$$P(\theta) = \frac{2}{v^2}\left[e^{-v} - (1 - v)\right] \tag{14}$$

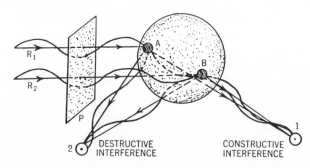

Figure 15 Light scattering from a particle comparable in size to the wavelength, showing the effect of phase difference in causing destructive interference.

where

$$v = \frac{8\pi^2}{3} \frac{n^2 \overline{r^2}}{\lambda^2} \sin^2 \frac{\theta}{2} \tag{15}$$

The significance of these equations is discussed further in Section F.

To take into account this effect describing light scattering in larger particles, Eq. 9 is modified to read

$$K \frac{c}{R_{90}} = H \frac{c}{\tau} = \frac{1}{\overline{M}_w P(\theta)} + 2A_2 c \tag{16}$$

Since most polymer solutions are not monodisperse, $P(\theta)$ is not given by Eq. 14, but by a summation of similar equations for different molecular sizes, weighted by the amounts of the various sized species present. Since this information is generally not known, it is customary to treat the data in a way that does not require explicit knowledge of $P(\theta)$. By expanding $1/P(\theta)$ as a function of the angle variable, Zimm (51) showed that a plot of $K(c/R_{90})$ or $H(c/\tau)$ against $\sin^2(\theta/2)$ is linear for small angles (and in most cases over the entire angular range). He combined the necessary extrapolations to $c = 0$ and $\theta = 0$ in what is known as a *Zimm plot* (Fig. 16), in which the intercept is $1/\overline{M}_w$, as indicated by Eq. 16. In addition to $\overline{M}_w$, the second virial coefficient can be determined from the slope of the lines of constant angle on a Zimm plot, and (as discussed in Section F) the average size of the scattering particles can be derived from the slope of the lines of constant concentration.

Equipment and Methods for Determining $\overline{M}_w$

Light scattering from polymer solutions is measured in photoelectric turbidimeters, several of which are described in the literature (51, 53–56). A simplified diagram of a light-scattering photometer is given in Fig. 17. Most modern research photometers feature automatic scanning and recording of the scattered light intensity as a function of angle.

The proper preparation of a sample for light-scattering measurement is of major importance for obtaining reliable results. Dust and other impurities may scatter much more light than the polymer molecules, and must be carefully removed, for example by filtration or ultracentrifugation.

Although light scattering is an absolute method for measuring $\overline{M}_w$, the photometer must be calibrated to read turbidity or R_θ directly. Some of the materials used for this calibration are reflecting standards, colloidal suspensions, and simple liquids. Tungstosilicic acid, $H_4SiW_{12}O_{40}$, $M = 2879$, has been recommended as a primary calibration standard in preference to pure liquids (too sensitive to impurities) or uniform particle-size latexes (too sensitive to residual polydispersity). The calibration of light-scattering photometers has been discussed in detail (57).

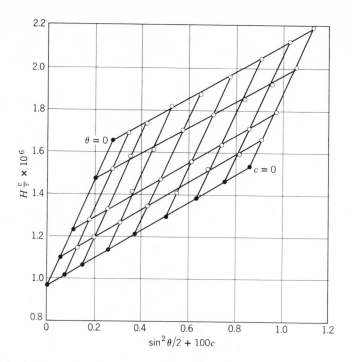

Figure 16 A Zimm plot showing the light scattering from a sample of polystyrene dissolved in butanone (52). The factor 100 multiplying c is an arbitrary constant introduced for convenience in plotting. This sample had $\overline{M}_w = 1{,}030{,}000$, $A_2 = 1.3 \times 10^{-4}$ ml-mole/g^2, and $(r^2)^{1/2} = 1120$ Å.

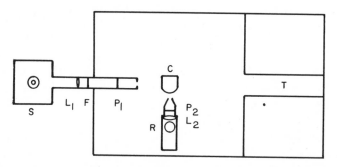

Figure 17 A simplified diagram of a typical light-scattering photometer. S is the light source, and L_1 a lens to provide a parallel light beam passing in succession through a polarizer P_1, a filter F isolating a single wavelength of light, and the scattering cell C before being absorbed in a light trap T. The phototube R views the center of the cell through a lens L_2 and a polarizer P_2, and is mounted on an arm rotating about the center of the cell to vary the angle of observation.

The light-scattering method is applicable to polymers of almost any molecular weight, from a few thousand to a few million, which can be dissolved in a solvent for which dn/dc is not too small. It has long been assumed that information on particle shape or on polydispersity could be obtained from light-scattering data, but recent work strongly suggests that neither of these can be done except under unusually favorable circumstances. Also, the method cannot be applied to copolymers without observing severe restrictions.

The measurement by light scattering of the sizes of particles considerably larger than polymer molecules requires special equipment and theories other than those mentioned here for its interpretation (58–60).

2. Equilibrium Ultracentrifugation (61–67)

Traditionally, ultracentrifugation has been more widely used for studying biological materials than random-coil polymers, but more recently its usefulness for synthetic materials has been well demonstrated.

Equipment

As indicated in Fig. 18, the analytical ultracentrifuge consists of an aluminum or titanium alloy rotor which is rotated at high speed in an evacuated chamber. The solution being measured is placed in a small cell near the outer edge of the rotor and viewed by an optical system which allows measurement of the concentration of the polymer along the length of the cell. Most modern ultracentrifuges use electric drive units and operate at speeds between 2000 and 60,000 rpm. The lower speeds are used for the thermodynamic methods described in this section, and the higher speeds for the transport method described in Section E.

Ultracentrifugation requires that the polymer and solvent be different in density (to ensure sedimentation) and in refractive index (which is commonly used to measure concentration). With suitable solvents, the method can be used for polymers with a wide range of molecular weights.

Sedimentation Equilibrium

When the ultracentrifuge is operated at low speeds for long times (often several days), a thermodynamic equilibrium is reached in which the polymer is distributed along the cell (in the direction of the centrifugal force, perpendicular to the axis of rotation) according to its molecular weight. The force of sedimentation on each species is just balanced by its tendency to diffuse back against the concentration gradient resulting from its movement

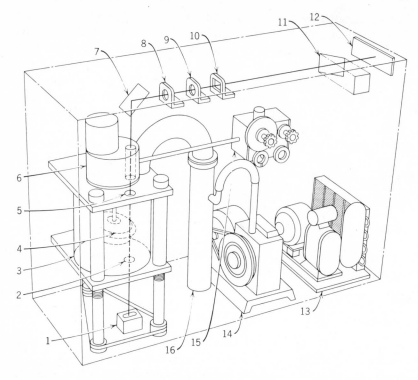

Figure 18 Schematic perspective of the "Spinco" analytical ultracentrifuge. The rotor is shown at 4, inside the vacuum chamber 3, with the electric drive unit 6 above. The optical system by which the cell is viewed includes lamp 1, various lenses, windows, and mirrors 2, 5, 7–11, and a photographic plate 12. A refrigeration unit is at 13, vacuum pumps at 14 and 16, and a speed-control device at 15. Courtesy Spinco Division, Beckman Instruments, Inc.

in the centrifugal field. It can be shown that, at $c = 0$,

$$\overline{M}_w = \frac{2RT}{(1 - \bar{v}\rho)\omega^2} \frac{\ln(c_b/c_m)}{r_b^2 - r_m^2} \tag{17}$$

where r is the distance from the axis of rotation and the subscripts m and b indicate the meniscus and the cell bottom, respectively. It is at these points that the polymer concentration, c, must be evaluated in order that all molecular species be included in the average $\overline{M}_w$. The quantity ω is the angular speed of rotation of the centrifuge, and the term $(1 - \bar{v}\rho)$, in which $\bar{v}$ is the partial specific volume of the polymer and ρ the density of the solution, describes the difference in density between the polymer and the solvent. The second virial coefficient can be determined from the dependence of the apparent molecular weight of the polymer on the concentration.

By treating the experimental data somewhat differently, making use of the refractive index difference between solution and solvent as a function of distance from the center of rotation, it is possible to determine the z-average molecular weight of the polymer,

$$\overline{M}_z = \frac{\sum_{i=1}^{\infty} N_i M_i^3}{\sum_{i=1}^{\infty} N_i M_i^2} \tag{18}$$

This is the only thermodynamic method available for determining $\overline{M}_z$.

Since the solute is distributed in the ultracentrifuge cell according to its molecular weight, the equilibrium method gives considerable information about the molecular weight distribution of the polymer.

Approach to Equilibrium

By making measurements only in the vicinity of the meniscus and the cell bottom at a sufficiently early time (often within 10–60 min of the start of centrifugation) that no redistribution of molecular species has taken place, it is possible to determine $\overline{M}_w$ and A_2 from thermodynamic equations. This method avoids the long times needed to redistribute the polymer species characteristic of the sedimentation–equilibrium process, but sedimentation equilibrium using short cells is often preferred.

Sedimentation in a Density Gradient

If sedimentation is carried out in a mixed solvent, the two solvent components can distribute so as to form a density gradient within the cell. The solute forms a band centering at the point where its effective density equals that of the solvent mixture. The method is valuable for determining compositional differences leading to differences in solute density.

The *sedimentation velocity* technique (a transport method), is described in Section E.

3. Comparison of Data

Data (68) comparing weight-average molecular weights obtained by light scattering and by approach-to-equilibrium ultracentrifugation are given in Table 6. In each case, the two methods are considered to agree within their combined experimental errors.

Table 6   Weight-average molecular weight by various methods

Sample	$\bar{M}_w$ by Light Scattering	$\bar{M}_w$ by Approach-to-Equilibrium Ultracentrifugation	Reference
Linear polyethylene	144,000	126,000	69a
Polymethyl methacrylate			
Fraction	148,000	159,000	69b
Whole polymer	490,000	524,000	69b
Polyethylene glycol	12,000	10,900	70
Polystyrene			
H	49,600	47,600	71
L	12,700	14,300	71
A-1	25,600	26,300	72
SM-1 (2-butanone)	27,900	28,200	73
SM-1 (cyclohexane)	28,900	28,200	73
S-111 (2-butanone)	20,400	22,200	73
S-111 (cyclohexanone)	26,500	22,200	73
705	179,300	189,800	74
706	257,800	288,100	74

E. TRANSPORT MOLECULAR WEIGHT METHODS

In addition to the direct methods of molecular weight measurement discussed in Sections B–D, extensive use is made of empirical correlations between certain transport properties and molecular weight. These transport properties, which are associated with the frictional drag on a polymer molecule passing through the solvent, include the intrinsic viscosity, the sedimentation velocity, and the diffusion rate. Basically, as developed further in Section F, they measure the size (end-to-end distance, for example) rather than the weight of the polymer molecule.

By far the best known and easiest to measure of the above-mentioned transport properties is the intrinsic viscosity; for these reasons, the discussion of these properties is developed in terms of this quantity.

1. The Intrinsic Viscosity (75–78)

The viscosity of dilute polymer solutions is measured in simple glass capillary viscometers (Fig. 19), through which the time required for a fixed volume of liquid to flow is measured. If the solution and solvent have the same density, and neglecting possible effects of shear rate, the flow times of these liquids are

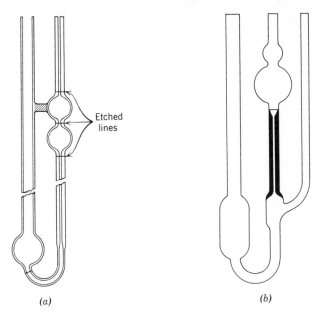

Etched
lines

(a)   *(b)*

Figure 19   Two common varieties of capillary viscometer: (*a*) Ostwald–Fenske, and
(*b*) Ubbelohde. The former requires a fixed volume of solution, whereas with (*b*)
successive dilutions of the polymer solution can be made in the viscometer.

proportional to their viscosities. Some common viscosity functions derived
from these data are listed in Table 7 (79).

The hypothesis of the long-chain nature of high polymers first received
widespread acceptance through the work of Staudinger in the late 1920s
and early 1930s, an important part of which was his use of viscosity as a
measure of polymer molecular weights (80). He postulated, and provided
convincing evidence to support, the relationship

$$\eta_{sp} = KcM \qquad (19)$$

where K is a constant for a given polymer type, solvent, and temperature.
In the ensuing decades, only two modifications to this equation have been
found necessary:

1.   The reduced viscosity η_{sp}/c has been replaced by the intrinsic viscosity
$[\eta]$, for it has been found that the former quantity is not independent of
concentration. Huggins (81) has shown empirically that

$$\frac{\eta_{sp}}{c} = [\eta] + k'[\eta]^2 c \qquad (20)$$

where k' is a constant for a given polymer type, solvent, and temperature.

Table 7   Nomenclature of solution viscosity (79)

Quantity	Symbol and Defining Equation
Relative viscosity	$n_r = \dfrac{\eta}{\eta_0} \cong \dfrac{t}{t_0}$
Specific viscosity	$\eta_{sp} = \eta_r - 1 = \dfrac{\eta - \eta_0}{\eta_0} \cong \dfrac{t - t_0}{t_0}$
Reduced viscosity	$\eta_{red} = \dfrac{\eta_{sp}}{c}$
Inherent viscosity	$\eta_{inh} = \dfrac{\ln \eta_r}{c}$
Intrinsic viscosity	$[\eta] = \left(\dfrac{\eta_{sp}}{c}\right)_{c=0}$
	$\quad = \left(\dfrac{\ln \eta_r}{c}\right)_{c=0}$

A similar equation can be written (82) for the inherent viscosity:

$$\frac{\ln \eta_r}{c} = [\eta] + k''[\eta]^2 c \tag{21}$$

Because k'' (often about -0.15) is usually smaller than k' (often about 0.35), the inherent viscosity usually varies less with concentration than the reduced viscosity.

2.   The proportionality has been found to be to a power of molecular weight somewhat lower than the first.

Empirically, then, it is found that

$$[\eta] = K'\overline{M}_v^a \tag{22}$$

known as the Mark–Houwink equation. The exponent a varies from 0.5 in a theta solvent to about 0.8; typical values of K' range between 0.5 and 5×10^{-4} when $[\eta]$ is expressed in the usual units of deciliters/gram (corresponding to concentration in grams/deciliter). K' and a are constants for a given polymer type, solvent, and temperature, and have been determined and tabulated (7, 82) for many systems. $\overline{M}_v$ is the *viscosity-average molecular weight* defined by the equation

$$\overline{M}_v = \left[\frac{\sum_{i=1}^{\infty} N_i M_i^{1+a}}{\sum_{i=1}^{\infty} N_i M_i}\right]^{1/a} \tag{23}$$

which is seen to depend on the value of a and therefore the conditions of measurement of $[\eta]$. For this reason, $\overline{M}_v$ cannot be located uniquely with respect to other averages, for example, on a molecular weight distribution curve such as that of Fig. 7. $\overline{M}_v$ is, moreover, not directly measurable by any other technique.

The intrinsic viscosity is a quantity that can be correlated *empirically* with molecular weight. The constants K' and a in Eq. 22 must be determined for each experimental system by measuring the intrinsic viscosity of polymer samples whose molecular weight is known as the result of other measurements. It is not possible to determine absolute values of molecular weight from $[\eta]$ without such calibration. Since $\overline{M}_v$ is not usually available (except for hypothetical monodisperse polymers for which all averages are equal), and because $\overline{M}_v$ is more nearly equal to $\overline{M}_w$ than to other average molecular weights, it is common to use values of $\overline{M}_w$, for example, from light scattering, to calibrate the viscosity method, and to rewrite Eq. 22

$$[\eta] = K''\overline{M}_w^a \tag{24}$$

whose use requires the added restriction that samples to be measured must have molecular weight distributions (specifically, values of $\overline{M}_v/\overline{M}_w$) similar to those of the calibrating polymers. (A further restriction, that the polymers be linear rather than branched, is discussed in Section F.)

A typical calibration curve for the viscosity method is shown in Fig. 20 (83).

2. Sedimentation Velocity

In the *sedimentation transport* or *sedimentation velocity* method, the ultracentrifuge described in Section D is operated at high speed so that the polymer molecules are transported to the bottom of the cell. The rate of sedimentation is given by the *sedimentation constant*

$$s = \frac{1}{\omega^2 r}\frac{dr}{dt} \tag{25}$$

Like the intrinsic viscosity, the value of s obtained by empirical extrapolation to $c = 0$ can be related empirically to molecular weight:

$$s_0 = K_s\overline{M}_s^{a_s} \tag{26}$$

where the constants K_s and a_s are determined like those of the corresponding viscosity equation, and $\overline{M}_s$ is a similarly defined, but little used, average molecular weight.

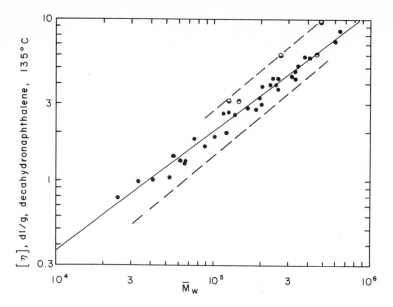

Figure 20 A calibration curve for the intrinsic viscosity–molecular weight relationship (Eq. 24), for fractionated samples of linear polyethylene dissolved in decahydronaphthalene and measured at 135°C (83).

The most fruitful result of the sedimentation transport method appears to be the examination of the distribution of sedimentation constants in a polydisperse sample. This can, with several approximations, be related to the distribution of molecular weights, and the experiment is one of the few that allow direct, if only semiquantitative, observation of this distribution.

3.   The Diffusion Coefficient (84)

If a boundary is artificially produced between a polymer solution and its solvent, the rate of diffusion of the polymer species across the boundary, resulting from Brownian motion, can be measured and expressed in terms of the *diffusion constant*, D. This quantity can be empirically related to molecular weight exactly like the intrinsic viscosity and the sedimentation constant:

$$D_0 = K_D \overline{M}_D^{a_D} \tag{27}$$

The method is relatively little used except in conjunction with the sedimentation velocity as described below.

4. Combinations of Methods

Each of the transport methods just described is related to a *frictional coefficient, f*. The equations for monodisperse polymers are

$$[\eta] = \frac{(f/K_f)^3}{M} \tag{28}$$

$$s_0 = \frac{M(1 - \bar{v}\rho)}{f} \tag{29}$$

$$D_0 = \frac{RT}{f} \tag{30}$$

If it is assumed that, for random-coil polymers, the frictional coefficient is the same for all three methods, the results of these measurements can be combined in pairs to allow direct determination of the molecular weight. The combination of sedimentation and diffusion is due to Svedberg (85) and is widely used for biological materials for which f is well defined and monodisperse species are often found:

$$\frac{D_0}{s_0} = \frac{RT}{M(1 - \bar{v}\rho)} \tag{31}$$

Mandelkern and Flory (86) suggested the similar combination of the intrinsic viscosity and the sedimentation constant:

$$\frac{s_0}{[\eta]^{1/3}} = \frac{(1 - \bar{v}\rho)M^{2/3}}{K_f} \tag{32}$$

It should be emphasized that the elimination of the need for absolute calibration of these methods by combining them in pairs does not make them any less transport methods, subject to the restrictions associated with these methods. In addition, even for a monodisperse solute the sedimentation equations have not been extended to take into account simultaneously the effects of diffusion in the ultracentrifuge cell, polymer–solvent interactions, and compressibility of solvent and solute in the high centrifugal fields required. For polydisperse samples, additional complications arise in assigning average molecular weights, since the methods being combined yield different averages, themselves relatively complex.

F. MEASUREMENT OF MOLECULAR SIZE

We come now to the consideration of methods for measuring the size—that is, the dimensions (in contrast to the mass)—of polymer molecules. It will be recalled that the random-coil nature of synthetic polymers, and their characterization in terms of size, was discussed in Section A.

1. Light Scattering

The most direct way of measuring the size of polymer molecules is by the angular dependence of light scattering. In Section D it was stated that, when scattering particles become comparable in size to (0.05–0.10 of) the wavelength, there is interference between light waves scattered from different regions of the particle. This results in a dependence of scattered intensity on the angle of observation, described by a function $P(\theta)$, which was given for monodisperse random coils in Eqs. 14 and 15. In Section D, $P(\theta)$ was used as a correction factor to the scattered intensity, allowing the calculation of true values of molecular weight after extrapolation to $c = 0$.

It should be clear, however, that the angular dependence of light scattering can equally well be used to evaluate the size of the scattering particles, as measured by the mean square end-to-end distance appearing in Eq. 15 or appropriate parameters for other types of particles such as rods, spheres, or disks.

Since most polymers are not monodisperse, it is convenient to evaluate an average end-to-end distance from the Zimm plot. It was shown by Zimm (51) that the z average of $\bar{r}^2$ results from this treatment:

$$\overline{r_z^2} = \frac{9\lambda^2}{8\pi^2 n^2} \times \frac{\text{initial slope}}{\text{intercept}} \tag{33}$$

where the initial slope of the Zimm plot is that of the $c = 0$ line as a function of $\sin^2(\theta/2)$. In a sense it is unfortunate that the z average is given by light scattering, since the comparison of these data with those obtainable in other ways is rendered more difficult.

2. The Intrinsic Viscosity (7, 76, 87)

The most commonly accepted theory relating the intrinsic viscosity, defined in Section E, to molecular size is that of Flory (88), who (like others before him) postulated that $[\eta]$ is proportional to the effective hydrodynamic volume of the molecule—that is, the apparent volume occupied by the dissolved random-coil molecule.

The nature of this dependence follows from Einstein's theory (89) of the viscosity of suspensions of spheres not penetrated by the solvent. He showed that the specific viscosity of such a suspension is proportional to the volume fraction occupied by the spheres, independent of their viscosity or mass. On assuming, for random-coil polymers, that the effective volume of the molecule is proportional to the cube of the root-mean-square end-to-end distance, and converting from volume fraction to the usual concentration units, it can be shown that

$$[\eta] = \Phi \frac{(\overline{r^2})^{3/2}}{M} \tag{34}$$

the key equation of Flory's theory. Here Φ is a universal constant for all polymer types and temperatures. It depends slightly on solvent power, ranging in value from about 2.5×10^{21} in theta solvents to about 2.8×10^{21} in good solvents, when $[\eta]$ is expressed in deciliters/gram and $(\overline{r^2})^{1/2}$ in centimeters.

Replacing $(\overline{r^2})$ by $\alpha(\overline{r_0^2})^{1/2}$, and recalling that $(\overline{r_0^2})^{1/2}$ is proportional to $M^{1/2}$, so that $\overline{r_0^2}/M$ is a function of chain structure only, independent of solvent, temperature, or molecular weight, we can write

$$[\eta] = \Phi\left(\frac{\overline{r_0^2}}{M}\right)^{3/2} M^{1/2}\alpha^3 = KM^{1/2}\alpha^3 \tag{35}$$

where $K = \Phi(\overline{r_0^2}/M)^{3/2}$ is a constant for a given polymer, independent of other parameters. It follows that in a theta solvent,

$$[\eta]_\theta = KM^{1/2} \tag{36}$$

The exponent of one-half has been verified many times, and this fact constitutes one of the most convincing evidences of the validity of Flory's viscosity theory. Values of K are near 1×10^{-3} for a number of polymer systems.

Further developments of the theory show that in good solvents α is approximately proportional to $M^{0.1}$, so that the total dependence of $[\eta]$ on M from Eq. 35 is about $0.5 + 3(0.1) = 0.8$, in good agreement with the empirical results of Section E.

3. The Dimensions of Branched Polymers

It has been shown (90) that the arrangement of a given number of polymer chain segments in a branched, rather than a linear, configuration, causes a reduction in the size of the random coil. Since branched polymers have many ends, it is usual to express their size in terms of the *radius of gyration S* rather than the end-to-end distance. (For linear polymers, $\overline{r^2} = 6\overline{S^2}$.) The relative size of linear and branched molecules of the same mass is given as a function of the number and type of branch points by a function

$$g = \frac{\overline{S^2} \text{ (branched)}}{\overline{S^2} \text{ (linear)}} \tag{37}$$

which decreases continuously from unity as branching increases. Unfortunately, the relation between g and $[\eta]$ is not well established (91). Though the ratio $[\eta]_{branched}/[\eta]_{linear}$ at constant molecular weight is certainly a qualitative indication of chain branching, it should not currently be given quantitative significance (92).

G. FRACTIONATION OF POLYMERS (93–98)

So far, no mention has been made of methods for separating the various molecular species making up a typical polymer. The purpose of this section is to provide a brief description of techniques for accomplishing this *fractionation* of a polymer according to molecular weight. It is assumed throughout that chemical heterogeneity is not present.

1. Fractionation Based on Solubility

Many of the common methods of fractionating polymers are based on the fact that the solubility of macromolecules is a function of molecular weight, with the highest molecular weight species being the first to precipitate out from solution as solvent power is reduced.

Precipitation Methods

Traditionally, fractionation has been carried out in the manner just described, with solvent power being controlled by temperature or by varying the ratio of solvent to nonsolvent in a two-component liquid mixture. Nonsolvent may be added to a dilute polymer solution (ca 1 g/l) until a slight turbidity develops. After a time, a precipitated phase consisting of a highly swollen relatively concentrated polymer solution settles out and can be removed. Addition of a second increment of nonsolvent allows the separation of a second fraction, and so on, until the sample is separated, typically, into 10–20 fractions. The process is both time-consuming and extremely inefficient. It takes about 1 day to remove each fraction, and theory shows clearly that the separation by molecular weight is far from complete. The calculated molecular weight distributions for a series of fractions obtained this way are shown in Fig. 21; the overlap in the various fractions is obvious.

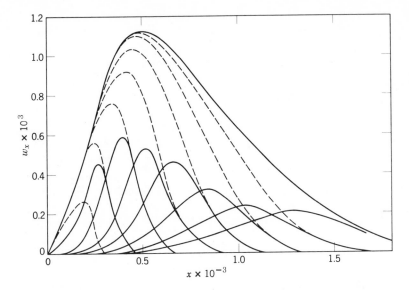

Figure 21   Molecular weight distribution curves calculated for the separation of a polymer (upper curve) into eight fractions (lower solid curves). The dashed curves represent the polymer remaining in solution after successive fractions are removed by precipitation (100).

Elution Methods

Somewhat more success has been achieved by reversing the fractionation technique so that species of increasing molecular weight are dissolved out of a swollen but high-viscosity polymer-rich phase, usually coated on the support of a chromatographic type column. Sand, "celite" (a diatomaceous earth), and glass beads are typical column packing materials.

In *solvent-gradient elution* fractionation, a liquid mixture of continuously increasing solvent power is pumped through the column at constant temperature. The lowest molecular weight species are dissolved and eluted first, with species of increasing molecular weight following in order. (If the column loading is too high or the solvent power is changed too rapidly, a "reversal" may occur in which some low molecular weight species are held back and eluted later than their normal location.)

In analytical application, a small (0.4 g) sample has been separated into some 400 fractions by this method (99). The technique has also been scaled up to the 1-lb level (101). The method of polymer deposition may (102) or may not (103) be important; there is some evidence that the efficiency of the separation is improved if the polymer is selectively precipitated, with species of lower molecular weight precipitating last.

A modification of this technique is *temperature-gradient elution* (104), in which a temperature gradient is imposed along the length of the column, so that each species undergoes a series of solution and precipitation steps as it is carried down the column by the gradient of solvent composition.

2.   Gel Permeation Chromatography

Gel permeation chromatography is a powerful new separation technique (similar in principle to but advanced in practice (105) over gel filtration as practiced by biochemists) which has found wide acceptance in the polymer field since its discovery (106) in 1961. The separation takes place in a chromatographic column filled with beads of a rigid porous "gel"; highly cross-linked porous polystyrene and porous glass are preferred column packing materials. The pores in these gels are of the same size as the dimensions of polymer molecules.

A sample of a dilute polymer solution is introduced into a solvent stream flowing through the column. As the dissolved polymer molecules flow past the porous beads (Fig. 22), they can diffuse into the internal pore structure of the gel to an extent depending on their size and the pore-size distribution of the gel. Larger molecules can enter only a small fraction of the internal portion of the gel, or are completely excluded; smaller polymer molecules penetrate a larger fraction of the interior of the gel.

The larger the molecule, therefore, the less time it spends inside the gel, and the sooner it flows through the column. The different molecular species are eluted from the column in order of their molecular size as distinguished from their molecular weight, the largest emerging first.

A complete theory predicting retention times or volumes as a function of molecular size has not appeared for gel permeation chromatography. A specific column or set of columns (with gels of differing pore size) is calibrated empirically to give such a relationship, by means of which a plot of amount of solute versus retention volume can be converted into a molecular size distribution curve. If the calibration is made in terms of a molecular size parameter, for example, $[\eta]M$, whose relation to size is given by Eq. 34, it can be applied to a wide variety of both linear and branched polymers.

As in all chromatographic processes, the band of solute emerging from the column is broadened by a number of processes, including contributions from the apparatus, flow of the solution through the packed bed of gel particles, and the permeation process itself (107). Corrections for this zone broadening can be made empirically (108); it usually becomes unimportant when the sample has $M_w/M_n > 2$.

Gel permeation chromatography has proved extremely valuable for both analytical and preparative work with a wide variety of systems ranging from low molecular weights to very high. The method can be applied to a wide variety of solvents and polymers, depending on the type of gel used. With polystyrene gels, relatively nonpolar polymers can be measured in solvents such as tetrahydrofuran, toluene, or (at high temperatures) *o*-dichlorobenzene; with porous glass gels, more polar systems including aqueous solvents can be used. A sample of a few milligrams suffices for analytical work, and the determination is complete in 2–4 hr in typical cases. An extensive bibliography on gel permeation chromatography has been compiled (109).

The results of careful gel permeation chromatography experiments for molecular weight distribution agree so well with results from other techniques

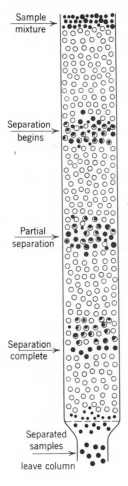

Sample mixture →

Separation begins →

Partial separation →

Separation complete →

Separated samples leave column →

Figure 22 Principle of the separation of polymer molecules according to size by gel permeation chromatography, as explained in the text (95).

that there is serious doubt as to which is correct when residual discrepancies occur. Figure 23 shows the extent of agreement between this method and a solvent-gradient elution fractionation, and Fig. 24 demonstrates the degree of fit between the experiment and a distribution curve calculated from polymerization kinetics.

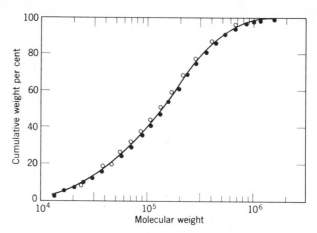

Figure 23   Typical cumulative molecular weight distribution curve (110). The polymer is polypropylene; solvent gradient elution data are represented by closed circles, and gel permeation chromatography data by open circles. Molecular weight is plotted logarithmically because of the very broad molecular weight distribution in the sample.

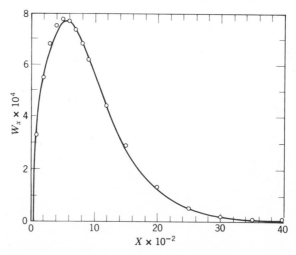

Figure 24   Fit of gel permeation chromatography data for polystyrene to a molecular weight distribution curve calculated from polymerization kinetics (111).

3. Treatment of Data

The process of fractionation provides no more than the separation of a polymer sample into a number of fractions. Until these are characterized for molecular weight, nothing is learned about the distribution of molecular weights in the sample. Usually the fractions are themselves characterized, but occasionally (as in the current practice of gel permeation chromatography) characterization may result from a calibration of the procedure or equipment.

When the weights and molecular weights of all the fractions are known, a cumulative molecular weight distribution curve like that of Fig. 23 is plotted. The familiar differential distribution curve (like that of Fig. 7) is obtained by differentiation. Experimental errors, resulting in the scatter of points in the integral curve, are magnified in this process, and care must be taken to avoid overinterpretation of the results.

4. Analytical Fractionation

Several techniques have been developed in which the direct result is a display related to the molecular weight distribution curve without the physical separation of fractions. The use of the analytical ultracentrifuge, as described in Sections D and E, is one example; others are described in the general references.

H. THE THERMODYNAMICS OF POLYMER SOLUTIONS (112–114)

The purpose of this short section is to indicate the existence of theories for the thermodynamics of solutions of high polymers, which give reasonably satisfactory explanations for the phenomena described in this chapter, and through which the equations presented have been derived.

These theories, originally developed independently by Flory (115) and Huggins (116), explain the gross deviations from ideal solution behavior, characteristic of polymer solutions, in terms of the difference in size of the molecules of polymer and solvent. These deviations are attributed to the small change in entropy (a measure of the degree of disorder in a system) when the polymer and solvent are mixed.

The polymer is assumed to be made up of a large number, x, of segments each equivalent in size to a solvent molecule. Since the x segments are joined together in a chain, they cannot be distributed as randomly in the solvent as could x molecules of another low molecular weight component. Thus the entropy change on mixing polymer with solvent is small compared to that in ordinary solutions.

The calculation of the entropy and the free energy of mixing for polymer solutions on the basis of this model is simple and straightforward. Many useful relationships, such as the equations for molecular weight determination by colligative property measurement presented in Section C, follow directly from the theory. The widespread conformity of these equations to physical reality is evidence for the satisfactory nature of the theory.

The relationship derived from the Flory–Huggins treatment for the free energy of dilution of a polymer solution would be

$$\mu_1 - \mu_1^0 = RT\{\ln(1 - \phi_2) + [1 - (1 - x)]\phi_2 + \chi\phi_2^2\} \tag{38}$$

where $\mu_1 - \mu_1^0$ is the free energy of dilution, ϕ_2 the volume fraction of polymer, x the number of segments in a polymer chain, and χ the interaction parameter (a measure of the solvent power of the system). Since osmotic pressure is related to the free energy of dilution, where $\overline{V}_1$ is the partial molar volume of the solvent, by

$$\pi\overline{V}_1 = -(\mu_1 - \mu_1^0) \tag{39}$$

the osmotic pressure may be substituted into Eq. 38.

$$\pi\overline{V}_1 = -RT\left[\ln(1 - \phi_2) + \left(1 - \frac{1}{x}\right)\phi_2 + \chi\phi_2^2\right] \tag{40}$$

In dilute solutions, $\phi_2 \ll 1$, so $\ln(1 - \phi_2) = -\phi_2 - \phi_2^2/2 - \phi_2^3/3 - \cdots$ and Eq. 40 may be rewritten as

$$\pi = -RT\left[\frac{\phi_2}{x\overline{V}_1} + \left(\frac{1}{2} - \chi\right)\frac{\phi_2^2}{\overline{V}_1} + \frac{\phi_2^3}{3\overline{V}_1} + \cdots\right] \tag{41}$$

To express the osmotic pressure in terms of concentration (weight of polymer per volume of solution) and molecular weight the appropriate substitutions (113) are made and Eq. 42 is obtained:

$$\pi = RT\left[\frac{c}{\overline{M}_n} + \frac{\bar{v}^2}{V_1}\left(\frac{1}{2} - \chi\right)c + \cdots\right] \tag{42}$$

where $\bar{v}$ is the partial specific volume of the polymer and V_1 the molar volume of the solvent. Upon rearranging,

$$\frac{\pi}{RTc} = \frac{1}{\overline{M}_n} + \left(\frac{\bar{v}^2}{V_1}\right)\left(\frac{1}{2} - \chi\right)c^2 + \cdots \tag{43}$$

It can now be understood from Eq. 43 why osmotic pressure data (as well as the other colligative properties) must be extrapolated to zero concentration. For most systems the higher concentration terms may be dropped and

Eq. 6 would be obtained, where the second virial coefficient A_2 is equivalent to $(\bar{V}^2/V_1)(\frac{1}{2} - \chi)$.

The original Flory–Huggins treatment neglected the fact that a very dilute polymer solution must be discontinuous in structure, including some regions containing only solvent. Flory and Krigbaum (117) extended the theory to this case, introducing the excluded volume effect described in Section A, and *see* *pg. 160* the concepts of the theta solvent and theta temperature. This treatment has required further modification with time, and considerable work is still needed in certain areas because of formidable mathematical difficulties.

Both the Flory–Huggins and the Flory–Krigbaum theories have serious limitations, many of which have been overcome in powerful new free-volume theories (118, 119), which take into account such factors as volume change on mixing polymer and solvent and the possibility of negative heats of mixing. The full consequences of the application of these theories are just beginning to appear in the literature.

GENERAL REFERENCES

1. P. J. Flory, *Principles of Polymer Chemistry*, Cornell University Press, Ithaca, N.Y., 1953.
2. P. W. Allen (Ed.), *Techniques of Polymer Characterization*, Butterworths, London, 1959.
3. D. McIntyre (Ed.), *Characterization of Macromolecular Structure*, National Academy of Sciences Publication 1573, Washington, D.C., 1968.
4. I. M. Kolthoff, and P. J. Elving (Eds.), *Treatise on Analytical Chemistry*, Wiley, New York, 1964.
5. H. F. Mark, N. G. Gaylord and N. M. Bikales (Eds.), *Encyclopedia of Polymer Science and Technology*, Wiley-Interscience, New York.
6. H. Tompa, *Polymer Solutions*, Butterworths, London, 1956.
7. M. Kurata, and W. H. Stockmayer, *Fortschr. Hochpolym.-Forsch.*, **3**, 196 (1963).
8. H. Morawetz, *Macromolecules in Solution*, Wiley-Interscience, New York, 1965.

SPECIFIC REFERENCES

9. P. J. Flory, in Ref. 1, Chap. 10.
10. J. Tompa, in Ref. 6, Chap. 8.
11. M. V. Volkenstein, *Configurational Statistics of Polymeric Chains*, trans. by S. N. Timasheff and M. J. Timasheff, Wiley-Interscience, New York, 1963
12. P. J. Flory, *Statistical Mechanics of Chain Molecules*, Wiley-Interscience, New York, 1969.
13. For a review of the excluded volume effect, see E. F. Casassa, *Ann. Rev. Phys. Chem.*, **11**, 447 (1960).
14. R. P. Smith, *J. Chem. Phys.*, **38**, 1463 (1963).
15. C. Tanford, *Physical Chemistry of Macromolecules*, Wiley, New York, 1961, p. 404.

16. L. H. Peebles, Jr., *Molecular Weight Distributions in Polymers*, Wiley-Interscience, New York, 1971.

17. G. F. Price, "Techniques of End-Group Analysis," Chap. 7 in Ref. 2.

18. M. Hellman, and L. A. Wall, "End-Group Analysis," Chap. 5 in G. M. Kline (Ed.), *Analytical Chemistry of Polymers*, Part III, Wiley-Interscience, New York, 1962.

19. F. W. Billmeyer, Jr., *Textbook of Polymer Science*, 2nd ed., Wiley, New York, 1971, Chap. 3B; see also F. W. Billmeyer, Jr., *J. Polym. Sci.*, **C8**, 161 (1956), *Polym. Eng. Sci.*, **6**, 359 (1966); *Appl. Polym. Symp.* **10**, 1 (1969).

20. R. U. Bonnar, M. Dimbat, and F. H. Stross, *Number-Average Molecular Weights*, Interscience, New York, 1958.

21. C. A. Glover, "Determination of Number-Average Molecular Weights by Ebulliometry," C. N. Reilley and F. W. McLafferty (Eds.), *Advances in Analytical Chemistry and Instrumentation*, Vol. 5, Wiley-Interscience, New York, 1966, pp. 1–67.

22. H. T. Hookway, "Number-Average Molecular Weights by Osmometry," Chap. 3 in Ref. 2.

23. W. R. Krigbaum and R.-J. Roe, "Measurement of Osmotic Pressure," Chap. 79 in Ref. 4, Part I, Vol. 7, 1967.

24. R. L. Parrette, *J, Polym. Sci.*, **15**, 447 (1955).

25. A. T. Williamson, *Proc. Roy. Soc. (London)*, **A195**, 97 (1948).

26. E. J. Newitt and V. Kokle, *J. Polym. Sci.*, **A2**, **4**, 705 (1966).

27. C. A. Glover and J. E. Kirn, *J. Polym. Sci.*, **B3**, 27 (1965).

28. R. M. Fuoss and D. J. Mead, *J. Phys. Chem.*, **47**, 59 (1943).

29. B. H. Zimm and I. Myerson, *J. Am. Chem. Soc.*, **68**, 911 (1946).

30. J. V. Stabin and E. H. Immergut, *J. Polym. Sci.*, **14**, 209 (1954).

31. F. B. Rolfson and H. Coll, *Anal. Chem.*, **36**, 888 (1964).

32. T. R. Reiff and M. J. Yiengst, *J. Lab. Clin. Med.*, **53**, 291 (1959).

33. R. E. Steele, W. E. Walker, and H. C. Ehrmantraut, paper presented at the Pittsburgh Conference on Analytical Chemistry, March, 1963.

34. P. M. Holleran and F. W. Billmeyer, Jr., *J. Polym. Sci.*, **B6**, 137 (1968).

35. V. Kokle, F. W. Billmeyer, Jr., L. T. Muus, and E. J. Newitt, *J. Polym. Sci.*, **62**, 251 (1962).

36. These theories are fully reviewed by H.-G. Elias, in Ref. 3, pp. 28–50.

37. W. J. Badgley and H. Mark, "Osmometry and Viscometry of Polymer Solutions," in *High Molecular Weight Organic Compounds* (*Frontiers in Chemistry*, Vol. VI), R. E. Burk and O. Grummitt (Eds.), Wiley-Interscience, New York, 1949, pp. 75–113.

38. G. Gee, *Trans. Faraday Soc.*, **40**, 261 (1944); data of A. Dobry, *J. Chim. Phys.*, **32**, 50 (1935).

39. R. Pasternak, P. Brady, and H. Ehrmantraut, paper presented at the ACHEMA 1961, Frankfurt-am-Main, 1961.

40. W. Simon, et al., *Chimia*, **14**, 301 (1960); *Microchem. J. (Symp. Ser.)*, **2**, 1069 (1962); *Tetrahedron*, **19**, 949 (1963); *Hevl. Chim. Acta*, **47**, 515 (1964); *Microchem. J.*, **10**, 495 (1966); *Helv. Chim. Acta*, **50**, 2193 (1967); *Anal. Chem.*, **41**, 90 (1969)

41. F. W. Billmeyer, Jr., and V. Kokle, *J. Am. Chem. Soc.*, **86**, 3544 (1964).

42. Chap. 2, Section F.5.

43. F. W. Billmeyer, Jr., "Principles of Light Scattering," Chap. 56 in Ref. 4, Part I, Vol. 5, 1964.

44. V. E. Eskin, *Sov. Phys. Usp.*, **7**, 270 (1964).

45. J. P. Kratohvil, *Anal. Chem.* **36**, 458R (1964); 38, 517R (1966).

46. D. McIntyre and F. Gornick (Eds.), *Light Scattering from Dilute Polymer Solutions*, Gordon and Breach, New York, 1964.

47. F. W. Peaker, "Light-Scattering Techniques," in Ref. 2, Chap. 5.

48. Lord Rayleigh, *Phil. Mag.*, [4], **41**, 107, 224, 447 (1871).

49. A. Einstein, *Ann. Phys.*, **33**, 1275 (1910); M. S. Smoluchowski, *Ann. Phys.*, **25**, 205 (1908); *Phil. Mag.* [6], **23**, 165 (1912).

50. P. Debye, *J. Appl. Phys.*, **15**, 338 (1944); *J. Phys. Coll. Chem.*, **51**, 18 (1947).

51. B. H. Zimm, *J. Chem. Phys.*, **16**, 1093, 1099 (1948).

52. F. W. Billmeyer, Jr., and C. B. de Than, *J. Am. Chem. Soc.*, **77**, 4763 (1955).

53. F. J. Baum and F. W. Billmeyer, Jr., *J. Opt. Soc. Am.*, **51**, 452 (1961).

54. B. A. Brice, M. Halwer, and R. Speiser, *J. Opt. Soc. Am.*, **40**, 768 (1950).

55. P. P. Debye, *J. Chem. Phys.*, **17**, 392 (1946).

56. D. W. Ovenall and F. W. Peaker, *Makromol. Chem.*, **33**, 222 (1959).

57. J. P. Kratohvil et al., *J. Polym. Sci.*, **57**, 59 (1962); *J. Polym. Sci.*, *A*, **2**, 303 (1964); *J. Colloid Sci.*, **20**, 875 (1965); *J. Colloid Interface Sci.*, **21**, 498 (1966).

58. H. C. van de Hulst, *Light Scattering by Small Particles*, Wiley, New York, 1957.

59. P. J. Livesey and F. W. Billmeyer, Jr., *J. Colloid Interface Sci.*, **30**, 447 (1969); see also F. W. Billmeyer, Jr., in Ref. 43.

60. M. Kerker, *The Scattering of Light and Other Electromagnetic Radiation*, Academic, New York, 1969.

61. E. T. Adams, Jr., "Molecular Weights and Molecular–Weight Distributions from Sedimentation-Equilibrium Experiments," in Ref. 3, pp. 84–142.

62. R. L. Baldwin and K. E. van Holde, *Fortschr. Hochpolym.-Forsch.*, **1**, 451 (1960).

63. H. Fujita, *Mathematical Theory of Sedimentation Analysis*, Academic, New York, 1962.

64. H. K. Schachman, *Ultracentrifugation in Biochemistry*, Academic, New York, 1959.

65. K. E. van Holde, "Measurement of Sedimentation," in Ref. 4, Part I, Vol. 7, 1967, Chap. 80.

66. J. W. Williams (Ed.), *Ultracentrifugational Analysis in Theory and Experiment*, Academic, New York, 1963.

67. R. C. Williams, Jr., and D. A. Yphantis, "Ultracentrifugation," in Ref. 5, Vol. 14, 1971, pp. 97–116.

68. Data taken in large part from T. Kotaka and H. Inagaki, *Bull. Inst. Chem. Res. (Kyoto Univ.)*, **42**, 176 (1964).

69. N. E. Weston and F. W. Billmeyer, Jr., (a) *J. Phys. Chem.*, **67**, 2728 (1963); (b) unpublished results.

70. H.-G. Elias, *Angew. Chem.*, **73**, 209 (1961).

71. H. Fujita, H. Inagaki, T. Kotaka, and H. Utiyama, *J. Phys. Chem.*, **66**, 4 (1962).

72. H. Inagaki and S. Kawai, *Makromol. Chem.*, **79**, 42 (1964).

73. Y. Toyoshima and H. Fujita, *J. Phys. Chem.*, **68**, 1378 (1964).

74. Data from the National Bureau of Standards, Washington, D.C.

75. J. W. Lyons, "Measurement of Viscosity," in Ref. 4, Part I, Vol. 7, 1967, Chap. 83.

76. G. Meyerhoff, *Fortschr. Hochpolym.-Forsch.*, **3**, 59 (1961).

77. W. R. Moore, "Viscosities of Dilute Polymer Solutions," *Progress in Polymer Science*, Vol. 1, A. D. Jenkins (Ed.), Pergamon, New York, 1967, Chap. 1.

78. J. B. Kinsinger, "Viscometry," in Ref. 5, Vol. 14, 1971, pp. 717–740.

79. L. H. Cragg, *J. Colloid Sci.*, **1**, 261 (1946); but see also *J. Polym. Sci.*, **8**, 257 (1952).

80. H. Staudinger and W. Heuer, *Chem. Ber.*, **63**, 222 (1930).

81. M. L. Huggins, *J. Am. Chem. Soc.*, **64**, 2716 (1942).

82. M. Kurata, M. Iwawa, and K. Kamada, "Viscosity—Molecular Weight Relationships and Unperturbed Dimensions of Linear Chain Molecules," in *Polymer Handbook*, J. Brandrup and E. H. Immergut (Eds.), Wiley-Interscience, New York, 1966, pp. IV-1–IV-72.

83. M. O. de la Cuesta and F. W. Billmeyer, Jr., *J. Polym. Sci.*, **A1**, 1721 (1963).

84. A. L. Geddes and R. B. Pontius, "Determination of Diffusivity," in *Technique of Organic Chemistry*, A. Weissberger (Ed.), 3rd ed., Part 2, Wiley-Interscience, New York, 1959, Chap. 16.

85. T. Svedberg and K. O. Pederson, *The Ultracentrifuge*, Clarendon Press, Oxford, 1940; Johnson Reprint Corp., New York, 1959.

86. L. Mandelkern and P. J. Flory, *J. Chem. Phys.*, **20**, 212, 1392 (1952).

87. P. J. Flory, in Ref. 1, Chap. 14.

88. P. J. Flory, *J. Chem. Phys.*, **17**, 303 (1949); P. J. Flory and T. G. Fox, *J. Polym. Sci.*, **5**, 745 (1950), *J. Am. Chem. Soc.*, **73**, 1904 (1951); T. G. Fox, J. C. Fox, and P. J. Flory, *J. Am. Chem. Soc.*, **73**, 1901 (1951).

89. A. Einstein, *Ann. Physik.*, [4], **19**, 289 (1906); **34**, 591 (1911).

90. B. H. Zimm and W. H. Stockmayer, *J. Chem. Phys.*, **17**, 1301 (1949).

91. W. H. Stockmayer and M. Fixman, *Ann. N.Y. Acad. Sci.*, **57**, 334 (1953); B. H. Zimm and R. W. Kilb, *J. Polym. Sci.*, **37**, 19 (1959).

92. W. W. Graessley, "Detection and Measurement of Branching in Polymers," in Ref. 3, pp. 371–388.

93. K. H. Altgelt, "Theory and Mechanics of Gel Permeation Chromatography," in *Advances in Chromatography*, Vol. 7, J. C. Giddings and R. A. Keller (Eds.), Dekker, New York, 1968.

94. M. J. R. Cantow (Ed.), *Polymer Fractionation*, Academic, New York, 1967.

95. J. Cazes, "Topics in Chemical Instrumentation. XXIX. Gel Permeation Chromatography," *J. Chem. Educ.*, **43**, A567, A625 (1966); "Current Trends in Gel Permeation Chromatography," *J. Chem. Educ.*, **47**, A461, A505 (1970).

96. G. M. Guzman, "Fractionation of High Polymers," in *Progress in High Polymers*, Vol. 1, J. C. Robb and F. W. Peaker (Eds.), Academic, New York, 1961, pp. 113–183.

97. R. W. Hall, "The Fractionation of High Polymers," in Ref. 2, Chap. 2.

98. J. F. Johnson, M. J. R. Cantow, and R. S. Porter, "Fractionation," in Ref. 5, Vol. 7, 1967, pp. 231–260.

99. V. Kokle and F. W. Billmeyer, Jr., *J. Polym. Sci.*, **C8**, 217 (1965).

100. P. J. Flory, in Ref. 1, p. 562; data of G. V. Schulz, *Z. Phys. Chem.*, **B46**, 137 (1940); **B47**, 155 (1940).

101. A. S. Kenyon, I. O. Salyer, J. E. Kurz, and D. R. Brown, *J. Polym. Sci.*, **C8**, 205 (1965).

102. A. S. Kenyon and I. O. Salyer, *J. Polym. Sci.*, **43**, 427 (1960).

103. R. T. Traskos, N. S. Schneider, and A. S. Hoffman, *J. Appl. Polym. Sci.*, **12**, 509 (1968).

104. C. A. Baker and R. J. P. Williams, *J. Chem. Soc.*, **1956**, 2352.

105. L. E. Maley, *J. Polym. Sci.*, **C8**, 253 (1965).

106. J. C. Moore, *J. Polym. Sci.*, **A2**, 835 (1964).

107. F. W. Billmeyer, Jr., et al., *J. Chromatog.*, **34**, 316, 322 (1968); *Anal. Chem.*, **41**, 876 (1969); *J. Chromatogr.*, **42**, 399 (1970); *Separation Sci.*, **5**, 291 (1970).

108. J. H. Duerksen and A. E. Hamilec, *J. Polym. Sci. C*, **21**, 83 (1968)

109. Waters Associates, Framingham, Mass.

110. P. Crouzet, F. Fine, and P. Magnin, *J. Appl. Polym. Sci.*, **13**, 205 (1969).

111. J. A. May, Jr., and W. B. Smith, *J. Phys. Chem.*, **72**, 216 (1968).

112. P. J. Flory, *Discuss. Faraday Soc.*, **49**, 7 (1970).

113. D. K. Carpenter, "Solution Properties," in Ref. 5, Vol. 12, 1970, pp. 627–678.

114. P. J. Flory, in Ref. 1, Chaps. 12, 13.

115. P. J. Flory, *J. Chem. Phys.*, **10**, 51 (1942).

116. M. L. Huggins, *J. Phys. Chem.*, **46**, 151 (1942).

117. P. J. Flory and W. R. Krigbaum, *J. Chem. Phys.*, **18**, 1086 (1950).

118. P. J. Flory, et al., *Macromolecules*, **1**, 279 (1969); *Trans. Faraday Soc.*, **64**, 2035, 2053, 2061, 2066 (1968). See also D. K. Carpenter, in Ref. 113.

119. D. Patterson, *Rubber. Chem. Technol.*, **40**, 1 (1967); *Macromolecules*, **2**, 672 (1969).

DISCUSSION QUESTIONS AND PROBLEMS

1. Discuss the value of knowledge of the molecular weight and distribution of a polymer to the plastics engineer.

2. Which methods would you use to obtain this information on a routine basis, as in process control? Why?

3. Which methods would you use to obtain this information for a new polymer type not previously studied? Why?

4. Of what value is the ability to make measurements by such absolute methods as membrane osmometry or light scattering? Review your answers to the previous questions assuming that these absolute methods were not available, and discuss.

5. You wish to determine $\overline{M}_n$ for a polymer, but find it too high to be measured by vapor-phase osmometry. Yet when using membrane osmometry, there is significant diffusion through the membrane. Outline two different ways of obtaining the desired information, and discuss their relative merits.

6. In 1974 Paul J. Flory received the Nobel prize in chemistry, in part for his calculations of the unperturbed dimensions of polymer molecules. Discuss the practical value of these calculated results to the scientist responsible for the molecular characterization of polymers.

CHAPTER 5

Polymer Morphology

P. H. GEIL
Division of Macromolecular Science
Case Western Reserve University
Cleveland, Ohio

A. MOLECULAR PACKING

As has been discussed in Chapter 1, it is known that many polymers, despite their long chains, can crystallize to at least some extent (1). Both rubber and polyoxymethylene, the macromolecules that served as a basis for much of the early research that led to a recognition of the long-chain nature of polymers, crystallize. X-ray diffraction patterns from these polymers, which since have been found to be typical of many other polymers also, indicated that long-range order existed in the polymers, that crystalline regions at least several hundred angstroms in size were present (Fig. 1). In addition, the patterns showed that some regions of the polymer did not have long-range order (> ca. 20 Å), but rather gave rise to diffraction similar to that obtained from molten polymer. The presence of both liquidlike or amorphous and crystal-

Figure 1   Flat plate X-ray diffraction "powder pattern" of an unoriented sample of polyethylene.

linelike diffraction from the same sample has given rise to the term semi-crystalline and led to the simplifying assumption, in many treatments of polymer physical properties, that "semicrystalline" polymers consist of two distinct phases, crystalline and amorphous (2). Recently, however, as a result of detailed studies of polymer fine structure or morphology, it appears to be more desirable in some cases to consider them as one-phase systems of varying degrees of order, from sample to sample, and from place to place within a sample (3). Following consideration of the ways an individual molecule packs within a crystal, we consider this fine structure, that is, the size, shape, and interrelationship of the crystals and amorphous regions, and its effect on physical properties.

1. Criteria for Crystallinity

To obtain X-ray diffraction patterns, such as those in Fig. 1, requires that there be a regular, perfect, repeating arrangement of individual atoms from portions of at least several molecules (for typical molecular weights) in all three dimensions for at least several hundred angstroms and possibly even further in one or more directions if some relaxation of the requirements of perfection is permitted; that is, there must exist a unit cell for the polymer. The existence of a unit cell places a critical requirement on the chemical structure of the individual polymer chain; its chemical repeat unit must have a regular configuration, which is identical from one unit to another, for at least those sequences that are present in a given crystal.

In addition, within the crystal, it must assume a regular conformation. Not only must the type and succession of atoms and bonds be the same (a regular configuration), but their three-dimensional arrangement (a regular conformation), as produced by rotation about the bonds, must also be regular (see Fig. 4). The former can be affected only by chemical means; if irregular, chemical bonds would have to be broken and reformed to permit a regular configuration. On the other hand, physical treatments, such as annealing or stress, are sufficient to permit changes in the conformation of a molecule.

2. Unit Cell, Helical Structure

Polymers crystallize with a wide variety of unit cells (4). Common to all of them, however, is a basic anisotropy (properties differ along different axes) owing to the difference in interatomic forces along the bonds within a chain and between atoms of different chains; thus no polymer crystallizes with a cubic unit cell although all other types are found. This anisotropy leads to unique physical properties obtainable by orienting polymer molecules so that their axes are oriented in the direction of maximum expected stress. In the next few paragraphs we consider the unit cells of polyethylene, polypropylene, and their copolymers. Within this system we find most of the features representative of the molecular packing in unit cells of polymers in general.

An X-ray diffractometer scan of a sample of linear polyethylene is shown in Fig. 2 (in comparison with the powder pattern in Fig. 1 a diffractometer scan corresponds to a plot of the intensity along one of the radii as a function of angle of diffraction θ, or its equivalent interplanar spacing d (in Å) from Bragg's law ($n\lambda = 2d \sin \theta$, where λ = wavelength and n = order of reflection)). Indicated on this pattern are the portions of the scattering attributed to crystalline and amorphous regions in the polymer and various background contributions. From patterns such as this and, in particular, fiber patterns (X-ray diffraction patterns using a drawn fiber of the polymer as a sample, the advantage being that the molecular axes are aligned in the draw direction, rather than all axes being randomly oriented as in a normal sample) it was determined that segments of the polyethylene molecules are packed in the crystal in a manner identical to that of the orthorhombic unit cell previously known for the short-chain paraffins.

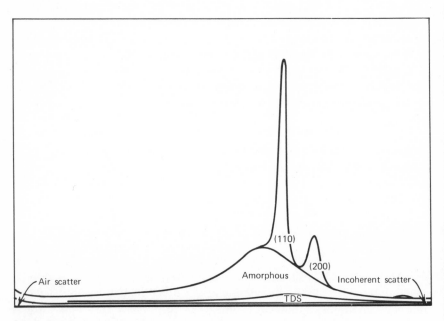

Figure 2   X-ray diffraction scan of a polyethylene sample. The thermal diffuse scatter (TDS) results from thermal motion of the atoms in the crystalline regions.

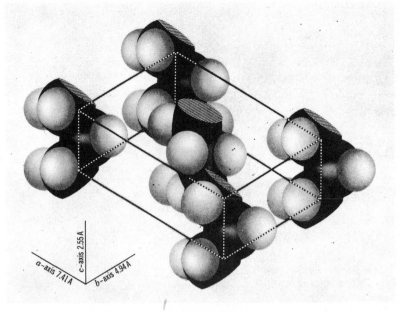

Figure 3   Diagram of unit cell of polyethylene.

In the polyethylene unit cell (Fig. 3) the molecule is fully extended in the planar zigzag conformation. In the paraffins, the c axis repeat distance is measured in terms of the length of the molecule, whereas for polyethylene, the c axis repeat distance is generally taken as one zigzag. It is found that, in most cases, molecular segments only on the order of 100 Å in length, rather than the entire molecule, have a regular conformation and are incorporated into a given crystal. Subsequent segments from the same molecule may be incorporated in the same or different crystals, the intersegmental portions of the molecule being of variable length and having a different conformation from the portions within the crystal. The explanation of this feature in terms of the fine structure of a polymer is the concern of much of the remainder of this chapter.

Polypropylene can be pictured as having a linear polyethylene backbone on which an H atom on every other C atom is replaced by a CH_3 (methyl) group. In contrast to polyethylene, which has an inherent regular configuration, the position of the CH_3 group (i.e., which one of the two H atoms on a given C atom it replaces) may yield a polymer with either a regular or irregular configuration. If irregular (atactic), it cannot crystallize. However, if it is either wholly isotactic or syndiotactic, or if sufficiently long blocks of either stereoregular form exist, it can and does crystallize. We consider only the isotactic form, the form in which the "same" H atom is replaced on each substituted C atom (Chapter 2, Fig. 8a). Because of the bulky size of the CH_3 group in comparison with an H atom, the backbone cannot exist in the planar zigzag form; it must twist. For polypropylene it is found that lowest energy state is attained for a regular twist such that each chemical repeat unit is rotated by $120°$; that is, a 3/1 helix (three chemical repeat units per turn) is formed which packs into a monoclinic unit cell (Fig. 4). Crystal structures can be built by packing any regular helix and, in other stereoregular polymers,

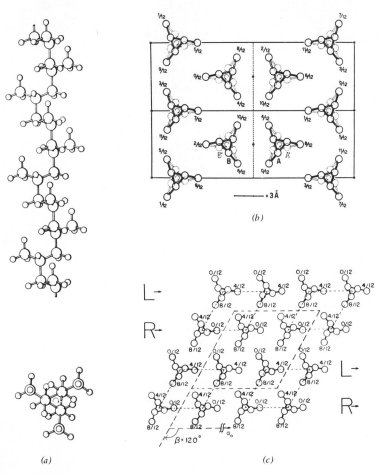

(a)   (b)

(c)

Figure 4   Diagram of (a) single molecule helix, and unit cells (b) monoclinic and (c) hexagonal of polypropylene.

a wide variety of helices are found, the helix formed being that yielding the best inter- and intramolecular packing of the side chains. In a number of polymers two or more different helices are found, most frequently being stable at different temperatures and resulting in different unit cells. In polybutene ($-CH_2CH_3$ side group rather than CH_3), for instance, an 11/3 helix forms during crystallization from the melt, which gradually transforms to a 3/1 helix if the sample is left to stand at room temperature. The resulting change in crystal structure causes a change in dimensions and properties which creates difficulties in commercializing the polymer (Chapter 6, Section D.2.)

Even with the same conformation (and configuration), polymers as well as low molecular weight materials can be packed into two or more different unit cells. Thus polypropylene can be crystallized into either the more usual monoclinic unit cell or, within a limited temperature range and particularly in the presence of certain foreign nuclei, into a hexagonal unit cell. In both cases the molecules are twisted into a 3/1 helix, the difference being the relative positions and hand (right or left twist) of neighboring helices. Polypropylene also can be quenched to a liquid crystal form, one in which each molecular segment is appropriately twisted but neighboring segments are apparently only aligned rather than being packed well in a lattice. Upon warming above ca. 40°C molecular motion becomes sufficient to permit conversion of this form to the monoclinic form. These basic morphological states have different properties, yet all are based on the same polymer.

Polyethylene can also exist in two crystal forms, monoclinic as well as orthorhombic. The monoclinic form, in which all the planar zigzags have the same orientation, can be formed from the orthorhombic by compressing it normal to the molecular axes. It is metastable at room temperature, melting about 40°C below the orthorhombic form.

In addition to the lack of a regular configuration caused by a nonregular tacticity, as mentioned above, configurational defects can result from branches, cross-links, and end groups (if large or numerous). All these defects result in disturbances within the crystal, if the associated segment is accepted, or in many cases in rejection from the crystal. For instance, if a small amount of propylene is copolymerized with ethylene, the resulting molecules have corresponding numbers of CH_3 branches randomly located along the chain. It has been found that the CH_3 group is small enough that a limited number ($< 13\%$) can be accepted in the polyethylene lattice, the lattice expanding slightly and proportionately in the a axis direction (5). If more than 13% is copolymerized, however, all long-range order is lost and an apparently homogeneous amorphous polymer results. A mixture of corresponding amounts of the homopolymers, on the other hand, is incompatible; the polymers segregate upon solidifying, each crystallizing into microscopic domains of pure polymer.

3. Amorphous Polymers

Polymer melts and wholly amorphous solid polymers (such as polystyrene and polymethyl methacrylate) are generally assumed to consist of randomly coiled and entangled chains (2). Recently, however, in the light of density considerations (6) and electron microscopic observations (7), there have been suggestions that order to varying degrees exists in the solid amorphous polymers and, very likely, in polymer melts also. One pictures small (< 100 Å) clusters of aligned molecular segments. These clusters would be nonpermanent in the molten state, continually forming and disappearing as a result of thermal motion, but would become permanent as the sample was cooled to a temperature below that at which segmental motion occurs within experimental times (the glass-transition temperature; see Chapter 6). This structure is thus expected in polymers with an irregular configuration and in quenched samples of polymers which have difficulty in reaching a regular conformation ("stiff chains"). In the case of the latter polymers, heating at temperatures between the glass-transition temperature and the melting point results in molecular motion and development of crystallinity.

B. POLYMER CRYSTALLIZATION

When a low molecular weight material, such as a metal, crystallizes from the melt, nucleation occurs at various points, from each of which a crystal or grain grows. Grains of zinc, for instance, can easily be seen on a piece of galvanized steel. In recent years control of grain crystal structure, size, shape perfection, and arrangement, that is, the morphology of the sample, has led to the development of metals and ceramics with significantly improved properties. Many of these advances have resulted from basic studies of the properties of individual single crystals.

Likewise, when a molten crystallizable polymer is cooled, crystallization spreads out from individual nuclei. However, instead of individual grains, a considerably more complex structure, termed a spherulite because of its gross spherical symmetry, develops from each nucleus.

Knowledge and control of the morphology of these structures, other than for their size, in order to optimize polymer physical properties is just beginning to be attempted. As we shall see, this is primarily because polymer crystals in objects of commercial interest have at least one and sometimes all dimensions restricted to the order of 100 Å, requiring electron microscopy for their observation and unique techniques for examination of their properties. In addition, it is only since 1957 that individual single crystals of polymers have been recognized and become available for study (9). We consider first the growth and structure of the single crystals and then the development and structure of the more complex spherulites. Following a brief discussion of the origin of the restricted size of polymer crystals, the remainder of the chapter is concerned with the effect of stress on a crystalline polymer and the morphology of an oriented polymer such as a fiber or film.

C. SINGLE CRYSTALS

1. Growth and Thickness

In order to obtain as perfect and large single crystals as possible one would desire to (1) keep the number of nuclei low; (2) keep the growth rate slow and molecular motion on the crystal face high, to permit the rearrangement of the molecules on the growth face if defects are introduced; and (3) in the case of polymers in particular, reduce the number of interactions between molecules or portions thereof still in the liquid state. This can be done most easily by crystallizing from dilute solution at relatively high temperatures, although for some polymers, even larger single crystals can be grown from the melt, on substrates at high temperature, than from solution. The lateral size of a polymer single crystal grown from solution appears to be restricted by hydrodynamic turbulence, which leads to defects known as screw dislocations (see below).

Figure 5 is an electron micrograph of crystals of linear polyethylene obtained by cooling a dilute solution (0.1 % in tetrachloroethylene). These crystals, which are $\sim 10\ \mu$ in lateral size, are not as large as can be grown (crystals as large as $100\ \mu$ have been grown), but they are relatively perfect for polymer crystals. One notes first that they are lamellar (platelike or disklike); the lamellae in this preparation were measured by electron microscopy and X-ray diffraction to average 104 Å thick. Crystals with larger lateral dimensions are found to be multilamellar.

In order to obtain these small but relatively perfect crystals it was necessary to increase the number of nuclei over those normally present in the solution; this was done by adding to the polymer solution a number of previously prepared crystals with the solution temperature such that the crystals almost dissolved. Upon cooling the remnants of these crystals served as seed nuclei for the growth of the crystals shown (10).

The temperature (and thus the rate of crystal growth which increases with decreasing temperature) of crystallization can be varied over a limited range for a given solvent. The thickness of the lamellar crystals has been found for most, but not all, polymers to increase with increasing temperatures. The maximum crystallization temperature for a homogeneous solution appears

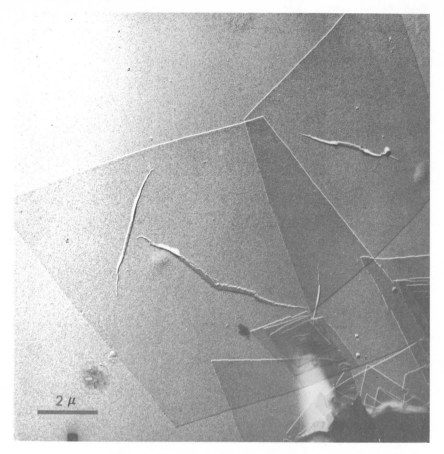

Figure 5   Single crystals of linear polyethylene.

to be determined by nucleation. As is seen later the size of the nucleus required to initiate crystal growth increases with increasing temperature; the high temperature limit appears to be determined by the maximum size of the nucleus which can be formed from a single molecule. At lower temperatures only a portion of a molecule is sufficient to form a stable nucleus. The lower temperature limit for crystallization is usually determined by the experimental technique, crystallization occurring so rapidly at intermediate temperatures that it is not possible to cool the solution to the desired temperature without the sample already having crystallized. For polyethylene crystallized from xylene, lamellae varying from a thickness of 92 Å at 50°C to 150 Å at 90°C have been reported; for other solvents the range is from about 60 to 210 Å.

When crystallized isothermally the lamellar thickness is remarkably uniform; small-angle X-ray measurements indicate that 95 % of the lamellae in a given sample may be within 5 % of the same thickness. If the temperature is changed during growth, the thickness of the new material added changes correspondingly; thus if growth occurs as a solution is cooled, the edges of the crystals are thinner than the central portion.

The lamellar nature of a polymer has been found to be fundamental. Growth of the crystal normal to the lamellar surface occurs not by the gradual addition of new molecules to the surface, as would occur in low molecular weight materials whose crystals are more or less isotropic, but rather by the formation of additional lamellae of the same thickness as the basal lamella. For low molecular weight materials the presence of screw dislocations can greatly increase the crystal growth rate. For polymer crystals it is found to be almost essential to have a screw dislocation present to permit growth of these additional lamellae. One source of the dislocation is tears in the growth face of the basal lamella produced by the hydrodynamic forces referred to previously. If a tear is formed, lateral growth can occur on each face of the tear, the now overlapping layers yielding a spiral growth of new lamellae on each side of the original lamella. The region at the apex of the tear has the characteristics of a screw dislocation. Figure 6 shows several spiral growths on a single crystal of nylon-6 crystallized from glycerin. In some cases the lamellae of the opposite spiral can be seen on the underside of the basal lamella.

With increasing growth rate (decreasing temperature) more and more defects become incorporated and remain in the crystal; in particular the planar growth faces observed in Fig. 5 become irregular, serrations and re-entrant faces develop, and dendritic or treelike growth occurs (Fig. 7). The apices of the reentrant faces serve as sources of spiral growths. While suspended in the solvent the lamellae in the various spiral growths splay; that is, they are separated from each other except in the vicinity of the screw dislocation. There is little cohesion between the faces of adjacent lamellae.

2. Folding

Polymer molecules are several thousand angstroms long; the crystals are tens of thousands of angstroms in lateral dimensions but only 100 Å thick. We inquire next concerning the arrangement of the molecules within the crystal, that is, the orientation of the unit cell with respect to the faces of the crystal. For this purpose diffraction patterns from an individual crystal are needed. Although too small to be examined using standard X-ray techniques, most electron microscopes are so constructed that electron diffraction patterns can be obtained (by adjusting the focal length of the lenses) from

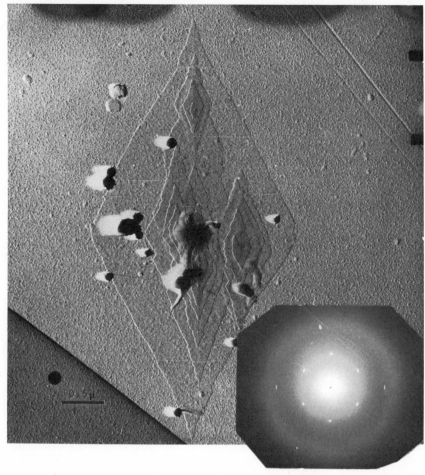

Figure 6   Single crystal of nylon-6 crystallized from glycerin (with diffraction pattern).

selected areas as small as 1 μ^2. Using suitable conditions to sufficiently reduce electron beam irradiation damage (which induces cross-linking and/or degradation leading to destruction of crystallinity) and heating, one can obtain electron diffraction patterns from individual crystals or portions thereof (Fig. 8). Besides confirming that the lamellar structures are single crystals, these diffraction patterns have shown the most unique feature of polymer crystals; the molecules several thousand angstroms long are oriented normal (or in some polymers at an acute angle) to the lamellar surfaces. A molecule can remain in crystalline order for only ~ 100 Å before it reaches

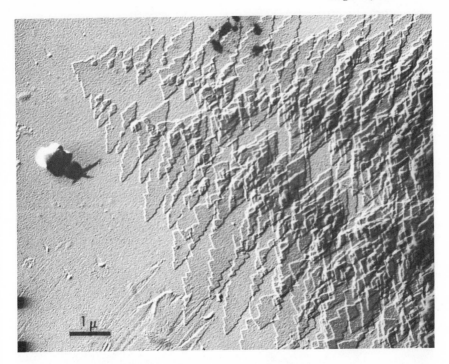

Figure 7   Dendritic crystal of polyethylene.

the surface of the crystal. In the single-lamella crystals it then must fold back on itself and reenter the crystal at some other point (11). As is seen below, this feature of chain folding is characteristic of crystallizable polymers, even in multilamellar crystals and in spherulitic structures formed during crystallization from the melt.

A problem of major concern at present is the nature of the surface of the crystals. Not only are there at least three different models for the conformation of the fold, but one also needs to consider whether the ends of the molecules are included within the crystal, forming a defect, or whether they are excluded from the crystal, forming cilia. Infrared studies of Keller and Priest (12) suggest that approximately 90% of the ends are excluded from polyethylene crystals; Krimm and Bank (13) have shown that these cilia can become included in neighboring crystals during annealing following sedimentation.

The question of the conformation of the fold (and the effect of the fold on the shape of the crystal) can, at present, be answered with reasonable confidence only for the case of solution-grown single crystals of polyethylene,

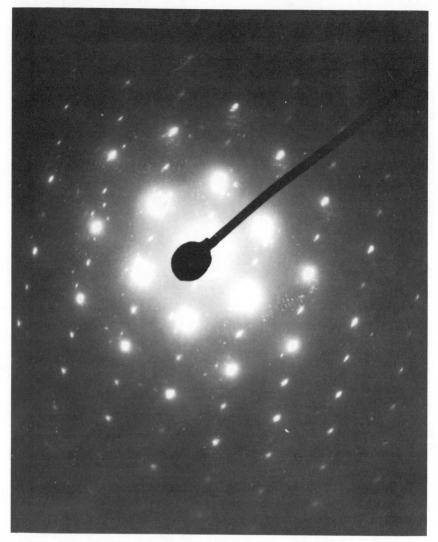

Figure 8   Electron diffraction from a polyethylene single crystal.

primarily because of the detailed studies of Keller and co-workers (14). These findings, however, are expected to apply also to solution-grown crystals of other polymers. If one reexamines Fig. 5, one notes that the two crystals are not single lamellae; one has a spiral growth, but each has a pleat of two extra layers along one of its diagonals. These pleats were the first indication that the crystals were not simple flat lamellae; it has been shown that instead polyethylene crystals, while in suspension, may be either pyramidal or corrugated. Figure 9 shows several crystals in which the corrugated shape has been at least partially retained by sedimenting on a soft substrate. The central portion of these crystals, or the entire crystal in the other cases, has the pyramidal morphology; the pyramids are hollow, the faces being 100-Å-thick lamellae. The crystal is seen to be divided into four sectors, each sloping away from the apex. It was postulated that in each sector the folds are made parallel to the sector's growth face. The corrugated texture is related to the pyramidal, the slope reversing along a series of lines [(310) planes of the lattice] to form the corrugations.

Electron microscopic studies of the results of deforming individual polyethylene crystals confirmed the idea that the folds are made in planes parallel to the growth faces. They thus change direction at the sector boundaries. (Further details of the mechanism of deformation are described below.) Within the crystal the only forces between any two adjacent segments

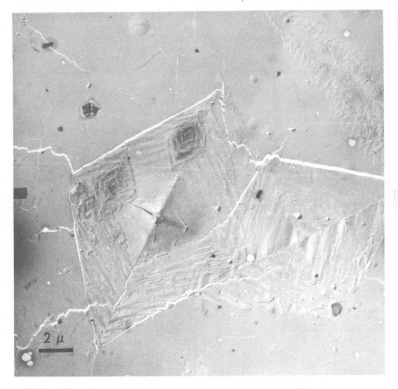

Figure 9   Single crystals of polyethylene deposited and shadowed on a glycerin substrate.

are low-strength van der Waals bonds. However, those segments connected by folds are also connected by the strong covalent bonds in the fold. Thus when a polyethylene crystal is deformed, cracks developing in a sector at an angle to its growth face would have to break covalent bonds; instead the molecules are pulled out across the cracks in the form of fibrils. No fibrils are formed when the cracks are parallel to the growth face. Thus it is concluded that the planes parallel to the growth faces (the 110 planes, it is known, for the polyethylene crystal in Fig. 5) are the fold planes, the planes into which the successive segments of a given molecule are folded.

The pyramidal morphology has been suggested to result from the packing of the folds on the surface of the crystal. Keller's results (14), we believe, suggest that the molecular segments lie down on the lateral face of the crystal with an irregular fold period (segment length between folds) but that the associated rough surface is a high energy state. During further growth, or perhaps even after growth has ceased, segmental motion within the crystal is sufficient to permit a smoothing of the surface, this smoothing occurring by the development of a uniform fold period. However, in polyethylene at least, the fold conformation is such that the tightest fold connecting two adjacent segments is somewhat larger than the cross-section of the segments themselves; thus if all the folds were on a smooth surface normal to the molecular axis direction, considerable surface stress would occur. Instead the lowest energy state turns out to be one in which the smooth surface lies at an angle to the molecular axes; adjacent folds are uniformly displaced or staggered. Keller has been able to show that in one particular preparation of crystals the folds are staggered both within a fold plane and between fold planes by definite amounts, the displacements involving a half or a full repeat distance along the c axis. A model of a presumed conformation for the folds and the displacement of the fold planes, but not the displacement of the folds within a fold plane, for a small portion of a crystal is shown in Fig. 10. In crystals grown at other temperatures the fold displacement can be different.

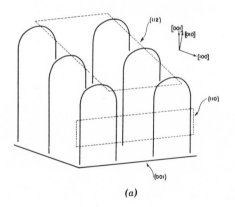

(a)

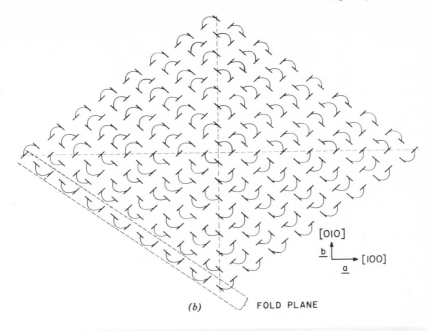

[OIO]

b

a

[IOO]

(b)   FOLD PLANE

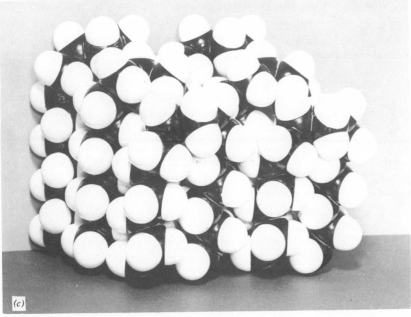

(c)

Figure 10 Diagrams (*a, b*) and photograph (*c*) of molecular models shadowing a possible fold conformation and packing of adjacent folds. In this model adjacent folds planes are displaded by $\frac{1}{2}$ the *c* axis repeat distance.

222 of the chain crystal

The pyramidal morphology implies that the lowest energy state for a folded chain crystal is one in which the fold period is uniform and in which the molecule reenters the crystal in the next lattice position in the fold plane. This has been described as the regular, adjacent reentry model. The presumed initial form of the crystal, in which the fold period is nonuniform, is described as the irregular, adjacent reentry model. Although it would appear difficult to explain the pyramidal crystals in terms of this model, Blackadder and Roberts (15) have postulated a related model in which the fold and a small portion of the attached segments lie at an angle to the main portion of the segments. The lateral spacing of the small segments is greater than the lattice spacing in the bulk of the crystal, permitting more flexibility and randomness; that is, an "amorphous" surface layer is formed. This would be in agreement with a number of physical measurements, including density, X-ray diffraction, and thermal measurements, suggesting that polyethylene crystals are only 85% or less crystalline. A third model that has been proposed for a lamellar crystal is the switchboard model. In this model the molecule leaving the surface of the crystal reenters not in the adjacent lattice position but rather at some apparently random position at least several lattice positions removed. For this model, there must thus be a relatively thick, noncrystalline layer on each surface of the crystal. Although this model does not appear to apply to solution grown crystals, particularly to the pyramidal crystals, it has been suggested to apply to crystallization from the melt.

3.   Thermal Properties

The thermal properties of polymer crystals can be studied either individually or as sedimented cakes of crystals. As is the case of all small crystals, the melting point is a function of crystal size. Thus the higher the temperature at which the crystal was grown (and therefore the thicker it is), the higher will be its melting point. A crystal of polyethylene with infinite thickness (and lateral size) would melt at a temperature of about 145°C; typical solution-grown folded chain polyethylene crystals melt at temperatures below 130°C. However, the melting point can be measured only if the heating rate is rapid enough; even at such relatively fast rates as 10°C/min it is found that the crystals thicken before melting. The amount of thickening, and thus the finally determined melting point, depends on the heating rate. The origin of this annealing effect lies in the energy required to form a fold. The most stable form for a polymer crystal would be one in which all the molecules are of the same length and are fully extended. As discussed below, however, those crystals that are found, for crystallization at a given temperature, are those that are nucleated and grow most rapidly; in most cases the chains are folded rather than extended. However, once crystallized, if the sample is heated to a suitably high temperature, molecular motion of a type similar to that occurring during smoothing of the fold surface occurs and the crystal thickens. This motion can occur at temperatures above the so-called α transition temperature (Chapter 6, Section A.6) of the polymer, $\sim 110°C$ for polyethylene (the temperature depends on both the crystal thickness and perfection).

Figure 11 is an electron micrograph of a polyethylene crystal which has been annealed at 120°C for 30 min. The thickness doubled at the same time as the holes developed. If a cake of the crystals is heated in air it is found that the thickness increases logarithmically with time. On the other hand, if the cake is heated rapidly, for instance, by immersion in a liquid bath, the increase in thickness occurs much more rapidly initially and then appears to approach

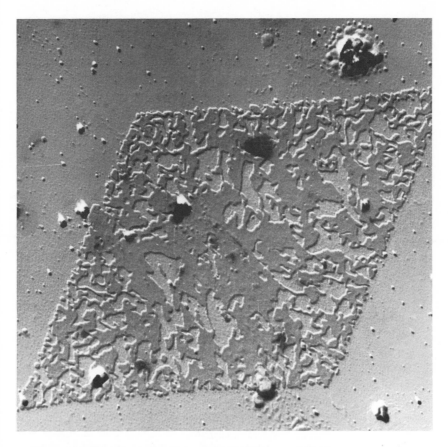

Figure 11   Single crystals of linear polyethylene annealed at 120°C for 30 min.

a limit. In the first case the thickening presumably occurs by molecular motion within the crystal; in the second case it is believed that an actual melting of the crystal occurs followed by recrystallization with a new, larger thickness. Since both features can be occurring simultaneously, caution is needed in interpreting measurements of the thermal characteristics of crystalline polymers. Although the same type of molecular motion is envisioned as giving rise to the thickening process when the crystals are annealed in air and to the smoothing of the surface when the crystals are in the solvent at the crystallization temperature (which is in effect an annealing process also), the latter crystals do not thicken. In fact if heated in the solvent to a higher temperature the crystals, in most cases, dissolve rather than thicken. No explanation for this effect is known at present.

D. SPHERULITES

1. Nucleation and Growth

Under some circumstances, usually slow growth rates in thin films (which reduce the number of nuclei per unit area of the film), one can grow single crystals from the melt in the form of single lamellae or stacks of lamellae originating from an apparent screw dislocation. Under these circumstances, despite the condensed, presumably entangled state of the macromolecules in the melt, the crystals formed appear to be as perfect, with as little connection between neighboring lamellae, as those formed during crystallization from solution. The thicker structures, for instance, fracture easily between the lamellae.

Under more rapid crystallization conditions, approaching those found in commercial practice in which the melt is often quenched, spherulitic crystallization occurs. As indicated previously, nucleation occurs at various points within the sample, usually on heterogeneous (foreign particle) nuclei. For polyethylene the heterogeneous nuclei are effective at supercoolings of 20–25°C, whereas homogeneous nucleation (formation of a nucleus by aggregation of sufficient molecular segments) requires supercoolings of 50–80°C. The heterogeneous nuclei are effective as a result of the adsorption on their surface of molecular segments in crystalline order at temperatures above the melting point. From each of the nuclei formed, regardless of its origin, a spherulite grows (Fig. 12). Although initially having a sheaflike structure (see below), after sufficient radial growth a nearly spherical growth front develops. The spherulites grow until they meet, completely filling the volume. Because the density increases during crystallization the sample shrinks toward each nucleus, leading to weak points and actual voids at the

Figure 12   Spherulites of isotactic polypropylene.

spherulite boundaries if the spherulites are large. Thus one method of toughening a sample is to add heterogeneous nuclei to reduce the spherulite size.

The spherulites of polypropylene in Fig. 12 are of particular interest with respect to understanding the mechanism of growth. As observed here between crossed polaroids, two types are seen. X-ray diffraction has shown that the difference is due to the crystallization of polypropylene in two unit cells. Although growing at the same temperature, all the polymer molecules in the brighter spherulites have crystallized in the hexagonal unit cell, whereas all the polymer in the darker spherulites is in a monoclinic unit cell. By varying the type of heterogeneous nuclei present, either one or the other type can be grown independently; the structure of the nucleus determines the unit cell throughout the entire spherulite. The birefringence pattern observed has been shown to result from an average tangential orientation of the molecular axes. Thus the molecules and unit cells have a different orientation along different radii, in contrast to the uniform orientation of the unit cell throughout a metal grain.

The explanation of the apparent far-reaching effect of the nucleus structure and the molecular orientation is found in electron micrographs of polymer spherulites. The spherulites are composed of aggregates of lamellar crystals, which, in the region beyond the sheaflike center, are radially oriented. As in the case of solution-grown crystals, electron diffraction has shown that the molecular axes are normal, or nearly so, to the lamellar surfaces and thus are (a) tangentially oriented in the spherulite, and (b) presumably folded. The latter conclusion is based on the assumption that most of the molecules return to the same lamellae rather than entering an adjacent one. This is discussed further below.

Observation of electron micrographs of a number of spherulites, of various polymers, shows that the sheaflike central region, which can be several orders of magnitude larger than the nucleus, is composed of a stack of nearly uniformly oriented lamellae. Presumably they all originated at the nucleus; some evidence suggests they result from a spiral growth process. However, as they grow outward from the nucleus they do not remain coplanar, but instead begin to diverge, twist, and branch. When this splaying and branching (which is needed to fill space with radially oriented lamellae) has proceeded sufficiently far, spherical symmetry is approached and the typical spherulitic morphology is observed.

2. Impurity Segregation Model of Growth

An explanation of the mechanism of spherulitic growth, which is also observed under some conditions for low molecular weight materials, has been suggested and tested by Keith and Padden (16). Spherulitic growth is found in impure systems of high melt viscosity. In the case of polymers, "impurities" is a broad term covering molecules of varying degrees of crystallizability. Such factors as tacticity, branches, copolymers, and molecular weight (ends) are of concern. High viscosity, and therefore a low diffusion rate, is inherent in polymers. Keith and Padden define a factor $\delta = D/G$, where G is the growth rate of the crystal–melt interface and D is the diffusion rate of the noncrystallizing molecules. They point out that as a growth face advances into the melt it builds up in front of it a zone of impurity rich melt, the impurities being rejected by the growing crystal. This zone, across which there is a gradient of impurity concentration, can be shown to be of the order of δ thick. Under the conditions of polymer crystallization, owing to the low thermal conductivity of polymers and the release of heat of crystallization, the crystal–melt interface usually is hotter than either the melt or the solid. The combined thermal and impurity effects create a situation in which a planar growth face is not stable; instead the growth face breaks down into a number of projections or cells, each of the order of δ in lateral size. The impurities are segregated in the intercellular regions as the cells continue to grow into the melt. Depending on the type of impurity, it may either remain noncrystalline or crystallize upon subsequent cooling. In a polymer the cells appear to be composed of one or a few lamellar ribbons. It is presumed that when the growth face of a lamella in the sheaf growing out of the central nucleus becomes larger than δ, it will split into two or more lamellae. These lamellae need not remain coplanar as they continue to grow, resulting in (1) a source of a spiral growth at the point of splitting and therefore still further lamellae, (2) the development of so-called small-angle branching in which the lamellar ribbons diverge and spherical symmetry is approached, and (3) retention of the segregated impurities in interlamellar regions.

The texture of the spherulite is thus a function of δ; D and G can be adjusted by controlling the temperature, and the molecular weight and type of impurities. For small δ (fast growth rate and/or slow diffusion rate) a finely textured spherulite forms in which the noncrystalline regions are dispersed throughout the sample. For large δ (high temperature and/or low molecular weight impurities) the lamellae are large and rather perfect, and the noncrystalline material is also aggregated in large regions between the coarse arms of the spherulite and at its boundaries. Figure 13 shows a spherulite of

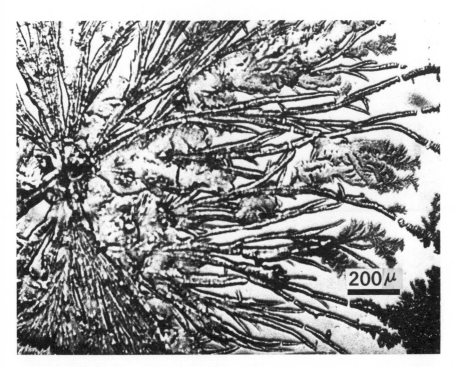

Figure 13   Polypropylene (monoclinic unit cell) spherulite grown in a thin film. Following completion of crystallization and cooling to room temperature the film was washed with benzene to remove most atactic and degraded (low molecular weight) material.

polypropylene which was grown for a period of time under conditions yielding a large δ (high temperature) followed by a period of growth with small δ (low temperature). The majority of the noncrystalline polymer (in this case atactic and degraded polypropylene) was removed by solvent to reveal the crystalline cellular arms. In the electron microscope each arm is seen to be composed of a complex array of lamellae.

Of primary concern in the practical use of a crystalline polymer is its mechanical strength; as described above a crystalline polymer is composed of spherulites, each of which can be looked on as a radial array of lamellae with interspersed noncrystalline regions. A cake of relatively perfect single crystal lamellae, when bent, fractures readily between the lamellae since the fold surfaces have little cohesive strength. On the other hand mats of less perfect lamellae, presumably those containing irregular fold surfaces or "cilia" (molecular ends extending from the surface) are found to be ductile (17). A question of concern is thus the origin of the cohesion of a polymer crystallized from the melt.

It is generally presumed, although direct evidence is lacking, that the surfaces of lamellae crystallized from the melt are more irregular than those crystallized from solution; even the switchboard model cannot be ruled out. However, the same "smoothing forces" should be present during melt crystallization and there is infrared spectroscopic evidence that the conformation of the folds in several different slowly cooled or annealed polymers is the same as in solution-grown crystals (18). Thus an interaction between the fold surfaces, such as an entangling of the irregular folds, would not appear to be sufficient in all cases to give rise to the cohesion. The major difference between crystallization from the melt and from solution is the considerably higher concentration of macromolecules; segments from more than one molecule would be expected to be in contact with even small areas of the growth face at all times. In addition neighboring lamellae may be much closer than in solution, close enough to each other to permit segments of one molecule to be incorporated simultaneously into two different lamellae. Following further crystallization the results would be (1) higher concentration of defects than in solution-grown crystals and (2) the presence of "tie" molecules, molecules that are folded into two different lamellae. The number of defects and tie molecules would be expected to increase with increasing molecular weight and increasing crystallization rates—both of these factors give rise to "tougher" samples.

It has not yet been possible to measure the number of tie molecules and defects in melt crystallized lamellae. Keith, Padden, and Vadimsky (19), however, in their studies of segregation during spherulite growth found considerably larger interlamellar links in samples containing large amounts of noncrystallizable impurity. These links are fibrils up to several hundred angstroms in diameter, extending taut between neighboring arms of a spherulite and across spherulite boundaries. They form in the noncrystalline regions and were revealed, by electron microscopy, following removal of the noncrystalline material. More recently Hase and Geil (20) have shown the presence of interlamellar links in swollen poly-4-methyl-1-pentene (Fig. 14). These links are formed during the swelling, presumably by a drawing out and coalescence of tie molecules and other interlamellar material.

As might be expected the effect of temperature is the same for melt crystallized lamellae as for solution crystallized. Annealing can result in an increase in lamellar thickness. In addition it can produce an apparent increase in crystallinity (crystal perfection?) and a decrease in toughness. All these effects result from molecular motion within the crystal, and can occur during growth as well as upon subsequent annealing. As a crystal grows it is found to also thicken; that is, while chains are folding against a growth face, those already folded are still able to and do readjust their positions within the lattice, removing defects and increasing the fold period.

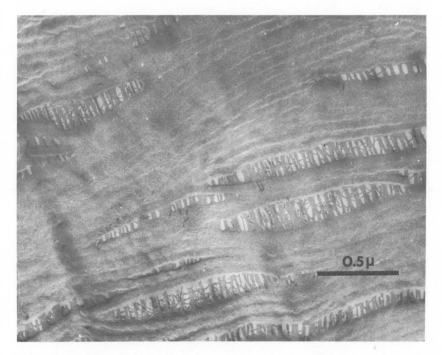

Figure 14   Fracture surface of swollen poly-4-methyl-1-pentene. The lamellae are approximately 400 Å thick and, in this region, are nearly perpendicular to the fracture surface.

E. THEORY OF POLYMER CRYSTALLIZATION

Once the orientation of the molecular axes in the lamellar crystals was determined, it was necessary to accept the concept that a polymer molecule, upon reaching the surface of the crystal, must fold back on itself and reenter the crystal. As indicated there is still considerable question, particularly during crystallization from the melt, of the distance between the points of exit and entry and of the conformation of the molecule in the fold, but there is little question of the necessity of gaining understanding of the morphology of individual lamellae and their interaction in order to be able to understand the properties of a crystalline polymer. A significant question in this respect is understanding why the thickness of the crystal varies as it does with the temperature of crystallization and subsequent annealing treatments. Two distinctly different theories have been advanced to answer this fundamental question: (1) the *thermodynamic* theory developed by Peterlin, Fischer, and Reinhold (21), and (2) the *kinetic* theory developed to its present state by Hoffman, Lauritzen, and Passaglia with related work by Price and Frank and Tosi (22).

1. Thermodynamic Theory

The thermodynamic theory is based on the large anisotropy in the interatomic forces in a polymer crystal. Peterlin, Fischer, and Reinhold suggested that the small interatomic forces permit polymer molecules in a crystal to undergo translational vibration as a unit along the chain axis or torsional vibration about the axis with an amplitude that depends on the length of the segment. They calculate a free energy term that increases with the length of the molecule that is in a crystal lattice. The longer the vibrating segment, the larger the amplitude of vibration and the more smeared the periodic potential in the lattice that holds the molecule in lattice register. The combination of this factor with the decrease in free energy resulting from crystallization yields a minimum in the total free energy density or molar free energy of the polymer for a crystal length that varies with temperature. This minimum is only a metastable equilibrium state, however, and, as in the kinetic theory, fully extended chain crystals in which all molecules in a given crystal are same length are the stable equilibrium state.

This theory has been successful in describing some aspects of the crystallization of polyethylene from dilute solution, but does not seem to apply to crystallization from the melt. In polyethylene, using reasonable values for the parameters required, no minimum is found at temperatures above about 110°C; high molecular weight polyethylene crystallizes from the melt in the form of folded chain lamellae at temperatures up to about 130°C.

2. Kinetic Theory

In the kinetic theory as originally developed by Hoffman and Lauritzen, and by Price, it is accepted that polymers crystallize by folding and use is made of the theory of nucleation as developed for small molecules to calculate the rate of primary nucleation (formation of a new crystal) and secondary nucleation (nucleation of a new ribbon of folded molecules on a completed growth face) as a function of the fold period. It is assumed that the free energy density within the crystal and the surface free energies are independent of the fold period. The primary nuclei found, they assume, are those formed most rapidly, that is, those having the lowest activation energy, even though ones of longer fold period would be thermodynamically more stable under their assumption of a fold period independent molar free energy. It takes energy to make a fold and thus the more folds per unit volume of crystalline material (the thinner the lamellae), the larger the total free energy.

The fold period of the secondary nuclei has been similarly evaluated and is about half that of the primary nucleus. Again, those secondary nuclei found are those that nucleate most rapidly. It is the fold period of these secondary nuclei that determines the character of the final crystals. In the original derivation, it is assumed that, once nucleated, a new ribbon of folded molecules (a fold plane) rapidly completes growth with the same fold period as that of its nucleus. Price, Frank and Tossi, and Lauritzen and Passaglia have all investigated the effect of permitting one or more variations in fold period during growth of a fold plane. It is found that the incorporation of the possibility of fluctuations in fold period does not invalidate the basic results of the simple theory, namely, the uniformity of the fold period and its particular dependence on supercooling. There is even some electron microscopic evidence for the difference in thickness of the primary and secondary nuclei.

As indicated, the thermodynamically stable crystal is one in which the molecules are fully extended, and also one in which all the molecules in a given crystal are of the same length. A mixture of chain lengths requires incorporation of the chain ends of short molecules in the lattice with accompanying "row" vacancies. Under normal crystallization conditions, extended chain crystals are found only for low molecular weight substances. In polyethylene it appears that during normal crystallization from the melt, the high molecular weight polymer crystallizes by chain folding, but that low molecular weight polymer (chain lengths on the order of less than two to three times the fold period) serves as Keith and Padden's impurities, is segregated, and subsequently crystallizes during cooling as extended chain paraffinlike crystals. It has been found that, when polyethylene is crystallized under pressure, all the chains can be included in extended chain crystals. At 5000 atm almost the entire sample can crystallize in this fashion. Recently it has been shown that the molecules first crystallize by folding and then rapidly "anneal" to the extended chain form (23). It appears that some fractionation takes place during the crystallization process, the longer chains being incorporated in the thicker crystals. Samples prepared in this way have crystallinities of 95–99 %, but are extremely brittle. There is little or no cohesion between or within the crystals. Cross-linking after crystallization, however, appears to yield exceptionally tough samples.

F. DEFORMATION AND ORIENTATION

Many practical applications of crystalline polymers take advantage of the high strength of the covalent bond; the molecular axes are aligned or oriented in the direction of the anticipated stress. Fibers, films, and pipes are examples in which, respectively, uniaxial, biaxial, and hoop alignment of the molecules are desired. However, molecular alignment alone, is not sufficient for maximum properties; it is essential to obtain a suitable morphology also. For instance, the molecular axes in a stack of polymer single crystals are all aligned parallel to the axis of the stack but the strength parallel to the axis is minimal. Unfortunately, as will be seen, the morphology of oriented polymers is not yet satisfactorily defined. In most cases the orientation is produced by drawing (deforming) the object after its solidification or by having crystallization occur while the melt is under stress, that is, from an "oriented melt." We first consider observations of the mechanism of deformation of crystalline polymers followed by a discussion of current concepts of the morphology of uniaxially oriented objects prepared by both methods. In both types of oriented samples, chain-folded lamellae are observed. At present there are no reported observations of the morphology of hoop oriented polymers. Biaxially oriented specimens of several polymers have been shown to consist of planar arrays of fibrils; considerably more research is needed.

1. Single Crystal Deformation

Deformation of polycrystalline metals occurs through various combinations of dislocation slip, twinning, grain boundary migration, and void formation. Much of the understanding has been obtained by studies of the deformation of single crystals. Studies of the deformation of polymer single crystals have proved similarly useful.

Deformation of polymer single crystals has been followed by depositing the crystals on a smooth plastic film, drawing it a predetermined amount, and then observing the results with electron microscopy and diffraction.

Two basically different types of deformation are observed; uniform and discontinuous, with the difference seeming to depend on the adhesion of the crystal to the plastic film, the type of crystal, temperature, and the number of lamellae (as in spiral growths) (3, 24).

Figure 15 shows a polyethylene crystal drawn on a Viton film. The deformation is localized along various cracks, across some of which fibrils have been drawn. In many cases all the fibrils are 100–200 Å in diameter; on this sample some of the oriented material is much wider. One notes that fibrils are present only in those cracks in a given sector that are at an angle to a growth face. This is in agreement with the suggestion that the molecules are folded (with adjacent reentry) in planes parallel to the growth faces. Although the fibrils "neck down" as they are drawn off the edge of the crack, there has

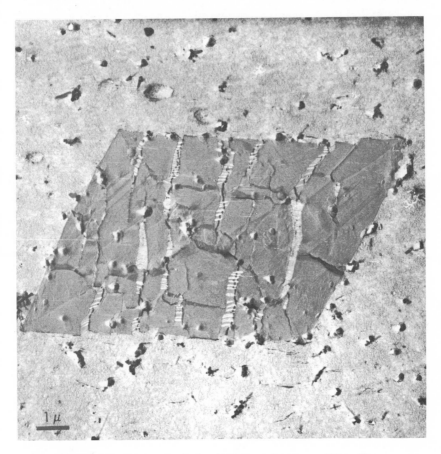

Figure 15   Single crystal of polyethylene drawn on a Viton film.

been no evidence reported concerning the mechanism of the deformation that is occurring. These observations would be in agreement with a simple unfolding of the molecules off the edge of the crack and a coalescence into a fiber, but the results of annealing experiments (see below) suggest a much more complicated mechanism may occur. In addition, observations of the discontinuous deformation of polyoxymethylene crystals suggest that the fibril formation may, in some cases, be accompanied by a localized melting and recrystallization.

Uniform deformation of polyethylene crystals occurs when single lamella crystals are drawn on a Mylar substrate. Depending on the direction of draw, the early stages of deformation ($< 25\%$ elongation) can occur by $\{110\}$ twinning or be accompanied by a phase change from orthorhombic to a monoclinic unit cell. The phase change appears to require a lateral compression as well as the elongation. With further draw, up to 100% deformation, the molecular axes within the crystal tilt (shown by electron diffraction and a thinning of the crystal) and the crystal appears to break up into ~ 100 Å diameter mosaic blocks. The results of still further elongation have not been reported; the Mylar substrate breaks.

Multilayered polyethylene crystals drawn on Mylar tend to deform discontinuously, fibers being drawn across cracks (Fig. 16). The difference is attributed to the difference in adhesion of the crystal to the Mylar and of one lamella to another. Annealing of these fibrils results in the formation of striations normal to the fibril axis (Fig. 17). These striations have been shown to be individual single crystals (25). The molecular axes remain parallel to the fibril axis, but the molecules are presumably folded to a major extent within the individual striations. In some cases further draw results in a separation of the crystals with no evidence of tie molecules, whereas in others the lamellae draw out into 100 Å diameter fibrils. In the former case the melting point may be sufficiently lowered by the small lateral size of the fibril crystals for melting to occur during the thermal treatment, whereas in the latter case there must be numerous tie molecules remaining between the lamellae.

2.   Deformation of Bulk Samples

When a bulk polymer (melt crystallized) is drawn, several different types of draw can also occur. Linear polyethylene, for instance, and most other crystalline polymers line draw ("neck") at room temperature; with increasing tension yield occurs at some point in the sample and a localized zone of deformation is formed which travels along the sample. The mechanism of deformation occurring in the necking zone is believed to be the same as that occurring throughout samples of the few polymers that deform uniformly.

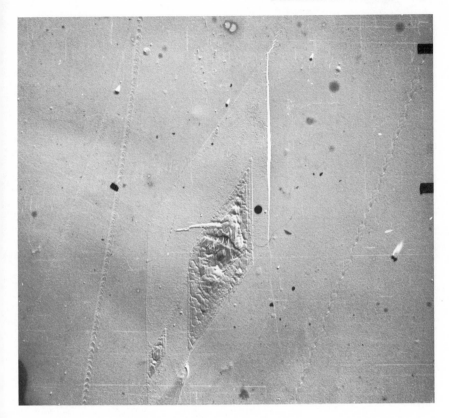

Figure 16   Single crystal of polyethylene drawn 100% on a Mylar film.

This has not been proved, however. Depending to some extent on the temperature (at lower temperatures) discontinuous draw has also been observed in a number of polymers. Usually the total amount of draw obtainable in these cases is limited, the sample failing in what appears macroscopically to be a brittle fashion.

In all the types of deformation, the fully drawn polymer has been found to consist of ~ 100 Å diameter fibrils. Small-angle X-ray diffraction indicates there is a more or less regular periodicity (~ 100 Å also) along the fibril axes. Annealing relaxed or taut (heat setting, as is often done commercially during fiber production) results in a perfecting of the periodicity, as shown by an increase in the intensity of the small-angle scattering. Infrared spectroscopic measurements have shown that regular folds are formed during the annealing (15), in agreement with the X-ray measurements and electron microscopic

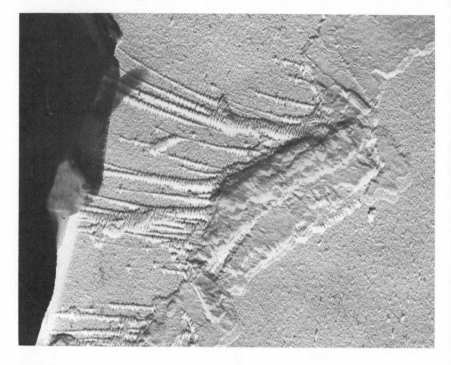

Figure 17 Fibers drawn across a crack in a polyethylene single crystal deformed on a Mylar film. The fibers have been annealed at 120°C for 30 mins.

observations of the development of broad lamellae normal to the fibril axes. The lamellae are considerably broader than the original fibrils.

3. Morphology of Oriented Polymers

On the basis of the above observations of the deformation of single crystals and bulk polymer it has been suggested that the fibrils are formed from mosaic blocks of folded chains "broken off" the lamellae in the original spherulite (26). The mechanism of their formation and alignment into the fibrils should depend on the relationship between the orientation of the lamella in the spherulite and the draw direction. As should be obvious, however, there are considerable topological problems in the rearrangement of the mosaic blocks and their interconnection both along and between fibrils, and it must be concluded that at present neither the mechanism of deformation of bulk polymer nor the morphology of uniaxially drawn material is well understood.

One of the most interesting recent developments has been the correlation of studies on solution crystallization during stirring (27) and crystallization from an oriented melt (28). The latter process occurs in all extrusion and injection molding processes used commercially. As in the case of stirrer crystallization, nuclei are formed parallel to the flow direction on which folded chain lamellae crystallize. The lamellae are thus oriented predominantly perpendicular to the flow direction. Keller and Hill have shown that most of the tensile strength of the resulting object is due to the fibrillar nuclei; the lamellae, as in a stack of single crystals, can bear little force even though the molecular axes within them are also in the flow direction. Thus although the tensile strength of some polymer fibers already exceeds that of steel wire on a weight basis, it is anticipated that considerable improvement in the strength of oriented structures should be possible by further control of their morphology.

REFERENCES

1. K. H. Meyer and H. Mark, *Ber. Deut. Chem. Ges.*, **61**, 593, 1939 (1928).

2. P. J. Flory, *Polymer Chemistry*, Cornell University Press, Ithaca, N.Y., 1953.

3. P. H. Geil, *Polymer Single Crystals*, Wiley-Interscience, New York, 1963.

4. See Appendix 1 in Ref. 3 and Section III-1, *Polymer Handbook*, J. Brandrup and E. Immergut (Eds.), Wiley-Interscience, New York, 1966.

5. P. R. Swan, *J. Polym. Sci.*, **56**, 409 (1962).

6. R. E. Robertson, *J. Phys. Chem.*, **69**, 1575 (1965).

7. G. S. Y. Yeh and P. H. Geil, *J. Macromol. Sci. (Phys.)*, **B1**, 235 (1967).

8. T. G. F. Schoon and O. Teichmann, *Kolloid-Z.*, **197**, 35 (1964); T. G. F. Schoon and R. Kretschmer, *Kolloid-Z.*, **197**, 45 (1964).

9. See Ref. 3 for historical treatment and general review. The other references cited are particularly important papers and recent results not treated in Ref. 3.

10. D. J. Blundell and A. Keller, *J. Macromol. Sci. (Phys.)*, **B2**, 301 and 337 (1968).

11. A. Keller, *Phil. Mag.*, **2**, 1171 (1957).

12. A. Keller and D. J. Priest, *J. Macromol. Sci. (Phys.)*, **B2**, 479 (1968).

13. M. I. Bank and S. Krimm, *Polym. Lett.*, **8**, 143 (1970).

14. D. C. Bassett, F. C. Frank, and A. Keller, *Phil. Mag.*, **8**, 1739 and 1753 (1963).

15. D. A. Blackadder and T. L. Roberts, *Makromol. Chem.*, **126**, 116 (1969).

16. H. D. Keith and F. J. Padden, *J. Appl. Phys.*, **34**, 2409 (1963); **35**, 1270 (1964).

17. P. J. Holdsworth and A. Keller, *J. Polym. Sci.*, A2, **6**, 707 (1968).

18. J. L. Koenig and M. L. Hannon, *J. Macromol. Sci (Phys.)*, **B1**, 119 (1967); J. L. Koenig and M. C. Agboatwalla, *J. Macromol. Sci. (Phys.)*, **B2**, 391 (1968).

19. H. D. Keith, F. J. Padden, and R. G. Vadimsky, *J. Polym. Sci.*, A4, 267 (1966).

20. Y. Hase and P. H. Geil, *Polym. J.*, **2**, 560 (1971).

21. A. Peterlin, E. W. Fischer, and Chr. Reinhold, *J. Chem. Phys.*, **37**, 1403 (1962).

22. For review see J. D. Hoffman, *SPE Trans.*, **4**, 315 (1964).
23. D. V. Rees and D. C. Bassett, *Polym. Lett.*, 7, 273 (1969); B. Wunderlich et al., *J. Polym. Sci.*, *A2*, **7**, 2043–2113 (1969).
24. H. Kiho, A. Peterlin, and P. H. Geil, *J. Appl. Phys.*, **35**, 1599 (1964).
25. P. H. Geil, *J. Polym. Sci.*, **A2**, 3835 (1964).
26. A. Peterlin, in *Man-Made Fibers*, Vol. 1, H. F. Mark, S. M. Atlas, and E. Cernia (Eds.), Wiley, 1967.
27. A. J. Pennings and A. M. Kiel, *Kolloid-Z.*, **205**, 160 (1965); A. J. Pennings, *J. Polym. Sci.*, **C16**, 1799 (1967).
28. A. Keller and M. J. Machin, *J. Macromol. Sci.* (*Phys.*), **B1**, 41 (1967); M. J. Hill (Machin) and A. Keller, *J. Macromol. Sci.* (*Phys.*), **B3**, 153 (1969).

DISCUSSION QUESTIONS AND PROBLEMS

1. What are the basic criteria for achieving crystallizability in polymers?

2. What are the effects of molecular weight (chain length), and temperature on degree and form of crystallization polymers?

3. Using the list of polymers and structures shown in Chapter 6 Table 3, indicate if they might be expected to be crystallizable or amorphous.

4. Polymers often crystallize with a chain-folding morphology. Review the evidence for this type of structure and indicate how an interpenetrating array of these chain-folded molecules could help to explain the increased strength and lower elongation usually associated with crystalline specimens.

5. Review the morphology of oriented structures and indicate how this explains the unusual properties of films and fibers. Include a consideration of spherulites.

CHAPTER 6 ⎯⎯⎯⎯⎯⎯⎯⎯⎯⎯⎯⎯

Transitions and Relaxations in Polymers

C. D. ARMENIADES
Department of Chemical Engineering
Rice University
Houston, Texas

ERIC BAER
Division of Macromolecular Science
Case Western Reserve University
Cleveland, Ohio

A. INTRODUCTION

The terms *transition* and *relaxation* are used to describe temperature or time-dependent changes in the macroscopic properties of polymers. For example, crystalline polymers show discontinuities in thermodynamic properties, such as density and enthalpy, when heated through their melting points. Glassy polymers, when heated through their softening temperatures, show discontinuities in their thermal expansion coefficients and specific heats. These phenomena are termed *transitions* and the temperatures at which they occur are *transition temperatures*. Thus we speak of melting as a transition, and we associate with it a particular temperature, the melting point (where the solid state is in equilibrium with the liquid state), which is a property of the crystalline structure under consideration. Similarly, we associate a glass transition and a glass-transition temperature (associated with "freezing in" or cessation of molecular motion) with noncrystalline structures.

It is also possible to observe significant changes in the physical properties of a polymer as a function of time. The elastic modulus and dielectric constant usually decrease with time (under constant mechanical force or electric field). These properties also show frequency dependence when the material is subjected to oscillatory stresses or an alternating electric field. A steep decrease in modulus over a narrow frequency (or time) range is termed a *relaxation* and the frequency associated with it is also a property of the structure under consideration. In this chapter we see that transitions and relaxations are interrelated phenomena and constitute macroscopic manifestations of molecular motion in a polymeric system.

1.   Structural Parameters and Molecular Motion

Molecular motion in a polymer sample is promoted by its thermal energy. It is opposed by the cohesive forces between structural segments (groups of atoms) both along the chain and between neighboring chains. These cohesive forces depend on the particular structure of the sample and can be discussed in terms of molecular parameters such as size and geometry of the molecular chains, rotational flexibility of the chain segments, and intermolecular forces.

Molecular geometry is of major importance in determining the *crystallizability* of a polymer, that is, its ability to form molecular arrangements with long-range three-dimensional order (as discussed in Chapter 5). Polymers with symmetric or stereoregular chains, such as polyethylene or isotactic polypropylene, are *crystallizable*. Those with irregular backbone chains or randomly placed side groups, such as butadiene–styrene copolymers or atactic polymethyl methacrylate, are *noncrystallizable*. The degree of order (% crystallinity) in a crystalline polymer sample can vary widely, depending on the crystallizability of the polymer and the process history of the particular sample. Likewise, the molecular chains in amorphous polymers are not completely random but possess varying amounts of short-range order.

The flexibility of molecular chains depends on the ease with which the different structural elements along the chain can rotate around covalent bonds. This rotation is opposed by the stiffness of the bond and bulkiness of the rotating group, which give rise to hindrance potentials restricting chain flexibility. For example, polyethylene has very flexible molecular chains with a hindrance potential of only 1.9 kcal/mole for rotation of the neighboring methylenes around the C—C bond. However, if every other hydrogen in the chain is replaced by a phenyl group, as in polystyrene, the hindrance potential becomes about 7 kcal/mole, resulting in a stiff chain.

The intermolecular forces determine the cohesion of molecular aggregates. In the case of hydrocarbon polymers the chains are held together by dispersion (van der Waals) forces only. Polar chains develop, in addition, induction forces between neighboring dipoles, which increase intermolecular cohesion. Certain polymers, such as nylon, cellulose, and proteins, can also form intermolecular hydrogen bonds. These cohesive forces have important effects on the properties of the bulk polymer.

The intermolecular forces and internal rotation barriers tend to restrict the polymer chains to positions of minimum potential energy. At $0°K$ the molecular structure is indeed immobile. However, at higher temperatures, the various structural segments acquire thermal vibrations, which grow with increasing temperature. As the kinetic energy of the vibrating segments approaches and exceeds the value of the various hindrance potentials to internal rotation, additional motions set in. Each of these molecular motions has a characteristic "natural frequency" determined by the temperature and moment of inertia of the participating groups. The onset of these molecular motions causes changes in the macroscopic physical properties of the polymer, giving rise to transition or relaxation phenomena.

2. Phenomenological Approaches

The association of each molecular motion with a specific temperature and frequency suggests two experimental approaches for their investigation. One could vary the temperature of the specimen and observe the resulting changes in thermodynamic properties (such as the specific volume or specific heat) at the onset of the various molecular motions. If the rate of temperature change is slow enough to allow the specimen to approach equilibrium, any observed discontinuities in these properties could be treated as thermodynamic transitions. This approach has been traditionally used to study melting phenomena and glass transitions. It is responsible for the use of the term "transitions" to characterize the macroscopic effects of large-scale molecular motions. However, the rate of change of the macroscopic properties in most systems is much too slow to allow equilibration within reasonable observation times. This is especially true in glass formation and in the changes of solid-state properties owing to restricted molecular motions. As a result the observed "transition" temperatures and the corresponding values of the measured variables become dependent on the rate of temperature change during the experiment and the phenomena do not represent true thermodynamic transitions.

The alternative approach to the study of molecular motions utilizes the association of characteristic frequencies, with different molecular motions, to create resonance phenomena with externally applied oscillations. Thus if a polymer specimen is subjected to a small oscillatory strain at constant frequency while its temperature is slowly increased, it reaches a temperature at which certain molecular motions come into resonance with the external oscillation. This resonance manifests itself by the transfer of energy from the external system to the oscillating structural elements in the polymer. It can be observed experimentally as a dispersion (energy loss) maximum accompanied by a relaxation (a steep decrease in modulus) in the temperature region where the molecular frequencies are in resonance with the external system. Most polymers show several dispersion and relaxation maxima, when tested over a sufficiently wide temperature range, each relaxation corresponding to resonance of the external oscillation with specific molecular motions at that temperature. Amorphous polymers show a very pronounced relaxation near their glass-transition temperatures. In addition they show smaller relaxations at lower temperatures, owing to motions of side chains or other conformational changes. These experimental techniques that employ oscillating disturbances are called dynamic methods. They utilize relaxation phenomena as macroscopic manifestations of molecular motion and have proved to be a very powerful tool for the study of transitions and molecular motions in solid structures.

It should now be evident to the reader that the terms "transition" and "relaxation" arise from the two different approaches to the macroscopic phenomena associated with molecular motions. In the following sections, we describe briefly the prevalent experimental methods and theories developed for the study of these phenomena. We then consider the type of transitions that occur in crystalline and amorphous polymers as well as in heterogeneous systems, which form the basis of many commercial plastics. Finally, we discuss the effects of transitions and relaxations on the mechanical and transport properties of bulk polymers. These effects are of fundamental importance in polymer processing and applications, since the major transition temperatures (melting, glass transition) are often used to assess the physical properties and processability of commercial plastics.

B. MEASUREMENT

1. Dilatometry

One of the most commonly used methods for detecting transitions is *dilatometry*, the measurement of volume change as a function of temperature. A typical apparatus contains the sample and a confining liquid, usually mercury. As the sample expands or contracts with temperature it displaces this liquid an amount corresponding to its volume change. The liquid displacement is monitored in a graduated capillary.

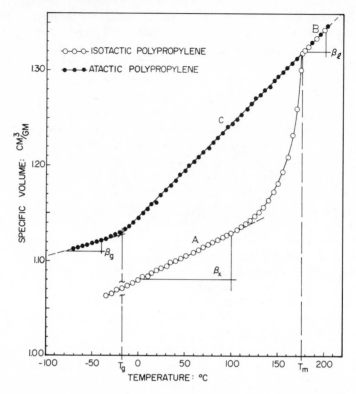

Figure 1 Specific volume of atactic and isotactic polypropylene as a function of temperature based on dilatometric measurements (data on atactic polypropylene below room temperature taken with an alcohol dilatometer).

Figure 1 shows a plot of specific volume as a function of temperature at constant pressure for atactic and isotactic polypropylene. The isotactic sample is highly crystalline and the slope of the isobar A gives the thermal expansion coefficient of the crystalline structure. As the polymer is heated to 176°C, the crystallites melt, causing an abrupt change in volume. Since melting involves a discontinuity in a primary thermodynamic variable (volume) it can be considered a *first-order transition* (4). Above 176°C both the isotactic and the atactic polymer are in the liquid state and the slope of the isobar B is the thermal expansion coefficient of the melt. The atactic specimen cannot crystallize and continues to contract linearly at temperatures well below the crystalline melting point (isobar C). Around −18°C, however, the slope of the isobar changes to a lower value and the specimen becomes a rigid glass. The temperature at which the two extrapolated slopes intersect is called the *glass-transition temperature*, T_g. This transition from a liquid to a glass involves a discontinuity in a second-order variable (dV/dT); hence it could be considered a *second-order transition*. The treatment of melting and glass formation as thermodynamic transitions implies the existence of phase equilibrium at these points. In practice, experimental limitations give kinetic values that depend on the rate of temperature change. This is especially true in the case of T_g, where very long times are needed for equilibration of the chains in the highly viscous glassy polymer. Hence the experimentally determined values of T_g and the specific volume of the glass are usually higher than the corresponding equilibrium values.

Recent developments in instrumentation, coupled with numerical differentiation, have extended the use of dilatometric techniques to the measurement of discontinuities in specific volume at temperatures much lower than T_g (5). A study of a series of methacrylate polymers was carried out using this technique (6). Discontinuities observed in the specific volume below T_g were attributed to the onset of restricted motions involving small molecular segments, such as the side chains, and can be correlated with relaxation phenomena, observed at these temperatures by other methods.

2.   Calorimetry and Thermal Analysis

Polymer transitions can also be studied from changes in thermal properties. Melting causes a discontinuity in enthalpy (H) with temperature, whereas the glass transition causes a discontinuity in specific heat $[C_p = (\partial H/\partial T)_p]$. Experimentally the most convenient and widely used thermal method is differential thermal analysis, DTA. Figure 2 shows a schematic diagram of a typical DTA apparatus. The sample and an inert reference substance, which does not have any transitions in the temperature range of interest, are heated simultaneously at exactly the same rate. Changes in the specific heat of the sample create a small difference between its temperature and that of the reference substance. This difference is measured accurately and recorded as a function of sample temperature. Figure 3 shows a typical "thermogram" of a semicrystalline sample of polyethylene terephthalate. Initial differentials in temperature and heating rate between sample and reference are compensated electronically so that the "base line" has zero slope. At 70°C the amorphous part of the sample undergoes a glass transition, which causes an abrupt increase in specific heat and the sample temperature starts to lag behind. Thus T_g shows up as a drop in base-line slope. The melting of crystalline regions at 250°C involves a large amount of heat absorption and shows up as a sharp drop in base line (endothermic peak). As in the case of dilatometric measurements, the melting and glass-transition temperatures measured by DTA are kinetic values and depend on the rate of temperature change.

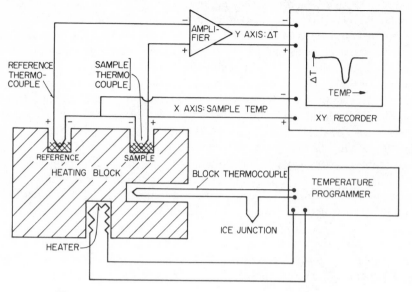

Figure 2 Basic arrangement for differential thermal analysis.

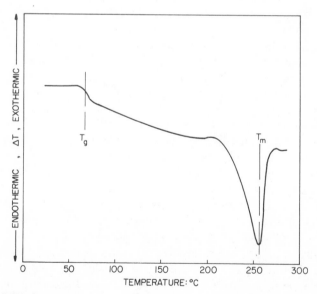

Figure 3 DTA thermogram of a semicrystalline sample of polyethylene terephthalate.

246

In its simplest form the DTA apparatus can only designate the temperatures where changes in specific heat occur, but cannot give quantitative information about C_p. Scanning calorimeters (7) are also available. These instruments can give quantitative information on the enthalpy changes involved in melting and crystallization processes.

3. Dynamic Methods

The dynamic methods utilize oscillatory disturbances, such as sinusoidal strains, alternating electric fields, or alternating magnetic fields, to create resonance with molecular motions. As a class they represent one of the most powerful and perhaps the most widely used approach to the study of transition and relaxation phenomena. We outline here the theoretical basis of the dynamic methods and subsequently give a brief description of the most prevalent experimental techniques and instrumentation.

If an external stress is applied to a perfectly elastic (Hookean) solid, the resulting strain is proportional to the stress, the proportionality constant being the elastic modulus of the solid. The energy of deformation is stored by the material, and is released upon removal of the stress, returning the specimen to its original dimensions. If the applied stress is periodic the resulting strain is completely in phase (see Fig. 4a) and the process involves no net loss of energy. If, on the other hand, an external stress is applied to a perfectly viscous (Newtonian) liquid, the liquid deforms permanently, dissipating in the process an amount of energy proportional to its viscosity. With a periodic stress the strain lags 90° behind the stress, as shown in Fig. 4a.

As discussed in Chapter 7, the response of polymers to external stress is partly elastic and partly viscous in nature. This viscoelastic behavior is due to interactions of the external stress with the molecular chains, which tend to rearrange into conformations of lower energy. In the case of periodic stresses, these molecular rearrangements cause a lag in strain as well as dissipation of a certain amount of energy. Figure 4b shows a typical viscoelastic response to a sinusoidal stress. The sinusoidal strain lags by a phase angle δ. The stress and strain can be expressed in complex form as follows:

$$\sigma^* = \sigma_0 \exp(i\omega t) \tag{1}$$

$$\gamma^* = \gamma_0 \exp[i(\omega t - \delta)] \tag{2}$$

where σ_0 and γ_0 are the stress and strain amplitudes and ω is the angular frequency. By dividing the complex stress (σ^*) by the strain amplitude, we obtain the complex modulus of the material. This can be a tensile, shear, or bulk modulus (E^*, G^*, or K^*) depending on the nature of the applied stress. It is convenient to resolve σ^* and the resulting complex modulus

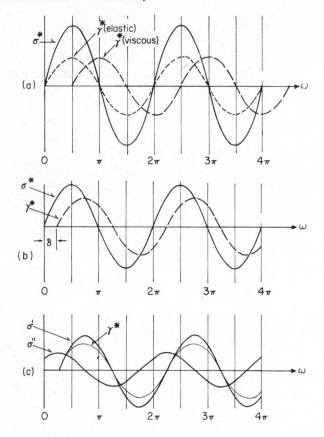

Figure 4 Response of various materials to a sinusoidal stress field. (*a*) Perfectly elastic and perfectly viscous response; (*b*) viscoelastic response; (*c*) resolution of stress into elastic and viscous components.

into two components, one in phase with the strain, the other 90° out of phase, as shown in Fig. 4*c* for the complex shear modulus:

$$G^* = G' + iG'' \qquad \text{where } G' = |G^*| \cos \delta \qquad \text{and } G'' = |G^*| \sin \delta \quad (3)$$

It can be easily shown that the in-phase component of the complex modulus gives the energy (W) stored elastically by the specimen at maximum deflection, whereas the out-of-phase component is a measure of the energy dissipated in the deformation process. This component is therefore termed "loss modulus." The ratio of the energy dissipated over a complete stress cycle (ΔW) to the maximum stored energy (W) is a characteristic quantity of the viscoelastic material and is called "internal friction." It is directly related to the components of the complex modulus:

$$\frac{\Delta W}{W} = 2\pi \frac{G''}{G'} = 2\pi \tan \delta \tag{4}$$

where $\tan \delta$ is referred to as the "loss tangent."

In principle the complex modulus can be related to molecular motions by a simple relaxation model (8), which assumes that certain molecular segments in the material under consideration can exist in two or more different conformational states, separated by an energy barrier ΔH. If the material is at a temperature T, where its thermal energy is sufficient to overcome ΔH, its molecular segments fluctuate between the two conformational states. These segmental motions are characterized by a natural relaxation time, τ, which shows an Arrhenius type temperature dependence

$$\tau = \tau_0 \exp\left(\frac{\Delta H}{kT}\right) \tag{5}$$

where τ_0 is the relaxation time at a convenient reference temperature, T_0. Under thermal equilibrium the average number of segments in each state remains unchanged and their distribution can be determined by Boltzmann statistics. If an external force is now applied to the material, so as to preferentially lower the energy of one conformational state with respect to the other, it disturbs the equilibrium, favoring a net conformational change toward the lower energy state. It is expected that the external oscillatory stress will be most effective in inducing these conformational changes when its frequency, ω, corresponds to their natural relaxation frequency $1/\tau$. At this frequency the molecular motion is in resonance with the external stresses and the material shows a maximum in internal friction. On the basis of this relaxation model the complex modulus of the material can be related to ω and τ by the following equations*:

$$G' = R_R + \frac{(G_u - G_R)\omega\tau}{1 + \omega^2\tau^2} \tag{6}$$

$$G'' = \frac{(G_u - G_R)\omega\tau}{1 + \omega^2\tau^2} \tag{7}$$

where G_u and G_R are the unrelaxed and relaxed moduli, corresponding to frequencies well above or below the resonance frequency $1/\tau$, at which the internal friction maximum occurs.

---

* The derivation of these equations is beyond the scope of this book. The interested reader is referred to special texts such as McCrum, Read, and Williams (Ref. 1, pp. 9–18).

In addition to mechanical relaxation, polymers with polar groups show also dielectric relaxation phenomena, when placed in an alternating electric field. The field causes reorientation of the dipoles in the polymer (polarization), which lags behind by a phase angle δ_e. Expressing the field (E) and polarization (D) in complex form, we have

$$E^* = E_0 \exp(i\omega t) \quad \text{and} \quad D^* = D_0 \exp(i\omega t - \delta_e) \tag{8}$$

We can now define a complex dielectric constant ε^*, which is a quantity analogous to the complex compliance ($J^* = 1/G^*$) in mechanical relaxation:

$$\varepsilon^* = \frac{D^*}{E^*} = |\varepsilon^*|\cos \delta_e + i|\varepsilon^*|\sin \delta_e = \varepsilon' + i\varepsilon'' \tag{9}$$

Using a dielectric relaxation model analogous to that for mechanical relaxation (9), we can also derive relations between the complex dielectric constant ε^*, the frequency of the applied field ω_e, and the dielectric relaxation time τ_e, corresponding to Eqs. 6 and 7.

The various mechanical and dielectric relaxation phenomena can be studied subjecting the polymer specimens to oscillatory strains or alternating electric fields of varying frequency and observing the corresponding changes in G^* or ε^*. This can be done easily in dielectric experiments. However, in the case of mechanical oscillation experiments the frequency variation, possible within a single apparatus is quite limited. For this reason in most dynamic mechanical measurements we utilize the temperature dependence of the relaxation time expressed by Eq. 5 and create resonance between ω and τ by varying the specimen temperature (which changes τ) at constant frequency or within a limited frequency range.

Mechanical Methods

Free Oscillation. Free oscillation methods are usually employed at low frequencies (0.1–10 Hz) and utilize torsional pendulum arrangements to measure shear modulus and internal friction. In the simplest arrangement (Fig. 5) the upper end of the specimen is clamped rigidly, whereas its lower end is attached to a moment of inertia member. When this member is rotated slightly and then released, it oscillates freely, twisting and untwisting the specimen. The internal friction of the polymer imposes a damping force on these oscillations; hence their amplitude (A) decays with time (Fig. 6). This amplitude decay is measured in terms of its logarithmic decrement (Δ):

$$\Delta = \ln \frac{A_1}{A_2} = \ln \frac{A_2}{A_3} = \cdots = \frac{1}{k}\ln\left[\frac{A_n}{A_{(n+k)}}\right] \tag{10}$$

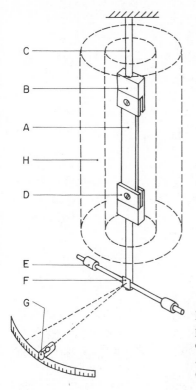

Figure 5 Schematic diagram of a simple torsion pendulum arrangement. A, Specimen; B, upper clamp; C, connecting rod to rigid mount; D, lower clamp; E, inertia arm; F, mirror; G, light source and scale; H, thermal jacket.

In the absence of external damping forces, the logarithmic decrement is a direct measurement of the specimen's internal friction and is therefore simply related to the complex modulus:

$$\Delta = \frac{1}{2}\frac{\Delta W}{W} = \pi \frac{G''}{G'} \tag{11}$$

For an isotropic specimen oscillating under the influence of a periodic shear field only, the shear modulus (G') can be computed using classical mechanics (10) from the oscillation frequency (f) and the moment of inertia of the system (I):

$$G' = kIf^2 \tag{12}$$

where k is a geometry factor, involving specimen dimensions. This formula does not consider the tensile force imposed on the specimen by the weight of the inertia member and may lead to substantial error in calculating G'.

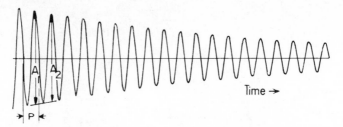

Figure 6   Typical decaying oscillation pattern.

The inverted pendulum arrangement, shown in Fig. 7, permits one to counterbalance the tensile load on the specimen and is used for precision measurements.

Resonance. Resonance methods are usually employed at intermediate frequencies (50–50,000 Hz) with rigid specimens having low internal friction ($\Delta \leq 0.2$). The specimen (in the form of a bar) is suspended at two points, one of which is subjected to vibrations of fixed amplitude, while the resulting motion at the other point is detected. As the excitation frequency is changed the motion at the detector point goes through a maximum at a resonance frequency, which depends on the density of the sample (ρ) and its geometry. The internal friction is measured usually from the half-width of the resonance peak and the resonance frequency (f_r), and the dynamic modulus is a function of f_r, ρ, and specimen geometry. Resonance methods can measure accurately very low levels of internal friction. They demand, however, considerable experimental skill, especially in mounting and exciting the specimen and in avoiding spurious resonances.

Forced Oscillation. Forced oscillation techniques utilize the phase angle δ between stress and strain to measure internal friction. With suitable instrumentation they can cover continuously a very wide frequency range (10^{-3}–10^4 Hz) and can measure internal friction levels down to 10^{-3}. The forced oscillations can be either transverse (with rigid rod specimens) or longitudinal (with rigid rods or fibers). One end of the specimen is subjected to a regulated forced oscillatory strain (γ^*), while the resulting stress (σ^*) at the other end is detected continuously. By converting stress and strain to electrical signals the complex modulus and loss tangent can be measured directly:

$$|E^*| = \frac{\sigma_{max}}{\gamma_{max}} \quad \text{and} \quad \tan \delta = |\sigma^* - \gamma^*| \tag{13}$$

A variety of forced oscillation instruments have been developed.

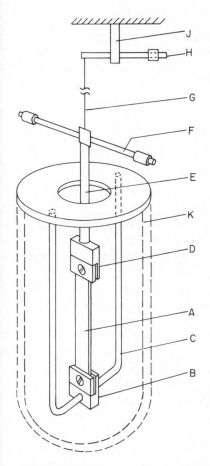

Figure 7 Schematic diagram of an inverted torsion pendulum arrangement. A, Specimen; B, lower clamp; C, rigid frame for lower clamp; D, upper clamp; E, connecting rod; F, inertia arm; G, suspension filament; H, adjustable counterbalance; J, knife edge support for counterbalance.

Wave Propagation. These methods measure the complex modulus of solid substances by utilizing its relation with the velocity of sound wave propagation in the solid (11):

$$v_l^* = \sqrt{\frac{K^* + 4/3G^*}{\rho}} \quad \text{and} \quad v_t^* = \sqrt{\frac{G^*}{\rho}} \qquad (14)$$

where v_l^* and v_t^* are the complex velocities for longitudinal and transverse waves, respectively, in a medium of density ρ. K^* and G^* are the complex bulk and shear moduli of the medium. Damping can be measured from the attenuation of the vibrations as the wave travels the length of the specimen. Wave propagation methods are normally used for relaxation measurements at high frequencies (10^5–10^7 Hz).

Dielectric Methods

These methods are applicable to polar polymers. They usually measure variations in the complex compliance of the specimen with frequency, since frequency can be varied easily over several decades with a single apparatus. At very low frequencies (10^{-4}–10^{-1} Hz), dielectric measurements can be performed quite satisfactorily by *dc transient current methods*, which measure the charging (or discharging) transient currents that accompany the sudden application of a dc potential across the sample. The duration of these transients can be related to the dielectric relaxation time, (τ_e). For a wide intermediate frequency range (10^{-2}–10^8 Hz) the *lumped circuit* methods are used predominantly. These measure the equivalent capacitance and resistance of the specimen, from which ε^* can be calculated as a function of frequency. At very high frequencies (10^8–10^{11} Hz) the effect of residual inductance in the measuring assembly makes it difficult to regard the sample as a resistance–capacitance arrangement. Hence the dielectric tests utilize *distributed circuit* techniques, which measure the attenuation factor α and phase factor β as functions of frequency. These quantities can be related to the complex dielectric constant of the specimen (ε^*). A description of the actual experimental systems and derivation of the equations used in dielectric measurements can be found in reviews by Westphal (12), DeVos (13), and Smyth (14).

Dielectric measurements are often easier to interpret structurally than dynamic mechanical data, since only the polar groups of the polymer chain are involved in dielectric relaxations. In addition, the theory of dielectric relaxation is quite well developed, to the point where the strength of experimentally observed relaxations can be quantitatively related to the number of relaxing structural units (15). This is not possible with dynamic mechanical measurements, where the mode of coupling of the macroscopic stress with the relaxing elements is not known.

The use of dynamic mechanical and dielectric measurements for relaxation studies is illustrated in Figs. 8 to 10. Figure 8 is a three-dimensional plot of the loss modulus, G'', of polyvinyl chloride as a function of temperature and frequency. It was constructed by combining dynamic mechanical measurements of G'' versus temperature at different frequencies. The G'' surface obtained in this manner shows two regions of high internal friction: a pronounced peak, β, around 100°C, which is attributed to the glass transition, and a smaller maximum, γ, which is associated with local motions of small chain segments in the glassy polymer. Corresponding regions of high loss are observed in the dielectric measurements for the same polymer, shown in Figs. 9 and 10.

It is noteworthy that the two loss maxima, β and γ, show different depen-

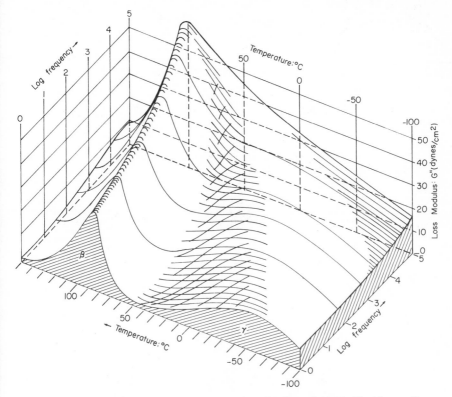

Figure 8 Three-dimesional plot of loss modulus (G'') for polyvinyl chloride as a function of temperature and frequency. Plot constructed by combining measurements of G'' vs. temperature obtained with three different instruments: torsion pendulum data at 1 Hz, forced oscillation data at 20,200 and 900 Hz, and sound wave propagation measurements at 9500 and 10^5 Hz.

dence on frequency. The temperature of the β peak changes only 25°C over five decades in frequency (see Fig. 8), while the γ peak changes by 100°C over the same frequency range. As a result the two loss maxima, which are well resolved at low frequencies, tend to merge at 10^5 Hz. The frequency dependence of the different loss maxima can be used to measure the activation energy, ΔH, associated with each relaxation process. According to Eq. 5 ΔH can be obtained directly from the slope of a semilogarithmic plot of frequency versus the reciprocal temperature of each loss maximum. Such a plot of the data from Figs. 8 to 10 is shown in Fig. 11. As expected, the dynamic mechanical and dielectric data for each loss peak fall on the same line,

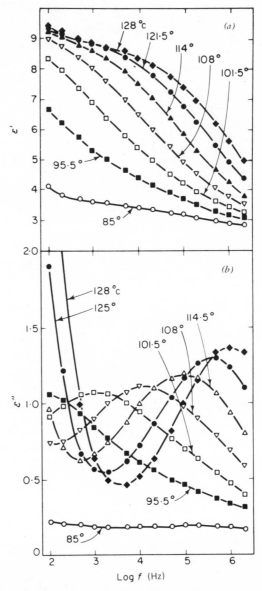

Figure 9 Frequency dependence of (*a*) ε′ and (*b*) ε″ for PVC at various temperatures in the α relaxation region [from Y. Ishida, *Kolloid Z.*, **168**, 28 (1960)].

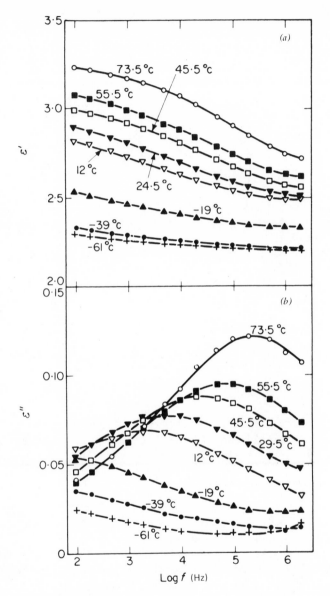

Figure 10 Frequency dependence of (*a*) ε' and (*b*) ε'' at various temperatures for PVC in the β relaxation region [from Y. Ishida, *Kolloid Z.*, **168**, 28 (1960)].

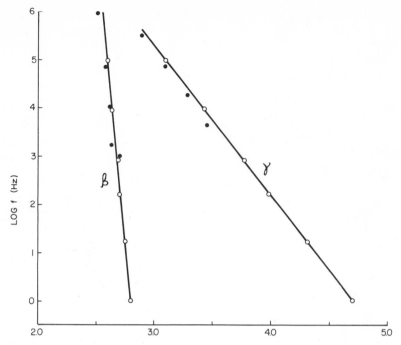

Figure 11 Plot of log f against reciprocal temperature for the mechanical (open circles) and dielectric (filled circles) loss maxima of PVC.

indicating that the same type of molecular motion gives rise to the corresponding mechanical and dielectric losses. The activation energies are 225 and 15 kcal/mole repeat units, respectively, for the β and γ relaxations.

Nuclear Magnetic Resonance

Nuclear magnetic resonance (NMR) is a spectroscopic technique, used primarily for structural analysis. Here we describe briefly its application to the study of molecular motions in polymers.

The data obtained from the study of polymers in the solid state are quite different from those obtained in dilute solution. In the solid state the resonant transition for a given nucleus (i.e., a proton) differs slightly for various protons, because the local field experienced by each proton is influenced by the neighboring nuclear and electron spins. Usually a distribution of local fields exists within a solid; hence resonance occurs over a range of the applied magnetic field (H_0), rather than a specific value, giving a "broad line" signal. However, the molecular motions in the solid create changes in the local fields. If these motions are rapid enough, they tend to average out the local fields and cause resonance at a single value of H_0, giving a "narrow line" signal. The rate of molecular motion necessary for this is of the order of 10^4–10^5 Hz. If the NMR experiment is performed on a solid at increasing temperatures, starting from absolute zero, we would expect the onset of each molecular motion to decrease the width of the broad line signal. Since the glass transition involves Brownian type motion of large molecular segments, it causes pronounced line narrowing. Quantitatively the second moment of the NMR signal (the mean square deviation of the signal from the center of the line) can be related on the basis of various models to the sixth power of the nuclear separation in the sample.

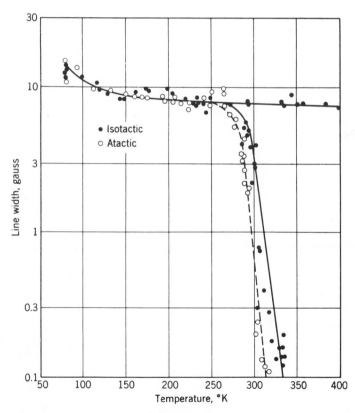

Figure 12   Variation of proton NMR line width with temperature in isotactic and atactic polypropylene [from W. P. Slichter and E. R. Mandell, *J. Appl. Phys.*, **29**, 1438 (1958)].

An example of the application of NMR line narrowing to the study of molecular motions is shown in Fig. 12. Atactic (noncrystallizable) polypropylene shows appreciable line narrowing around $300°$K. This corresponds to the glass transition at $-18°$C determined dilatometrically in Fig. 1. The semicrystalline sample of isotactic polypropylene shows two line widths above $300°$K. The narrow line is attributed to the protons in the noncrystalline chains, which undergo a glass transition similar to the atactic polymer. The broad line is attributed to the protons of the crystalline structures, which do not possess large-scale motion at these temperatures.

An alternative NMR technique for studying molecular motions is to observe the temperature dependence of the spin-lattice relaxation time, T_1. This is the time constant for the exponential decay of spin excitation, if the resonant field $\mathbf{H}_1$ is suddenly turned off. The excited protons release their excess energy to the surrounding matter by means of magnetic coupling with other protons and electrons. This coupling is facilitated by molecular motions at the resonance time scale ($10^7–10^8$ Hz). Thus T_1 shows minima at temperatures where such motions occur. Figure 13 shows a plot of T_1 versus temperature for atactic polypropylene. The minimum at $70°$C corresponds to the glass-transition motions. This is observed here at higher temperatures

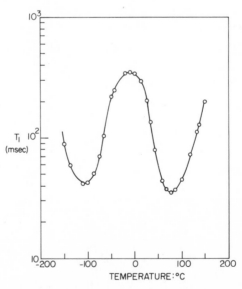

Figure 13 Temperature dependence of spin-lattice relaxation time in atactic polypropylene [from W. P. Slichter, *J. Polym. Sci.*, **C14**, 33 (1966)].

than the line narrowing data because the frequency of the associated molecular motion is higher by about three orders of magnitude. The second minimum of T_1 around $-110°C$ is attributed to motion of the pendant methyl group. This motion can be detected in Fig. 12 as a small decrease in line width around $100°K$.

NMR techniques are not as widely used as the other methods for the study of transitions. This is partly due to the rather complex instrumentation and mathematics required for quantitative interpretation of the data. However, the ability of these techniques to measure directly molecular motions at high frequencies makes them a valuable supplement to the other dynamic methods. It should be also noted that the fraction of nuclei participating in NMR transitions is extremely small (about 6 ppm of the nuclei precessing at a given ω_0) and the energy of the transition is only a few small calories per mole. Thus the transitions take place without appreciable changes in the overall temperature of the sample and the technique comes very close to the "ideal" experiments of classical physics, which let an observer make a measurement without perturbing the system.

A recapitulation of the various dynamic methods used in the study of relaxation phenomena is presented in Table 1. This table is by no means exhaustive. Many other techniques are described in the listed references and continue to appear in the current literature. The selection of the proper method for a given study depends on many factors, such as the polymer to be studied, temperature and frequency of interest, and the type of information desired.

4.   Stress Relaxation and Creep

Stress relaxation or creep phenomena (Chapter 7) can be utilized to study transitions in polymers. In a stress relaxation experiment a constant strain, γ_0, is applied suddenly to the specimen and the resulting stress, $\sigma(t)$, is measured as a function of time. The time-dependent modulus, $E(t)$, is given by the ratio $\sigma(t)/\gamma_0$. In creep experiments the specimen is subjected to a constant stress, σ_0, and the resulting strain, $\gamma(t)$, is measured as a function of time. The time-dependent compliance, $J(t)$, is given by the ratio $\gamma(t)/\sigma_0$. Stress relaxation experiments can be performed in tension, flexure, or shear on commercial force-elongation equipment. The specimen is clamped between a movable crosshead, which imposes the strain, and a load cell, which senses the stress. The decay of stress at a fixed strain is recorded as a function of time. Simple creep tests can be performed by hanging a weight on the specimen and observing the change in its length with time.

A plot of $E(t)$ or $J(t)$ over an extensive range of time can be obtained from a series of stress relaxation or creep experiments at different temperatures by

Table 1 Dynamic methods for relaxation studies

Method	Frequency Range (Hz)	Specificity	References
Dynamic mechanical			
Free oscillation	10^{-1}–10	Measures G', Δ	McCrum, Read, and Williams (1), pp. 192–205
			Nielsen (2), pp. 140–153
Resonance	50–5×10^4	Measures E', E''/E'	J. D. Ferry, *Viscoelastic Properties of Polymers*, Wiley, New York, 1961, pp. 82–146
Forced oscillation	10^{-3}–10^4	Measures E^*, $\tan\delta$	
Wave propagation	10^5–10^7	Measures K^*, G^*, Δ	
Dielectric			
Dc transient current	10^{-4}–10^{-1}	Applicable to polar polymers only; measures τ_e	Westphal (12)
			DeVos (13)
			Smyth (14)
Lumped circuits	10^{-2}–10^8	Measures ε', ε''	
Distributed circuits	10^8–10^{11}	Measures ε', ε'' from α, β	
Nuclear magnetic resonance			
Line narrowing	10^4–10^5	Measures mobility of hydrogens in polymer chain	W. P. Slichter, *Rubber Chem. Technol.*, **34**, 1574 (1961)
Spin-lattice relaxation	10^7–10^8		J. G. Powles, *Polymer*, **1**, 219 (1960)

use of the time–temperature superposition principle (16). Such a "master curve" is usually sigmoidal in shape, the inflection corresponding to the relaxation time or temperature associated with the glass transition. An example of this technique is given in Fig. 14, where a master stress relaxation curve for isobutylene at 25°C is constructed from stress relaxation data at different temperatures, by shifting the different data curves along the log time axis till they superimpose on the curve at 25°C (reference temperature). The master curve indicates a relaxation at $-70°C$, which corresponds to the dilatometrically observed glass transition of the polymer. At the reference temperature of 25°C, where the polymer is in the rubbery state, the relaxation time associated with the molecular motions, which start occurring at T_g, is quite small, 10^{-9} hr. If, however, the master curve is constructed with T_g as the reference temperature the corresponding relaxation time becomes about 10^{-2} hr, corresponding to the time scale of dilatometric experiments.

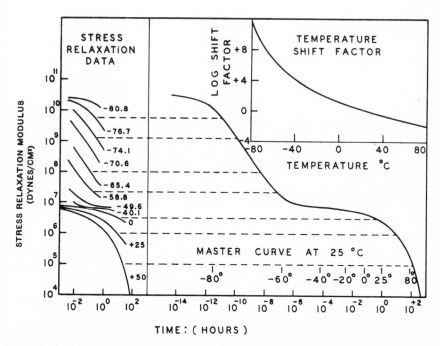

Figure 14 Time–temperature superposition principle illustrated with polyisobutylene data. The reference temperature of the master curve is 25°C. The inset graph gives the amount of curve shifting required at the different temperatures [from E. Castiff and A. V. Tobolsky, *J. Colloid Sci.*, **10**, 375 (1955); *J. Polym. Sci.*, **19**, 111 (1956)].

5. Other Methods

Certain transitions in crystalline polymers, such as melting, crystal–crystal transitions, and crystalline transformations are often investigated by means of techniques normally used for structural studies. For example, the melting of crystalline bulk can be monitored with an optical microscope equipped with a hot stage, using polarized light. Spherulitic structures, when viewed with crossed polaroids, usually appear as circularly birefringent areas with a Maltese cross. This is due to the twisted lamellar conformation of the molecular chains. As the polymer sample in the microscope is heated to the melting temperature, its structural order is destroyed and birefringence disappears. Optical microscopy can thus be used to correlate melting temperature with the size and perfection of the spherulitic structures, since small or imperfect crystallites should melt at lower temperatures. It can also be used to observe the nucleation and growth of spherulites during crystallization from the melt (Chapter 5).

Crystal–crystal transitions and crystalline transformations can be studied effectively by means of X-ray diffraction, by comparing the diffraction patterns of the structure before and after the transition or by monitoring the intensity changes of certain key diffraction lines during the transition. The diffraction patterns are obtained photographically or by means of electronic counter systems.

6. Classification of Transitions and Relaxations

To date there is no universally accepted scheme for classifying transitions and relaxations in polymers. This is partly because of the diversity of experimental methods used for the study of these phenomena and partly to their complex nature.

The early literature follows the thermodynamic approach, which classifies transitions according to their occurrence in noncrystalline or crystalline structures, which are considered as separate systems. Noncrystalline polymers are characterized by a glass transition at a temperature T_g. Transitions in the glass, below T_g, are numbered in order of decreasing temperature, T_{gg1}, T_{gg2}, etc. Transitions in the (liquid) polymer melt above T_g are numbered in order of increasing temperature, T_{ll1}, T_{ll2}, etc. Crystalline polymers are characterized by a melting transition at a temperature T_m. In addition they may show crystal–crystal transformations or transitions below T_m. These are named on the basis of the crystalline structures involved. This method of classification provides a convenient frame of reference for discussing transitions, if their structural origin is known. Most polymer samples, however, contain complex organizations of molecular chains with varying order. They are therefore likely to show transition and relaxation phenomena associated

Table 2 Classification of relaxation processes

Symbol	Temperature Range (°K)	ΔH (kcal/mole) repeat unit	Appearance	Structural Origin
α_c	0.8–0.9 T_m	~80 ~15 ~60	Multiple peak; in crystalline structures	α_c–A: chain fold motion coupled with interior α_c–B: chain twisting in interior α_c–C_f: chain fold motion, uncoupled α_c–C_c: interior chain motion, uncoupled
β	near T_g	Varies (WLF eq.)	Sharp peak; in noncrystalline structures	Large-scale segmental motion (50–100 C atoms) associated with the glass transition
γ_c	0.6–0.8 T_g	~14	Composite peak; in crystalline as well as disordered structures	γ_c: motion of crystal defects (chain, ends, edge dislocations, raw vacancies)
γ_a		~18		γ_a: small-scale segmental motion (3–5 units) in amorphous regions ("crankshaft" motion)
$\gamma_{s.c.}$	Varies widely	10–20	Sharp peak; indirectly affected by crystallinity	Side chain motion
δ	~0.1 T_m	< 5	Usually broad, weak maximum	Not well defined. Probably local in-chain motions affected by molecular organization

265

with crystalline as well as noncrystalline structures. The determination of the structural origin of these phenomena is by no means easy. Hence in many instances we have data showing transitions or relaxations, which cannot be unequivocably assigned to crystalline or amorphous structures. This problem is encountered quite often with dynamic mechanical and dielectric measurements showing multiple relaxation maxima.

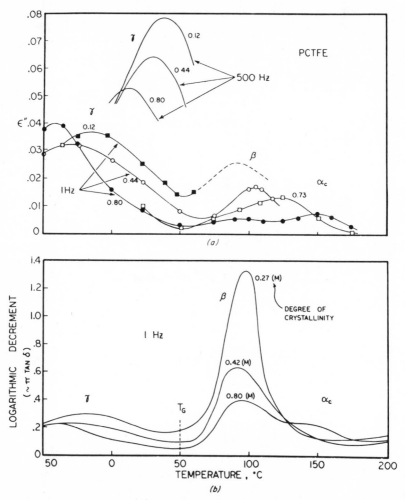

Figure 15 (*a*) ε'' versus temperature at 1 Hz for PCTFE specimens of 12, 44, 73, and 80% crystallinity [from A. H. Scott et al., *J. Res. Natl. Bur. Std.*, **66A**, 269 (1962)]. (*b*) Logarithmic decrement versus temperature at 1 Hz for PCTFE specimens of 27, 42, and 80% crystallinity [from N. G. McCrum, *J. Polym. Sci.*, **60**, 53 (1962)].

Several authors have attempted to solve this problem by labeling the experimentally determined relaxation maxima for a given polymer sample using Greek letters α, β, γ, etc. in the order of decreasing temperature (or increasing frequency). This has created some confusion in the case of polymers, which can give semicrystalline, as well as noncrystalline samples. The former show melting–crystallization and crystal disordering relaxations that are absent in noncrystalline samples where the first relaxation (α) would be due to the glass transition. This difficulty can be overcome by assigning certain letters to specific types of relaxations and by using subscripts to indicate their molecular origins, when known. This scheme of classification is shown in Table 2 and is followed in the presentation of all relaxation data in this text.

As an example of the treatment of experimental relaxation measurements on a polymer that can give samples with different crystalline contents, we consider briefly the analysis and classification of the dynamic mechanical and dielectric relaxation data of polychlorotrifluoroethylene (PCTFE), shown in Fig. 15. The 80% crystalline sample shows a well-defined loss maximum, α, at 150°C in both mechanical and dielectric measurements. Based on microscopic observations, showing this sample alone to have well-formed, lamellar spherulites, Scott and co-workers (17) attributed this loss to chain-fold motions in the lamellar surface (α_c). All samples show a sharp loss peak, β, at 100°C, which diminishes significantly with increasing crystallinity, indicating that it originates in the amorphous polymer. This loss is accompanied by a large decrease in shear modulus in the 27% crystalline sample and is associated with the glass transition of PCTFE. There is also a broad loss maximum, γ, which decreases in intensity and shifts to lower temperatures with increasing crystallinity. On the basis of this behavior, McCrum (18) considered this to be a composite peak, owing to overlapping relaxations at -40 and -10°C. The -40 relaxation was assigned to crystalline structures (γ_c), whereas the relaxation at -10°C, which increases with decreasing crystallinity, was attributed to local motions in the amorphous polymer (γ_a).

C. TRANSITIONS IN NONCRYSTALLINE POLYMERS

1. The Glass Transition

Macroscopically the glass transition in a polymer is manifested by drastic changes in many of its physical properties. The most familiar, and perhaps the most important technologically, is the change from a viscous liquid or rubber, at temperatures above T_g, to a relatively brittle solid below. On a molecular scale the glass transition is associated with cooperative motion of large chain segments (20–50 consecutive carbon atoms). This motion takes place above T_g, where the material is liquidlike; it cannot occur below T_g, where the material behaves as a solid. The glassy state can therefore be defined as a form of matter that has physical properties similar to a crystalline solid but has the molecular disorder of a liquid. These peculiar characteristics are shared by all glasses, regardless of their molecular composition, and lend support to the view, expressed by some authors, that glasses constitute a fourth state of matter, separate from liquids and crystalline solids.

Alternatively a glass can be thought of as a supercooled liquid, where the molecular motions have been frozen in. Indeed a number of crystallizable polymers can be made into glasses, by cooling the melt rapidly enough to prevent ordering of the chains in a crystal lattice. Noncrystallizable polymers will, of course, form glasses regardless of cooling rate. However, the cooling rate determines the amount of disorder frozen in a glass, hence it affects its specific volume as well as the observed glass-transition temperature. Because of the "frozen in" molecular disorder the experimentally observed glassy state is inherently a nonequilibrium state and its properties are rate dependent.

The various theories of the glass transition have followed either a thermodynamic approach, based on entropy considerations of the ideal glassy state, or a kinetic approach, based on the relaxational phenomena, which accompany the glass transition. These theories are considered briefly in the following section, together with theories based on the concept of "free volume." Subsequently we discuss the effect of structure on the glass-transition temperature.

Theoretical Treatment

Free Volume Theories. The "free volume" of a substance can be thought of as the difference between its specific volume and the space actually occupied by the molecules. It can be considered as the "elbow room" within the material, which accommodates molecular motion. This concept has proved very useful in treating transition and relaxation phenomena and is used for this purpose by a number of authors, who have developed different operational definitions.

Fox and Flory (9) define free volume, V_f, as the difference between the specific volume, $\overline{V}$, and the occupied volume, V_0, which is expressed as

$$V_0 = V' + \beta_g T \tag{15}$$

where V' is the extrapolated volume of the glass at absolute zero and β_g the thermal expansion coefficient of the glass (see Fig. 16). For a large number of glass-forming polymers, the free volume fraction, $f = V_f/\overline{V}$, is found to be

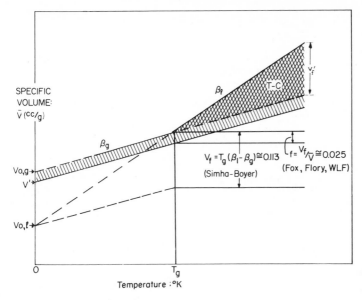

Figure 16 Schematic diagram of free volume as defined by Fox and Flory (19) or Williams, Landel, and Ferry (WLF), Simha and Boyer (20), and Turnbull and Cohen (T–C) (21).

a constant equal to 0.025. A polymer is therefore expected to become a glass upon cooling, when its free volume reaches this lower limit, at which the free volume stops changing with temperature. The glassy state can thus be considered as an *isofree volume state*.

Doolittle proposed the following semiempirical expression for the viscosity of polymers, based on free volume:

$$\eta = A \exp\left(\frac{BV_0}{V_f}\right) \qquad (16)$$

This relation can be used to derive the WLF equation (see Chapter 7), which is the most successful semiempirical expression for the treatment of relaxation phenomena near T_g. We choose T_g as the reference temperature and approximate $V_0/\overline{V}_f$ with $1/f$ at a temperature T, near T_g where $V_f \ll \overline{V}_0$. Taking the logarithms of both sides of Eq. 21, we have

$$\ln \eta_T = \ln A + B\left(\frac{V_0}{V_f}\right) = \ln A + \frac{B}{f} \qquad (17a)$$

$$\ln \eta_{T_g} = \ln A + B\left(\frac{V_0}{V_{f_g}}\right) = \ln A + \frac{B}{f_g} \qquad (17b)$$

where η_T and η_{T_G} are the viscosities at T and T_g, respectively. Subtracting the two equations we have:

$$\ln \frac{\eta_T}{\eta_{T_g}} = B\left(\frac{1}{f} - \frac{1}{f_g}\right) \tag{18}$$

We now make use of the (assumed) linear change in free volume fraction with temperature above T_g:

$$f = f_g(T - T_g) \tag{19}$$

where β_f is the difference in thermal expansion coefficients above and below T_g. Inserting Eq. 19 in 18 and rearranging, we have the WLF equation:

$$\ln \frac{\eta_T}{\eta_{T_g}} = \frac{B}{2.303 f_g} \frac{T - T_g}{(f_g/\beta_f) + T - T_g} \tag{20}$$

Using the principle of corresponding states Simha and Boyer (20) have suggested an alternative method for calculating the free volume fraction v_f at T_g:

$$v_f = T_g(\beta_l - \beta_g) = 0.113 \tag{21}$$

where β_l is the thermal expansion coefficient above T_g. This finding implies that the free volume of a polymer at T_g is 11.3% of the total volume. The difference between this value and the WLF treatment arises from differences in the definition of free volume and in the method of extrapolation to absolute zero (see Fig. 16).

Turnbull and Cohen (21) offer another definition of free volume based on the potential function of molecular segments, considered to be in a "lattice" formed by their nearest neighbors. They define as "excess volume," $\Delta \bar{v}$, the difference between the specific volume per molecule, $\bar{v}$, and the molecular volume v_0, based on the most stable van der Waals radii. The potential energy of the molecular system is considered to be at a minimum when the total excess volume is distributed uniformly among the molecular segments. At low temperatures, where $\Delta \bar{v}$ is small, its nonuniform redistribution causes a significant increase of the potential energy. At high temperatures $\Delta \bar{v}$ is large enough so that part of it can be redistributed without appreciable change in energy. This part of the excess volume is defined as the free volume, v_f':

$$\Delta \bar{v} = v_f' + \Delta v_c \tag{22}$$

v_f' becomes zero at T_g; hence the glass contracts thermally like a crystalline solid, the decrease in $\Delta \bar{v}_c$ being caused by the anharmonicity of the diminishing thermal vibrations of the molecules. The Turnbull–Cohen (T–C) free volume is shown in the crosshatched area in Fig. 16.

A fourth definition of "free volume" was proposed recently by Litt and Tobolsky (22). They define "fractional unoccupied volume," as

$$\bar{f} = \frac{V_a - V_c}{V_a} \tag{23}$$

where V_a and V_c are the specified volumes of the glassy and crystalline phases of the polymer. This is based on the premise that at any temperature the crystalline state is the one with the lowest entropy and closest packing; therefore Eq. 23 should give a measure of the volume fraction available for molecular rearrangement. In this treatment the "free volume" becomes a property of the specific polymer, rather than a universal constant, and can be related to its toughness in the glassy state.

Kinetic Theories. Kinetic theories consider the glass transition as a rate phenomenon. This approach is supported by experimental data, where T_g and the specific volume of the glass decrease with decreasing cooling rates. It is also supported by dynamic measurements, which show the temperature of the relaxations associated with T_g to be a function of the test frequency. Indeed all experimental measurements of the glass transition are associated with kinetic phenomena, since they involve a perturbation of the polymer system by temperature changes or stress fields. These perturbations change the conformational energetics of the polymer chains, giving rise to molecular rearrangements. A number of theories have been proposed to correlate these molecular motions with the changes in macroscopic properties observed in the experiment.

One of the first kinetic treatments of the volumetric phenomena at the glass transition was advanced by Alfrey, Goldfinger, and Mark (23). These authors separated the volume contraction of a polymer upon cooling into two components: an instantaneous component, due to the decrease in amplitude of the anharmonic thermal vibrations of the molecular segments, and a delayed component, due to conformational rearrangements of the chain segments to states of lower energy. These molecular motions are associated with a natural relaxation time τ, which is temperature dependent. At low temperatures τ becomes increasingly large and the approach to equilibrium slows down beyond the cooling rates of the experiment. Thus nonequilibrium conformations become "frozen in," to a degree dependent on cooling rate.

An alternate treatment of the volumetric phenomena near T_g was advanced by Kovacs (24). He considers an activation energy for viscous flow and arrives at an expression similar to the WLF equation, which is in good agreement with experimental measurements. Kovac's treatment provides a theoretical framework for predicting volume recovery phenomena, that is, the slow approach to the equilibrium volume in rapidly quenched polymers.

The kinetics of the glass transition have also been treated by means of barrier theories. These consider a distribution of polymer chain segments over several conformational states, separated by energy barriers. At temperatures above T_g these molecular segments possess sufficient thermal energy to overcome the barriers and fluctuate between the different conformations in a state of dynamic equilibrium. These fluctuations are characterized by a distribution of relaxation times, which is temperature dependent. Variation of specimen temperature or application of an external stress field changes the conformational energetics of the polymer and disrupts the dynamic equilibrium of its chain segments, causing a net flux toward conformations of lower energy. Volkenstein and Ptitsyn (25) have used this approach to develop mathematical expressions relating the observed T_g with the rate of temperature variation. Norwick (8) and Fröhlich (9) used barrier theories to treat dynamic mechanical and dielectric relaxation phenomena, respectively, and derived expressions relating the natural relaxation time with the external stress frequency and internal friction or dielectric behavior of the polymer (see Section B.3). These relations are not limited to the glass transition but can be also used for the treatment of other relaxation phenomena.

Thermodynamic Theories. The thermodynamic theories consider the "ideal" glass transition as a true second-order transition possessing equilibrium properties. This ideal state requires infinite time for its realization and cannot be obtained experimentally. Although a number of thermodynamic theories have been advanced since 1920 to explain vitrification of inorganic glasses (26–28), the theory of Gibbs and DiMarzio (29), was the first one developed specifically for polymers. This theory considers the glass-transition process as a consequence of conformational entropy changes with temperature. As the temperature is lowered the number of available conformational states decreases, causing the observed slowdown in molecular reorganization near the transition temperature. Eventually a thermodynamic second-order transition temperature, T_2, is reached, where the equilibrium conformational entropy becomes zero. Using statistical mechanics on a "quasi-lattice" model, Gibbs and DiMarzio predict that T_2 lies about 50°C below the experimentally determined T_g.

More recently, Adam and Gibbs (30) sought to bridge that gap between the equilibrium and rate theories, by relating the relaxation behavior near T_g with the static properties of the ideal glass at T_2. The resulting expression is similar in form to the semiempirical WLF equation and agrees with the Gibbs–DiMarzio theory.

We emphasize here again the fact that the kinetic and thermodynamic theories of the glass transition consider different macroscopic aspects of the same molecular phenomenon. In particular the strictly thermodynamic theories depend on equilibrium considerations, which may be impossible to achieve experimentally. It is noteworthy that treatments which combine the kinetic and thermodynamic aspects, such as the work of Adam and Gibbs (30), result in expressions similar to the semiempirical WLF equation and are in substantial agreement with experimental data.

Effect of Structure

In the introductory discussion we considered briefly the interaction of structural parameters with the molecular motions, which give rise to the various transitions and relaxations. Since the glass transition involves large segmental chain motions, the temperature at which it occurs depends on the geometry and flexibility of the molecular chains and the interchain forces of the polymer under consideration. In the case of very short chains, where the entire molecule may participate as a unit in the micro-Brownian motion characteristic of the glass transition, T_g is also affected by molecular weight. However, most commercial polymers have an average chain length at least an order of magnitude greater than the 20–50 carbon atom segments involved in glass-transition motions and their glass-transition temperatures are independent of molecular weight. Table 3 lists these temperatures for a number of polymers. They cover a range of almost 400°C, from -123°C for polydimethylsiloxane to 264°C for polyacenaphthalene.

Molecular Flexibility and Geometry. Among linear polymers, molecular flexibility is the most important factor in determining T_g. Polymers with low hindrance potentials to internal rotation, such as linear polyethylene or polyoxymethylene, show very low T_g (-120 and -85°C, respectively), whereas stiffer chain polymers, such as polyethylene terephthalate and polycarbonate require higher temperatures for the onset of molecular motions necessary for the glass transition (T_g is 70 and 150°C, respectively).

Bulky side groups decrease the mobility of the chain; hence they raise T_g. For instance, substitution of alternate hydrogens in the polyethylene chain with methyl groups (polypropylene) or with phenyls (polystyrene) increase T_g from -120 to -18 and $+100$°C, respectively. Further increase in the bulkiness of the phenyl group in polystyrene by ring substitution causes additional increases in T_g. On the other hand flexible side groups, such as aliphatic chains, tend to lower T_g. This is illustrated in the series of alkyl acrylate and methacrylate polymers, where T_g decreases as the ester alkyl chain length increases. The configurations of these side groups are such that they tend to push apart neighboring backbone chains and thus they

Table 3 Glass-transition temperatures of several polymers[a]

Polymer	Repeat Unit	T_g (°C)

Polyethylene	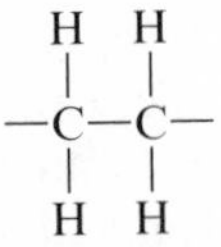 H H
	  | |
	—C—C—
	  | |
	H H
		-125

Polypropylene	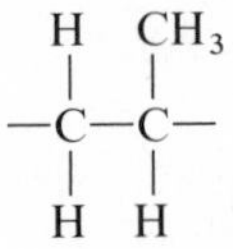 H CH$_3$
	  | |
	—C—C—
	  | |
	H H
		$-10, -18$

Poly-1-butene	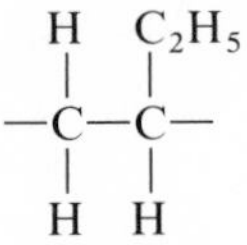 H C$_2$H$_5$
	  | |
	—C—C—
	  | |
	H H
		-25

Poly-1-pentene	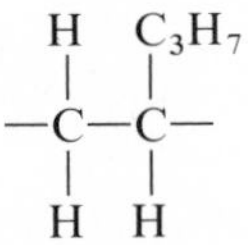 H C$_3$H$_7$
	  | |
	—C—C—
	  | |
	H H
		$-40, -24$

Poly-1-hexene	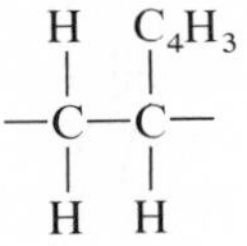 H C$_4$H$_3$
	  | |
	—C—C—
	  | |
	H H
		-50

Poly-1-octene	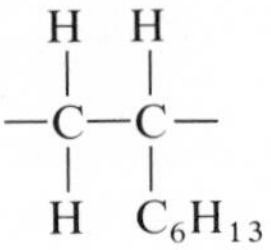 H H
	  | |
	—C—C—
	  | |
	H C$_6$H$_{13}$
		-65

Poly-1-dodecene	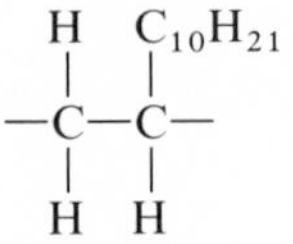 H C$_{10}$H$_{21}$
	  | |
	—C—C—
	  | |
	H H
		-25 (?)

Poly-4-methyl-1-pentene	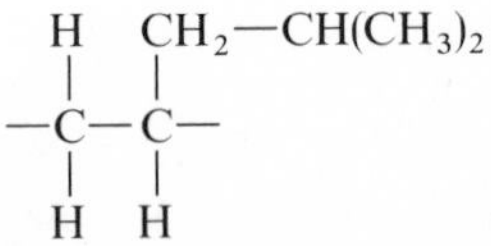 H CH$_2$—CH(CH$_3$)$_2$
	  | |
	—C—C—
	  | |
	H H
		$+18$ (cryst.)
		29 (amorph.)

Polyisobutylene	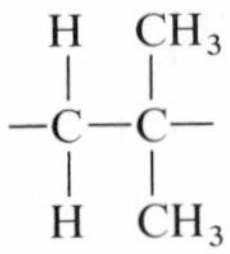 H CH$_3$
	  | |
	—C—C—
	  | |
	H CH$_3$
		$-75, -60$

Table 3 (*continued*)

Polymer	Repeat Unit	T_g (°C)
Polychloroprene	$\begin{array}{ccccc} H & Cl & H & H \\ | & | & | & | \\ -C & -C= & C & -C- \\ | & & & | \\ H & & & H \end{array}$	-50
Polyisoprene	$\begin{array}{ccccc} H & CH_3 & H & H \\ | & | & | & | \\ -C & -C= & C & -C- \\ | & & & | \\ H & & & H \end{array}$	-73
Polybutadiene	$\begin{array}{ccccc} H & H & H & H \\ | & | & | & | \\ -C & -C= & C & -C- \\ | & & & | \\ H & & & H \end{array}$	-90
Polyoxymethylene	$\begin{array}{c} H \\ | \\ -C-O- \\ | \\ H \end{array}$	$-50\,(-85)$
Polyvinyl methyl ether	$\begin{array}{cc} H & O-CH_3 \\ | & | \\ -C & -C- \\ | & | \\ H & H \end{array}$	$-20, -10$
Polyvinyl ethyl ether	$\begin{array}{cc} H & O-C_2H_5 \\ | & | \\ -C & -C- \\ | & | \\ H & H \end{array}$	-25
Polyvinyl-*n*-butyl ether	$\begin{array}{cc} H & O-C_4H_9 \\ | & | \\ -C & -C- \\ | & | \\ H & H \end{array}$	-52
Polyvinyl isobutyl ether	$\begin{array}{cc} H & O-CH_2-CH(CH_3)_2 \\ | & | \\ -C & -C- \\ | & | \\ H & H \end{array}$	$-5, -18$
Polyvinyl *t*-butyl ether	$\begin{array}{cc} H & O-C(CH_3)_3 \\ | & | \\ -C & -C- \\ | & | \\ H & H \end{array}$	$+88$

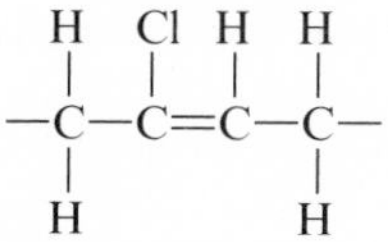

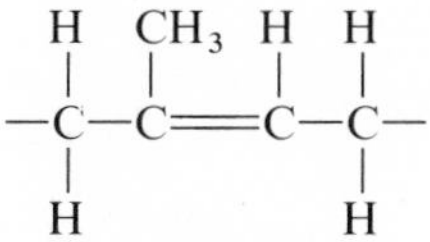

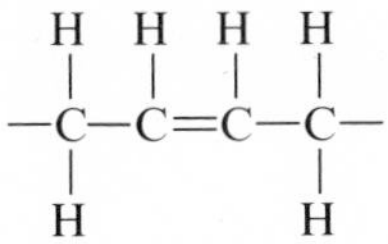

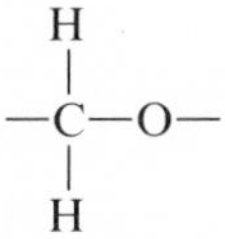

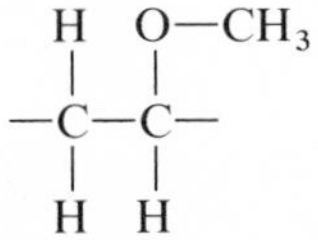

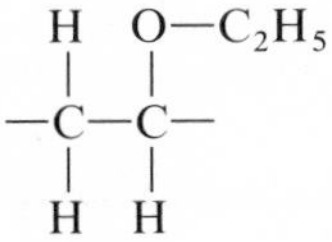

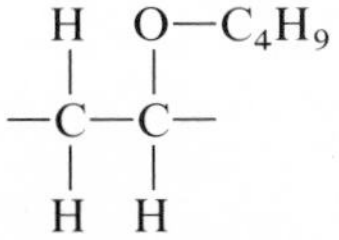

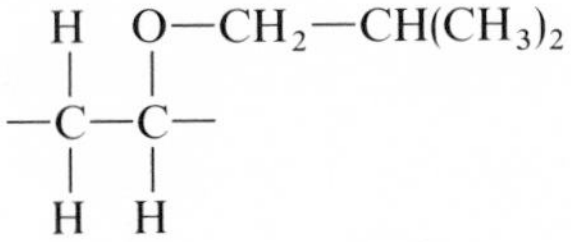

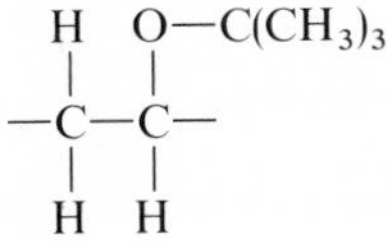

Table 3 (*continued*)

Polymer	Repeat Unit	T_g (°C)
Polydimethyl siloxane	$$\begin{array}{c} CH_3 \\ | \\ -Si-O- \\ | \\ CH_3 \end{array}$$	−123
Polyvinyl fluoride	$$\begin{array}{cc} H & F \\ | & | \\ -C- & C- \\ | & | \\ H & H \end{array}$$	−20
Polyvinyl chloride	$$\begin{array}{cc} H & Cl \\ | & | \\ -C- & C- \\ | & | \\ H & H \end{array}$$	87
Polyvinylidene fluoride	$$\begin{array}{cc} H & F \\ | & | \\ -C- & C- \\ | & | \\ H & F \end{array}$$	−35
Polyvinylidene chloride	$$\begin{array}{cc} H & Cl \\ | & | \\ -C- & C- \\ | & | \\ H & Cl \end{array}$$	−17
Polychlorotrifluoro-ethylene	$$\begin{array}{cc} Cl & F \\ | & | \\ -C- & C- \\ | & | \\ F & F \end{array}$$	45
Polytetrafluoroethylene	$$\begin{array}{cc} F & F \\ | & | \\ -C- & C- \\ | & | \\ F & F \end{array}$$	126
Polyacrylonitrile	$$\begin{array}{cc} H & CN \\ | & | \\ -C- & C- \\ | & | \\ H & H \end{array}$$	104, 130
Polymethacrylonitrile	$$\begin{array}{cc} H & CH_3 \\ | & | \\ -C- & C- \\ | & | \\ H & CN \end{array}$$	120

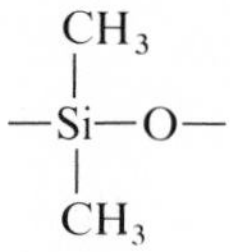

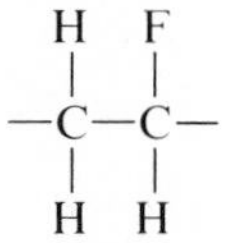

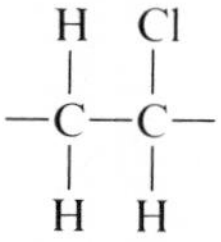

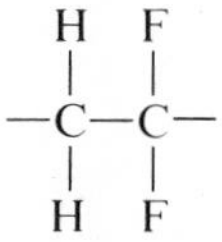

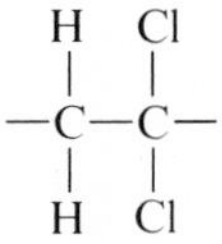

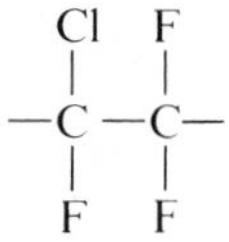

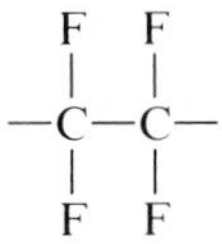

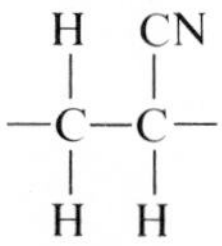

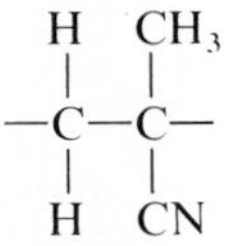

Table 3 (*continued*)

Polymer	Repeat Unit	T_g (°C)
Cellulose nitrate	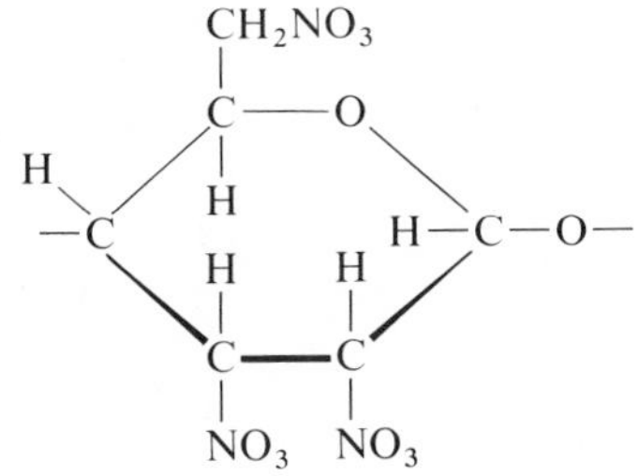	53?

Polymer	Repeat Unit	T_g (°C)
Cellulose triacetate		105, 69?
Cellulose (2.3) acetate		120
Cellulose tributyrate		120
Cellulose (2.6) butyrate		125
Ethyl cellulose		43
Polyvinyl alcohol	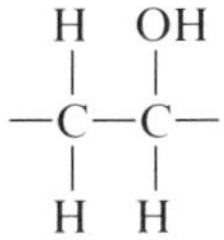	85
Polycarbonate	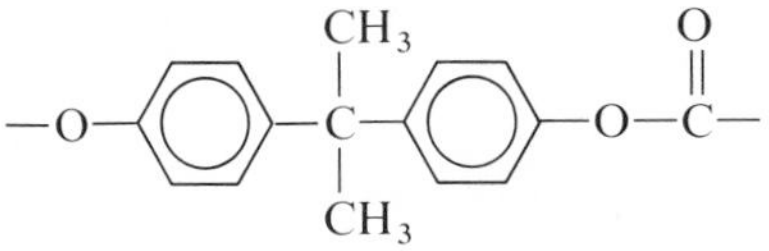	150
Polyethylene terephthalate	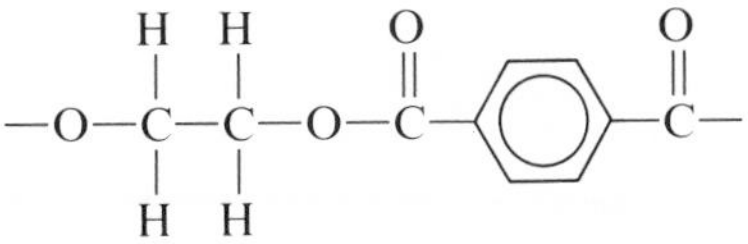	69
Polyethylene adipate	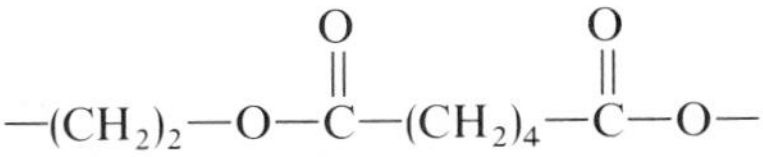	−70
Polytetramethylene sebacate	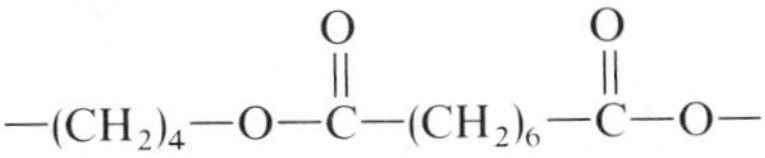	−57
Polyhexamethylene adipamide (nylon-6,6)	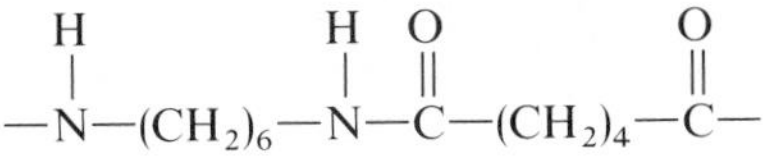	50
Polyhexamethylene sebacamide (nylon-6,10)	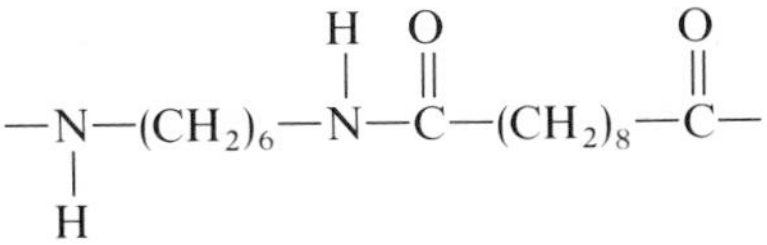	40

Table 3 (*continued*)

Polymer	Repeat Unit	T_g (°C)
Polycaprolactam (nylon-6)	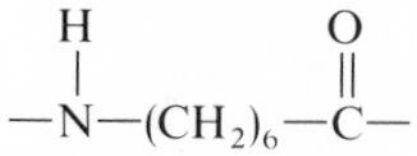	50
Polystyrene	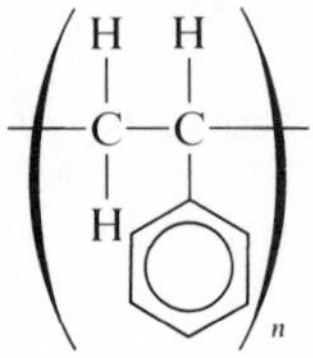	100, 105

Polycaprolactam repeat unit:

$$-\overset{\displaystyle H}{\underset{\displaystyle |}{N}}-(CH_2)_6-\overset{\displaystyle O}{\overset{\displaystyle \|}{C}}-$$

Polystyrene repeat unit:

Ring modified polystyrenes

	T_g (°C)
o-Methyl	115, 125
2,4-Dimethyl	119, 129
2,5-Dimethyl	122
p-Chloro	128
2,5-Dichloro	130, 115
3,4-Dichloro	138, 103
2,6-Dichloro	167, 132

Acrylates and Methacrylates repeat units:

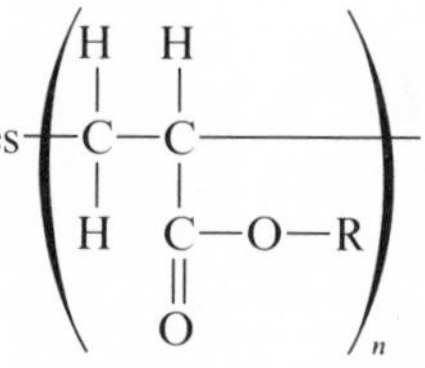

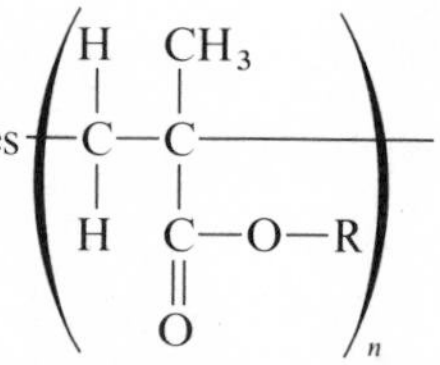

R	T_g (°C) of Acrylate	T_g (°C) of Methacrylate
Methyl	+3	+105, 120 (syndiotactic) +45 (isotactic)
Ethyl	−20	+65
n-Propyl	−44	+35
n-Butyl	−56	+21
n-Hexyl	—	−5
n-Octyl	—	−20
n-Dodecyl	—	−65
n-Octadecyl	—	−100
H (acid)	+106	

[a] See also Table 1, Chapter 7.

enhance backbone motions. This effect is similar to that produced by a plasticizer (see Section E.2) and is known as "internal plasticization."

Molecular symmetry tends to lower the glass-transition temperature. For example, polyvinyl chloride has a T_g of 87°C, whereas for polyvinylidene chloride T_g is -10°C. Similarly T_g of polypropylene is -18°C versus -65°C for polyisobutylene. In these cases the increase in molecular symmetry more than compensates for the additional side groups.

Intermolecular Forces. In general the presence of polar groups, hydrogen bonding, or other factors that increase the intermolecular forces tends to raise T_g. This effect can be seen by comparing the glass transition temperatures of polypropylene, polyvinyl chloride and polyacrylonitrile (-18, 87, and 103°C). If we consider the steric effects of $-CH_3$, $-Cl$, and $-CN$ to be similar, the steep increase in T_g can be attributed to the polarity of the $-Cl$ and $-CN$ substituents. Hydrogen bonding has similar effects. The difference of 107°C between the T_g's of polytetramethylene sebacate and nylon-6,6 can be attributed in part to the hydrogen bonding in the latter polymer. Similarly, polyacrylic acid, which has strong hydrogen bonds, shows a glass transition about 100°C higher than polymethyl methacrylate.

Ionic bonding is especially effective in raising T_g. For instance, addition of metallic ions in polyacrylic acids causes a drastic increase in the glass transition temperature of the system, the effect depending on the valency of the ion. Substitution of Na^+ for the acid hydrogen increases T_g from 106 to 280°C, and Cu^{2+} raises it further, to about 500°C.

2. Relaxations Below T_g

Although the onset of the glassy state marks the loss of large-scale molecular mobility, local motions such as hindered rotation or libration of small molecule segments continue to occur at temperatures far below T_g, giving rise to low-temperature relaxation phenomena. The exact molecular mechanisms of these relaxations are not yet clearly understood. Experimental evidence shows that various types of side group as well as main chain motion give rise to relaxations below T_g.

An example of side group relaxation is given by the dynamic mechanical behavior of the alkyl methacrylate ester polymers, shown in Fig. 17. The broad maximum in G'' (γ) around 10°C is associated with motions of the ester side group as a unit. The temperature of this loss maximum is not affected by the length of the alkyl side chain. This alkyl relaxes independently of the oxycarbonyl group and of the main chain, giving rise to the γ loss maximum shown by PnPMA and PnBMA at -190°C. Measurements at lower temperatures have shown the corresponding γ' loss for PEMA to occur at 41°K (31), whereas the ester methyl in PMMA is believed to relax

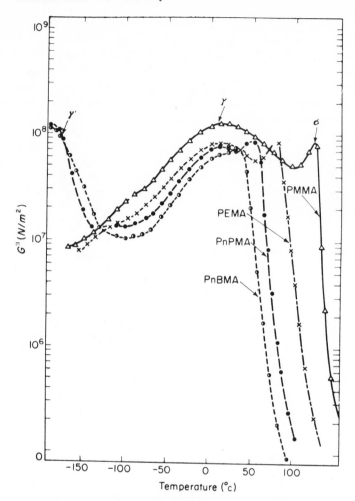

Figure 17   Temperature dependence of G'' at 1 Hz for PMMA, polyethyl methacrylate (PEMA), poly-n-propyl methacrylate (PnPMA) and poly-n-butyl methacrylate (PnBMA) [from Hijboer, in *Physics of Noncrystalline Solids*, North-Holland, Amsterdam, 1965, p. 231].

below 4°K. The β loss maximum in Fig. 17 is associated with the glass transition of these polymers. These data also show the effect of increasing length of the ester alkyl in lowering T_g.

Various mechanisms of "crankshaft" rotation of backbone chain segments have been proposed by Schatzki (32), Wunderlich (33), and Boyer (34) to explain the relaxation phenomena observed in a number of linear polymers at 150–200°K. These mechanisms involve simultaneous rotation of two to four carbon atoms about collinear bonds and appear sterically possible in chain segments of at least four sequential carbon atoms without bulky side groups. The loss maxima observed in linear polyethylene and various polyesters and polyamides in this temperature range have been explained in terms of this motion. Recent experimental evidence, however, indicates that the crankshaft model is rather simplistic and that several other mechanisms of restricted backbone segment motion may also be operative below T_g.

A number of linear polymers have recently shown evidence of mechanical relaxations of low activation energy (about 4 kcal/mole) at temperatures below 80°K. These relaxations are sensitive to molecular organization (crystallinity, orientation) and may be related to the mechanism of low temperature deformation of the polymer at large strains. There is evidence that these cryogenic relaxations may be due to the motion or aggregation of defects in the molecular organization of the polymer.

3. Relaxations Above T_g

Noncrystalline polymers become melts or rubbers above their glass transition. In this state they possess motions of large molecular segments, involving 50–100 carbon atoms. These segments are still very small fractions of the total molecular length. Hence in a homogeneous melt a few degrees above T_g the average position of the molecular chains remains unchanged. It is postulated that when the melt is heated sufficiently above T_g, it reaches a temperature where the increased free volume and thermal energy permit the entire chain to move as a unit. The onset of this motion has been designated by Boyer (36) as T_{ll}, liquid–liquid transition (see Section B.6).

There is considerable experimental evidence that atactic polystyrene melts show a T_{ll} 60–100°C above their T_g. As expected this transition shows a dependence on molecular weight, increasing from 137°C for samples with $M_n = 3000$ to 192°C for samples with $M_n = 392,000$. Similar transitions have been observed dilatometrically (37) in melts of atactic and isotactic polypropylene, poly-1-butene, and poly-1-pentene at 190 to 230°C.

D. TRANSITIONS IN CRYSTALLINE POLYMERS

1. Melting

Theoretical Treatment

Melting (fusion) is the change of a crystalline structure into a liquid. In classical thermodynamics the process is considered to be a first-order transition, characterized by a discontinuity in the primary thermodynamic variables of the system (such as specific volume of enthalpy) at constant pressure. The temperature at which the crystalline and liquid phases are in equilibrium is the (equilibrium) melting point, T_m°. This temperature is independent of the relative amounts of the two phases and is defined solely by the ratio of the enthalpy and entropy differences (ΔH_m and ΔS_m) between the ordered (crystalline) and disordered (liquid) phase:

$$T_m^0 = \left(\frac{\Delta H_m}{\Delta S_m}\right)_p \tag{24}$$

ΔH_m and ΔS_m are referred to as the enthalpy and entropy of melting. This concept of fusion has been extended to the melting of polymeric crystalline structures. However, the macromolecular nature and polydispersity (distribution of molecular weights within a sample) of polymers cause considerable deviations from ideality in their melting behavior.

As discussed in Chapter 5 polymers crystallize by chain folding in the form of lamellar structures, the lamellar thickness depending primarily on the kinetics of the crystallization process and the subsequent thermal history of the sample. These structures have at least two types of defects, created by the presence of the chain folds and chain ends. In addition polymer samples always contain a certain amount of noncrystallizable "impurities," such as catalyst fragments, branches, or atactic polymer, which are attached to crystallizable chain segments. Thus a certain amount of defects and disordered material is always present even in the most carefully crystallized samples. Since the presence of defects increases the entropy of the crystalline state, the actual melting point, T_m, of any polymer crystal is substantially lower than its equilibrium melting point, T_m^0. The latter would be determined from Eq. 24 on the basis of a perfect crystal of infinite size, made from a pure, linear polymer of infinite chain length.

Using statistical mechanics based on a lattice model Flory (38) derived a relation between the actual and equilibrium melting points (T_m and T_m^0) for a homogeneous polymer of degree of polymerization x:

$$\frac{1}{T_m} - \frac{1}{T_m^0} = \frac{R}{\Delta H_m}\left(\frac{1}{xW_a} + \frac{1}{x - Z + 1}\right) \tag{25}$$

In Eq. 25 W_a is the weight fraction of noncrystalline material, Z is a parameter associated with crystallite size, R is the gas constant, and ΔH_m is the heat of fusion per mole of (crystalline) chemical repeat unit. Since Z and x cover a range of values within a polymer sample, the melting process is expected to take place over a finite temperature range, rather than at a single temperature. This "melting" range increases with increasing breadth of molecular weight distribution. The broadening of the melting range by the polydispersity of the sample is illustrated in Fig. 18, which compares dilatometric melting data of a sample of commercial linear polyethylene with similar data for a narrow molecular weight fraction.

Application of Equation 25 to the calculation of T_m^0 from experimental data is quite cumbersome since it requires knowledge of ΔH_m, W_a, Z and x. T_m^0 is therefore usually calculated from simpler relations, based on the kinetic theory of crystallization. This theory predicts a linear relation between the

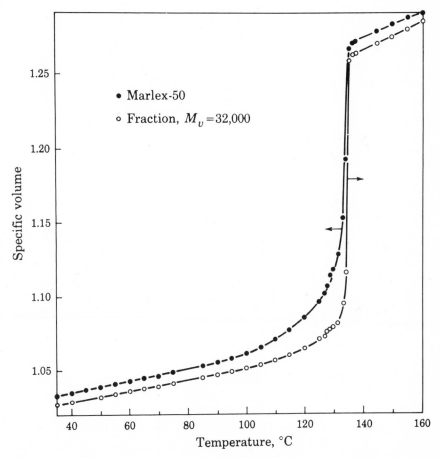

Figure 18 Specific volume–temperature relations for linear polyethylene samples. Solid circles, unfractionated Marlex-50; open circles, fraction $M_v = 32,000$. Samples initially crystallized at 131.3°C for 40 days. [from Chiang and Flory, *J. Am. Chem. Soc.*, **83**, 2857 (1961). Reprinted with permission of the copyright owner, the American Chemical Society.]

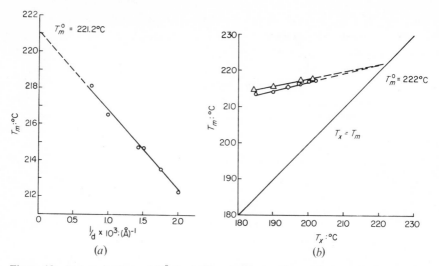

Figure 19   Determination of T_m^0 for bulk polychlorotrifluoroethylene (a) from a plot of reciprocal lamellar thickness (measured by means of low-angle X-ray scattering) versus observed melting temperature for a number of samples crystallized at different temperatures; (b) from a plot of T_m vs. T_x.

observed melting temperatures (T_m) and the reciprocal lamellar thickness ($1/d$) of polymer samples, crystallized isothermally at different temperatures (T_x). It also predicts a linear relation between T_m and T_x for these samples. The equilibrium melting point can, therefore, be obtained from plots of T_m versus $1/d$, or T_m versus T_x, by extrapolating to equilibrium conditions, where $d \to \infty$ and $T_m = T_x$. Figure 19 shows an example of this technique with samples of bulk-crystallized polychlorotrifluoroethylene.

It should be noted here that, despite the availability of many different experimental techniques for observing melting phenomena (dilatometry, thermal analysis, optical microscopy, X-ray diffraction, and others) the precise determination of the melting temperature of a given polymer structure is very difficult, owing to the interference of rate effects. Use of slow rates can cause annealing of the structure, giving a spuriously high T_m. Rapid rates may superheat the sample or cause thermal gradients within it. We should also keep in mind that the various methods use different physical criteria for determining melting and may therefore differ by a few degrees in their values of T_m even with identical heating rates.

Effect of Structure

Several attempts have been made to correlate the molecular structure of polymers with the melting points of their crystalline structures (39–41).

The correlation, however, remains at best qualitative. The factors affecting melting temperature are quite complex, involving, in addition to molecular mobility, the order–disorder energetics of the crystal lattice as well as the degree of order in the melt. Table 4 lists the melting temperatures of several polymers.

In general, polymers with rigid molecules, such as poly-p-xylylene and poly(α-vinylnaphthalene), would be expected to melt at higher temperatures than flexible chain polymers, such as polyethylene or polybutadiene. However, the melting temperature cannot be predicted or even extrapolated from simple considerations of the molecular structure. A case in point is the following series: polyoxymethylene (T_m 181°C), polyethylene oxide (66°C), and polypropylene oxide (75°C). The effect of small side groups is also hard to predict, as evidenced by the series polyethylene (137°C), polypropylene (176°C), and poly-1-butene (126°C). Here the unexpectedly high melting point of isotactic polypropylene seems to be due to its low entropy of fusion (see Eq. 24), which results from the presence of residual order in the melt.

The effect of intermolecular forces on the melting temperature seems to be similar to that observed for T_g. Polar chains form crystalline structures with high T_m, as evidenced by the fluorinated hydrocarbons and polyacrylonitrile. The hydrogen bonded polyamides also show high melting temperatures.

Several authors have attempted to derive simple relations between the melting and glass transition temperatures of a polymer. Boyer (42) and Beaman (43) suggested a linear relation of the type

$$T_m = KT_g \tag{26}$$

where K has a value of 2 for symmetric chains and 1.4 for asymmetric chains. This expression, although obviously oversimplified, represents a reasonable approximation for many polymers, as can be readily seen from the values of the T_m/T_g ratio listed in Table 4.

2. Crystalline Transitions and Transformations

A number of crystalline polymers are *polymorphic*; that is, they can form several crystallographically different structures. Each of these crystalline forms is usually associated with a temperature and pressure range within which it is stable. There are also crystalline structures, which are thermodynamically unstable, but which can exist because of kinetic barriers to the formation of the more stable forms. Changes in the solid state from one structure to another can be induced by suitable changes in the temperature and/or

Table 4 Melting temperatures of polymers

Polymer	$T_m(°C)$	$T_g(°C)$	$T_m/T_g{}^a$
Linear polyethylene	137	−120	2.67
Polypropylene	176	−18	1.76
Poly-1-butene	126	−25	1.61
Poly-1-pentene	75	−40	1.49
Poly-3-methyl-1-butene	310	+50	1.80
Poly-4-methyl-1-pentene	250	+29	1.73
Poly-4-methyl-1-hexene	188		
Poly-5-methyl-1-hexene	130		
Polyisoprene, cis (natural rubber)	28 (36)	−73	1.50
Polyisoprene, trans (gutta-percha)	74	−67	1.68
1,2-polybutadiene (syndiotactic)	154		
1,2-polybutadiene (isotactic)	120		
1,4-trans-polybutadiene	148 (92)		
Polyisobutylene	128 (105)	−70	1.97
Polyvinyl cyclohexane	305, 375	+90	1.59
Polystyrene (isotactic)	240	+100	1.38
Poly-o-methylstyrene	>360		
Poly-m-methylstyrene	215	+72	1.41
Poly-2,4-dimethyl styrene	310 (350)	+119	1.48
Poly-2,5-dimethylstyrene	340	+122	1.55
Poly-3,5-dimethylstyrene	290		
Poly-3,4-dimethylstyrene	240	+102	1.36
Poly-o-fluorostyrene	270		
Poly-p-fluorostyrene	265		
Poly-2-methyl-4-fluorostyrene	360		
Poly-α-vinylnaphthalene	360	135	1.55
Poly-p-xylene	375		
Polyoxymethylene	181	−50	2.03
Polyethylene oxide	66	−56	1.56
Polypropylene oxide	75	−62	1.64
Polyvinyl methyl ether	144	−20	1.64
Polyvinyl ethyl ether	86	−25	1.44
Polyvinyl-n-propyl ether	76		
Polyvinyl isopropyl ether	190		
Polyvinyl-n-butyl ether	64	−52	1.52
Polyvinyl isobutyl ether	115 (165)	−5	1.44
Polyvinyl-tert-butyl ether	260	+88	1.47
Polyvinyl benzyl ether	162		
Polyisopropyl acrylate (isotactic)	162	+11	1.53
Polytertiary butyl acrylate	193		
Polymethyl methacrylate (isotactic)	160	+45	1.36
Polymethyl methacrylate (syndiotactic)	>200	+105	

Table 4 (*continued*)

Polymer	$T_m(°C)$	$T_g(°C)$	$T_m/T_g{}^a$
Polyethylene terephthalate	267	+69	1.57
Polytrimethylene terephthalate	233		
Polytetramethylene terephthalate	232		
Polypentamethylene terephthalate	134		
Polyhexamethylene terephthalate	160		
Polyoctamethylene terephthalate	132		
Polynonamethylene terephthalate	85		
Polydecamethylene terephthalate	138		
Polyethylene isophthalate	240		
Polytrimethylene isophthalate	132		
Polytetramethylene isophthalate	152		
Polyhexamethylene isophthalate	140		
Polyethylene sebacate	76		
Polytetramethylene sebacate	64	−57	1.56
Polydecamethylene sebacate	80		
Polyethylene adipate	50	−70	1.59
Polytrimethylene adipate	38		
Polydecamethylene adipate	80		
Polytrimethylene succinate	47		
Polycaproamide (nylon-6)	225 (215)	+50	1.54
Nylon-11	194		
Polyhexamethylene adipamide (nylon-6, 6)	265	+50	1.66
Polyhexamethylene sebacamide (nylon-6, 10)	277	+40	1.59
Nylon-9, 9	175		
Nylon-10, 9	214		
Polydimethylsiloxane	−80	−123	1.28
Polydecamethylene sebacamide (nylon-10, 10)	210 (216)		
Cellulose triacetate	306	+105	1.53
Cellulose tripropionate	234		
Cellulose tributyrate	183 (207)		
Polyvinyl chloride	212	+87	1.34
Polyvinylidene chloride	198	−35	1.97
Polychloroprene	80	−50	1.58
Polyvinyl fluoride	200	−20	1.86
Polychlorotrifluoroethylene	220	+45	1.55
Polytetrafluoroethylene	327	+126	1.50 (3.75)
Polyacrylonitrile	317	+104	1.56
Polycarbonate (bisphenol A)	220 (267)	150	1.16 (1.27)

^{a} Ratio calculated from T_g and T_m temperatures in °K.

pressure of the crystalline sample. When these changes are thermodynamically reversible they are termed enantiotropic transitions. When they occur irreversibly they are termed monotropic transitions or transformations.

A polymer that shows well defined enantiotropic transitions is polytetrafluorethylene. These transitions have been studied by means of pressure dilatometry (45–47). They appear as discontinuities in the volume–temperature plot at different pressures as shown in Fig. 20a. Figure 20b is a phase diagram of the polymer. It shows the existence of three enantiotropic crystalline structures. The Form II to Form I transition, which occurs at 19°C at atmospheric pressure, has been studied extensively by Clark and Muus (48) using X-ray diffraction. They attribute it to a dynamic disordering

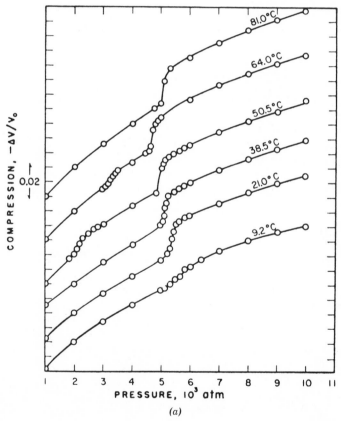

(a)

Figure 20   (*a*) Compression curves of polytetrafluoroethylene as a function of pressure.

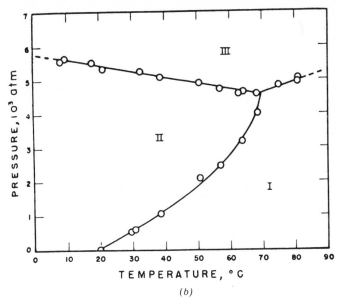

Figure 20   (*b*) phase diagram of polytetrafluoroethylene [from C. E. Weir, *J. Res. Natl. Bur. Std.*, **50C**, 95 (1953)].

of the crystal lattice, consisting of librations of chain segments about their long axis, accompanied by a slight untwisting of the helical conformation of the molecule.

Several polymers show monotropic transitions from metastable to stable structures. A good example is the Form II to Form I transformation of isotactic poly-1-butene. This polymer crystallizes from the melt in a tetragonal structure, Form II, characterized by an 11_3 helix. This structure is metastable and transforms spontaneously to a denser, stable, rhombohedral structure, Form I, characterized by a 3_1 helix. The transformation involves an increase in crystalline density of 7.3%. The kinetics of the process are controlled by temperature and pressure. Figure 21 is a plot of the transformation rate, expressed in terms of the half time of the process, at different temperatures and pressures. The rate reaches a maximum around room temperature and is drastically accelerated by pressure. At atmospheric pressure the transformation becomes sufficiently sluggish below -20 and above $50°C$ to allow prolonged storage of the polymer at these temperatures in the metastable Form II. Isotactic polypropylene shows a monotropic transition. It involves the transformation of the γ form, obtained by crystallization of the melt under pressure, to the more common α structure. The transformation occurs

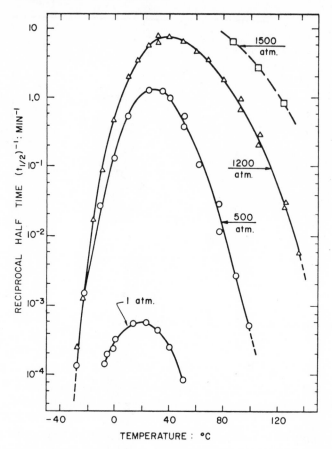

Figure 21   Reciprocal half times $(t_{1/2})^{-1}$ of the Form II to I transformation of poly-1-butene as a function of temperature at various pressures [from C. D. Armeniades and E. Baer, *J. Macromol. Sci. (Phys.)*, **B1** (2), 309 (1967)].

spontaneously, when the γ form is heated at atmospheric pressure to 150°C (49).

Certain polymers show crystalline transformations as a result of tensile or shear strains. Linear polyethylene, for instance, transforms from its orthorhombic structure to a monoclinic form, when subjected to compressive forces normal to the direction of the molecular chains (50). Polyvinylidene fluoride, when drawn uniaxially above 50°C, transforms from the orthorhombic form II, where the chains have an alternating *trans*-gauche conformation (51), to a planar zigzag conformation (Form I).

3.   Relaxations Below T_m

Several crystalline polymers show relaxation phenomena below their melting temperatures. These are usually observed with dynamic methods as internal friction maxima at temperatures slightly below T_m (α_c peaks) and in the range of 0.3 to 0.5 T_m (γ_c peaks).

The α loss maximum appears usually as a composite peak, associated with several relaxation processes, which are made possible by the increased mobility of the crystalline structures, as the melting temperature is approached. These processes are as follows (52, 53): motion of chain folds on the lamellar crystal surface ($\alpha_c - C_f$); chain twisting in the crystal interior ($\alpha_c - C_c$); motion resulting from coupling of the above relaxations ($\alpha_c - A$); and relaxations, which are associated with irreversible structural changes due to annealing (α_f). The occurrence and relative strength of these relaxations depend on the molecular organization of the crystalline structures. Hence the temperature, magnitude, and shape of the α loss are strongly affected by the process history of the crystalline sample and changes with annealing. Often the α loss is studied using single crystal mats, rather than bulk crystallized polymers, in order to minimize the amount of noncrystalline material.

The γ_c loss maximum is attributed to motion of crystal defects, such as chain ends, edge dislocations, or row vacancies (52). This relaxation has activation energies, similar to those of the γ_a process in noncrystalline polymers and may involve similar mechanisms of molecular motion. Figure 22 shows an example of the α_c and γ_c loss maxima observed on isothermally crystallized mats of polyethylene single crystals, which were annealed at different temperatures. The γ_c maximum increases in temperature and strength with annealing. This behavior is attributed to the increased number of crystal defects introduced by the annealing process, as the chain ends are pulled in by the thickening crystals, creating row vacancies. The α_c maximum is attributed by Sinnott (54) to fold motions on the crystal surface coupled with the interior ($\alpha_c - A$). This interpretation is supported by the decrease in relaxation strength with annealing, since the total fold area decreases as the lamellae thicken. The coupling of the fold motion with the crystal interior is indicated by the increase in peak temperature with the number of $-CH_2-$ groups between folds. On the other hand, Takayanagi and Matsuo (53) attribute this loss maximum to chain motions in the crystal interior only ($\alpha_c - C_c$) and explain its behavior with annealing on the basis of a crystal mosaic model.

As the reader can surmise from this brief discussion, the area of crystalline relaxations in polymers is not well understood at present. It is the subject of considerable research and much discussion among various investigators.

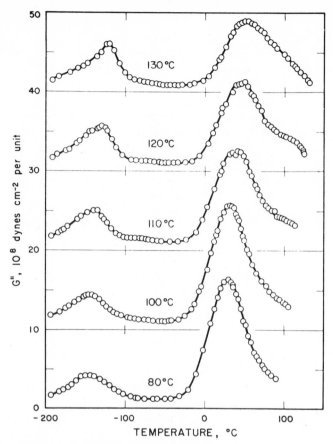

Figure 22   Temperature dependence of loss modulus (G'') for single crystals of poly-ethylene annealed at the temperatures indicated. Data for each annealing temperature above 80°C displaced vertically by 10 units above that for the precedng annealing temperature [from K. M. Sinnott, *J. Polym. Sci.* **C14**, 141 (1966)].

E.   TRANSITIONS IN HETEROGENEOUS SYSTEMS

Many commercial plastics are not homogeneous on a molecular scale. They may be copolymers or terpolymers; they may be physical blends of several polymers; they may also contain plasticizers or other additives. The transition and relaxation behavior of these heterogeneous systems is often different from that of the pure homopolymer. In most cases it can be related rather simply to the behavior of the various components.

1. Copolymers

Random

Random copolymers have glass transitions at temperatures intermediate between those of the two homopolymers. The T_g of the copolymer can be related to its composition by either of the equations:

$$T_g = v_A T_{gA} + v_B T_{gB} \tag{27a}$$

$$\frac{1}{T_g} = \frac{W_A}{T_{gA}} + \frac{W_B}{T_{gB}} \tag{27b}$$

where T_{gA} and T_{gB} are the glass-transition temperatures of the homopolymers A and B, v_A and v_B are their respective volume fractions in the copolymer, and W_A and W_B are their respective weight fractions.

In addition to shifting the glass transition, copolymerization also tends to broaden the temperature range over which it occurs. This is due to differences in chemical composition among the copolymer chains in the same sample. These differences tend to be more pronounced in systems, where the co-monomers have different reactivity ratios, unless steps are taken in the poly-merization process to counteract this tendency.

The melting behavior of random copolymers depends on the crystalliza-bility of the comonomers. In most cases the two comonomers cannot be accommodated in the same crystalline lattice. As a result copolymers have a lower degree of crystallinity and more defective structures than the cor-responding homopolymers. Flory (55) has proposed an equilibrium theory of crystallization and melting of random copolymers. It considers successive crystallization of the different uninterrupted crystallizable sequences according to their length. The crystalline copolymer is expected to have a lower melting temperature and a broader melting range than the correspond-ing homopolymer. This broadening of the melting range can be observed in the case of branched polyethylene, which can be considered a random co-polymer of ethylene with n-alkylenes. Indeed it has a lower melting tempera-ture and broader melting range than the linear polymer. The degree of crystallinity of branched polyethylene is also lower.

Ordered

Ordered copolymers show different transition behavior from that of random systems. Alternating copolymers may be regarded as homopolymers having the two-monomer sequence as the chemical repeat unit. They show a single, characteristic glass transition and, if crystallizable, they usually pack in a lattice different from that of either homopolymer. Melting of this structure gives a single sharp melting temperature.

The glass-transition behavior of block and graft copolymers is similar to that of miscible polyblends (Chapter 3). Two distinct relaxations usually occur near the glass-transition temperatures of the two types of homopolymer sequences. Block and graft copolymers can crystallize if they have sufficient uninterrupted homopolymer sequences to form a lattice. The melting range of these structures is usually sharper than that of random copolymers.

2.   Polyblends and Plasticized Systems

Polyblends are physical mixtures of two or more homopolymers. Their properties and transition behavior depend on the mutual solubility of the components. If the components are in molecular contact, the behavior of the polyblend is similar to that of a copolymer with the same composition. If the components of the polyblend are only partly miscible, the polyblend behaves like a heterogeneous copolymer, showing a broad transition temperature range. In the cases where the components of the polyblend are totally immiscible, the system is composed of distinct grains of each component. It shows two transitions, corresponding to the glass transitions of the two components (Chapter 3).

As an example of a polyblend having a crystallizable component we can consider commercial polypropylene, which consists predominantly of isotactic chains, with a small amount of noncrystallizable atactic polymer. Because of their different steric configurations the atactic chains constitute indeed a second component. During crystallization the isotactic chains form lamellar spherulites, whereas the atactic material segregates in the inter-lamellar or interspherulitic regions. A solid polypropylene sample with 10% or more atactic content will show, therefore, two transitions: the glass transition of the atactic chains below 0°C and the melting of the isotactic crystalline structures around 176°C.

Plasticizers are usually liquids, that are mechanically blended with polymer in order to improve their ductility in the glassy state or the processability of the melt. The addition of plasticizer lowers the glass-transition temperature of the system. This effect can be estimated by the use of Eqs. 27, where the plasticizer content and its T_g are substituted in place of component B. The addition of plasticizer also broadens the temperature range of the glass-transition relaxation. The extent of this broadening depends on plasticizer content as well as the mutual miscibility between the polymer and plasticizer molecules (Chapter 3, Section A.2).

F.   EFFECT OF TRANSITIONS ON PHYSICAL PROPERTIES

Both major transitions in polymers (glass transition and crystalline melting) involve large-scale changes in the conformational mobility of the molecular chains. These are manifested macroscopically as drastic changes in the physical properties of the polymer. In mechanical behavior these transitions mark a very steep drop in modulus and increase in ductility. The elastic modulus decreases from 10^9 to 10^{10} dyne/cm^2, in glassy or crystalline polymers, to 10^7 to 10^4 in rubbers and melts. Although most glasses and many of the crystalline polymers fail under stress at deformations of only a few percent, cross-linked rubbers can sustain deformations of several hundred percent, and melts show liquidlike flow. For these reasons T_g and T_m constitute in effect the usage limits of polymer systems.

In the case of elastomers a low T_g is desirable, because it extends the temperature range of their usage above T_g. Silicone rubber ($T_g = -123°C$) is, for this reason, a better gasket material at cryogenic temperatures than natural rubber ($T_g = -60°C$). In some cases the glass-transition temperature can be lowered as much as 100°C by efficient use of plasticizers. A good example of plasticizer effect are the various PVC resins. The pure polymer has a glass transition at 87°C and is quite brittle at room temperature. However, with the addition of sufficient plasticizer, it becomes tough and flexible at room temperature.

The plastics industry has used for a long time the empirical "brittleness test," which determines the temperature where the impact strength of a polymer falls below an arbitrary value. Several attempts have been made to correlate the "brittleness temperature," determined by this test, with T_g, but they have met with very limited success. Implicit in this approach is the assumption that all glassy polymers are brittle. This generalization is not correct. Examination of Table 5, which contains impact strength data for several glassy polymers, shows values covering two orders of magnitude, from the very brittle polystyrene to the extremely tough polycarbonate and PPO.

It is, of course, quite desirable to relate the mechanical behavior of glassy polymers to their structure and transitions (Chapter 7). Several authors have suggested qualitative relations between impact strength and relaxations in the glass, based on the observation that impact-resistant polymers show pronounced relaxations below T_g, which are associated with backbone chain motions (56, 57). Nylon-6, PET, and polycarbonate in Table 5 show this behavior. It is postulated that the molecular motions associated with these relaxations contribute to the toughness of the glassy polymer, by dissipating part of the impact energy. A notable exception is PPO, which has only a weak relaxation around 0°C, but retains its high impact strength down to

Table 5 Impact strength of glassy polymers at room temperature

Polymer	Impact Strength (ft-lb/in. notch)	T_g (°C)	Relaxation Below T_g(°C) and its Structural Origin	Fractional Unoccupied Volume (f)
Polystyrene (PS)	0.25–0.40	100	−50, T_g of rubber particles	0.05^a
Rubber modified polystyrene	1.0	100	10, methyl ester side groups	—
Polymethyl methacrylate (PMMA)	0.4	105	−80, cyclohexyl side groups	—
Polycyclohexyl methacrylate (PCHMA)	< 0.5	105	−50, very weak, crankshaft	
Polyvinyl chloride (PVC)	0.4–0.8	87	−130, strong, (—CH$_2$)$_n$ crankshaft	0.07
Amorphous nylon-6	0.9–2.0	50		0.10
Polyethylene terephthalate (PET)	5–8	69	−70, very strong	0.09
Bisphenol A polycarbonate	12–16	150	−100, very strong	0.09
Poly(2,6-dimethylphenylene oxide) (PPO)	18–24	210	0, very weak	0.09

^{a} Based on isotactic polymer.

$-100°C$ (58). It must also be noted that relaxations, owing to side chain motions, do not seem to affect impact strength. For instance, PMMA and PCHMA, which show strong side chain relaxations at 10 and $-80°C$, are quite brittle at room temperature. It appears that the side group relaxations are not involved in the stress transfer mechanism under load and can not participate effectively in the dissipation of the impact energy.

Boyer discusses the difference in impact strength among glassy polymers in terms of "degrees of glassiness in the glassy state" (34) arising from small differences in the free volume of the glass. Litt and Tobolsky attempted a similar correlation of impact strength with fractional unoccupied volume, $\bar{f}$ (see Section C.1) by observing that polymers with high $\bar{f}$ show high impact strength. This can be seen in Table 5. The concept of fractional unoccupied volume is also used to explain the mechanical behavior of certain glassy polymers, such as PET and polycarbonate under tensile strain. These polymers can be made to yield and cold draw under uniaxial strain at room temperature, instead of exhibiting brittle failure. The yielding behavior is explained in terms of a local increase in free volume during drawing, since Poisson's ratio of the glass is less than 0.5. This free volume increase is equivalent to lowering the T_g of the polymer to the test temperature and allows long-range motion of the chains to take place.

In semicrystalline polymers, the α_c relaxations of the crystalline structures usually cause a substantial decrease in modulus and constitute in effect the upper temperature limits of dimensional stability in the bulk polymer. The mechanical properties of semicrystalline polymers are also affected by the presence of noncrystalline chains. At temperatures above the glass transition these serve as internal plasticizers, decreasing the modulus and increasing the toughness of the polymer. This effect disappears below T_g. For example, polypropylene and poly-1-butene show impact strengths of 1.5 and 18 ft-lb/in. notch, respectively, at room temperature. These values drop to 0.4 and 0.8 ft-lb/in. notch below $-20°C$. The plasticizing effect of the non-crystalline chains is only one of several factors affecting the mechanical properties of semicrystalline polymers. These properties depend primarily on molecular organization.

The effects discussed in this chapter are not the only macroscopic manifestations of molecular motions in polymers. A number of physical properties, such as thermal and electrical conductivity, optical properties, chemical luminescence, fluorescence, and gas permeability are affected significantly by transitions. The effect of T_g on gas permeability is of considerable practical importance since a large number of polymeric systems are used as coatings or protective materials in the form of paint or plastic film. In general the temperature coefficient of diffusion of the glassy polymer is lower than the corresponding quantity in the rubbery state.

GENERAL REFERENCES

1. N. G. McCrum, B. E. Read, and G. Williams, *Anelastic and Dielectric Effects in Polymeric Solids*, Wiley, New York, 1967.
2. L. E. Nielsen *Mechanical Properties of Polymers*, Reinhold, New York, 1962.
3. M. C. Shen and A. Eisenberg, "Glass Transition in Polymers," in *Progress in Solid State Chemistry*, Vol. 3, Pergamon, New York, 1966, p. 407.

SPECIFIC REFERENCES

4. W. J. Moore, *Physical Chemistry*, Prentice-Hall, Englewood Cliffs, N.J., 1962, p. 107.
5. H. C. Hershey, J. L. Zakin, and R. Simha, *IEC Fundam.*, **6**, 413 (1967).
6. R. A. Haldon and R. Simha, *J. Appl. Phys.*, **39**, 1890 (1968).
7. Bacon Ke, in *Newer Methods of Polymer Characterization*, Bacon Ke (Ed.), Wiley-Interscience, New York, 1964, p. 347.
8. A. S. Norwick in *Progress in Metal Physics*, Wiley-Interscience, New York, 1953, p. 1.
9. H. Fröhlich, *Theory of Dielectrics*, Oxford University Press, 1949.
10. S. Timoshenko and J. N. Goodier, *Theory of Elasticity*, Wiley, New York, 1951, p. 275.
11. R. Resnik and D. Halliday *Physics*, Vol. 1, Wiley, New York, 1961, p. 425.
12. W. H. Westphal, in *Dielectric Materials and Applications*, A. R. von Hippel (Ed.), Wiley, New York, 1954.
13. F. C. DeVos, Thesis, Leyden, 1958.
14. C. P. Smyth in *Dielectric Behavior and Structure*, McGraw-Hill, New York, 1955.
15. Ref. 1, pp. 102–127.
16. Ref. 2, pp. 89–92.
17. A. H. Scott, D. J. Scheiber, A. J. Curtis, J. J. Lauritzen, Jr., and J. D. Hoffman, *J. Res. Natl. Bur. Std.*, **66A**, 269 (1962).
18. N. G. McCrum, *J. Polym. Sci.*, **60**, 53 (1962).
19. T. G. Fox and P. J. Flory, *J. Appl. Phys.*, **21**, 581 (1950).
20. R. Simha and R. F. Boyer, *J. Chem. Phys.*, **37**, 1003 (1962).
21. D. Turnbull and M. H. Cohen, *J. Chem. Phys.*, **31**, 1164 (1959); **34**, 120 (1961).
22. M. Litt and A. V. Tobolsky, *J. Macromol. Sci.*, **B1**, 433 (1967).
23. T. Alfrey, G. Goldfinger, and H. Mark, *J. Appl. Phys.*, **14**, 700 (1943).
24. A. J. Kovacs, *J. Polym. Sci.*, **30**, 131 (1958).
25. M. V. Volkenstein and O. B. Ptitsyn, *Sov. Phys.—Tech. Phys.*, **1**, 2138 (1957).
26. G. N. Lewis and G. E. Gibson, *J. Am. Chem. Soc.*, **42**, 1529 (1920).
27. F. E. Simon, *Ergeb. Exact. Naturwiss.* **9**, 244 (1930).
28. W. Kauzmann, *Chem. Rev.* **43**, 219 (1948).
29. J. H. Gibbs and E. A. DiMarzio, *J. Chem. Phys.*, **28**, 373 (1958).
30. G. Adam and S. H. Gibbs, *J. Chem. Phys.*, **43**, 139 (1965).
31. K. M. Sinnott, *J. Polym. Sci.*, **42**, 3 (1960).
32. T. F. Schatzki, *J. Polym. Sci.*, **57**, 496 (1962).
33. B. Wunderlich, *J. Chem. Phys.*, **37**, 2429 (1962).

34. R. F. Boyer, *Rubber Rev.*, **34**, 1303 (1963).
35. C. D. Armeniades, I. Kuriyama, J. M. Roe, and Eric Baer, *J. Macromol. Sci.*, **B1**, 777 (1967).
36. R. F. Boyer, *J. Polym. Sci.*, **C14**, 267 (1966).
37. H. N. Beck and A. A. Hiltz, *SPE Trans.*, **5**, 15 (1965).
38. P. J. Flory, *J. Chem. Phys.*, **17**, 223 (1949).
39. C. N. Bunn, *J. Polym. Sci.*, **16**, 323 (1955).
40. L. Mandelkern, *The Crystallization of Polymers*, McGraw-Hill, New York, 1964.
41. M. Dale, *Fortschr. Hochpolymer. Forsch.*, **2**, 221 (1960).
42. R. F. Boyer, *Changements des Phases*, Societe de Chimie Physique, Paris, 1952.
43. R. G. Beaman, *J. Polym. Sci.*, **9**, 472 (1953).
44. R. J. Beecraft and C. A. Swenson, *J. Appl. Phys.*, **30**, 1793 (1959).
45. H. A. Rigby and W. C. Bunn, *Nature*, **1949**, 583.
46. C. E. Weir, *J. Res. Natl. Bur. Std.*, **50C**, 95 (1953).
47. C. W. F. T. Pistorius, *Polymer*, **5**, 315 (1964).
48. E. S. Clark and L. T. Muus, *Z. Krist*, **117**, 119 (1962).
49. J. L. Kardos, A. Christiansen, and Eric Baer, *J. Polym. Sci.*, **A2**, 777 (1966).
50. P. H. Geil, *J. Polym. Sci.*, **A2**, 3813 (1964).
51. J. B. Lando and W. W. Doll, *J. Macromol. Sci.*, **B2**, 205 (1968).
52. J. D. Hoffman, G. Williams, and E. Passaglia, *J. Polym. Sci.*, **C14**, 173 (1966).
53. M. Takayanagi and T. Matsuo, *J. Macromol. Sci.*, **B1**, 407 (1967).
54. K. M. Sinnott, *J. Polym. Sci.*, **C14**, 141 (1966).
55. P. J. Flory, *Trans. Faraday Soc.*, **51**, 848 (1955).
56. A. J. Staverman and J. Heijboer, *Kunstoffe*, **50**, 23 (1960).
57. Ref. 2, p. 180.
58. R. F. Boyer, *Polym. Eng. Sci.*, **8**, 161 (1968).

DISCUSSION QUESTIONS AND PROBLEMS

1. Give phenomenological and molecular definitions of the terms *transition* and *relaxation*. Cite examples of transition and relaxation phenomena in nonpolymeric materials.

2. In a dynamic mechanical experiment, where:

$$\sigma^* = \sigma_0 \sin(\omega t) \qquad \text{and} \qquad \gamma^* = \gamma_0 \sin(\omega t - \delta)$$

show that the amount of energy (W_{st}) stored elastically at maximum specimen deflection and the amount of energy (ΔW) dissipated over a complete cycle are related by the expression:

$$\frac{\Delta W}{W_{st}} = 2\pi \tan \delta = 2\pi \frac{G''}{G'}$$

Hint: Calculate the work performed on the specimen in a quarter cycle.

3. The table below summarizes the results of various dynamic mechanical measurements on atactic polystyrene. Calculate the apparent activation energy of the β and γ relaxations. If the dilatometrically observed T_G for polystyrene is 100°C, what can be said about the natural frequency of the microbrownian motion of polystyrene to melt at temperatures slightly above T_g?

Experimental Method	Frequency (HZ)	β Relaxation Temperature (°C)	γ Relaxation Temperature (°C)
Torsion pendulum	0.1	100	30
Torsion pendulum	1.0	108	42
Forced oscillations	10	119	54
Vibrating reed	160	129	68

4. Construct a plot of T_m vs T_g for the polymers listed in Table 3. Discuss possible structural reasons for the observed deviations from the Boyer and Beaman relation (Eq. 26).

CHAPTER 7

Mechanical Properties of High Polymers

J. A. SAUER
K. D. PAE
Rutgers University,
The State University of New Jersey
New Brunswick, New Jersey

A. INTRODUCTION

In recent years polymers have been called upon to serve our technological society in hundreds of applications. These have included bearings, synthetic fibers, tires, gears, machine components, packaging units, structural panels and bodies, floor coverings, wire insulation, pipe, adhesives, and various types of composite members involving polymer matrices plus fiber type of reinforcement. Polymers are increasingly used because their unique properties, as well as light weight, make them more suitable in many applications than metallic or ceramic materials. In many of these applications the mechanical properties of the polymer are of prime importance. In other applications, the choice of a particular plastic or polymer may be made largely because of other characteristics, such as excellent electrical properties, low thermal conductivity, or good chemical resistance. Here too, however, it is essential that the polymer part retain its mechanical stability and integrity in order that it continue to execute its function over a useful service life without failure or fracture.

By mechanical properties of polymers we mean those parameters or characteristics of the material that determine its response to applied stresses or strains. The nature of this response depends markedly on temperature, time, and pressure as well as on the structure, preparation conditions, and past history of the polymer. Under some conditions, such as low applied strains and relatively low temperatures, polymers may behave as linear elastic materials. Under higher applied stresses and strains and at normal temperatures polymers may show yield phenomena, plastic deformation, and cold drawing. Under other conditions, as when the ambient temperature is above T_g and the polymer is amorphous, polymers may show nonlinear but recoverable behavior over wide ranges of strain, or they may exhibit viscous flow. Also, over wide ranges of strain magnitude, temperature, and test frequencies, high polymers show viscoelastic behavior that is sometimes linear and sometimes nonlinear. Finally, at relatively modest temperatures but above the melting point, crystalline polymers also show flow behavior, and this may be Newtonian or non-Newtonian.

In view of the above, it is evident that an engineer or designer who wishes to use polymers wisely, intelligently, and economically must know how their properties are affected by time, temperature, pressure, and environment, and he must be able to apply and utilize various theories of theoretical and applied mechanics, such as linear elasticity theory, linear viscoelasticity theory, rubberlike elasticity theory, and theories of yielding and of fracture. Combined knowledge of material behavior, both experimental and theoretical, is essential for establishing meaningful design criteria and for making the most efficient use of the qualities and properties of each material. Many of these matters are considered later in some detail. However, it is appropriate first to discuss, in a general way, the chemical and physical structure of high polymers and to outline the various commercial forms of polymers that are available for engineering uses.

1.   Nature and Types of Polymers

Molecular and Crystal Structure

Polymers differ appreciably from metals and ceramics in chemical structure, and these differences cause their mechanical behavior to differ also in many important respects. As discussed in earlier chapters of this book, the fundamental units of solid polymers are not isolated atoms or simple molecules but are macromolecules. These macromolecules contain many monomer units joined together by strong primary bonds of the shared electron valence type. The molecular weight of a thermoplastic polymer chain may reach hundreds of thousands or even millions, and hence a single polymer chain may contain up to 100,000 monomer units linked strongly together.

The intermolecular forces in most polymers are usually secondary forces of the van der Waals type. However, in highly cross-linked rubbers or three-dimensional thermosetting polymers, chains may be connected into a network structure by means of primary bonds and, in this case, the entire bulk polymer is essentially one molecule and the MW is effectively infinite. Intermolecular forces of the van der Waals type, such as polar forces, dispersion forces, and hydrogen bond forces, are much weaker than primary valence types of forces. In many thermoplastic or rubberlike polymers, rotation about segmental bonds can occur relatively easily even at modest temperatures. As a result, the long macromolecular chains have considerable flexibility and many different chain conformations are possible. The degree of mobility in any given polymer depends on the temperature as well as on its specific chemical and physical structure.

Polymers are frequently classified as either amorphous polymers or crystalline polymers (Chapter 5). In amorphous polymers, like atactic polystyrene or polymethyl methacrylate, there is no long-range order and X-ray diffraction gives only broad amorphous halos. In crystalline polymers, like polyethylene and polypropylene, X-ray diffraction gives both a diffuse background, characteristic of noncrystalline regions, and sharp peaks, characteristic of fairly well-defined crystalline regions. The degree of crystallinity of the sample is affected by many variables, including chain structure, tacticity, ease of chain packing, preparation conditions, annealing, and orientation. It, in turn, affects the mechanical properties, for the crystalline regions tend to give stiffness and strength to the polymer and the noncrystalline regions are usually responsible for the flexibility and toughness of polymers. Unstretched natural rubber is amorphous, but upon stretching 300% or more, the chains move into a more ordered array and X-ray studies indicate the stretched rubber is crystalline.

The individual crystals in crystalline polymers tend to be very small and, if permitted to grow slowly, usually develop in the form of platelets (1, 2). An example of the presence of crystalline platelets in a bulk specimen of polypropylene crystallized by slow cooling from an elevated temperature while under high pressure is shown by the electron micrograph of Fig. 1 (31). Similar but better-defined platelets are obtained when crystals are grown from solution (32). These platelets consist of folded chains with the chain axis generally at right angles to the surface of the platelet. The platelet thickness is usually of the order of 100 to 150 Å, and the length and width, which depend greatly on the cooling rate from the melt or solution, are of the order of microns. There is also a tendency for the crystallites to grow in dendritic fashion and to form fairly complex spherulitic aggregates. The spherulites may vary in size from microns to hundreds of microns, depending on the nucleation and growth rates. Interspherulitic and intercrystalline

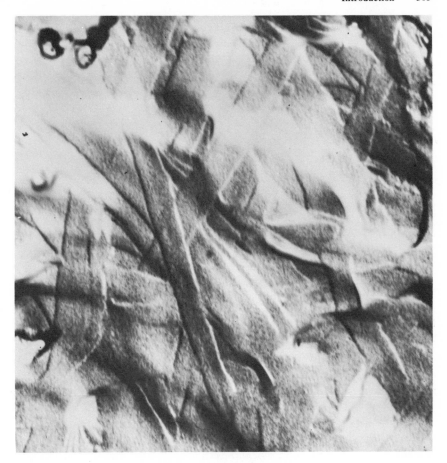

Figure 1 Electron micrograph of replica of fracture surface of a polypropylene specimen crystallized by slow cooling while under high pressure (31).

tie molecules form part of the amorphous phase and these, as well as the nature of the inter- and intramolecular force fields, are very important to mechanical behavior of the bulk polymer.

Rubbers and Synthetic Elastomers

In rubbers and elastomeric materials, the intermolecular forces and the barriers to rotation are so low that, even at room temperature, rotation about bonds is essentially free and hence the polymer chains tend to adopt the more probable curled-up state rather than the less probable extended state. When natural or synthetic rubbers are stretched, the chain network is easily extended to very large deformations (hundreds of percent) and the modulus is very low ($\sim 10^7$ dyne/cm^2 or ~ 150 psi). However, because of purposely introduced chemical cross-links across chains or of physical entanglements owing to high molecular weight of the chains, permanent plastic deformation or flow is hindered or avoided. On removal of the stretching force, there is an immediate contraction of the polymer to the initial dimensions as the stressed chains return to their equilibrium arrangement. This phenomenon of rubber, or entropy, elasticity is unique to polymers and is a direct result of the long-chain nature of polymeric materials. An equilibrium theory of rubber elasticity, based on statistical mechanics and the concept of free rotation, has been developed (3, 4). A short discussion of this theory, which is a non-linear one, and of the relation between the modulus of a rubber, the temperature, and the material parameters is given in Section B.

Thermoplastics

In typical thermoplastics, like polyethylene (PE), polypropylene (PP), nylon, polycarbonate (PC), polystyrene (PS), and polymethyl methacrylate (PMMA), the intermolecular forces and the barriers to rotation about C—C bonds are greater than in the rubbers or elastomeric polymers. These increased forces may arise because of better molecular packing and a high degree of crystallinity, as in PE and PP; from the presence of side chains and increased steric forces, as in PS and PMMA; or from increased dipolar forces and/or hydrogen bond forces, as in PC and nylon.

In thermoplastics, the stiffness or elastic modulus is some 1000 times greater than that of the rubber materials. At room temperature the tensile modulus usually falls in the range of 0.4 to 4.0 $\times$ 10^{10} dyne/cm^2 ($\sim 60,000-600,000$ psi). However, if noncrystalline polymers, like atactic PS and PMMA, are raised in temperature to and above their glass-transition temperature values, then the modulus falls, over a relatively narrow temperature region, from values near 10^{10} dyne/cm^2 to the usual 10^7 dyne/cm^2 characteristic of rubbers or synthetic elastomeric materials.

Despite the comparatively high room-temperature modulus of most thermoplastic polymers, there is still a considerable amount of internal molecular motion and chain flexibility in these materials (5–11). For example, if the temperature of a PE specimen is lowered from room temperature to liquid nitrogen temperature, the modulus increases by a factor of three or more. Thus in addition to the primary glass-to-rubber transition, most thermoplastic materials also show one or more secondary relaxation regions in which, as the temperature is increased, some small side chain units or localized main chain units acquire sufficient thermal energy to reorient over a potential barrier. These secondary transitions are important both from the viewpoint of their relation to polymer structure and molecular mobility and also because they have a bearing on deformational response and mechanical behavior. As a result, both primary and secondary relaxations in polymers have been widely studied in recent years, using various techniques such as specific volume, dielectric measurements, nuclear magnetic resonance, and dynamic mechanical behavior (Chapter 6).

Thermosetting Plastics

Thermosetting polymers include the phenolic resins, the epoxy resins, the melamine–formaldehyde resins, some polyester resins, and many others (discussed in Chapter 3). The thermosetting resins differ from the thermoplastics already discussed in that the polymer chains are no longer linear but form a tight three-dimensional network. They also differ from the rubbers and elastomers in that the degree of cross-linking between chains is very much greater. As a result, they have a room-temperature modulus of about 10^{11} dyne/cm^2 (1–2 million psi), or three to ten times that of most thermoplastics.

Because of the high degree of cross-linking and the rigid three-dimensional network structure of thermosetting polymers, these materials have low flexibility and tend to be brittle if used in the pure state. However, they are not generally used this way, but rather as adhesives for laminates or as a matrix in combination with reinforcing fillers such as organic or inorganic fibers, fabric or woven mats, or asbestos. Because of their network structure, thermosetting resins are much less affected by temperature changes than thermoplastic ones. However, they cannot be reprocessed, for at elevated temperatures they tend to decompose rather than melt and flow.

Fibers and Oriented Films

In many injection-molded or extruded polymers, the orientation of the polymer chains or of the crystals, if these are present, is frequently random, and hence the mechanical properties of the bulk polymer may be essentially independent of direction of loading, even though the chains themselves are anisotropic. Even in highly crystalline polymers, with spherulitic structure, the properties may not vary much with orientation, for these properties are highly complicated averages taken over regions of varying degrees of order and orientation.

The situation is different in spun synthetic or natural fibers or in drawn films. In these forms, the molecular axis of the polymer chains tends to align along the draw direction, and the greater the draw ratio, the greater the degree of alignment or orientation. As a result, the observed mechanical properties are significantly different in the longitudinal or draw direction than in a transverse direction, and the values of the measured properties may differ significantly from those of isotropic cast or molded thermoplastics.

The stiffness or tensile modulus of natural fibers like hemp or flax may approach 10^{12} dyne/cm^2 in the chain direction and similar values are approached for highly drawn synthetic polymers like PE. In fact, if the experimentally determined modulus is based on X-ray recorded changes in d spacings of the crystallographic lattice, there is comparatively good agreement between the experimental values and theoretical values based on assumed intermolecular potentials or force constants obtained from spectroscopic studies (12). For example, the calculated lattice modulus values for linear polyethylene, which crystallizes in the zigzag extended chain arrangement, range from 1.8 to 3.5×10^{12} dyne/cm^2 as compared to an X-ray measured lattice value of 2.4×10^{12} dyne/cm^2. For polymers that crystallize in a helical chain conformations, the computed modulus values are somewhat lower, but theoretical and experimental values are still in close agreement. For example, in Teflon, with only a slight twist, because there are 13 monomer units per turn of the helix, the computed theoretical value and the experimental lattice value are 1.6×10^{12} dyne/cm^2 and 1.5×10^{12} dyne/cm^2, respectively; in polypropylene, with three monomer units per turn, the computed theoretical values are in the range 1 to 4.9×10^{11} dyne/cm^2, and the observed lattice modulus is about 4.1×10^{11} dyne/cm^2.

Transverse modulus values in fibers or uniaxially drawn films are more difficult to calculate and there is quite a wide spread in the theoretical values that are available, depending on the nature of the analysis and the assumptions involved. However, theoretical and experimental values are in fair agreement and both show that transverse moduli are only of the order of 1/50 to 1/100 of the longitudinal modulus (12).

Many polymer films are not uniaxially but biaxially drawn. This is done so that the anisotropy in the plane of the film is greatly reduced. In this case the longitudinal and transverse strengths and moduli are comparable in value and there is little tendency for the material to tear in any given direction, as can happen in uniaxially oriented films. The strength and stiffness are, of course, reduced compared to values in the longitudinal direction of uniaxially drawn film, but they are still considerably larger than those recorded for an isotropic cast or molded specimen of the same material.

2. Glass and Melting Transitions and Effects of Temperature

The mechanical characteristics of a given polymer are dependent upon the ambient temperature and the location of this ambient temperature relative to the glass-transition temperature, T_g, and the melting temperature, T_m. A tabulation of T_g values for a representative group of common amorphous

Table 1 Glass-transition temperatures for various amorphous polymers

Polymer	Designation	Glass-Transition Temperature, T_g ($^\circ$C)
Silicon rubber		-123
Polyisobutylene	PIB	-75
Natural rubber	NR	-72
Poly(n-butyl vinyl ether)	PnBVE	-55
Poly(ethyl vinyl ether)	PEVE	-40
Poly(isobutyl vinyl ether)	PiBVE	-20
Poly(n-butyl acrylate)	PnBA	-55
Poly(n-propylene acrylate)	PnPA	-45
Polyethyl acrylate	PEA	-20
Polymethyl acrylate	PMA	10
Polypropylene, atactic	PP	-20
Polyvinyl propionate	PVPr	12
Polyvinyl acetate	PVA	30
Polyethyl methacrylate	PEMA	65
Polymethyl methacrylate	PMMA	105
Polystyrene	PS	100
Polycarbonate	PC	150
Poly(n-vinyl carbazole)	PnVCb	211

polymers is given in Table 1 and a tabulation of both T_g and T_m values for a representative group of partially crystalline polymers is given in Table 2. (See Chapter 6 for a more complete tabulation.) In some highly crystalline polymers, like PE and POM, the location of T_g is controversial and several different values have been suggested.

From Tables 1 and 2 we can draw the following general conclusions as to the mechanical behavior of any given polymer at any given temperature.

1. If the ambient temperature is below the T_g of a given polymer we can expect that polymer to exhibit a tensile modulus value in the neighborhood of 10^{10} to 10^{11} dyne/cm^2. The material will show elastic or linear viscoelastic behavior at low strains and either plastic deformation or fracture at high strains, depending on its specific structure, the rate of loading, and the state of stress applied.

2. If the polymer is amorphous and the ambient temperature is well above T_g, then we can expect viscous flow or high elasticity behavior, depending on the molecular weight and/or the absence or presence of cross-linking.

Table 2   Glass-transition and melting temperatures for partially crystalline polymers

Polymer	Designation	Glass-Transition Temperature, T_g (°C)	Melting Temperature, T_m (°C)
Polyethylene	PE	−125, −33	137
Polyoxymethylene	POM	−70, 0	181
Poly(cis-isoprene)	PcI	−70	28
Poly-1-octene, isotactic	POc	−65	−38
Poly-1-pentene, isotactic	PPen	−40	70
Poly-1-butene, isotactic	PB	−25	125
Polypropylene, isotactic	PP	−20	165
Polyvinylidene chloride	PVC²	−15	198
Poly(4-methyl-1-pentene)	P4MP1	20	250
Polymethyl methacrylate, isotactic	PMMA(iso)	45	160
Polymethyl methacrylate, syndiotactic	PMMA(syn)	115	200
Polychlorotrifluoroethylene	PCTFE	45	220
Polytetrafluoroethylene	PTFE	−33, 126	327
Polyvinyl chloride	PVC	87	212
Nylon-6,10	6-10	40	227
Nylon-6,6	6-6	50	265
Polyethylene terephthalate	PET	70	267
Polyacrylonitrile	PAN	104	

* See also Table 3, Chapt. 6.

Amorphous polymers that have negative T_g values (on the Celsius temperature scale) show rubberlike behavior at room temperature if their molecular weights are sufficiently high to give physical entanglements.

3.   The transition from glass like to rubbery behavior can occur in widely different temperature ranges, depending on the structure of the polymer. It occurs near −123°C for silicon rubber where the barrier-to-bond rotation is known to be low, and above 200°C for poly(n-vinyl carbazole), where rotational barriers are high owing to the presence of a very stiff side chain.

4.   The precise location of T_g, and hence also the mechanical behavior, depends on many factors. With increase of length of a flexible side branch, as in the series from polymethyl acrylate, PMA, to poly(n-butyl acrylate), PnBA, the glass-transition temperature, and hence the accompanying three-decade change in modulus, moves progressively to lower temperatures. This

behavior is also evident on comparing PMMA to polyethyl methacrylate, PEMA.

5. Bulky side groups tend to increase steric effects and chain stiffness and hence raise T_g. This is evident from the data of Table 1 by comparing poly(isobutyl vinyl ether), PiBVE, to poly(n-butyl vinyl ether), PnBVE, or by comparing PS to amorphous PP. The increased polarity of the side group is also a factor in this latter case.

6. Partially crystalline polymers possess essentially two major softening regions, one in the neighborhood of T_g, where chain cohesion in the amorphous phase is sufficiently overcome to allow micro-Brownian motions, and one near T_m, where cohesion in the crystalline phase is overcome. In the temperature range below T_g we expect elastic or viscoelastic behavior with modulus values above 10^{10} dyne/cm^2, and above T_m we expect flow or rubber-like behavior and low modulus values of the order of 10^7 dyne/cm^2 or less, depending on the molecular weight. When the ambient temperature lies between T_g and T_m, as it does for many of crystalline polymers, then the polymer is fairly stiff and strong because of the rigidity of the crystalline regions and, at the same time, very tough because of the flexibility of the polymer chains in the amorphous regions. Modulus values lie somewhere between 10^7 and 10^{10} dyne/cm^2, depending on the specific temperature of measurement and the degree of crystallinity of the sample.

This effect is illustrated graphically by Fig. 2, where dynamic mechanical measurements of modulus, taken at audio frequencies, are presented for crystalline and amorphous PP, for poly-1-butene, PB, and for poly-1-pentene, PPen (33, 34). The amorphous PP shows only a single modulus transition in the vicinity of its T_g (see Table 1), whereas the partially crystalline PP shows two dispersion regions in which the modulus drops fairly sharply, one just above T_g and near room temperature, and the other above 400°K and near the melting point. PB has a somewhat lower T_g and a lower T_m than polypropylene, as Table 2 shows, and hence the two modulus transition regions for this polymer occur at somewhat lower temperatures than for PP. Figure 2 also shows that both PB and PPen show a low-temperature relaxation region near 150°K. This is an example of a secondary relaxation, or so-called γ relaxation. The γ relaxation is here associated largely with onset of re-orientational motion of the short side chain branches, and it plays a part in the mechanical behavior of the bulk polymer, for it contributes to chain flexibility.

7. In crystalline polymers where both T_g and T_m are above room temperature, the polymer has high rigidity at room temperature with modulus values of 10^{10} dyne/cm^2 or higher. However, whether the polymer is brittle in behavior or not depends on whether additional secondary low-temperature relaxations, preferably involving some backbone chain motion, are present.

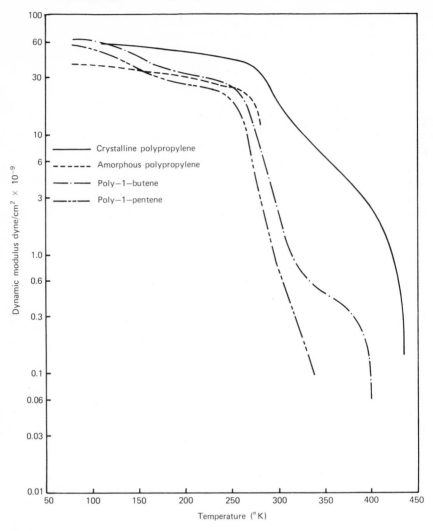

Figure 2   Temperature variation of elastic modulus of polypropylene (crystalline and amorphous), polybutene, and polypentene (33,34).

Where significant low temperature relaxations are not present, as in isotactic PS, the polymer is brittle and shows essentially linear elastic behavior to fracture. Where significant low-temperature relaxations are present, as in PC, polyethylene terephthalate (PET), and the nylons, the polymer yields prior to fracture, and significant plastic deformation and hence toughness are present.

Strong polar forces also have an influence, because by raising the melting temperature (compare the melting temperatures of nylon or PET with those of nonpolar PE, PP, or PB), they considerably extend the temperature range in which the polymer exhibits these desirable properties of both strength and toughness.

3.   Modes of Failure and Significant Mechanical Parameters

In subsequent sections we discuss various theories of mechanical behavior of polymers and give examples of how properties are affected by physical and chemical structure and other variables. In this section we outline the various modes by which polymer components or specimens may fail, either in the laboratory or in service, and define the pertinent mechanical parameters.

Static Strength Considerations

Polymers may fail in specific applications simply because they do not possess the necessary strength to carry the design load, or occasional overloads. To assess this quality of load-carrying ability, the mechanical properties of most interest are the strength of the polymer at yield, σ_y, and the ultimate strength, σ_u, that the polymer can withstand just prior to final fracture. It is customary to determine these stresses by dividing the yield or ultimate load by the original area of the specimen. The stresses so obtained are then referred to as nominal stresses to distinguish them from true stresses, which are values determined by dividing the loads by the instantaneous specimen areas. It is also desirable to know the values of these strength parameters under various conditions of loading, such as uniaxial tension, uniaxial compression, and simple shear.

Polymers are also frequently subjected in various applications to combined stresses. Under these conditions, yield or failure is not governed simply by stress or strain characteristics alone. Hence suitable theories of yielding under combined stresses must be developed that will accord with laboratory tests and that will provide a logical basis for design.

Stiffness Considerations

Polymers may fail to perform satisfactorily in a given application because of lack of rigidity and stiffness. This quality is becoming more important as polymers are considered for structural, load-bearing applications. It is essential, therefore, to measure the resistance of polymers to applied tensile or compressive strains, to shear strains, and to volume strains. The appropriate mechanical parameters of interest are the Young's modulus, E, the shear modulus, G, and the bulk modulus, K. If the material obeys the laws of linear elasticity and is isotropic, these three parameters, as well as Poisson's ratio v, are all related and determination of any two enables one to compute the others. The appropriate equations are (12, 13):

$$G = \frac{E}{2(1 + v)} \tag{1}$$

$$K = \frac{E}{3(1 - 2v)} \tag{2}$$

and

$$E = \frac{9KG}{3K + G} \tag{3}$$

Strain Considerations

In some applications, the polymer may have sufficient stiffness and strength but its usefulness may be limited by the amount of strain or deflection that it can carry prior to failure or to impairment of its function. Therefore one would like to know the magnitude of the strain, ε_y, that the polymer develops at the yield conditions as well as the magnitude of the strain, ε_u, that the polymer shows at final fracture.

The various strain, stress, and modulus parameters discussed above can be readily determined from a stress–strain curve. For a nylon-6,6 sample containing 2.5% moisture and tested at 73°F, a typical stress–strain curve is shown in Fig. 3 (35). The stress at the well-defined maximum in the curve, here 8500 psi, is taken as the yield stress, and the slope of the initial portion of the curve gives the elastic modulus. For this material, the modulus varies with amount of water present in the polymer. For 2.5% water it is about 175,000 psi and it rises to about 400,000 psi at a water content of 0.2%.

The stress–strain curve of Fig. 3 also enables one to obtain—for the given rate of loading, temperature, and pressure conditions prevailing during the test—other important mechanical property parameters. Thus the strain at yield is here 25%, the ultimate elongation is 300%, and the nominal ultimate strength is 11,000 psi.

Some polymers do not show a yield drop. Even polymers that do show a yield drop under one set of test conditions may not do so for other test conditions of temperature, pressure, and strain rate. When no yield maximum and subsequent drop are present, it is customary to define a yield stress by a so-called offset method. A detailed discussion of this subject follows, in Section D.

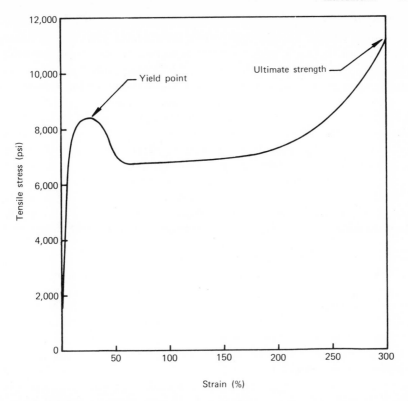

Figure 3 Tensile stress–strain curve of nylon-6,6 taken at a temperature of 73°F and a sample moisture content of 2.5 % (35).

Effects of Impact Loading

Polymers may fail in service because of being subjected to impact loads rather than to static loads. To assess the ability of polymers to withstand dynamic effects, the most significant parameter would be the area under the stress–strain curve obtained for the appropriate rate of loading. However, a more conventional parameter that is used for measuring the relative toughness of polymers is impact energy. This latter can be measured in various ways, but it is essentially a measure of the amount of energy it takes to break a standard size specimen under specified loading conditions. In the conventional Izod impact test, the recorded energy is that required to break a $\frac{1}{2} \times \frac{1}{2}$ in. rectangular specimen containing a V-notch and supported as a cantilever beam with the clamp just below the notch line.

Effects of Time

Polymers can fail because of excessive deformation with time. Polymers are much more susceptible to effects of time and temperature than are metals. Hence it is essential to study in the laboratory creep and stress relaxation behavior. Creep refers to continued deformation with time under constant stress or load conditions, and important mechanical parameters are the creep rate at a given stress level and time and the time to rupture. Stress relaxation refers to the reduction of initial stress with time under constant elongation conditions, and an important mechanical parameter is the relaxation modulus $E_r(t)$. This quantity is defined as the ratio of the value of the relaxed stress, at some given time t, to the initial elongation.

Both creep and stress relaxation are consequences of the viscoelastic nature of polymer materials, and insight into time-dependent behavior can be obtained from analysis of comparatively simple models.

Effects of Repeated Loading

Polymers are subject, in many applications, to alternating or repeated loads; under these conditions, they, like metals and other materials, may fail at stress levels much lower than those expected on the basis of static stress–strain curves. Hence it is essential to have laboratory data concerning the fatigue resistance of polymers as well as to be aware of the variables that affect the initiation and subsequent growth with time of microcracks. It is customary to represent fatigue data by an $S-N$ curve, which is a plot of the log of the number of cycles of alternating load to fracture, N, versus the maximum stress level in the specimen, S. An important mechanical parameter in fatigue testing is the endurance strength, or limit, S_e. This value is usually taken as the stress below which no failure occurs regardless of the number of load cycles, or as the stress at which the $S-N$ curve intersects the 10^7 line.

B. ELASTICITY OF RUBBERLIKE MATERIALS

Rubber elasticity is unique to macromolecular substances. All linear polymers that are lightly cross-linked, or are of sufficiently high molecular weight to produce physical entanglements between chains, exhibit the phenomenon of rubber elasticity in an appropriate range of temperatures. In this section, examples of rubber elastic behavior in various polymers are presented. Consideration is also given to the effects of various variables, such as temperature, pressure, and degree of cross-linking, on the mechanical properties of rubberlike materials. First, however, it is appropriate to discuss the theory of rubber elasticity, which is based on the concepts of equilibrium thermodynamics and statistical mechanics.

1. Kinetic Theory of Rubber Elasticity

Contributions to the theory of rubber elasticity have been made by many scientists, including Kuhn, Guth and James, Treloar, and Flory. The reader is referred to various textbooks for a fuller treatment of the subject (3, 4, 14, 15).

The necessary conditions for rubber elasticity are that we have a three-dimensional network of chains, that chain units are flexible and individual chain segments rotate freely, that no volume change accompanies deformation, and that the process is reversible. It is also usually considered, as a result of the free rotation assumption, that the internal energy, U, of the system does not change with deformation.

In the general case, an expression for the equilibrium tensile force, f, can be obtained from consideration of the first two laws of thermodynamics. From the first law, for any infinitesimal change of state,

$$dQ = dU + dW \qquad (4)$$

where dU is the internal energy change and dW the work done by the system. However, from the second law we may write

$$dS = \frac{dQ}{T} \qquad (5)$$

The change in internal energy then becomes

$$dU = T \, dS - dW \qquad (6)$$

If now the system is given a small displacement dl the work done by the system is

$$dW = p \, dV - f \, dl \qquad (7)$$

and the internal energy change becomes

$$dU = T \, dS + f \, dl - p \, dV \qquad (8)$$

Consider now the change of U with respect to l for constant temperature and constant volume conditions. This gives

$$\left(\frac{\partial U}{\partial l}\right)_{T,V} = T\left(\frac{\partial S}{\partial l}\right)_{T,V} + f \qquad (9)$$

or

$$f = \left(\frac{\partial U}{\partial l}\right)_{T,V} - T\left(\frac{\partial S}{\partial l}\right)_{T,V} \qquad (10)$$

The equilibrium tensile force is thus made up of two parts. One part arises from change of internal energy with deformation and the second part arises from the decrease of entropy with increased deformation.

An ideal rubber is defined as one in which

$$\left(\frac{\partial U}{\partial l}\right)_{T,V} = 0 \tag{11}$$

When this condition holds the only contribution to the force, f, arises from the entropy term; that is,

$$f = -T\left(\frac{\partial S}{\partial l}\right)_{T,V} \tag{12}$$

For a nonideal rubberlike material one would expect that the internal energy, U, of the elastomer would change somewhat as the polymer is stretched, since changes in length and lateral contractions would cause changes in both intermolecular and intramolecular distances. However, in loose network structures such as typical rubbers, it appears that this effect is small and hence, in simple theoretical treatments, the change of internal energy with deformation is generally neglected.

The quantity that does change significantly as a rubberlike material is deformed is its entropy, S. The long flexible polymer chains of a rubber tend, at any given temperature, to adopt a most probable distribution, and this distribution is a highly coiled one. Upon stretching, the chain segments tend to uncoil and straighten, and the new distribution changes to a less probable one. If the applied force is now removed, the system wants to return to the more probable coiled-up state, that is, the entropy of the system increases. In fact, the essential problem of rubber elasticity theory is the statistical mechanical problem of determining the change in entropy in passing from an undeformed state to a deformed state.

Consider a unit cube of rubberlike material. The retractive force f is now the same as the internal stress σ. Also, it is customary in rubber elasticity theory to represent deformation by a so-called extension ratio, α. This ratio is defined in terms of the elongation, e, and the nominal unit strain, ε, as follows:

$$\alpha = \frac{l}{l_0} = \frac{l_0 + e}{l_0} = 1 + \varepsilon \tag{13}$$

where l_0 is the original length in the direction of the applied load and l is the final length after the load is applied. In our case of a unit cube with $l_0 = 1$, dl is the same as $d\alpha$ and Eq. 12 may now be written as

$$\sigma = -T\left(\frac{\partial S}{\partial \alpha}\right)_{T,V} \tag{14}$$

The most probable state of a system is that which has the greatest number of ways, Z, of being realized. The entropy S is related to this number of complexions, Z, by the Boltzmann relation

$$S = k \ln Z \qquad (15)$$

The problem is to determine, from this relation, the change in entropy that occurs when the specimen is extended. This has been done and it is now known that the change in entropy, for a given extension ratio α, is given by

$$\Delta S = -\tfrac{1}{2} N k \left(\alpha^2 + \frac{2}{\alpha} - 3 \right) \qquad (16)$$

where N is the number of freely orienting chain segments that are present.

On substituting this expression for the entropy change into Eq. 14 we find for the stress–extension ratio relation

$$\sigma = N k T \left(\alpha - \frac{1}{\alpha^2} \right) \qquad (17)$$

This relation is frequently given in a slightly different form in order to show the dependence of the stress on the density, ρ, and on the average molecular weight between cross-links, M_c. To obtain this other form, note that we can express the number of chains in unit volume by $N = (N_0/M_c)\rho$, where N_0 is Avogadro's number, (N_0/M_c) is the number of chains per gram, and ρ is the mass in grams of unit volume. Since $N_0 k = R$, the gas constant, Eq. 17 becomes

$$\sigma = \rho \frac{RT}{M_c} \left(\alpha - \frac{1}{\alpha^2} \right) \qquad (18)$$

Equations 17 and 18 are remarkable in the sense that they indicate that the stress extension properties of any rubber are independent of its structure. The only factors that matter, other than the temperature, are the density, ρ, and the average molecular weight, M_c, between cross-links or entanglement points.

From Eq. 18 we can obtain an equation for the elastic modulus of a rubber in the low-strain region. In terms of the strain ε, Eq. 18 can be written

$$\sigma = \frac{\rho R T}{M_c} (3\varepsilon - 3\varepsilon^2 + \cdots) \qquad (19)$$

and hence for small strains, where we can neglect the square of ε compared to ε, we get

$$E = \frac{d\sigma}{d\varepsilon} = \frac{3\rho R T}{M_c} \qquad (20)$$

Equation 20 predicts that the modulus of a rubber should be directly proportional to the absolute temperature, and this behavior is found experimentally. In fact, from the observed temperature dependence of the modulus and the known density, one can use Eq. 20 to estimate the average molecular weight between cross-links and hence to determine the cross-link density.

Although Eqs. 17 and 18 show that the tensile stress–strain relation in an ideal rubber is a nonlinear one, it has been shown that the shear stress–shear strain relation is linear (14). The value of the shear modulus can be obtained from Eq. 20 by noting that for an ideal rubber Poisson's ratio is $\frac{1}{2}$ and hence, from elasticity theory (Eq. 1), $G = E/(2)(1 + v) = E/3$.

Therefore the shear modulus of a rubber is given by

$$G = \frac{\rho RT}{M_c} \tag{21}$$

and this equation is applicable for both small and large strains.

More sophisticated treatments have indicated that Eqs. 17 and 18 should be slightly modified to take proper account of the effects of finite molecular weight. According to one such modification the appropriate stress–extension equation is

$$\sigma = \frac{\rho RT}{M_c}\left(\alpha - \frac{1}{\alpha^2}\right) \cdot \left(1 - \frac{2M_c}{M_n}\right) \tag{22}$$

where M_n is the number-average molecular weight of the polymer prior to cross-linking. For most conditions $M_c \ll M_n$ and hence Eq. 22 reduced to Eq. 18.

2.   Experimental Data and Effects of Temperature

Despite the many assumptions that underlie the derivation of the stress–extension relation given by Eq. 18, the experimentally observed stress–strain curves are in quite good agreement with the theory for both natural and synthetic rubbers.

This is shown, for example, in Fig. 4, where the results of theory (dashed line) and experiment (unbroken line) are compared for the case of a vulcanized hevea rubber (17). The agreement is good except at very high strains, where the experimental results and the predicted performance begin to depart. The additional stiffening may be due to the fact that many rubbers crystallize when highly stretched. Below the strain value at which crystallization develops ($\sim 300\%$), the retractive force in the rubber is almost entirely that to be expected based on entropy effects alone.

Equation 20 indicates that the modulus of a rubber should increase linearly with increasing temperature. This effect can be shown by irradiation

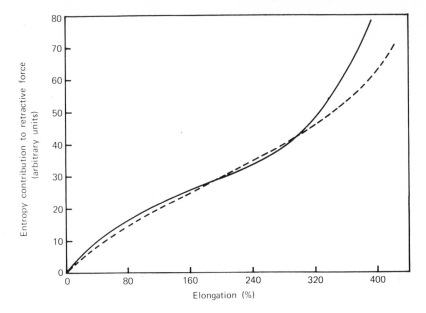

Figure 4 Comparison of theory (dashed line) and experiment (solid line) for a vulcanized hevea rubber (17).

of a material not normally considered as a rubber, such as high-density polyethylene, and then observing the modulus behavior in the region above the melting point of the unirradiated material. The effect of the radiation is to produce cross-links, as well as to reduce crystallinity, and when the cross-link density gets sufficiently high the chains are linked into a flexible three-dimensional network and the material shows rubberlike behavior above its melting temperature.

The modulus versus temperature curve for five samples of PE subjected to various pile irradiation dosages are shown in Fig. 5 (36). Note that above 410°K the unirradiated material simply softens and converts to a viscous liquid. However, for the material subjected to a dose of 0.6×10^{18} nvt in the Brookhaven reactor, modulus measurements were made up to 450°K and, as the data show, in the region from above 410 to 450°K the modulus–temperature relation is a linear one. In fact, from the observed slope and from use of rubber elasticity theory, the percent cross-linking was estimated as 5.6% for a dose of 0.6×10^{18} nvt and 11% for a dose of 1.1×10^{18} nvt.

Radiation also has an effect on the mechanical properties of the polyethylene below the melting temperature, and these changes are largely a result of loss of crystallinity owing to the irradiation. This is shown, for

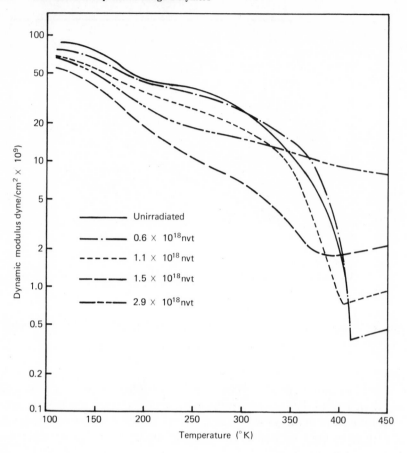

Figure 5   Modulus versus temperature for samples of irradiated polyethylene (36).

example, by the drop in modulus in the room-temperature region and below. Note, however, that for the very highest dosage of 2.9×10^{18} nvt the modulus again starts to increase. Evidently at the very high dosages the additional stiffening effect of an increasing degree of cross-linking between chains more than counterbalances the increased flexibility due to loss of crystallinity.

When the temperature of a natural or synthetic rubber decreases the modulus is reduced, in accordance with Eq. 20, so long as we remain in the temperature region where the assumptions involved in the derivation of the equations are applicable. However, as we reach or pass below the T_g of that sample, the rubber converts to a glassy polymer and the modulus rises rapidly in a narrow temperature interval by some three decades. The temperature at which this change from glassy to rubbery behavior occurs depends on the degree of cross-linking that is present.

This effect is shown in Fig. 6, taken from the data of Schmeider and Wolf (37). Here measurements are given as a function of temperature of the shear modulus of rubber samples having various amounts of sulfur added as a cross-linking agent. With 2 % sulfur the material behaves like a typical rubber,

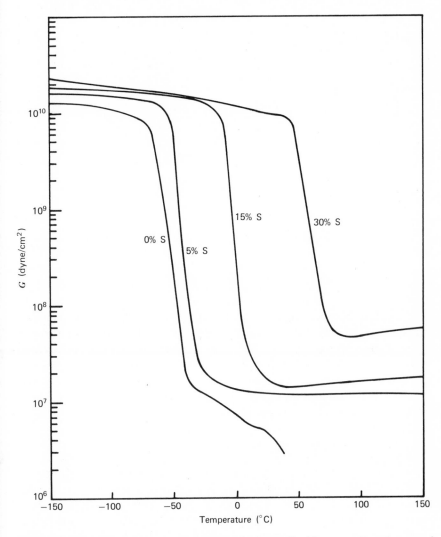

Figure 6   Shear Modulus versus temperature for natural rubber at various degrees of vulcanization (37).

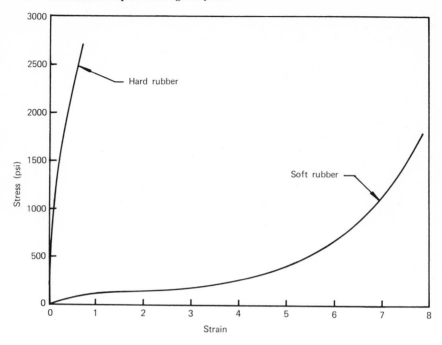

Figure 7   Stress–strain curves of soft rubber and hard rubber (38).

with the glass–rubber transition occurring at about $-60°$C. However, as the percent of sulfur is increased, the transition shifts up the temperature scale. For a hard rubber with about 30% sulfur the T_g is well above room temperature and the shear modulus at room temperature is about 10^{10} dyne/cm^2.

The distinction between a hard rubber, which at room temperature is really a rigid plastic rather than a rubber, and a soft rubber is even more vividly illustrated by comparison of their respective stress–strain curves. Such a comparison is given in Fig. 7 (38), and one notes a striking difference in their respective mechanical responses to applied stress.

3.   Effects of Hydrostatic Pressure

If the temperature of a rubber is decreased, the rubber passes through a rubber–glass transition region and, at temperatures below T_g, the modulus of the polymer reaches values of 10^{10} dyne/cm^2 or higher. Imposition of hydrostatic pressure can have a similar effect, even at room temperature, on the mechanical characteristics of a rubber. In Fig. 8 we show how the elastic modulus and yield stress of a commercial vulcanized rubber vary with

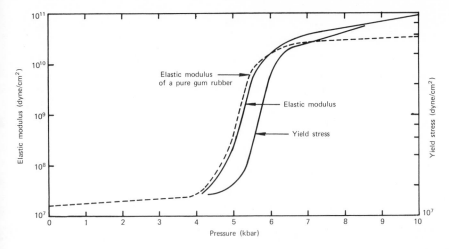

Figure 8   Variation of modulus and yield stress of vulcanized rubber with pressure (39).

magnitude of hydrostatic pressure (39). From Fig. 8 we note the following:

1.   At low pressures, up to about 4 kbars, the modulus is of the order of 10^7 dyne/cm^2 and the material behaves as a pure gum rubber (dotted line).

2.   In the region from 4 to 6 kbar there is a dramatic change in properties, with the modulus increasing by about three decades and the yield stress rising from less than 2×10^7 to about 5×10^8 dyne/cm^2.

3.   Above 6 kbar the material behaves like a typical thermoplastic, with a modulus above 10^{10} dyne/cm^2. Also its stress–strain curve is similar to that of other thermoplastics, with essentially a linear elastic region up to about 2% strain and then a region of decreasing slope beyond.

4.   In the region from 6 to 10 kbar both the elastic modulus and yield stress continue to rise with increasing pressure; similar examples of this behavior are demonstrated later for various thermoplastics. The ratio of yield stress to Young's modulus appears to be a constant (0.012) essentially independent of pressure.

From the test results reported above and those discussed earlier it is evident that a rubber can undergo a temperature-induced glass transition at atmospheric pressure or can experience a pressure-induced glass transition at room temperature.

4.   Constitutive Relations for Large Strains

The kinetic theory of rubber elasticity agrees quite well with experimental data on vulcanized rubbers, as Fig. 4 indicates, so long as the extension ratios are not too large. However, for large extension ratios (>3) the theory no longer gives a good representation of the experimental data. This is a problem of some concern, for rubberlike materials are frequently used in engineering applications at large extension ratios, as, for example, in high-altitude balloons.

To overcome the discrepancy at large strains, many scientists and engineers have proposed alternative forms of the constitutive relations to Eq. 17. In all the proposed theories, the strain energy density, W, is taken to be some function of the strain, or extension ratio invariants I_1, I_2, and I_3. If we let α_1, α_2, and α_3 be the specimen extension ratios along three Cartesian coordinate axes, then the strain or extension ratio invariants are

$$I_1 = \alpha_1^2 + \alpha_2^2 + \alpha_3^2$$
$$I_2 = \alpha_1^2\alpha_2^2 + \alpha_2^2\alpha_3^2 + \alpha_3^2\alpha_1^2 \qquad (23)$$
$$I_3 = \alpha_1\alpha_2\alpha_3$$

It is usually assumed that rubberlike materials are incompressible. If this is the case, then the third invariant, I_3, reduces to unity. Hence the strain energy can be written

$$W = f(I_1 I_2) \qquad (24)$$

Alexander (40) investigated experimentally, for a neoprene synthetic rubber, the validity of the assumption of constancy of volume over extension ratios ranging from 1.0 to 4.2. Over this wide range of applied strains, he found that the measured value of I_3 was very close to one and hence the assumption of incompressibility, at least for the elastomer tested, is confirmed.

The specific nature of the strain energy function W is not known, but various simple forms of this function have been suggested by different authors. Some of these suggested forms are (40) as follows.

Statistical theory

$$W = C_{10}(I_1 - 3) \qquad (25)$$

Mooney theory

$$W = C_{10}(I_1 - 3) + C_{01}(I_2 - 3) \qquad (26)$$

Three-Term theory

$$W = C_{10}(I_1 - 3) + C_{20}(I_1 - 3)^2 + C_{01}(I_2 - 3) \qquad (27)$$

Rivlin–Saunders theory

$$W = C_{10}(I_1 - 3) + f(I_2 - 3) \qquad (28)$$

Hart–Smith theory

$$W = G\left(\int \exp[k_1(I_1 - 3)^2]\, dI_2 + k_2 \ln \frac{I_2}{3} \right) \qquad (29)$$

Here the C's the k's, and G are all experimental parameters to be determined. Similarly, in the Rivlin–Saunders theory the unknown function $f(I_2 - 3)$ is to be determined experimentally.

Once a specific form for the strain energy density function has been assumed, one obtains the desired constitutive relations by use of the Rivlin equation (41):

$$\sigma_i = 2\left[\alpha_i^2 \frac{\partial W}{\partial I_1} - \frac{1}{\alpha_i^2} \frac{\partial W}{\partial I_2} \right] + p \qquad (30)$$

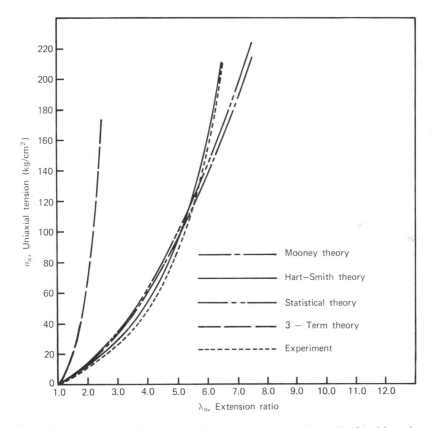

Figure 9 Comparison of experimental tensile stress–strain data (40,42) with various proposed theories for a rubber with 8% sulfur.

Here p is the hydrostatic pressure and i takes on values 1, 2, and 3. The desired constitutive equations for any specific theory can then be obtained by substituting into Eq. 30 the various assumed forms of the strain energy function given by Eqs. 25 to 29.

To illustrate how the various theories compare with one another and with experimental data, we present in Fig. 9 results obtained by Treloar (42) for uniaxial tension. The three-term theory is a poor representation of the data in this case, but is a good representation of the data for biaxial loading (42). The more complex Hart–Smith theory gives a good representation of the experimental data for both uniaxial and biaxial tension. For a synthetic neoprene rubber, Alexander (40) found that a theory combining features of both the Rivlin–Saunders theory and the Hart–Smith theory gave an excellent representation of the stress–strain response up to extension ratios of eight or higher.

C. ELASTICITY AND DEFORMATION BEHAVIOR OF SOLID POLYMERS

Polymers are now being used in so many different machine and structural components that their mechanical behavior under a wide variety of stress conditions, both simple and combined, must be investigated and established if these macromolecular materials are to be used efficiently and economically. For simple tension, compression, and flexure, standardized ASTM procedures are available for determination of mechanical properties. For parts subject to more complex stress states, it is necessary to utilize the information gathered from the simple tests together with criteria for yielding and fracture that are applicable for polymers. This matter is treated more fully in Section D.

1. Mechanical Properties in Uniaxial Tension

A standard method of determining the tensile properties of plastics is specified by ASTM Standard D638-64T. The specimen may be rectangular or circular in cross section. It usually has a reduced section at the center and larger dimensions at the ends where the specimen engages the grips of the apparatus that applies the tensile load. The speed of testing should be controlled so that comparisons of data obtained in different laboratories are valid. A recommended testing speed for many polymers is 1/20 in./min.

The elongation or strain in the specimen at various values of the applied load, or stress, is measured over a constant gauge length of the specimen by means of mechanical gauges, optical gauges, or electrical strain gauges. The experimental data are plotted in the form of a nominal stress–nominal strain plots. If the cross-sectional area is monitored one can plot the data in the form of true stress versus true strain curves.

The nominal stress, σ, is given by

$$\sigma = \frac{P}{A_0} \tag{31}$$

where P is the applied load, usually in pounds, and A_0 is the original cross-sectional area, usually given in square inches.

The elongation, e, is given by

$$e = l - l_0 \tag{32}$$

where l refers to the instantaneous length of the gauge section at any moment and l_0 refers to the original length.

The uniaxial tensile strain, ε, is defined as the elongation per unit of length, and hence

$$\varepsilon = \frac{l - l_0}{l_0} \tag{33}$$

A typical tensile stress–strain curve for a thermoplastic polymer has been given in Fig. 3. Over some finite range of strain the polymer exhibits Hookean behavior; that is, the stress–strain relation is essentially linear. Hence over this range we may write

$$\sigma = E\varepsilon \tag{34}$$

where E, the so-called Young's modulus of elasticity, is given by the slope of the initial portion of the stress–strain curve.

In general, as a polymer specimen is extended in one direction it will contract in a transverse direction. The ratio of this lateral strain, ε_T, to the longitudinal strain, ε, is Poisson's ratio, v. Therefore

$$v = -\left(\frac{\varepsilon_T}{\varepsilon}\right) \tag{35}$$

and the minus sign is introduced simply to keep v positive. The lateral strain can be measured by use of suitable strain gauges or directly by monitoring the lateral dimensions. If the specimen is circular and D represents the instantaneous value of the specimen diameter and D_0 the initial value, then the nominal value of the lateral strain is given by

$$\varepsilon_T = \frac{D - D_0}{D_0} \tag{36}$$

To make a true stress, σ', versus true strain, ε', plot, rather than a nominal stress versus nominal strain plot, the actual cross-sectional area must be measured during the loading period. If we let A be the instantaneous value of this area, then

$$\sigma' = \frac{P}{A} \tag{37}$$

The true strain ε' can be obtained, provided the specimen cross-sectional area remains uniform over the gauge length, directly from values of the nominal strain; that is,

$$\varepsilon' = \int_{l_0}^{l} \frac{dl}{l} = \ln\left(\frac{l}{l_0}\right) = \ln\left(\frac{l_0 + e}{l_0}\right) = \ln(1 + \varepsilon) \tag{38}$$

The yield stress is taken as the stress at which the slope of the $\sigma-\varepsilon$ curve in the vicinity of the linear region is zero. If no maximum is present, it is customary to define an arbitrary yield stress by an offset method. For metals, an offset strain of 0.002 is used, but in polymers it is usual to take the offset strain to be 0.02. When a maximum is present in the $\sigma-\varepsilon$ curve, the strain at yield can be much larger than that obtained using the offset method. For example, for the nylon-6,6 sample shown in Fig. 3 it is 25%.

The ultimate strength, σ_u, generally refers to the maximum nominal stress. Somewhat higher values are obtained if true stresses, based on instantaneous area values, are used. However, if necking has occurred prior to fracture, as is frequently the case, then even this value, σ'_u, is only approximate, for once necking has occurred, the stress state is no longer uniaxial tension.

The percent ductility of the specimen is simply the strain, ε_u, at fracture multiplied by 100. In polymers tested at room temperatures and at usual rates of straining, the elongation at fracture may vary from very low values, $\sim 1.5\%$, as for some types of polystyrene, to very high values, $\sim 500\%$ or more, as for some polyethylenes.

Stress–Strain Curves of Polymers in Tension

Because of variations in external factors, such as strain rate, temperature, pressure, and environment, and because of variations in internal factors, such as chemical structure, crystallinity, molecular weight, molecular weight distribution, and presence of water or plasticizer, it is not possible to give precise values of the static mechanical properties. We therefore present examples of typical stress–strain curves for various commercial polymers, mostly obtained in our laboratory at an elongation rate of 0.1 in./min, and then give a table of mechanical property data for typical polymers.

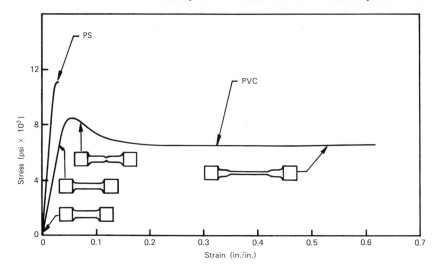

Figure 10   Tensile stress–strain data for two amorphous polymers: polyvinyl chloride (PVC) and polystyrene (PS). Sketches show modes of specimen deformation at various points on stress–strain curve.

To illustrate the broad range of tensile properties available in plastics, we present in Fig. 10 tensile stress–strain data obtained in our laboratory on two typical amorphous polymers having varying degrees of ductility, namely, polyvinyl chloride (PVC), and polystyrene (PS), and in Fig. 11 the tensile stress–strain data for two typical crystalline polymers, namely, polyethylene (PE) and polytetrafluoroethylene (PTFE).

From the tensile stress–strain curves it may be noted that the various thermoplastics differ considerably from one another in elastic modulus, in yield stress, and in elongation to fracture. Nevertheless, over small ranges of strain, both amorphous and crystalline polymers show essentially linear elastic behavior. Here the strain, which arises from bond angle deformation and bond stretching, is essentially instantaneously recoverable. As the strain is further increased, the phenomenon of strain softening sets in, that is, the instantaneous modulus of the material, the slope of the σ–ε curve, reduces. In molecular terms this strain is associated with some uncoiling and straightening of polymer chains, and it is largely recoverable. In many polymers a yield maximum is soon reached. At this point no increase in applied stress is required to increase the deformation further. In fact, the nominal stress, after reaching a maximum value, the so-called yield stress, or upper yield stress, then falls. At this stage it is thought that some slippage

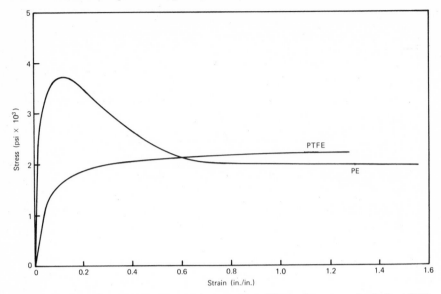

Figure 11   Tensile stress–strain data for two crystalline polymers: polyethylene (PE) and polytetrafluoroethylene (PTFE).

of chains past one another has occurred. Hence, upon unloading, recovery is slow and probably incomplete.

Until the yield stress is reached deformation is essentially homogeneous throughout the specimen, but, at the upper yield stress, an instability sets in and, as shown schematically in Fig. 10, the specimen begins to neck down at some particular point along its gauge length. For some polymeric materials like PVC and PE, this necked region soon stabilizes at some reduced diameter and deformation proceeds at essentially constant nominal stress, the so-called lower yield or drawing stress. Gradually the necked region propagates along the gauge length, as indicated for PVC in Fig. 10. In some cases, as in PTFE, there is no localized necking but only a gradual uniform reduction in cross-sectional area. In other materials, like PS, there is little yielding and fracture occurs at low strains.

For polymers that show a yield maximum, marked local deformation occurs in the vicinity of the necked region. If the material is transparent, deformation bands are visible. What occurs after the yield stress is reached depends on how effectively the polymer can strain-harden. As necking starts, the local stress at the necked region rises because of the reduced cross-sectional area. As the polymer is strained and the neck is developing, the material in the vicinity of the necked region is being extensively deformed and polymer chains in the amorphous regions are undergoing conformational changes and marked reorientation. This is shown, for example, by development of birefringence or by changes in the X-ray diffraction pattern. As deformation proceeds, the strength and stiffness of the original necked region rises. If orientation stiffening is great enough to overcome the increased stress owing to area reduction, then further extension of the specimen occurs only by propagation of the neck along the specimen, rather than by further contraction at the neck. If, however, the increased stress at the necked section rises faster than the increase in yield strength of the material owing to orientation, then the neck continues to deepen and finally local fracture occurs at that point.

As Fig. 11 indicates, many of the partially crystalline polymers exhibit the process of cold drawing at room temperature, and deformations of several hundred percent or more are possible without fracture. As cold drawing proceeds the initial spherulitic structure is broken down and converted to a fibrillar structure, and more and more of the polymer chains become aligned in the draw direction. Hence both modulus and strength in this direction increase. In polymers that cannot be cold-drawn without fracture similar increases in strength and stiffness can be realized by carrying out the drawing at elevated temperatures, preferably above T_g, and then rapidly quenching the polymer while in the oriented state.

Mechanical Property Data for Polymers Subjected to Tension

There are available today so many homopolymers, homopolymers with specific additives, copolymers, block copolymers, reinforced polymers, etc., that no simple tabulation can do justice to the subject. Nevertheless it is helpful to see the range of tensile mechanical property data that polymers exhibit at room temperature and to note what structural features give rise to strength, to stiffness, and to ductility. With this in mind, and recognizing the approximate nature of the values given, we present in Table 3 mechanical property data of various thermoplastic and thermosetting materials subjected to simple tension. The reader is referred to other sources for more complete data and for data on polymers not shown in Table 3 (11, 12, 18–20).

From inspection of Table 3 one notes the wide variety of mechanical properties available in the polymer field. Some conclusions concerning mechanical properties in tension are as follows:

1.   Nonpolar polymers like PE and PTFE tend to have relatively low modulus values for plastics, but great ductility and toughness.

2.   Polypropylene has somewhat higher strength and stiffness than PE, possibly owing to increased steric forces arising from the presence of a methyl group in each repeat unit.

Table 3 Tensile mechanical property data for various plastics

Polymer	Designation	Elastic Modulus (10^5 psi)	Yield Strength (10^3 psi)	Ultimate Strength (10^3 psi)	Elongation to Fracture (%)
Polyethylene, low ρ	PE	0.2–0.4	1–2	1.5–2.5	400–700
Polyethylene, high ρ	PE	0.6–1.5	2.5–5	2.5–5.5	100–600
Polytetrafluoroethylene	PTFE	0.6	1.5–2	2–4	100–350
Polychlorotrifluoroethylene	PCTFE	1.5–3	4–5	4.5–6	80–250
Polypropylene	PP	1.5–2.25	3–4	3.5–5.5	200–600
Nylon-6,6	6-6	1.8–4	8.5–11.5	9–12	60–300
Polycarbonate	PC	3.5	8–10	8–10	60–120
Polymethyl methacrylate	PMMA	3.5–5	7–9	7–10	2–10
Polyoxymethylene	POM	4	7–8	9–10	20–80
Polystyrene	PS	4–5	—	5.5–8	1–2.5
Polyvinyl chloride, rigid	PVC	3–6	8–10	6–11	5–60
Phenolic, cast		4–5		6–9	1.5–2
Phenolic, mineral-filled		10–20		4–7	0.2–0.5
Melamine, fabric-filled		12–16		6–9	0.4–0.6
Epoxy, glass-filled		30		15–60	3–4

3. Substitution of Cl for H, as in PVC, and Cl for F, as in PCTFE, gives greater strength and modulus owing to increased polar forces.

4. Strongly polar polymers like nylon, polyoxymethylene (POM), and PC have good combinations of stiffness, strength, and ductility.

5. Amorphous polymers like PS and PMMA, which have relatively stiff chains owing to bulky and polar side groups, also have high rigidity and strength, but not great toughness.

6. Where very high stiffness is desired, the filled thermosetting polymers are frequently the choice, because their modulus is usually four to five times that of even the best engineering types of thermoplastics.

7. High modulus and high strength can be realized from combinations of thermosetting resins and glass fibers.

8. The values given in the table are applicable only to a given set of conditions. Wide variations can occur owing to changes in temperature, pressure, and strain rate; examples of these effects are given later. Variations may also arise, even for a given polymer, owing to changes in crystallinity, molecular weight, and processing conditions.

2. Mechanical Properties in Uniaxial Compression

Test specimens for compression testing of polymers are usually in the form of cylinders of diameter d and length l. Standard test procedures are described in ASTM Standard D695-63T; the recommended testing speed is 1/20 in./min.

The nominal compressive stress is given by

$$\sigma = \frac{P}{A_0} \tag{39}$$

and the nominal strain, which in this case is a contraction, by

$$\varepsilon = \frac{l - l_0}{l_0} \tag{40}$$

The compressive strains can be measured by simply monitoring the distance between platens of the testing machine or by use of appropriate strain gauges on the specimen itself.

Compressive stress–strain data are used primarily to determine modulus of elasticity and compressive yield stress. The modulus is given by the slope of the initial linear portion of the σ–ε curve. The yield stress is determined by an offset method when there is no apparent yield maximum.

As load is increased and the specimen contracts axially, the diameter of the specimen increases due to Poisson's effect. Though in principle this lateral expansion should be uniform throughout the equally stressed test specimen, it is generally not so because of friction at the contact surfaces. Thus a compressive specimen tends to take on a barrel shape under increased loading. Because of this, as well as the possibility of failure by buckling, compressive tests are not usually used to determine ultimate or fracture stresses or ultimate elongations. However, they are useful for providing information about yielding, for even polymers like PS, which are brittle in tension tests, show yielding and deformation bands when tested in compression.

Stress–Strain Curves of Polymers in Compression

Stress–strain data are presented in Fig. 12 for two typical amorphous polymers, PVC and cellulose acetate (CA), and in Fig. 13 for two typical crystalline polymers, PTFE and polychlorotrifluoroethylene (PCTFE).

Both the amorphous polymers show a yield maximum. This represents a true softening process; it is not a result of localized necking or of the use of nominal stresses and strains instead of true stresses and strains. The two

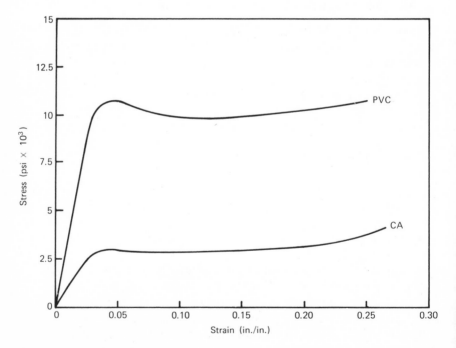

Figure 12 Compressive stress–strain data for two amorphous polymers: polyvinyl chloride (PVC) and cellulose acetate (CA).

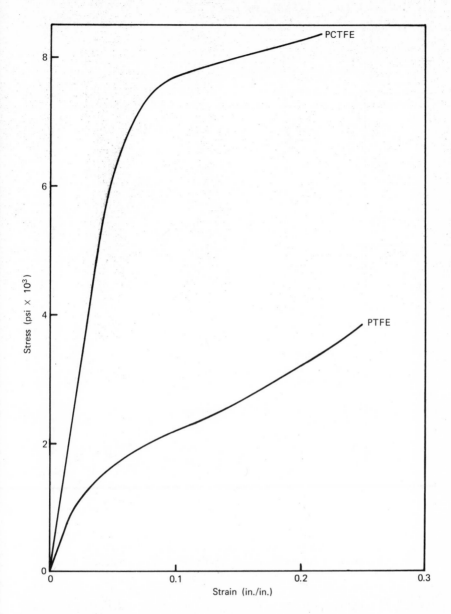

Figure 13 Compressive stress–strain data for two crystalline polymers: polytetra-fluoroethylene (PTFE) and polychlorotrifluoroethylene (PCTFE).

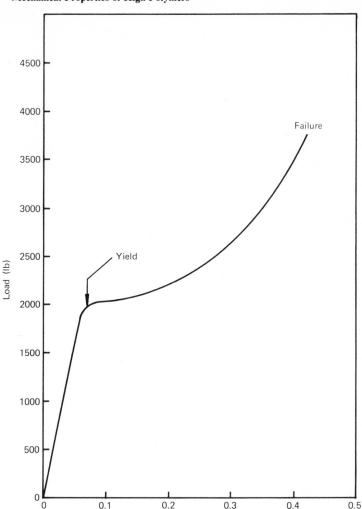

Figure 14   Compressive stress–strain curve of cast phenolic (20).

crystalline polymers, on the other hand, do not show a definite yield maximum; hence for these materials it would be necessary to define onset of yielding arbitrarily in terms of a 2 % offset method or by noting the point of intersection of extrapolated elastic and plastic portions of the curve.

When comparison is made of tensile yield stress and compressive yield stress for the same polymer, it is found that the compressive yield stress is higher, usually by about 20%. This fact alone indicates that the process of yielding in polymers is not purely a result of shear stress, for if this were true, there should be no difference between yielding in tension and yielding in compression. In view of the difference in these two quantities that does exist, it is a necessary consequence that theories of yielding and plastic flow which are independent of the hydrostatic stress, such as those normally used for metals, are not applicable to polymers and therefore new yield criteria will have to be established. We return to this point later.

Even a thermosetting polymer with poor ductility in tension (see Table 3) shows considerable plastic flow in compression. This behavior is demonstrated in Fig. 14 for a general-purpose cast phenolic resin. There is now a yield point and the strain to fracture is appreciable, about 40%. Under tension, however, the same material will break at about 2% strain and there is no yield.

Mechanical Property Data for Polymers Subjected to Compression

Data are presented in Table 4 with regard to the important mechanical parameters obtained from a compressive test. Strain at yield is given instead of strain at fracture, because compressive tests are seldom taken to fracture.

Thermoplastic polymers show yielding in compression at stress values that vary from about 1500 to possibly 15,000 psi. The filled or reinforced polymers do not show a yield drop and they can take quite high compressive stresses, in the 20,000–60,000 psi range, without fracture.

The compressive modulus values of polymers are in general agreement with values determined from a tension test, provided that both sets of values are determined at low strains and at common rates of straining. Thus compressive modulus values of polymers cover an equally wide range as the tensile modulus values, with most thermoplastics falling in the range from 50,000 to 500,000 psi and most filled thermosetting polymers in the range from 1,000,000 to 2,000,000 psi.

3. Mechanical Properties in Flexure

Standardized procedures for determining flexural properties of plastics are given in ASTM Standard D790-66. The specimen is usually in the form of a rectangular bar which is loaded at the center and simply supported at or near the ends.

Measurements are made of the deflection, δ, of the beam at the center, usually with a suitable dial gauge, as the load P is applied by the test machine. The experimental data are then generally plotted in the form of a load-elongation plot.

Table 4 Compressive mechanical property data for various plastics

Polymer	Designation	Compression Modulus (10^5 psi)	Stress at Yield (10^3 psi)	Strain at Yield (%)	Flexural Modulus (10^5 psi)	Flexural Strength (10^3 psi)
Polyethylene, low ρ	PE	0.5-1	3-5		0.1-0.6	3-6
Polyethylene, high ρ	PE	0.7-0.9	1.5-2		1-2	
Polytetrafluoroethylene	PTFE	1.8	5-8		0.5-2	7-10
Polychlorotrifluoroethylene	PCTFE	1.5-2.5	5-8	8	1.8	5-8
Polypropylene	PP	2-4	8-13		1.5-2	8-14
Nylon-6,6	6-6	3.5	10-12	4-20	2-4	10-13
Polycarbonate	PC	3.5-5	11-14		3.5	12-15
Polymethyl methacrylate	PMMA	4-4.5	10-16	5-10	3.5-5	13-14
Polyoxymethylene	POM	4-5.5	11-16		4-4.5	8-15
Polystyrene	PS	3-5	10-11	4-6	4-5	13-15
Polyvinyl chloride, rigid	PVC	3-5	14-18	5-6	4-6	11-17
Phenolic, cast				4-6	3-5	6-11
Phenolic, mineral-filled		5-20	15-30^a	4-6^a	10-20	13-17
Melamine, fabric-filled		10-15	25-32^a		16-20	20-80
Epoxy, glass-filled		25-30	30-70^a		25-45	

^{a} At fracture.

340

The important mechanical properties in a flexural test are the modulus of elasticity, E_b, and the flexural strength, S_b. For a simply supported beam of span length L subjected to a central load the deflection at the center is given by

$$\delta = \frac{PL^3}{48E_b I} \tag{41}$$

where I is the moment of inertia of the beam cross-section and, for a rectangular beam of width b and depth d, is given by $bd^3/12$. Thus the modulus of elasticity in bending E_b may be calculated from the initial slope,

$$m = \frac{P}{\delta} \tag{42}$$

of the load deflection curve. In terms of this slope, and the dimensions, it is given by

$$E_b = \frac{mL^3}{4bd^3} \tag{43}$$

At any value of P within the linear portion of the load deflection curve, the maximum fiber stress is given by the usual strength of materials formula:

$$S = \frac{Mc}{I} = \frac{3}{2}\left(\frac{PL}{bd^2}\right) \tag{44}$$

where M, the bending moment at the center, is $PL/4$ and c, the distance from the neutral axis to the outermost fiber, is $d/2$.

To obtain the so-called flexural strength, S_b, one simply substitutes into Eq. 44 the value of the load, P_b, at break or fracture. The resulting value is clearly not an accurate measure of ultimate strength because of use of an equation in a nonlinear region of the loading curve where it is not applicable. Nevertheless it is a convenient measure for comparing the relative strengths in flexure of beams of different materials.

In addition to listing mechanical property data obtained from compressive tests, Table 4 also gives values of flexural modulus and flexural strength. Not surprisingly, because flexure of beams involves simultaneous imposition of tension and compressive stresses, the flexural modulus is in general agreement with values determined from these other tests. However, the flexural strength value, which, as noted above, is of a somewhat fictitious value because it is determined by applying an equation that is not applicable in the nonlinear region, is somewhat higher than recorded tensile strength values.

4. Mechanical Properties in Shear

Static Properties

It is more difficult to perform a simple shear test than it is to carry out a simple tension or flexure test. This is probably why, even today, there is no standard ASTM test for determining the static shear modulus G. ASTM does have a recommended procedure for determining the shear strength of plastics, and this is specified in ASTM Standard D732-46 (1961). This is a punch type of test which is applicable to sheets or molded discs from 0.005 to 0.5 in. in thickness. The ultimate shear strength, τ_u, is then given by

$$\tau_u = \frac{P}{A_s} \tag{45}$$

where A_s is the shear area of the specimen subjected to the shear load P.

To determine the static shear modulus, torque tests can be carried out directly on solid cylinders, or on specimens in the form of thin tubes. For a solid cylindrical specimen of length L and radius r, the shear stress is calculated from

$$\tau = \frac{Tr}{J} \tag{46}$$

where T is the applied torque, r the radius, and J the polar moment of inertia of the specimen. The angle of twist, θ, is also related to the applied torque by the equation

$$\theta = \frac{Tr}{GJ} \tag{47}$$

Thus one can determine the shear modulus G directly from the initial slope of the linear portion of the torque twist curve or one can first plot a shear stress, τ, shear strain, γ, curve. The shear strain γ is given by

$$\gamma = \frac{r\theta}{L} \tag{48}$$

On eliminating T in the above equation we note

$$\tau = G\frac{r\theta}{L} = G\gamma \tag{49}$$

and hence G is simply the initial slope of the shear stress–shear strain curve.

If the polymer is isotropic, then the shear modulus is related directly to Young's modulus and Poisson's ratio:

$$G = \frac{E}{2(1 + v)} \tag{50}$$

Thus from experimentally observed values of G and E one can compute Poisson's ratio v. Poisson's ratio for polymers seems to fall in the range from about 0.30 to 0.50. For materials with $v = 0.3$ Young's modulus is about 2.6 times the shear modulus, whereas for rubbers, with $v = 0.5$, E should be three times the shear modulus.

These relations do seem to be obeyed for most polymers even over a wide range of temperatures. For example, from measurements of both E and G for polypropylene over the temperature range from 4.2 to $100°\text{K}$, it has been estimated that v is about one-third, and this value is in good agreement with direct measurements of this quantity made at room temperature.

Mechanical Properties from Torsion Pendulum

A convenient and frequently used method for measuring the shear modulus of polymers is a dynamic one. The specimen is in the form of a thin strip of length L, width b, and thickness t. One end is attached to a fixed clamp and the other to an inertial disk, having a moment of inertia I (Chapter 6). Procedures for operation of this torsional pendulum are detailed in ASTM Standard D2236-64T. The specimen is given an initial torsional displacement and then allowed to decay freely. From the frequency of the recorded decay curve and the dimensions of the specimen one can calculate the real part of the shear modulus, G'. If the dimensions are in centimeters, G' is in dyne/cm^2. We have

$$G' = \frac{64\pi^2 LI}{bt^3\mu} f^2 \tag{51}$$

In this equation μ is a shape factor whose value, given in the specification, depends on the ratio of specimen width to thickness. Because these tests are usually carried out at low frequency, ~ 1 cps, there is little difference between G' and the static shear modulus G, and for all practical purposes one may assume they are the same. Another advantage of this test is that one also gets a measure of the viscoelastic behavior of the test specimen. Usually one records the decay curve and obtains the logarithmic decrement, Δ, which is simply defined as

$$\Delta = \ln \frac{A_1}{A_2} \tag{52}$$

where A_1 and A_2 are the ratios of the values of two successive amplitudes. From the log decrement, Δ, one can also readily compute the imaginary part of the modulus, the so-called loss modulus G''. This is given by

$$G'' = \frac{\Delta G'}{\pi} \tag{53}$$

The torsional pendulum type of test is used to explore the properties of polymers under very small strain, in the region where the mechanical behavior is independent of strain and obeys the laws of linear viscoelasticity. However, it is a very convenient method for exploring the temperature dependence of the storage and loss modulus and hence for investigating, by mechanical means, the relaxation behavior of the polymer and its relation to chemical and physical structure.

5. Stress–Strain Relations for Isotropic Solids under Combined States of Stress

In the general case of loading, there are six independent stresses, σ_i, three normal stresses, σ_1, σ_2, and σ_3, and three shear stresses, σ_4, σ_5, and σ_6. There are also six independent strains, three normal strains, $\varepsilon_1, \varepsilon_2$, and ε_3, and three engineering shear strains, $\varepsilon_4, \varepsilon_5$, and ε_6. For a completely anisotropic material, such as a triclinic single crystal, there would then be, in principle, assuming linear behavior, 36 elastic constants and the generalized stress–strain relations could be written

$$\sigma_i = C_{ij}\varepsilon_j \qquad (i, j = 1, 2, \ldots, 6) \tag{54}$$

Because of symmetry conditions, these reduce to 21 (13). To describe completely the mechanical deformation behavior of a triclinic polymer single crystal, such as, say, the γ phase of polypropylene, one would first have to determine 21 different elastic constants. This is a prohibitive task and one that is currently beyond our capabilities, especially because the desired single crystals can be obtained only in the form of very thin platelets with lateral dimensions of the order of microns and thicknesses of the order of angstroms.

Fortunately, many bulk polymers are on a macroscopic scale, isotropic despite the fact that individual crystallites are anisotropic. For an isotropic solid the generalized stress–strain relations greatly simplify, and they may be written as follows (12, 13):

$$
\begin{aligned}
\sigma_1 &= \sigma_{xx} = C_{11}\varepsilon_{xx} + C_{12}(\varepsilon_{yy} + \varepsilon_{zz}) \\
\sigma_2 &= \sigma_{yy} = C_{11}\varepsilon_{yy} + C_{12}(\varepsilon_{xx} + \varepsilon_{zz}) \\
\sigma_3 &= \sigma_{zz} = C_{11}\varepsilon_{zz} + C_{12}(\varepsilon_{xx} + \varepsilon_{yy}) \\
\sigma_4 &= \tau_{xy} = C_{44}\gamma_{xy} \\
\sigma_5 &= \tau_{xz} = C_{44}\gamma_{xz} \\
\sigma_6 &= \tau_{yz} = C_{44}\gamma_{yz}
\end{aligned}
\tag{55}
$$

Thus the 36 elastic constants in the generalized constitutive equations have reduced to only three for the case of an isotropic material; furthermore, these three are related, as

$$C_{44} = \tfrac{1}{2}(C_{11} - C_{12})$$ (56)

It is customary to write the stress–strain equations in terms of Lamé's rigidity parameters, λ and μ. We have then

$$\lambda = C_{12}$$
$$C_{11} = 2G + \lambda$$ (57)

and

$$\mu = G = C_{44}$$

If we introduce the concept of volume strain, ε_v, defined for the case of small strains as

$$\varepsilon_v = \varepsilon_{xx} + \varepsilon_{yy} + \varepsilon_{zz} = \varepsilon_{kk}$$ (58)

then the six stress–strain equations above can be written in terms of one tensor equation, namely,

$$\sigma_{ij} = \lambda \varepsilon_{kk} \delta_{ij} + 2G\varepsilon_{ij} \qquad (i, j = x, y, z)$$ (59)

where δ_{ij} is the Kronecker delta. It has the value one if $i = j$ and zero if $i \neq j$, and the off-diagonal elements of the strain tensor are one-half the corresponding engineering shear strains.

The physical meaning of the parameter G is clear from the above. It simply represents the ratio of a simple shear stress, such as σ_{xy} or τ_{xy}, to a simple shear strain such as γ_{xy}, and hence it is simply the shear modulus. It is not easy to attach a similar physical meaning to λ. Hence it is customary to use parameters other than λ in the stress–strain relations. These may be the tension modulus, E, the Poisson's ratio, v, or the bulk modulus, K. The relations between these different parameters can be readily obtained by applying Eq. 55 above to special cases, such as simple hydrostatic pressure or simple tension. In this manner one can show

$$\lambda = K - \tfrac{2}{3}G$$ (60)

$$E = 3K(1 - 2v)$$ (61)

$$v = \frac{E - 2G}{2G}$$ (62)

Hence, regardless of the state of stress, only two elastic constants or parameters are needed to define the stress–strain behavior.

D. PLASTICITY AND THEORIES OF YIELDING OF POLYMERS

1.   Idealization of Stress–Strain Curves

Figures 10 and 11 give typical tensile stress–strain curves for three polymers that show yielding. If we neglect the yield drop, we can represent the behavior of these and similar polymers by the idealized stress–strain groups of Figs. 15 and 16. In these diagrams point A represents the proportional limit; to this point the stress is directly proportional to strain and upon unloading the stress–strain relation follows along path AO. Point B is a later point on the stress–strain curves and unloading from B may lead to permanent strain.

If loading occurs farther into the plastic range, as point C, and then unloading, strain is generally irrecoverable and the amount of strain remaining upon unloading, OD, is termed the plastic strain. In many polymers, because of their viscoelastic nature, strain recovery continues for some time after unloading. Upon reloading from point D subsequent yielding occurs at E and the flow or yield stress may then remain constant (Fig. 15) or increase (Fig. 16) to F (final fracture).

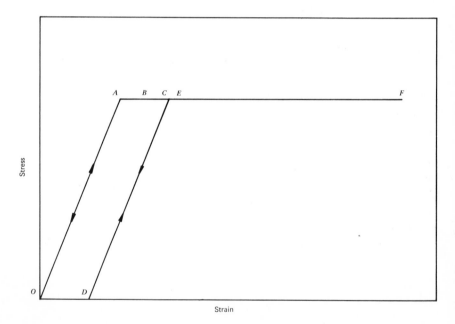

Figure 15   One type of idealized stress–strain relation for a solid polymer, showing both unloading and reloading paths.

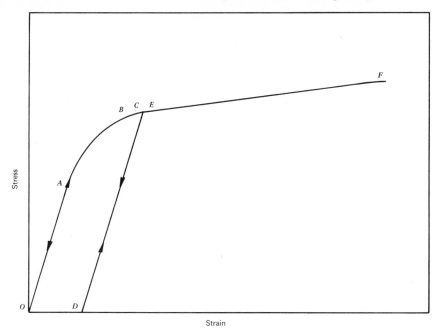

Figure 16   A second type of idealized stress–strain relation for a solid polymer, showing both unloading and reloading paths.

Materials exhibiting stress–strain curves of the general type of Fig. 15 are called elastic-perfectly plastic materials, and if they exhibit the behavior similar to that of Fig. 16 they are considered elastic-strain hardening materials (21, 22).

2. Loading Surface

The existence of a yield point in uniaxial states of stress, such as those shown in Figs. 15 and 16, suggests the possible existence of an initial yield surface in a nine-dimensional stress space which bounds the totality of all possible elastic states. The observation of a subsequent higher yield point for strain-hardening materials, as shown in Fig. 16, implies that subsequent yield surfaces may also exist. It can also be assumed that unloading from multi-axial states of stress takes place elastically, as in the case of uniaxial states of stress.

The mathematical formulation of an initial and subsequent yield surface can be made by introducing the concept of a loading function which, in general, is a complicated function of those variables that affect the yielding behavior. For time-independent and isothermal processes of deformation, the loading function may be expressed as (21–23)

$$f = f(\sigma_{ij}, \varepsilon_{ij}^p, K_{ij})$$ (63)

where σ_{ij} is the stress tensor, ε_{ij}^p are the plastic strains, and K_{ij} is a strain-hardening parameter. For a perfectly plastic material, K_{ij} is equal to zero. in other words, the yield surface remains fixed in the stress space as the material is loaded, unloaded, and reloaded as shown for a perfectly plastic material in Fig. 15. The function f is defined in such a way that $f = 0$ refers to a plastic state, $f < 0$ represents only elastic states, and no meaning is associated with $f > 0$.

Consider now the initial yield surface. A material is loaded from a virgin state through an entirely elastic path up to an initial yield point. No plastic deformation has yet taken place and, of course, no strain hardening is associated with the loading program. Therefore Eq. 63 for the initial yielding becomes

$$f(\sigma_{ij}) = 0$$ (64)

If the material is initially isotropic, the initial yield function must be independent of coordinate transformations and thus it may be expressed in terms of stress invariants:

$$f(J_1, J_2, J_3) = 0$$ (65)

where

$$J_1 = \sigma_{kk}$$ (66)

$$J_2 = \tfrac{1}{2}(\sigma_{kk}\sigma_{ll} - \sigma_{kl}\sigma_{lk})$$ (67)

$$J_3 = |\sigma_{kl}|$$ (68)

Before discussing specific forms of yield conditions and their possible shapes, we consider first some pertinent experimental findings.

Whitney and Andrews (43) have shown that the uniaxial tensile and compressive yield stresses for glassy polymers are not the same. In fact, the compressive yield stress is significantly higher than the tensile one. They also pointed out that the relative plastic volume change at yield is positive (expansion) in the uniaxial compression test. Bowden and Jukes (44) have reported that, in tests made under biaxial tension and compression on polymethyl methacrylate sheet tested in plane strain, the compressive yield stress decreased with applied tension more rapidly than would be expected if the shear yield stress of the material were independent of pressure.

Pae, Mears, and Sauer (45, 46) have shown that, for both crystalline and amorphous polymers, the yield stresses in uniaxial tension or compression increase markedly with increase of hydrostatic pressure. Similar observations have been reported by Baer and Radcliffe et al. (47, 48). Rabinowitz, Ward, and Parry (49) have observed that the torsional yield stress of crystalline and amorphous polymers also increases with increasing hydrostatic pressure. These experimental findings indicate that yielding in both crystalline and amorphous polymers is strongly dependent upon the hydrostatic component of stress.

3.   Pressure-Independent Yield Criteria

In most metals it has been observed that yielding is little affected by hydrostatic pressure (24, 50). For pressure-independent yielding, the loading function may be written as a function of the invariants of the stress deviator: that is,

$$f(J'_2, J'_3) = 0 \tag{69}$$

where

$$J'_2 = \tfrac{1}{2}(\sigma'_{ij}\sigma'_{ij}), \qquad J'_3 = \tfrac{1}{3}(\sigma'_{ij}\sigma'_{jk}\sigma'_{ki}) \tag{70}$$

and

$$\sigma'_{ij} = \sigma_{ij} - \sigma\delta_{ij} \tag{71}$$

In these equations σ'_{ij} are the components of the stress deviator, $\sigma = \tfrac{1}{3}\sigma_{kk}$, and δ_{ij} is the Kronecker delta. The first invariant of the stress deviator $J'_1 = \sigma'_{ii}$ is not included in Eq. 69 since it is identically zero. The most widely used pressure-independent yield conditions are those proposed by Tresca (1864) (51) and von Mises (1913) (52).

Tresca's yield condition, expressed for the general case of triaxial loading in terms of J'_2 and J'_3, is given by the rather cumbersome relation (53)

$$4J'^3_2 - 27J'^2_3 - 36k^2J'^2_2 + 96k^4J'_2 - 64k^6 = 0 \tag{72}$$

where k is the yield stress in simple shear. In terms of the principal stresses, $\sigma_1, \sigma_2,$ and σ_3, this equation becomes

$$[(\sigma_3 - \sigma_1)^2 - 4k^2][(\sigma_1 - \sigma_2)^2 - 4k^2][(\sigma_2 - \sigma_3)^2 - 4k^2] = 0 \tag{73}$$

From the above a simple Tresca yield condition may be derived as

$$\frac{\sigma_1 - \sigma_3}{2} = k \tag{74}$$

in which $\sigma_1 > \sigma_2 > \sigma_3$.

According to this theory, yielding is a result only of the presence of shear stresses, and it takes place only when the maximum value of the shear stress in a stressed element, $(\sigma_1 - \sigma_3)/2$, reaches the value, k, required to cause yielding in a simple shear. For the case of biaxial stresses with $\sigma_3 = 0$, the maximum shear stress will be either $\sigma_1/2$ or $\sigma_2/2$, provided both σ_1 and σ_2 are positive. When this value reaches the critical value, k, which is equivalent to one-half the yield stress, $\sigma_y/2$, that one would get in a uniaxial tension case, yielding will occur. If, however, one of the principal stresses is positive and one negative, then the maximum shear stress is $(\sigma_1 + \sigma_2)/2$ and when it reaches the value $\sigma_y/2$, failure occurs.

The failure envelope in σ_1, σ_2 space is shown by the hexagon of Fig. 17. The theory predicts, for example, that in simple shear (where $\sigma_1 = -\sigma_2$) failure occurs when both σ_1 and σ_2 are only one-half the expected value, σ_y, that produces yielding in uniaxial tension or uniaxial compression loading.

Although the simple shear theory has been widely used for yielding in metals, it is completely inapplicable for yielding in polymers, because it

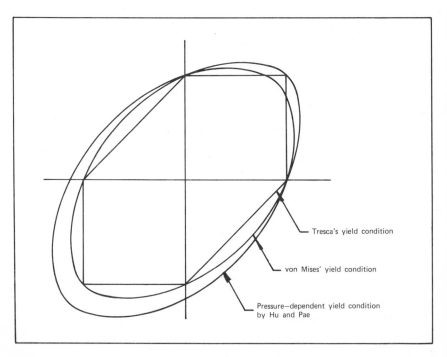

Figure 17   Yield conditions of Tresca, von Mises, and Hu and Pae, for various biaxial stress conditions.

implies equal yield stress values in tension and compression, and also that hydrostatic pressure has no effect on yielding. As already noted, the first of these is not true in polymers and, as we show below, hydrostatic pressure has a very large effect. Hence the Tresca, a simple shear theory, must *not* be used in the design of polymers for biaxial or triaxial stress applications.

The von Mises yield conditions can be simply stated in terms of the second invariant of the stress deviator. It is

$$J_2' = k^2 \tag{75}$$

Expressed in terms of the principal stresses, it is given by

$$(\sigma_1 - \sigma_2)^2 + (\sigma_2 - \sigma_3)^2 + (\sigma_3 - \sigma_1)^2 = 6k^2 \tag{76}$$

and, in terms of component stresses, by

$$(\sigma_x - \sigma_y)^2 + (\sigma_y - \sigma_z)^2 + (\sigma_z - \sigma_x)^2 + 6(\tau_{yz}^2 + \tau_{zx}^2 + \tau_{xy}^2) = 6k^2 \tag{77}$$

In these equations the parameter k is again the yield stress in simple shear. Von Mises suggested this yield condition purely on the basis of mathematical reasoning that led him to believe that when J_2 reached a certain value, yielding would occur. In later years, Eq. 75 was given physical interpretations by Hencky (1924) (53) and Nadai (1937) (54). Hencky pointed out that Eq. 75 implies that yielding begins when the elastic energy of distortion reaches a certain critical value. The elastic energy of deformation can be divided into two parts: one due to normal stresses associated with volume changes, and the other due to distortion, in turn due to shear stresses. Thus in the von Mises yield condition, a hydrostatic pressure does not cause yielding, since it produces only elastic energy associated with volume change. The elastic distortion energy can be expressed as a function of J_2' as

$$E \text{ (distortion)} = \frac{J_2'}{2G} \tag{78}$$

Nadai gave an alternative physical interpretation, namely, that yielding begins when the shear stress acting over the octahedral planes reaches a certain value. The octahedral shear stress can be expressed in terms of J_2' as

$$\sigma_{\text{oct}} = (\tfrac{2}{3}J_2')^{1/2} \tag{79}$$

Therefore the von Mises yield condition can now be written in terms of the octahedral shear stress as

$$\sigma_{\text{oct}} = \sqrt{\frac{2}{3}}\, k \tag{80}$$

A two-dimensional diagram of the von Mises yield condition is shown in Fig. 17.

The von Mises theory predicts a somewhat broader range of possible elastic stress states than does the Tresca theory, so that use of the simpler Tresca theory is always conservative. For the case of simple shear, the von Mises theory predicts yielding will occur at values of the principal stresses equal to $0.577\,\sigma_y$, as compared to $0.50\,\sigma_y$ for the Tresca theory. Actual data for metals subject to shear usually fall between these two sets of values.

However, this theory, like the Tresca theory, assumes that the shear stress in tensile and compressive yield are the same and that hydrostatic pressure has no effect on yielding. Hence the von Mises theory also is inapplicable to polymers.

4. Pressure-Dependent Yield Criteria

For polymers, crystalline and amorphous alike, any appropriate yield criterion, in view of the experimental data, must be dependent upon the hydrostatic component of stresses. Two possible yield criteria are now introduced. One is the theory put forward by Coulomb and sometimes referred to as the Mohr–Coulomb yield condition. According to this condition, yielding occurs when the shearing stress, τ, plus some constant times the normal stress, σ_n, acting on the same plane, reaches a certain critical value, that is,

$$\tau + \mu\sigma_n = \tau_0 \tag{81}$$

Here μ can be considered as a material constant which physically corresponds to an internal friction coefficient, and $\tau_0 = k$ is the yield stress in simple shear, when no normal stresses are acting. If we let $p = -\sigma_n$ be the normal pressure on the yield plane, then the shear stress at yielding is given by

$$\tau = \tau_0 + \mu p \tag{82}$$

and is proportional to the normal stress or pressure.

One immediate consequence of this theory is that the yield conditions for tensile loading and compressive loading should be different, for in one case the normal stress on the yielded plane will be tensile or positive, whereas in the other case the normal stress will be compressive or negative. Let us assume that yielding occurs under uniaxial loading on a plane inclined at 45° to the load axis. Then in the case of tension the maximum shear stress will be $\frac{1}{2}\sigma_{yT}$ and the normal stress on the same plane also $\frac{1}{2}\sigma_{yT}$. Equation 81 then becomes

$$\frac{\sigma_{yT}}{2} + \mu\frac{\sigma_{yT}}{2} = \tau_0 \tag{83}$$

whereas in the case of compression we get

$$\frac{\sigma_{yc}}{2} - \mu \frac{\sigma_{yc}}{2} = \tau_0 \tag{84}$$

Equating the two τ_0 values we find

$$\frac{\sigma_{yc}}{\sigma_{yT}} = \left(\frac{1 + \mu}{1 - \mu}\right) \tag{85}$$

Thus the yield stress in compression should be higher than the yield stress in tension, with the difference depending upon the value of the parameter μ. For many polymers μ seems to take on a value of about 0.05. Hence one would expect the compressive yield in these polymers to be of the order of 11% higher than the tensile yield.

The above is only approximate. If the Coulomb criterion is to be used, then the maximum value of the quantity $\tau + \mu\sigma_n$ will not occur at 45° to the tensile axis but at some other angle. If this is taken into account, and if we allow a hydrostatic pressure p to be superimposed on a uniaxial tension or compression, then it can be shown that Coulomb's criterion leads to the following equations for the dependence of the observed yield stress on pressure:

$$\sigma_{yT} = \sigma_T^0 + \frac{2 \sin \alpha}{1 + \sin \alpha} p = \sigma_T^0 + kp \tag{86}$$

$$\sigma_{yc} = \sigma_c^0 + \frac{2 \sin \alpha}{1 - \sin \alpha} p = \sigma_c^0 + k'p \tag{87}$$

when α and μ are related by

$$\tan \alpha = \mu \tag{88}$$

Thus if uniaxial stress–strain curves are obtained while a specimen is subjected to varying hydrostatic environments, this theory predicts a linear relation between the observed yield stress and the pressure.

For the case of tensile and compressive loadings the slopes k and k' of the yield stress–pressure relations are approximately given by

$$k = \frac{2\mu}{1 + \mu} \quad \text{and} \quad k' = \frac{2\mu}{1 - \mu} \tag{89}$$

because experiments show that α is a small angle and $\sin \alpha \cong \tan \alpha$. Hence the simple Coulomb yield criterion predicts that both the compressive and tensile yield stresses should increase linearly with increase of hydrostatic pressure, and that the slope of this line should be greater for compression than for tension.

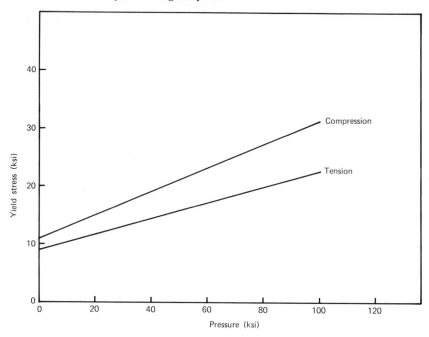

Figure 18 Effects of pressure on compressive and tensile yield stress for polyvinyl chloride (55).

Extensive tests on polymers carried out under a hydrostatic environment in our laboratories have shown that all these predictions are realized. It is therefore advocated that either the Mohr–Coulomb criterion or some more sophisticated theory that does take into account the hydrostatic component of stress be used for polymers rather than the Tresca or von Mises yield criteria.

To illustrate the marked pressure dependence of the yield stress of polymeric materials, we present in Fig. 18 data obtained on specimens of commercial PVC subjected to either uniaxial tension or compression plus hydrostatic pressure. From these data it may be noted that all three of the predictions of the Coulomb theory have been verified: (1) at atmospheric pressure, yield is higher in compression than in tension; (2) in both tension and compression, yield stress rises linearly with increasing pressure; and (3) the slope of the curve is higher for compression than for tension.

Similar behavior has been noted for a variety of other polymers, including cellulose acetate, polycarbonate, polyethylene, and polytetrafluoroethylene, and hence the simple Coulomb criterion of yielding can be considered as a suitable yield criterion for polymers. It is also possible to devise other yield criteria, such as a modified von Mises theory in which the failure criterion is taken to be that the octahedral shear stress reach some critical value which is itself a function of pressure. However, the Coulomb theory has the advantages of simplicity and adequate representation of the data.

Another example to illustrate the effect of combined stresses on polymers is taken from the work of Whitney and Andrews (43) on polystyrene (PS). These authors did not investigate the effects of hydrostatic pressure per se, but they did investigate the behavior of PS under five different stress fields, namely, uniaxial tension, uniaxial compression, torsional shear, biaxial tension, and biaxial compression. Their results, in terms of a principal stress $\sigma_1\sigma_2$ plot, are shown in Fig. 19. They concluded that the Tresca yield condition was not applicable and that their data were in accord with the Coulomb–Mohr yield criterion.

Another, more general yield condition, as formulated by Hu and Pae (56), is based on inclusion of a hydrostatic stress term directly in the loading

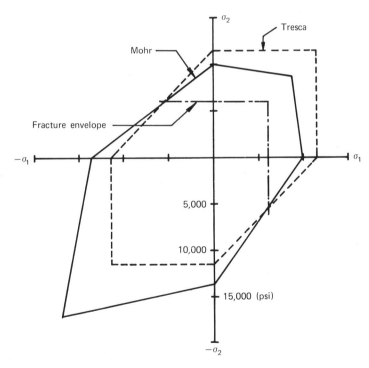

Figure 19   Yield behavior of polystyrene under various biaxial stress conditions (43).

function; that is, it is assumed that

$$f(J_1, J'_2, J'_3) = 0 \tag{90}$$

One can also assume that the plastic deformation of polymers is caused by the stress deviators but is influenced by the hydrostatic stress. The yield function may then be written in the general form

$$g(J'_2, J'_3) = \sum_{i=0}^{\infty} a_i J_1^i \tag{91}$$

where the a_i are parameters to be determined by experiment. In some cases a simpler form of Eq. 91 can be applied, in which the g function is simply J'_2 itself. Then we have

$$J'_2 = \sum_{i=0}^{\infty} a_i J_1^i \tag{92}$$

Also, for some polymers, it appears that, over a restricted pressure range, we need keep only the first term in the series expansion. The yield condition may then be written

$$J'_2 = a_0 + a_1 J_1 \tag{93}$$

In other cases, or for larger ranges of the pressure variable, it may be necessary to retain terms to the second power in J_1. Equation 92 then becomes

$$J'_2 = a_0 + a_1 J_1 + a_2 J_1^2 \tag{94}$$

where a_0, a_1, a_2, etc., are material constants that have to be determined by experiment. The above equations may be physically interpreted as saying that when the octahedral shear stress reaches a critical value which itself is influenced by hydrostatic pressure, yielding will occur. Figure 17 shows how the yielding envelope looks in a $\sigma_1 \sigma_2$ plot. If we take $a_1 = a_2 = 0$ and $a_0 = k^2$, then Eqs. 93 and 94 reduce to the von Mises yield condition. One of the possible variations of Eq. 92 or 93 is also

$$\tau_{\text{oct}} = k - \alpha p \tag{95}$$

which is an equation frequently used by polymer scientists. Both Eqs. 93 and 94 have been successfully applied to describe the behavior of various polymers under different loading conditions.

5. Crazing and Normal Stress Yielding

Many transparent amorphous glassy polymers exhibit two distinct modes of yielding. The type of yielding discussed in the preceding section involves an applied shear stress, although the yield phenomenon itself is influenced, as we have seen, by the normal stress component acting on the yield plane. The second type of yielding involves yielding under the influence of the largest principal stress. This type of yielding is frequently referred to as crazing, or normal stress yielding.

Crazing can be induced by stress or by combined stress and solvent action. It shows generally similar features in all polymers in which it has been observed and studied, such as polystyrene, polymethyl methacrylate, and polycarbonate. Crazing appears to the eye to be a fine, microscopic network of cracks almost always advancing in a direction at right angles to the maximum principal stress. Crazing generally originates on the surface at points of local stress concentration. In a static type of test, it appears that for crazing to occur the stress or strain must reach some critical value. However, crazing can occur at relatively low stress levels under long-time loading. Examples of crazing or normal stress yielding, as it has been called, are shown in Fig. 20 for several amorphous polymer samples (57).

It is known from extensive electron microscopic examination of crazed areas by Spurr and Niegisch (58), Kambour (59), and others that molecular chain orientation has occurred in the crazed regions and that oriented fibrils extend across the craze surfaces. This observation helps to explain the phenomenon already noted many years ago (60), namely, that crazing is associated with local orientation and that a specimen could be completely crazed across its entire cross section and yet continue to carry appreciable load without failure. From X-ray studies taken in the vicinity of the crazed areas it was inferred that molecular orientation was occurring and that it was an important feature of the crazing process. However, until electron microscopic studies became available, knowledge of the detailed structure of crazes and their mode of propagation was lacking.

Figure 20   Examples of crazing and normal stress yielding in a polymethyl methacrylate specimen (left) and in polystyrene specimens (right) subjected to different applied tensile stresses and durations of load application (57).

Recently the effect of biaxial stresses on the onset of crazing in PMMA has been studied for tubular specimens subjected to axial load and internal pressure and for tensile specimens containing a small hole. In both these instances a biaxial stress field is present, and Sternstein and his co-workers (61) observed the stress conditions prevailing at the onset of crazing. Their results, taken at three different temperatures, are presented in Fig. 21 for various values of principal stresses. It may be observed that there is a small effect of the second principal stress on the onset of crazing, although in all cases the crazing crack was in the direction of the minimum principal stress. Thus the 45° line in Fig. 21 represents points of instability because, for slight increases in either σ_1 or σ_2, the crack direction changes by 90°. Sternstein concluded that some chain conformational mobility was essential to development of crazing and that, if this were present, localized chain orientation could occur by normal stress yielding. In PMMA he noted that the higher the value of the first stress invariant $I_1 = \sigma_1 + \sigma_2 + \sigma_3$ the greater was the value of the stress bias $|\sigma_1 - \sigma_2|$ required to produce crazing. Under

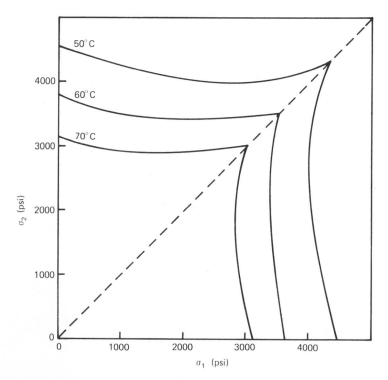

Figure 21 $\sigma_1-\sigma_2$ envelopes for craze formation in polymethyl methacrylate (61).

biaxial tensile stresses the yield envelope for normal stress yielding was everywhere inside that for shear yielding. However, for those states in which one principal stress is tensile and one compressive, the situation is different. In conditions approaching that of pure shear, a large increase in stress bias is required to get normal stress yielding and hence shear yielding occurs first. In particular, a stress field consisting of a simple tension and a hydrostatic pressure should lead to ductile or shear yielding in PMMA rather than crazing or normal stress yielding.

This latter effect, namely, the suppression of crazing by imposition of high pressure, has been observed to occur in several polymers, including polycarbonate, polystyrene, and polypropylene.

E. VISCOELASTICITY OF SOLID POLYMERS

One of the most striking characteristics of polymers is the marked dependence of their properties on time. This dependence is manifested in various ways. First of all, if a constant stress, or load, is applied to a polymer sample, the deformation continues to increase with time. This is the phenomenon known as creep. Second, if a polymer sample is deformed to and held at some constant value of strain, then the applied stress required to maintain that given amount of strain decreases with time. This is the phenomenon of stress relaxation. Third, if a sinusoidally varying stress is applied to a polymer, the resulting strain has an in-phase component and an out-of-phase component. The phase angle between stress and strain will be a measure of the mechanical strain energy that is converted to heat and, as such, it is a measure of what is normally called internal friction.

All these characteristics, namely, creep, stress relaxation, and internal friction, are expected manifestations of a linear viscoelastic material. Hence it is helpful, in visualizing or describing the behavior of a polymer, to compare it with the predicted behavior of simple viscoelastic models. Though no polymer can be completely and adequately represented by a simple model, nevertheless an understanding of the behavior of simple models is very instructive.

1. Mechanical Models

The simplest model of a viscoelastic material is a so-called Maxwell model. This consists, as shown in Fig. 22a, of a spring representing the elastic element in series with a dashpot representing the viscous element. Another simple model is the Voight or Voight–Kelvin model (Fig. 22b), in which the spring is in parallel with the dashpot. The Maxwell model is useful to depict stress

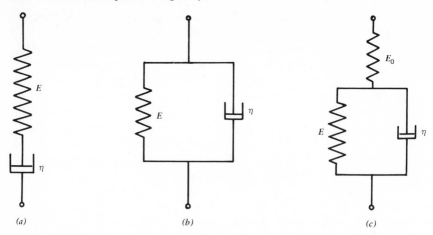

Figure 22   Mechanical models of spring and dashpot used in viscoelasticity theory. (*a*) Maxwell model. (*b*) Voight model. (*c*) Standard linear solid (three-parameter) model.

relaxation and the Voight model is useful to depict anelastic or creep deformation. However, neither of these simple models adequately shows the viscoelastic features of real polymeric materials. Therefore a third model, the so-called three-parameter model, or standard linear viscoelastic solid, is often used. This model, as Fig. 22*c* shows, consists of a spring in series with the elements of a Voight model. In these models we use E_0 to represent the instantaneous elastic response, η to represent the dashpot viscosity, and E to represent the elastic element in parallel or series with the dashpot.

We now derive the appropriate equations for creep and stress relaxation for the three-parameter model. Assume first that the model is subjected to a static stress S. Under the action of this stress there is an elastic strain ε_e arising from deformation of the spring. Since the spring is assumed to be linear or Hookean, we have

$$S = E_0 \varepsilon_e \tag{96}$$

The stress S also acts on the Voight element, with part of the stress being carried by the elastic element, characterized by E, and part by the dashpot element, characterized by η. Let the anelastic strain associated with the Voight element be ε_a. Then the stress due to the elastic element is $E\varepsilon_a$ and the stress due to the dashpot, assuming Newtonian behavior, is $\eta(d\varepsilon_a/dt)$. Hence the total stress, S, may be written

$$S = E\varepsilon_a + \eta \frac{d\varepsilon_a}{dt} \tag{97}$$

Now the total strain, ε, is the sum of the elastic and anelastic strains. Hence

$$\varepsilon = \varepsilon_e + \varepsilon_a \tag{98}$$

and

$$\frac{d\varepsilon}{dt} = \frac{d\varepsilon_e}{dt} + \frac{d\varepsilon_a}{dt} \tag{99}$$

On substituting for ε_e from Eq. 96 and for $d\varepsilon_a/dt$ from Eq. 97, we obtain

$$\frac{d\varepsilon}{dt} = \frac{1}{E_0}\frac{ds}{dt} + \frac{S}{\eta} - \frac{E\varepsilon_a}{\eta} \tag{100}$$

but ε_a can be written

$$\varepsilon_a = \varepsilon - \varepsilon_e = \varepsilon - \frac{S}{E_0} \tag{101}$$

Therefore Eq. 101 becomes

$$\frac{d\varepsilon}{dt} = \frac{1}{E_0}\frac{ds}{dt} + \frac{S}{\eta} - \frac{E\varepsilon}{\eta} + \frac{ES}{\eta E_0} \tag{102}$$

and, collecting terms, we have

$$\frac{d\varepsilon}{dt} + \frac{E\varepsilon}{\eta} = \frac{1}{E_0}\frac{ds}{dt} + \left(\frac{E_0 + E}{\eta E_0}\right)S \tag{103}$$

Equation 103 is a simplified version of a more general representation of linear viscoelasticity in which the stress and strain are related by a differential equation of the form:

$$A_0 S + A_1 \frac{ds}{dt} + A_2 \frac{d^2 S}{dt^2} \cdots = B_0 \varepsilon + B_1 \frac{d\varepsilon}{dt} + B_2 \frac{d^2 t}{dt^2} \cdots \tag{104}$$

Although Eq. 103 is a much simpler form of Eq. 104, it does show the principal viscoelastic behavior features observed in real polymeric materials. We now apply Eq. 103 to the phenomenon of creep.

Predicted Creep Behavior

If a tensile load is placed on a polymer specimen, the elongation will continue to increase with time. The standard linear solid also shows this feature. The appropriate equations follow immediately from Eq. 103 by letting $S = S_0 = $ a constant. Equation 103 then reduces to

$$\frac{d\varepsilon}{dt} + \frac{E}{\eta}\varepsilon = \frac{1}{\eta}\left(\frac{E_0 + E}{E_0}\right)S_0 \tag{105}$$

This equation can be readily solved by operator or other methods. Using the operator method, we can write

$$(D + a)\varepsilon = b \tag{106}$$

where $D = d/dt$, $a = E/\eta$, and b is a constant given by the right-hand side of Eq. 105.

The solution of the operator equation is

$$\varepsilon = \frac{b}{a} + Ce^{-at} = \left(\frac{E_0 + E}{E_0 E}\right)S_0 + Ce^{-at} \tag{107}$$

where C is a constant to be determined by the boundary conditions. In a creep experiment, the stress is held constant. Hence the initial strain is simply the instantaneous elastic strain, that is,

$$\varepsilon]_{t=0} = \varepsilon_0 = \frac{S_0}{E_0} \tag{108}$$

Placing this condition into Eq. 107, we find

$$\frac{S_0}{E_0} = \left(\frac{E_0 + E}{E}\right)\frac{S_0}{E_0} + C \tag{109}$$

Therefore

$$C = -\frac{S_0}{E} \tag{110}$$

and the strain at any time t is then given by

$$\varepsilon = \left(\frac{E_0 + E}{E_0 E}\right)S_0 - \frac{S_0}{E}e^{-at} \tag{111}$$

$$\varepsilon = \frac{S_0}{E_0} + \frac{S_0}{E}(1 - e^{-at}) \tag{112}$$

This equation thus predicts that there will be an instantaneous elastic response, given by S_0/E_0, and thereafter the strain will continue to increase, but at a steadily decreasing rate until a final asymptotic strain value, ε_F, is reached, where

$$\varepsilon_F = \frac{S_0}{E_0} + \frac{S_0}{E} = S_0\left(\frac{E + E_0}{E_0 E}\right) \tag{113}$$

Equation 112 does not provide for a deformation which continues to increase linearly with time. Such a plastic or viscous strain can be accommodated by making the three-parameter model into a four-parameter model.

This is done by adding an additional linear dashpot, with viscosity η_0, in series with the linear spring.

The net result of this will simply be to add an additional deformation term which is linear in time. The equation for creep deformation for the four-parameter model thus becomes

$$\varepsilon = \frac{S_0}{E_0} + \frac{S_0}{E}(1 - e^{-at}) + \frac{S_0}{\eta_0}t \qquad (114)$$

Equations 112 and 114 are usually written more directly in terms of creep compliance, D, and creep retardation time, τ_s, defined as follows:

$$D(t) = \frac{\varepsilon(t)}{S_0} \qquad \text{and} \qquad \tau_s = \frac{1}{a} = \frac{\eta}{E} \qquad (115)$$

The expressions for creep deformation and creep compliance as a function of time then become, where D_0 and D are the reciprocals of E_0 and E respectively:

For the two-parameter Voight model (here $E_0 \to \infty$):

$$\varepsilon(t) = \frac{S_0}{E_0}\left(1 - \exp\left(-\frac{t}{\tau_s}\right)\right) \qquad \text{or} \qquad D(t) = D_0\left(1 - \exp\left(-\frac{t}{\tau_s}\right)\right) \qquad (116)$$

For the three-parameter model:

$$\varepsilon(t) = \frac{S_0}{E_0} + \frac{S_0}{E}\left(1 - \exp\left(-\frac{t}{\tau_s}\right)\right)$$

or $\qquad\qquad\qquad\qquad\qquad\qquad\qquad\qquad\qquad\qquad\qquad\qquad\qquad$ (117)

$$D(t) = D_0 + D\left(1 - \exp\left(-\frac{t}{\tau_s}\right)\right)$$

For the four-parameter model:

$$\varepsilon(t) = \frac{S_0}{E_0} + \frac{S_0}{E}\left(1 - \exp\left(-\frac{t}{\tau_s}\right)\right) + \frac{S_0}{\eta_0}t$$

or $\qquad\qquad\qquad\qquad\qquad\qquad\qquad\qquad\qquad\qquad\qquad\qquad\qquad$ (118)

$$D(t) = D_0 + D\left(1 - \exp\left(-\frac{t}{\tau_s}\right)\right) + \frac{t}{\eta_0}$$

Predicted Stress Relaxation Behavior

The situation considered here is one in which an initial deformation is applied and then maintained constant. In this case let the deformation be constant at a value ε_0. Equation 103 then becomes

$$\frac{1}{E_0}\frac{ds}{dt} + \left(\frac{E_0 + E}{\eta E_0}\right)S = \frac{E}{\eta}\,\varepsilon_0 \tag{119}$$

or

$$\frac{ds}{dt} + \left(\frac{E_0 + E}{\eta}\right)S = \frac{E}{\eta}\,S_0 \tag{120}$$

since $E_0\varepsilon_0$ is simply the initial value of the stress, S_0. Equation 120 is also of the form

$$(D' + a_1)S = b_1 \tag{121}$$

where $a_1 = (E_0 + E)/\eta$ and $b_1 = (E/\eta)S_0$ are constants. Therefore the solution is

$$S = \frac{b_1}{a_1} + C_1 \exp(-a_1 t) = \left(\frac{E}{E_0 + E}\right)S_0 + C_1 \exp(-a_1 t) \tag{122}$$

The boundary condition for determining C_1 is $S]_{t=0} = S_0 = E_0\varepsilon_0$. Therefore

$$S_0 = \left(\frac{E}{E_0 + E}\right)S_0 + C_1 \tag{123}$$

and hence

$$C_1 = S_0\left(\frac{E_0}{E_0 + E}\right) \tag{124}$$

The expression for the stress as a function of time now becomes

$$S = \left(\frac{E}{E_0 + E}\right)S_0 + \left(\frac{E_0}{E_0 + E}\right)S_0 \exp(-a_1 t) \tag{125}$$

Now let S_∞ be the final relaxed value of the stress at infinite time. Then from Eq. 125

$$S_\infty = \frac{ES_0}{E_0 + E} = \frac{EE_0}{E_0 + E}\,\varepsilon_0 = E_r\varepsilon_0 \tag{126}$$

where E_r, referred to as the relaxed modulus, is given by

$$E_r = \frac{EE_0}{E_0 + E} \tag{127}$$

The coefficient of the exponential term in Eq. 125 can also be written in terms of S_∞ or E_r. Note that

$$S_0 - S_\infty = E_0 \varepsilon_0 - E_r \varepsilon_0 = E_0 \varepsilon_0 - \left(\frac{EE_0}{E_0 + E} \right) \varepsilon_0$$

$$= E_0 \varepsilon_0 \left(1 - \frac{E}{E_0 + E} \right) = S_0 \left(\frac{E_0}{E_0 + E} \right) \tag{128}$$

Now define a relaxation time, τ_ε, by

$$\tau_\varepsilon = \frac{1}{a_1} = \frac{\eta}{E_0 + E} \tag{129}$$

Then, in view of Eqs. 126, 128, and 129, Eq. 125 can be written

$$S(t) = S_\infty + (S_0 - S_\infty) \exp\left(-\frac{t}{\tau_\varepsilon} \right) \tag{130}$$

or, in terms of strain and modulus,

$$S(t) = E_r \varepsilon_0 + (E_0 - E_r)\varepsilon_0 \exp\left(-\frac{t}{\tau_\varepsilon} \right) \tag{131}$$

It is customary in stress relaxation experiments to define an instantaneous relaxation modulus, $E(t)$, as the ratio of the relaxed stress, $S(t)$, to the initial strain. Hence the relaxation modulus at any time t is given by

$$E(t) = E_r + (E_0 - E_r)\exp\left(-\frac{t}{\tau_\varepsilon} \right) \tag{132}$$

The three-parameter model thus predicts that, as a function of time, the stress decays in exponential fashion from an initial value S_0 to a final value S_∞, and the relaxation modulus similarly decays from an initial value E_0 to a final value E_r.

Dynamic Loading–Internal Friction

If a sinusoidally time-varying stress is applied to the model of the standard linear solid, it is convenient to write the solution of Eq. 103 for this case in terms of the real and imaginary components of a complex tensile compliance $D^* = D' - iD''$ or, in terms of the real and imaginary components of a complex modulus, $E^* = E' + iE''$.[1] The resulting strain is also sinusoidal, but it lags behind the stress by a phase angle, δ, where

$$\tan \delta = \frac{D''}{D'} = \frac{E''}{E'} \tag{133}$$

---

[1] Equivalently, if the applied stress is a shear stress rather than a tensile stress, one can use the shear compliance $J^* = J' - iJ''$ and the shear modulus $G^* = G' + iG''$ instead.

The solutions of Eq. 103, expressed in terms of the storage modulus, E', and the internal friction parameter, tan δ, are (5, 15)

$$\tan \delta = A \frac{\omega\tau}{1 + \omega^2\tau^2} \tag{134}$$

and

$$E' = E_0 - (E_0 - E_r)\frac{1}{1 + \omega^2\tau^2} \tag{135}$$

where the relaxation time, τ, is the geometric mean of the retardation time, τ_s, and the relaxation time, τ_ε, that have already been introduced; that is

$$\tau = \sqrt{\tau_s\tau_\varepsilon} \tag{136}$$

and ω is the circular frequency in radians per second. A is a parameter that can be written either directly in terms of the parameters of the model or in terms of the unrelaxed, E_0, and relaxed, E_r, moduli. It is given by

$$A = \frac{E_0}{[E(E_0 + E)]^{1/2}} = \frac{E_0 - E_r}{(E_0 E_r)^{1/2}} \tag{137}$$

Equations 134 and 135 show that the internal friction, tan δ, and the storage modulus, E', are functions of the quantity $\omega\tau$. As $\omega\tau$ is varied the internal friction increases, reaching a maximum value of $A/2$ at $\omega\tau = 1$, and then decreases again as $\omega\tau$ gets large. Hence the model predicts that the internal friction, either as a function of frequency at constant τ or as a function of τ at constant frequency, should be a bell-shaped curve with maximum occurring at $\omega\tau = 1$.

Equation 135 indicates that as $\omega\tau$ approaches zero the modulus takes its relaxed value E_r. As $\omega\tau$ increases the modulus rises, undergoes an inflection at $\omega\tau = 1$, and then approaches its asymptotic, or high-frequency, value of E_0, as $\omega\tau$ further increases. The influence of temperature on the relaxation process is obtained by noting that relaxation times usually vary exponentially with temperature; that is,

$$\tau = \tau_0 \exp\left(\frac{\Delta H}{kT}\right) \tag{138}$$

where ΔH is an activation energy for the process. The product $\omega\tau$ can thus be increased by either increasing ω or by decreasing temperature. Hence as a function of temperature, rather than of frequency, Eqs. 134 and 135 predict that the internal friction should pass through a maximum, and that the modulus should fall from a high unrelaxed value, E_0, at low temperatures, through an S-shaped drop in the vicinity of those temperatures for which $\omega\tau = 1$, and should finally attain the relaxed value, E_r, at high temperatures.

2. Comparison of Predicted and Observed Behavior

In the preceding section we have derived equations, based on relatively simple models of a linear viscoelastic material, that predict the following phenomena: under constant stress, deformation and compliance will continue to increase with time, as shown by Eqs. 116 to 118; under constant strain, the stress and the relaxation modulus continue to decrease with time, as shown by Eqs. 130 to 132; and under sinusoidally applied stresses or strains, the storage modulus and the internal friction will change with frequency, as indicated by Eqs. 134 and 135. Let us now examine the observed behavior of real polymeric materials to see to what extent they show the mechanical behavior expected of the simple linear viscoelastic models.

Creep Phenomena

All polymers show creep; that is, the deformation continues to increase with time under a constant applied load. When the load is first applied there is an "instantaneous" increase of strain, as predicted by Eqs. 117 and 118. The magnitude of this strain increment increases linearly with the applied stress, S_0, and is inversely proportional to the low strain elastic modulus, E_0, of the polymer. Also, in all real polymers subject to a constant stress, and provided the applied stress is not too large, the deformation increases with time but at a steadily decreasing rate. This, too, is predicted by the second term of Eqs. 117 and 118 although the change of strain with time does not always have the simple exponential character of these equations. Another experimental finding is that in many polymers the strain, after prolonged times, becomes linear in time or the strain rate becomes effectively constant. This effect is not predicted by the simple three-parameter model of Fig. 22c but it is predicted, as the last term of Eq. 118 shows, by a four-parameter model, in which a dashpot of viscosity η_0 is placed in series with the linear spring of modulus E_0.

As an illustration of the experimentally observed creep behavior of a specific polymer, we present in Fig. 23 creep strain–time curves, obtained at 77°F and 50% rh, for polystyrene specimens subjected to varying stress magnitudes (62). These creep curves show all three features mentioned above and hence are in general accord with expectations based on the simple four-parameter model. Although temperature noticeably enhances creep strains, it is clear from Fig. 23 that appreciable creep deformations may occur at moderate stress values and at temperatures far removed from the softening temperature, T_g, which, for polystyrene, is about 100°C. The stress dependence of the creep deformation is evident from the data but the creep rate is also markedly affected. For example, the observed "constant" creep rate at long time, for an applied stress of 3250 psi, is 60×10^{-7} in./in./hr,

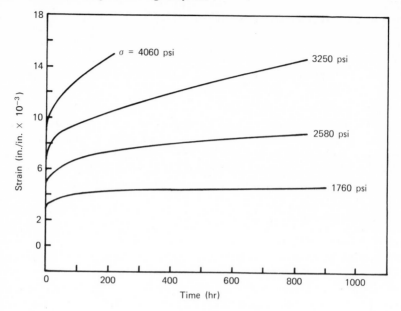

Figure 23   Creep curves for polystyrene at various tensile stresses (62).

whereas the observed creep rate, at the stress level of 1760 psi, is only 3×10^{-7} in./in./hr. Thus a stress increase of less than two times has caused a 20-fold increase in creep rate.

The deformation time dependence of polymers can also be expressed in terms of a creep modulus, or apparent modulus, $E_a(t)$. This creep modulus is simply the ratio of the initial applied stress, S_0, to the creep strain $\varepsilon(t)$ at the time and temperature in question. The creep modulus is a significant parameter to the designer of polymer structures or components who has to ensure that the polymer part will not undergo excessive deformation over its useful life. This modulus is, of course, a function of polymer composition and structure as well as of stress level, time, and temperature.

Values of the creep modulus of a number of commercial polymers, measured at room temperature and at a stress level of 1500 psi, unless stated otherwise, are given in Table 5 (19). The values of creep modulus are given for each polymer for various times ranging from 10 hr to, in many cases, 10,000 hr or more. As a rough guide to the effect of time on the value of the modulus, it may be noted from the data of this table that for each decade increase in time there is a drop in modulus of about 20%. Nevertheless, it is clear from the data of the table that some polymers, even after 10,000 hr at a stress level of 1000 psi, still retain comparatively high modulus values of the

Table 5   Creep modulus of various polymers at stress level of 1500 psi[a]

Material	Type	Creep Modulus (10³ psi; 73°F)				Comments
		10 hr	100 hr	10³ hr	10⁴ hr	
POM	Delrin 500	360	280	240	170	
ABS	Kralastic MH	278	263	240	—	
Nylon-6,6	Zytel 101	160	140	120	100	Tested at 1400 psi; equil. at 50% rh
Nylon-6	Plaskon 8202C	102	85	69	—	Equil. at 50% rh
PC	Lexan 141-111	260	225	197	190	Tested at 130°F
PE	Marlex 6050	36	28	24	—	Tested at 1000 psi
PMMA	Plexiglas G	410	318	245	—	Tested at 1000 psi
PPO	Noryl SE 1	416	395	320	269	
PP	Profax 6400	77	58	46	37	
PS	Bakelite TMDA 2120	100	37	11.3	—	Impact-modified PS
PSF	Bakelite P 3500	345	340	325	>280	Tested at 4000 psi
PTFE	Halon G 700	20	15	11	—	Tested at 1000 psi
PVC	Plaskon 2007	139	111	—	—	
PVC	Bakelite QMDA 2201	—	250	183	>120	
Phenolic, mineral-filled	Bakelite BMG 0750	—	4200	4000	>3400	Tested at 2030 psi

[a] See Ref. 19.

order of 200,000 psi or higher. Polymers in this class include Delrin, poly-carbonate, polyphenylene oxide, polysulfone, and the thermosetting polymers like mineral-filled phenolic.

Temperature has a marked effect on the creep behavior of polymers. With increasing thermal agitation and increasing molecular mobility creep deformation is accelerated and creep modulus reduced. The combined effects of both temperature and time for a given polymer, Delrin 500, are evident from Table 6 (19). From the data of this table, we may note that a polymer specimen may acquire a given creep modulus under different conditions. For example, under an applied stress of 1000 psi, a modulus value of 110,000 psi is reached after 100 hr at 185°F; however, if the specimen is held at 212°F, this modulus value is reached after 1 hr. For the effect of temperature and time on the creep modulus of other polymers the reader is referred to the original data.

The creep modulus of plastics can be significantly increased by appropriate reinforcement. This is demonstrated for a number of different commercial polymers in Table 7 (19). Note that even for relatively flexible polymers like polyethylene and polypropylene, where the normal creep modulus, as Table 5 indicates, is below 100,000 psi when tested at a stress level of 1000 psi, dramatic increases in creep modulus can be realized upon addition of some 30–40% of glass fiber. Table 7 shows also that many of the glass-filled thermoplastics such as POM, nylon, ABS, and rigid PVC maintain high creep moduli values of 10^6 psi or so even after long-time use at applied stresses of 5000 to 10,000 psi; thus the fiber reinforced thermoplastics begin to approach the stiffness qualities of thermosetting polymers.

The amount of reinforcement directly controls the increase in stiffness or creep modulus that results. This effect can be appreciated by comparing the Delrin data in the first two lines of Table 7 with that in the first line of Table 5.

Table 6   Effects of temperature on creep modulus of Delrin 500 at 1500 psi stress level[a]

Test Temp. (°F)	Creep Modulus (10^3 psi)			
	1 hr	100 hr	10^3 hr	10^4 hr
73	390	280	240	170
115	250	160	—	—
185	160	110	75	—
212	110	75	—	—

[a] See Ref. 19.

Table 7   Effects of reinforcement on creep modulus of polymers[a]

Material	Filler	Stress (psi)	Creep Modulus (10^3 psi)		
			1 hr	100 hr	10^3 hr
POM—Formaldafil G80/20	20% Glass fiber	4000	700	540	—
POM—Formaldafil G80/40	40% Glass fiber	4000	1250	950	—
ABS—LNPAF 1008	40% Glass fiber	5000	—	1650	1600
Nylon-6—Nylafil G13/40	40% Glass fiber	10000	1110	865	—
Nylon-6,6—Nylafil G1/30	30% Glass fiber	10000	1250	830	—
PP—Profil G60/40	40% Glass fiber	2000	970	840	—
PS—Styrafil G35/35	35% Glass fiber	8000	1350	1250	—
PE—Ethofil G90/40	40% Glass fiber	2000	840	680	—
PVC—LNP VF 1007	35% Glass fiber	10000	—	1540	1400
Epoxy—Plaskon Epiall 1914	Glass-filled	4000	1450	1250	—
Phenolic—Plaskon Phenall 8000	Glass-filled	4000	4250	3000	—

[a] See Ref. 19.

From this comparison we see that—even though the applied stress level has risen from 1500 to 4000 psi—the creep modulus at 100 hr is almost doubled by addition of 20% glass fiber and nearly doubled again by addition of another 20%.

Stress Relaxation Phenomena and Regions of Viscoelastic Behavior

All real polymeric materials exhibit stress relaxation. To a first approximation, it is found that the relaxed stress, $S(t)$ of Eq. 131, or the relaxation modulus, $E(t)$ of Eq. 132, gives an approximate representation of the actual data in that the measured stress or modulus continues to decrease with increase of time from a high, or unrelaxed value, to a low, or relaxed value. However, as noted in the case of creep phenomena, the actual change with time may not have the simple exponential form involving but one relaxation time constant. The relaxation modulus $E(t) = S(t)\varepsilon_0$, determined from a stress relaxation type experiment run at constant strain ε_0, is rather analogous to the creep modulus, $E_a(t)$, discussed in the preceding section, and which is obtained from a deformation test run at constant stress.

To take into account time-dependent effects in polymer design problems, it is appropriate, as a first approximation, to replace the static elastic modulus, E, that appears in the various equations of elasticity by either the apparent

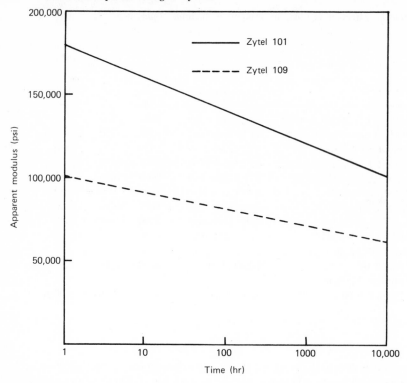

Figure 24 Variation with time of the apparent modulus of two nylons (35). Moisture content 2.5%.

creep modulus, $E_a(t)$, or the stress relaxation modulus, $E(t)$, appropriate for the desired lifetime. These moduli values of course vary with polymer structure as well as with time. Figure 24 shows the variation of apparent modulus with log time for two different polyamides; Zytel 101, a nylon-6,6 type polymer, and Zytel 109, a nylon copolymer (35). If the experimental variation is linear in log time it may be possible to predict long-time performance by extrapolation from the experimental measurements. From Fig. 24 it is seen that the reported modulus value of Zytel 101, after 1 hr, is approximately 180,000 psi. If this material were to be used in a part designed to function satisfactorily for many years, then design could be based on the extrapolated value of the data at 100,000 hr. In this case, one would use a modulus value of about 80,000 psi in the design equations to properly allow for long-time deformation effects.

Stress relaxation data are frequently used to explore the various regions of viscoelastic behavior that may be encountered by polymers (15). This could be done in principle by holding the temperature constant and varying the measuring time. However, if the temperature were less than T_g, the experimental times would be prohibitively long. It is more convenient in practice to hold the measuring time constant, at 10 sec, say, and vary the temperature instead. Stress-relaxation data for polystyrene, taken after 10 sec, are given as a function of temperature in Fig. 25 (15). From this type of data Tobolsky (15) has identified five different regions of viscoelastic behavior:

The Glassy Region. This region occurs at low temperatures, at $T < T_g$. Here long-range chain motions, the so-called convolutions of Alfrey (25), are no longer possible in the course of the experiment, for thermal agitation energy is not great enough to surmount the barriers to motion and to overcome entanglements. The modulus is high, of the order of 10^{10} dyne/cm^2, ($\sim 10^5$ psi). In partially crystalline polymers, like annealed isotactic polystyrene, ordered crystalline regions, as well as amorphous disordered regions are present. As a result, as Fig. 25 shows, the observed modulus is now

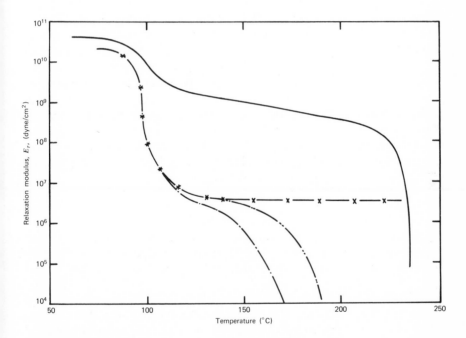

Figure 25 Relaxation modulus, E_r, versus temperature for different types of polystyrene specimens (15). — · — · —· Atactic sample, $\bar{M}_n = 1.4 \times 10^5$. — · · — · · —· · · Atactic sample, $\bar{M}_n = 2.17 \times 10^5$. – × – × – × Atactic sample, lightly cross-linked. ———— Isotactic sample, partially crystalline.

somewhat higher than that of the atactic polymer because of the better packing of the chains in the crystalline portion.

The Transition Region. This occurs at temperatures comparable to T_g and is marked by a drop in modulus of some three decades over a relatively narrow temperature interval. Thermal agitation energy is now sufficiently great to overcome barriers to segmental motions, even in an experimental time of 10 sec, and hence the modulus drops sharply as the polymer segments respond to the applied stress by rapid micro-Brownian movements.

For partially crystalline polymers, the magnitude of the modulus drop depends on the level of crystallinity in the sample. As Fig. 25 shows for isotactic polystyrene and as Fig. 2 shows for the crystalline polyolefins, a modulus drop still occurs in the vicinity of T_g but it is much less than three decades. This is because intermolecular cohesion has been lost only in the amorphous or disordered regions where chain packing is poor. In the crystalline regions, where the chains are packed in ordered arrays, the intermolecular forces are high and they provide rigidity to the system.

The Rubbery Plateau Region. This is the region where $T > T_g$. In this region the modulus tends to become constant with temperature at a value of about 10^7 dyne/cm. For a lightly cross-linked material, this region may extend for 100 to 200 degrees above T_g. This is the region where rubber elasticity is present. The cross-links, if present, prevent permanent chain slipping but local segmental diffusional motions are very rapid, and hence the polymer can quickly respond to applied stress or recover rapidly when the stress is removed. The extent of this rubbery plateau region depends on the molecular weight. For uncross-linked material, only chain entanglements are available to give a quasi-network structure; hence the higher the molecular weight, the further up the temperature scale this region extends.

The Rubbery Flow Region. This region is not present in the cross-linked material but in the uncross-linked polymers it occurs directly above the rubbery plateau region where the modulus begins to fall below the 10^7 value with increase of temperature. What is happening here is that in the time scale of the experiment, the entanglements are not holding and chains can slip by one another.

The Liquid Flow Region. This region occurs at temperatures above that at which the rubber flow region starts. The modulus falls at an increasingly faster rate with increasing temperatures. In the time scale of the experiment there is an increasing degree of viscous flow and a decreasing amount of elastic recovery. For the partially crystalline polymer, this region is not reached until the melting point is exceeded, which for isotactic PS is about

240°C. At temperatures between T_g and T_m the crystallites themselves act as cross-links, and hence crystalline polymers in this region are both strong and tough, for stiffness and strength arise from the ordered crystalline regions and high elastic deformations and rapid recovery from the amorphous regions.

When stress relaxation experiments are performed on a different time scale, say, 1000 sec instead of 10 sec, then all the viscoelastic regions shift to lower temperatures but the general form of the relaxation modulus–temperature relation remain the same. One can also shift the T_g location, and the accompanying modulus transitions, to lower temperatures by the addition of plasticizers or diluents; examples of this, as well as of the effect of various other variables on viscoelastic behavior, are given in the next section.

Dynamic Mechanical Phenomena

As noted in our discussion of viscoelastic models under dynamic loading, the loss modulus and the internal friction of a viscoelastic polymer would show a maximum as a function either of temperature or of frequency and, in this same region, the storage modulus would show dispersion.

Most polymers do show such behavior and many examples have been given in the literature (5, 8). In general, however, the polymer may exhibit more than one dispersion region or internal friction maximum, and furthermore each of the loss peaks is usually considerably broader than one would expect for a single relaxation-time process.

As one example of the type of information obtained from dynamic mechanical tests, we present in Fig. 26 data of Chung (63) on loss modulus values determined by means of a torsion pendulum (Chapter 6) over the temperature range from 80 to 300°K. The measurements were made on poly-p-xylylene and on another polymer, poly(chloro-p-xylylene), which has a similar chemical structure, except that one of the H atoms on the benzene ring is replaced by a Cl atom. The results show that, in both polymers, there is a broad maximum in the loss modulus (and internal friction) and that for the Cl-substituted polymer the maximum occurs at 254°K instead of at 159°K. This shift is the result of both dipolar and steric affects introduced by substitution of the larger and polar Cl atom for an H atom.

The breadth of the loss peaks in Fig. 26 is due to the wide distribution of relaxation times in all real polymers rather than a single relaxation time. Although not shown in the figure, the modulus falls more rapidly with temperature in the vicinity of the loss peaks than elsewhere, and this is the type of behavior we would expect for a relaxation type of transition. It has also been shown by various experimenters that test results, if the applied sinusoidal stresses are kept small, are independent of amplitude (5, 10), and this, too, follows from analysis of the mechanical models.

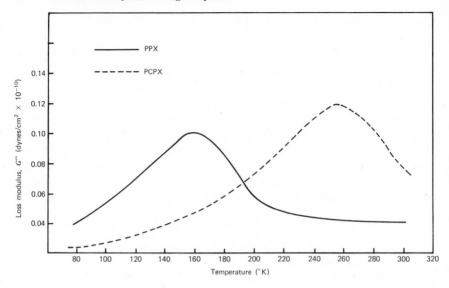

Figure 26 Loss modulus, G'', versus temperature for poly-p-xylylene (PPX) and poly-(chloro-p-xylylene) (PCPX) (63).

The loss peaks shown in Fig. 26 are associated with stress-biased reorientational motions of short chain segments involving the phenylene ring and probably neighboring CH_2 units. Both polymers also show another loss maximum, indicative of a primary glass–rubber relaxation process, in the region above room temperature, and this is associated with onset of micro-Brownian chain motion in the amorphous regions.

Another example, which shows how the dynamic modulus of several polyolefin polymers varies with temperature, has been given in Fig. 2. In this figure the large drop in the 250–270°K region is indicative of the glass–rubber transition. In addition, in both polypentene and polybutene there is a low-temperature drop in modulus at about $-120°C$. This is indicative of a secondary relaxation process and is here attributed largely to reorientational motion of the side chain units.

3. Generalized Models and Distributed Relaxation Times

Although the essential features of the viscoelastic behavior of real polymers are in approximate accord with the predictions of the phenomenological theory based on a three-(or four)parameter model, one cannot expect that models having but one relaxation time constant can truly represent actual polymers, with their broad molecular weight distribution and their highly complex internal structure. Hence more elaborate models have been introduced to describe viscoelastic materials. These models usually consist either of a discrete series of Maxwell units placed in parallel or a discrete series of Kelvin–Voight units placed in series. Although these two generalized models are essentially equivalent, the first is more convenient for describing stress relaxation phenomena and the second is more useful for describing creep phenomena. Also it is convenient to introduce the concept of a continuous distribution of relaxation times; in this case, we may speak of the polymer having a modulus relaxation function, $E(\tau)$, or if log times are used a logarithmic relaxation function, $H(\tau)$, and, correspondingly, a compliance function, $D(\tau)$, or a logarithmic retardation function $L(\tau)$.

For the generalized Maxwell model with a discrete set of relaxation times $\tau_1, \tau_2, \ldots, \tau_N$ the stress relaxation modulus is given by

$$E(t) = \sum_{i=1}^{N} E_i \exp\left(-\frac{t}{\tau_i}\right) \tag{139}$$

where E_i and τ_i are the modulus and relaxation time, respectively, of the ith element. If there is a continuous distribution of relaxation times, then Eq. 139 takes the form

$$E(t) = \int_0^\infty E(\tau)\exp\left(-\frac{t}{\tau}\right) d\tau \tag{140}$$

Also, if at very long times the material tends to approach an equilibrium or relaxed value of the modulus, E_r, say, then this fact can be accommodated by adding to the generalized Maxwell model one linear spring element and replacing $E(t)$ above by $E(t) - E_r$. It is also usual practice to describe the viscoelastic material behavior in terms of a relaxation distribution function $H(\tau)$, where $H(\tau) = \tau E(\tau)$.

In terms of H Eq. 140 then becomes

$$E(t) = \int_{-\infty}^\infty H(\tau)\exp\left(-\frac{t}{\tau}\right) d(\ln \tau) \tag{141}$$

Thus if the relaxation distribution function H is known for the material in question then the relaxation modulus at any time is given by the solution of the above integral equation instead of by solution of the much simpler Eq. 132 involving only one relaxation time constant.

For describing creep behavior it is more usual to utilize the generalized Voight–Kelvin model and to express behavior in terms either of the shear creep compliance, $J(t)$, or the tensile creep compliance, $D(t)$, where these compliances are simply defined as the corresponding shear or tensile strains divided by the appropriate constant value of the applied stress.

For a discrete model of relaxation times, we have

$$D(t) = \sum_{i=1}^{N} D_i\left(1 - \exp\left(-\frac{t}{\tau_i}\right)\right) \tag{142}$$

where D_i and τ_i are the compliance and the relaxation time, respectively, of the ith element; and for a continuous set of relaxation times, Eq. 142 becomes

$$D(t) = \int_0^{\infty} D(\tau)\left(1 - \exp\left(-\frac{t}{\tau}\right)\right) d\tau \tag{143}$$

If we wish to allow for an initial elastic response we can do so by adding a simple spring to the model and replacing $D(t)$ in Eqs. 142 and 143 by $D(t) - D_0$. Here D_0 is the initial compliance of the spring corresponding to the glassy state compliance of the system. Also if we wish to allow for permanent flow we can add a simple dashpot to the model, and in this case we add an extra term, t/η_0, on the right side of Eq. 143 where η_0 is the viscosity of the added dashpot.

The above results may also be expressed in terms of the retardation function $L(\tau)$, where $L(\tau) = \gamma D(\tau)$. The analogous equation to Eq. 143 then becomes

$$D(t) = \int_{-\infty}^{\infty} L(\tau)\left[1 - \exp\left(-\frac{t}{\tau}\right)\right] d(\ln \tau) \tag{144}$$

For a dynamic experiment, it is customary to describe behavior in terms of the real and imaginary parts of a complex modulus

$$E^*(\omega) = E'(\omega) + iE''(\omega) \tag{145}$$

or in terms of the real and imaginary parts of a complex compliance

$$D^*(\omega) = D'(\omega) - iD''(\omega) \tag{146}$$

In these equations $D^*(\omega) = 1/E^*(\omega)$ and the internal friction, $\tan \delta$, is given by either the ratio of the modulus or compliance components; that is,

$$\tan \delta = \frac{E''(\omega)}{E'(\omega)} = \frac{D''(\omega)}{D'(\omega)} \tag{147}$$

For the generalized Maxwell model, it is more appropriate to use the moduli rather than the compliances. (The reverse is true for the generalized Kelvin–Voigt model.) It can then be shown (10) that these moduli components are related to the modulus distribution function $E(\tau)$ or to the relaxation-time distribution function $H(\tau)$ by the equations

$$E'(\omega) = \int_0^{\infty} E(\tau)\frac{\omega^2\tau^2}{1 + \omega^2\tau^2} d\tau = \int_{-\infty}^{\infty} H(\tau)\frac{\omega^2\tau^2}{1 + \omega^2\tau^2} d(\ln \tau) \tag{148}$$

$$E''(\omega) = \int_0^{\infty} E(\tau)\frac{\omega\tau}{1 + \omega^2\tau^2} d\tau = \int_{-\infty}^{\infty} H(\tau)\frac{\omega\tau}{1 + \omega^2\tau^2} d(\ln \tau) \tag{149}$$

Hence if $E(\tau)$, $D(\tau)$, $H(\tau)$, or $L(\tau)$ is known, then the stress relaxation modulus, the creep compliance, and the dynamic behavior can be predicted from solutions of these integral equations. One difficulty is that the various functions are generally known only over a limited, instead of an infinite, time scale. For evaluating the functions H and L from available creep, stress relaxation, and dynamic data, various approximation procedures have been developed (10, 64, 65). The first approximation to $H(\tau)$, known as $H_1(\tau)$, can be evaluated directly from stress relaxation data by simply noting the slope of the experimental modulus–ln time curves. The applicable equation is given by

$$H_1(\tau) = -\left\{\frac{d[E(t)]}{d(\ln t)}\right\}_{t=\tau} \tag{150}$$

Similarly the first approximation to the retardation spectrum, $L_1(\tau)$, is obtained from experimental creep data by noting the slope of the compliance–ln time curve. Higher-order approximations can be obtained from dynamic or internal friction data, as well as from creep and stress relaxation data.

The dependence of the two functions, H and L, on time have been given for various polymers (9, 10). For amorphous polymers, such as polyisobutylene, PIB, which has been widely studied, the distribution of relaxation times can be approximated by two idealized distributions, a wedge-shaped distribution and a box-shaped distribution. The wedge type, which is operative at the low end of the relaxation time spectrum (10^{-9}–10^{-3} sec), is essentially independent of molecular weight and is associated with the transition region of the viscoelastic spectrum where the polymer passes from a glassy state, with limited chain mobility, to the rubbery state, where segmental mobility is rapid.

The box type, having a uniform distribution of relaxation times, is located at the upper end of the relaxation time spectrum (10^{-3}–10^5 sec). This box-like region shifts up the time scale with increasing molecular weight and hence it is separated from the wedge-shaped region at high molecular weight but partly overlaps the wedge region at low molecular weight. The boxlike region of the relaxation spectrum is associated with the rubbery flow region of viscoelasticity, discussed earlier, where the kinetic units that are involved are themselves large (several tens of chain links) and hence relaxation times are also large.

It should be noted that a molecular theory of viscoelasticity, based on free rotation about chain bonds and the concept of a Gaussian distribution of segment lengths assumed to behave as elastic springs with spring constants proportional to temperature, has been developed by Rouse (66), Bueche (14), and others (15). From the theory, one can determine the relaxation spectrum H and the retardation spectrum L. Theory predicts that ln H versus ln t

should be a straight line with slope of $-\frac{1}{2}$, and although the theory was developed for dilute polymer solutions, it does adequately predict, even for undiluted amorphous polymers like PIB, the form and slope of the wedge-shaped region referred to above.

The relaxation spectra of all polymers are affected by temperature. As temperature is decreased, the regions of intermediate and large time are appreciably shifted up the time scale whereas the region of small times is little affected. Hence the overall shape of the spectrum is altered and the separate regions may become more difficult to distinguish.

4. The Boltzmann Principle and Linear Viscoelasticity

To predict the general behavior of linear viscoelastic materials for various stress conditions one needs to know not simple material constants, as for the elastic case, but rather continuous time functions, such as the tensile compliance function, $D(t)$ [or the retardation spectrum function, $L(t)$], and the relaxation modulus $E(t)$ [or the relaxation spectrum function, $H(t)$]. Also it is necessary to utilize the Boltzmann superposition principle. This principle states that the behavior of the material at any time is determined by its entire past history or, in other words, that the net effect of a sum of causes is simply the sum of the effects of each cause taken separately. This is a consequence of the linear nature of the differential equations that describe the behavior of the Maxwell or Kelvin–Voight models.

Suppose, for example, application of a given stress σ_1 induces strain $\varepsilon_1(t)$ and application of a second stress σ_2 induces strain $\varepsilon_2(t)$. Then according to Boltzmann's theorem application of the combined stresses would induce strain $\{\varepsilon_1(t) + \varepsilon_2(t)\}$.

More generally, if we apply, say, a shear stress s_1 at time τ_1, another shear stress, s_2 at time τ_2, etc., then the total shear strain $\gamma(t)$ will be

$$\gamma(t) = s_1 J(t - \tau_1) + s_2 J(t - \tau_2) + s_3 J(t - \tau_3) + \cdots \qquad (151)$$

where $J(t)$ is the appropriate shear compliance function. If we now alter the stress continuously, the total strain will depend on the entire past history of the material and we may write

$$\gamma(t) = \int_{-\infty}^{t} J(t - \tau)\frac{ds(\tau)}{d\tau}\,d\tau \qquad (152)$$

One could just as easily express the stress at any time in terms of the past history of the strain. In this case the shear modulus function enters instead of the shear compliance function and the analogous equation to Eq. 152 becomes

$$S(t) = \int_{-\infty}^{t} G(t - \tau)\frac{d\gamma(t)}{d\tau}\,d\tau \qquad (153)$$

The functions $J(t)$ and $G(t)$ are not independent. In fact, it is a consequence of the Boltzmann superposition principle and of the assumed linear nature of the relation between stress and strain that all the various functions used to describe viscoelastic behavior under particular types of loading conditions are interrelated (26, 67).

It should be noted that the above considerations are applicable essentially only at low strains, that is, in the region of linear behavior. At high stress or strain values, many actual polymers exhibit nonlinear viscoelastic behavior. Under such conditions use of relations derived on the basis of linear visco-elasticity can at best be considered as rough first approximations.

5. Time–Temperature Equivalence and the WLF Equation

It has been observed by various investigators, especially by Leaderman (68), Tobolsky (15), and Ferry (10), that for amorphous polymers, mechanical relaxation data obtained at various temperatures—whether by creep, stress relaxation, or dynamic experiments—can be superimposed by an appropriate shift of the time scale. Thus the effect of temperature is simply to change the log time scale by a constant factor that depends on the temperature itself. This implies that the relaxation or retardation times involved vary with temperature in the same manner. The general phenomenon is known as time–temperature superposition. By utilizing appropriate shift factors, a_T, one can obtain from either mechanical or dielectric relaxation data a so-called master curve, appropriate to some arbitrary reference temperature. An illustration of time–temperature superposition based on stress–relaxation data on polyisobutylene (PIB) taken at a variety of temperatures is given in Chapter 6, Fig. 14. Many other examples, using dynamic as well as transient data, are given in the literature (5, 10). The method is also referred to frequently as the method of reduced variables as the master curve is a plot (generally on a log scale) of the specific viscoelastic function being measured against a reduced time variable (t/a_T) or against a reduced frequency (ωa_T). The shift factor, a_T, is essentially the ratio of the relaxation time, $\tau(T)$ at the temperature T to the relaxation time, $\tau(T_0)$, at some reference temperature, T_0.

Williams, Landel, and Ferry (69) have shown that for a wide variety of amorphous polymers there is essentially a universal relation relating the horizontal shift factor, a_T, to the temperature. This relation, which holds over a wide range of temperatures (from T_g to $T_g + 100°C$), is given by

$$\log(a_T) = -\frac{C_1(T - T_0)}{C_2 + T - T_0} \tag{154}$$

Here T_0 is a reference temperature (chosen generally as $T_g + 50°C$) and C_1 and C_2 are constants that apparently vary little from material to material, provided temperature has not caused any change in structure of the polymer. When the reference temperature is taken as T_g, Williams, Landel, and Ferry have shown that for a wide variety of polymers and glass-forming liquids, the equation for the shift factor may be written:

$$\log a_T = -\frac{17.4(T - T_g)}{51.6 + T - T_g} \tag{155}$$

Many amorphous polymers follow this WLF empirical equation quite closely, especially when appropriate corrections are made to the data to take care of both density changes and rubberlike behavior. The WLF equation is not expected to hold for crystalline-type polymers and, even for amorphous polymers, the presence of nearby secondary relaxations may influence the data and prevent superposition. The shift factor, a_T, is related to the temperature dependence of a segmental friction coefficient and hence to the viscosity or mobility of the polymer molecules.

Although the WLF equation is empirical in nature, its general form can be obtained from "free" volume considerations (69–72). The fractional free volume, f, is usually considered to be constant at a value f_g, below the glass-transition temperature. The "free" volume above T_g is taken as

$$f = f_g + (T - T_g)\Delta\alpha \tag{156}$$

where $\Delta\alpha$ is the difference of the expansion coefficients above and below T_g. The viscosity in the liquid state is expected to be a function of the free volume that is present at that temperature. As the free volume is decreased, as by cooling, molecular motion becomes more difficult. Doolittle (71) assumed a linear relation between the logarithm of the viscosity and the inverse free volume, and this leads to the equation

$$\ln a_T \cong \ln\left(\frac{\eta}{\eta_g}\right) = \left(\frac{1}{f} - \frac{1}{f_g}\right) \tag{157}$$

Upon substituting the expression for the free volume from Eq. 156 into 157, we get

$$\log a_T = -\frac{C_1(T - T_g)}{C_2 + T - T_g} \tag{158}$$

where

$$C_1 = \frac{1}{2.30 f_g} \tag{159}$$

and

$$C_2 = \frac{f_g}{\Delta\alpha} \tag{160}$$

On comparing Eqs. 158 to 160 with the empirical Eq. 155 we see that the fractional free volume at T_g is approximately 0.025, and that the change in expansion coefficient is 0.00048/deg C. As discussed by Gordon (70) and others, observed values of the quantity $\Delta\alpha$ are in reasonable agreement with the value cited above for a variety of glass-forming polymers and other materials; also for many different polymers, such as PMMA, PVC, and polyvinyl acetate (PVA), the fractional free volume falls very close to 0.025 (73).

From the form of the WLF equation, it is evident that the temperature dependence of the shift factor for the glass–rubber relaxation does not obey an Arrhenius type relation. In fact, if $\ln a_T$ is plotted against $1/T$ the slope at different points on the curve varies markedly. One example of this is PMMA, for which the apparent activation energy varies from a value of about 24 kcal/mole near 75°C to some 200 kcal/mole near 100°C, and then decreases again at higher temperatures. For some nonpolar polymers, like PIB, the variation of a_T with temperature is in approximate accord with an Arrhenius process, with the apparent activation energy near 16 kcal/mole. In general, however, the WLF equation is applicable only to the glass–rubber relaxation process whereas secondary relaxations tend to follow the Arrhenius relation.

The time–temperature superposition principle is of considerable practical importance because to obtain the viscoelastic functions of a rigid polymer, like PS or PMMA, by conducting measurements as a function of time at only one temperature is essentially an impossible operation; tests would have to be run for years in order to obtain needed data on rubberlike elasticity. Similarly in an elastomer, as natural rubber, one would have to make measurements at extremely short time scales to observe the "glassy" visco-elastic regions of these materials. However, by making viscoelastic measure-ments over a limited time scale but at various temperatures, and then superimposing the data, one can essentially separate the complex time–temperature dependence to obtain, for some given reference temperature, the separate time dependence of the various viscoelastic functions over wide ranges of time. The temperature dependence then shows up separately in the variation of the shift factor with temperature.

More general forms of the WLF equation have been given in the literature. That of Eisenberg (74) enables one to take account of free volume changes that arise from changes of pressure, molecular weight, and diluent concentration. Also, for some systems where structural changes may be introduced by changes of temperature, it may be necessary to obtain master curves to include vertical shifts along the modulus or compliance axis as well as horizontal shifts along the time axis.

Time–stress superposition methods have also been used for polymers subject to high stresses where nonlinear effects arise and must be considered (75). In these cases it has been shown that both horizontal and vertical shift factors arise, and that both are themselves functions of the applied stress. For low-density PE it was found, analogous to time–temperature superposition, that the horizontal shift factor a_σ varies with applied stress as

$$\ln a_\sigma = \frac{a_1(\sigma - \sigma_0)}{a_2 + (\sigma - \sigma_0)} \tag{161}$$

where σ_0 is a reference stress and a_1 and a_2 are parameters. Both time–temperature superposition and time–stress superposition have been used to predict long-time creep of PE with good results.

An interesting observation by Smith (76) is that even ultimate properties, such as tensile strength at break or ultimate strain, for materials such as GRS rubber can be superimposed by use of appropriate shift factors. He ran tests at various temperatures from -68 to $93°C$ and at various strain rates. The data were plotted at each temperature versus the reciprocal of the strain rate, and by appropriate shifts along the horizontal axis master curves were obtained. Smith also showed that these experimental shift factors varied with temperature in the same fashion as predicted by a WLF-type equation, with the reference temperature taken as $263°K$ (or about $50°C$ above T_g). Thus the variation with time and temperature of the ultimate properties of these amorphous rubbers are evidently controlled by the same internal friction or segmental mobility factors that govern the low strain viscoelastic behavior.

F. EFFECTS OF EXTERNAL AND INTERNAL FACTORS ON MECHANICAL BEHAVIOR

The mechanical behavior of polymers is affected by many factors, some of which have already been discussed in preceding sections. The important factors involved include both external variables and internal variables. External variables are factors that have an appreciable effect on mechanical properties and behavior but that are not directly concerned with the structure and composition of the polymer. The principal external variables are time, temperature, and pressure. Another external variable that may have a significant effect is chemical environment, such as the presence of radiation or of some other atmosphere or medium that causes changes in physical or chemical behavior.

Internal variables, on the other hand, are those that produce changes in mechanical properties by directly producing changes in the chemical and physical structure of the polymer. Some of the important internal factors are chemical structure and composition, degree of crystallinity, polar nature of substituents, molecular weight, the presence of low molecular weight diluents, such as water, monomer, and plasticizer, and the degree and extent of copolymerization. We now consider the effect of these different variables on mechanical behavior and, in each instance, present one or more illustrative examples.

1. Influence of Strain Rate

With increasing strain rate, polymers tend to become less ductile and more brittle. As the rate of straining increases most polymers show a rise in modulus, a rise in yield stress, and usually a drop in elongation to fracture. These effects are demonstrated in Fig. 27, which shows tensile stress–strain measurements of Hall (77) obtained on polyacrylonitrile fibers under different rates of straining. It will be noted that the change in strain rate covers more than six decades and that over this range the yield stress has increased about five times.

As a second example of the influence of strain rate on mechanical behavior, we offer the results obtained by Hsiao and Sauer (78) on an amorphous polymer, polystyrene. These investigators explored the effects of strain rate on the compressive stress–strain behavior of this material and their data are shown in Fig. 28. Under tension loading, PS exhibits brittle behavior and usually fractures without yielding at about 1.5% strain. However, under compressive loading PS is ductile and a definite yield point is present, as Fig. 28 shows. With increasing strain rate, the compressive yield stress rises from a value a little above 10,000 psi at the lowest strain rate to a value above 13,000 psi at the highest strain rate. From their data Hsiao and Sauer concluded that, with a fourfold increase in strain rate, the compressive modulus of PS increases by 2–3% and the compressive yield strength by about 8–10%.

In testing polymers at high rates of loading, appreciable temperature rises may occur in the specimen owing to work expended in viscous and plastic processes. When this happens, the effects observed on mechanical behavior are combined effects of both temperature and strain rate. Increasing temperature tends to soften the material, whereas increasing strain rate tends to make the material more rigid. To illustrate the amount of temperature rise that may be involved, reference is made to the studies of Nakamura and Skinner (79) on the tensile behavior of PE. The temperature rises that occurred in ASTM type dumbbell-shaped specimens for three different rates

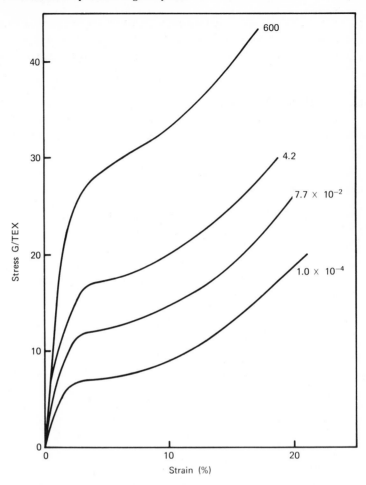

Figure 27 Effect of strain rate on tensile stress–strain curves of polyacrylonitrile fibers (77). Tex = weight in grams of 1000 m of yarn. G = applied force/g.

of straining were recorded. The temperature rises were found to be 11, 16.5, and 18.2°C for straining rates of 2, 5, and 10 in./min, respectively. Such appreciable rises in temperature could have a marked effect on the mechanical behavior.

2. Influence of Temperature

The mechanical properties of polymers are in general much more influenced by changes in temperature than are the properties of metals or ceramics

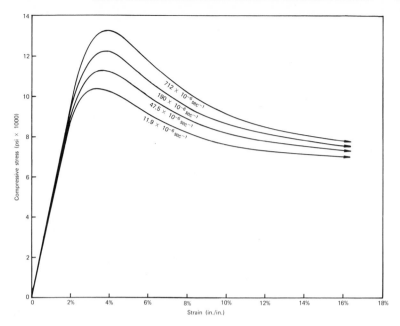

Figure 28   Effect of strain rate on compressive stress–strain curves of polystyrene (78).

(12, 16). As briefly discussed in Section A, and illustrated for several poly(α-olefins) in Fig. 2, the modulus or stiffness of a polymer tends to decrease with increasing temperature but not at a steady rate. As a temperature is approached at which some type of molecular motion is unfrozen and a relaxation process is initiated, there is a marked drop in modulus. For a secondary relaxation process, such as that involving onset of side branch motions and centered at about − 120°C in polybutene and polypentene, a modulus change of perhaps 10–50% occurs over a relatively narrow range of temperatures. For a primary relaxation process, involving large-scale main chain motions, such as that occurring near − 20°C in the polyolefins, the modulus may change by approximately one to three decades, depending on the degree of crystallinity of the polymer. If no internal molecular relaxation process is encountered over the temperature scale investigated, the modulus tends to decrease slowly with temperature, because normal thermal expansion occurs and causes greater separation of atoms and reduced intermolecular forces. As an example of the significance of this effect alone, we note that in PS, which has no appreciable relaxation transitions in the region from liquid nitrogen temperature to room temperature, the modulus nevertheless decreases by about 20%. Values of elastic and shear moduli at various

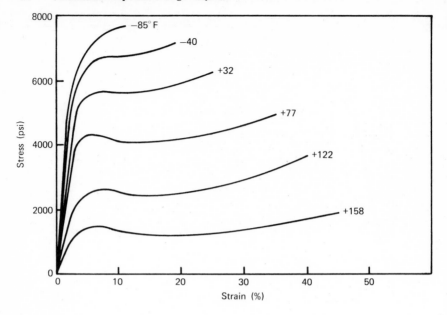

Figure 29   Effect of temperature on tensile stress–strain curves of cellulose acetate (20).

temperatures ranging from 4.2 to 473°K have been tabulated for a variety of polymers (11, 12, 80).

The influence of temperature on the stress–strain behavior of polymers is in general opposite to the effects of strain rate. With increasing temperature, the initial slope of the stress–strain curve decreases, the yield stress is reduced and yielding becomes more pronounced, and the polymer becomes more ductile. These effects are demonstrated, for samples of cellulose acetate, in Fig. 29, where the stress–strain response of this polymer is given for various temperatures from −85 to 158°F (20). This polymer exhibits ductile behavior at room temperature and at elevated temperatures and a definite yield point and yield drop are present. However, as the temperature is lowered to −85°F, yielding is suppressed, strength is increased, and ductility is cut to about 12%.

3.   Influence of Pressure

In polymers, imposition of hydrostatic pressure has a very large effect on mechanical behavior and on the nature of stress–strain curves. To illustrate, we present tensile stress–strain data obtained in our laboratory at various values of hydrostatic pressure for one amorphous polymer, PVC, and for one crystalline polymer, polypropylene. The respective stress–strain curves are shown in Figs. 30 and 31 for values of hydrostatic pressure ranging from atmospheric to 100,000 psi. In both materials—and this is typical of other polymers and in contrast to the behavior of metals—there is a dramatic increase in modulus and yield stress with increase of pressure (55).

The effect of pressure on elongation to fracture and on the nature of yielding varies considerably from polymer to polymer. In PVC the material at first becomes more brittle with increasing pressure and the elongation falls from about 100% at atmospheric pressure to about 20% at a pressure of 60 ksi. Thereafter, the elongation increases with increasing pressure. In the more brittle polymers such as polystyrene and polyimide, a pressure-induced brittle–ductile transition has been observed as the hydrostatic pressure is increased above some critical value. This increase of ductility with increasing pressure in polymers of low ductility is a phenomenon somewhat similar to that observed by Bridgman and others in tests on brittle materials like marble (24).

However, in polymers that show cold-drawing to large extension at atmospheric pressure, such as CA, PE, and PP, the effect of increasing pressure is to restrict cold-drawing and hence to decrease the nominal

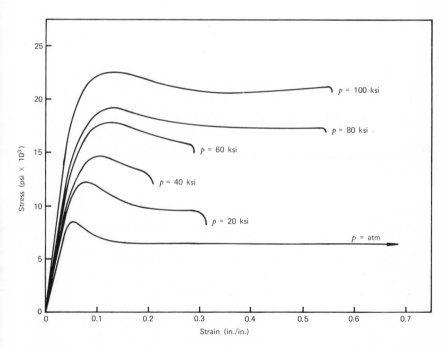

Figure 30   Effect of pressure on tensile stress–strain curves of polyvinyl chloride.

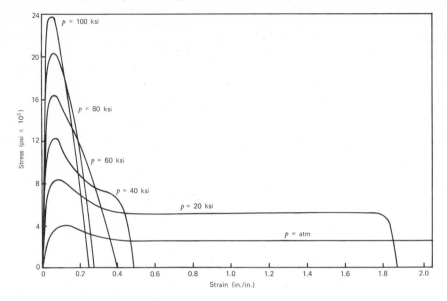

Figure 31   Effect of pressure on tensile stress–strain curves of polypropylene.

strain to fracture. This effect may arise from the combination of many causes, two of which are noted here. First, with increase of pressure the amount of free volume is reduced and hence chain mobility is lowered. Second, under increasing pressure relaxation transitions shift to higher temperatures. Hence increase of pressure at a given temperature produces a somewhat comparable effect on mechanical behavior as a reduction in temperature at a given pressure.

In PP the nominal ductility reduces from more than 200 % at atmospheric pressure to about 25 % at a pressure of 100,000 psi. However, at pressures above 40,000 psi the T_g of this polymer would be expected to shift from $-20°C$ to well above room temperature (55).

As an illustration of the dramatic increase in the elastic modulus of polymers that occurs with increase of hydrostatic pressure, we refer to Fig. 32, which shows the results obtained on PTFE samples. The modulus increased by a factor of six or more for an increase in pressure of only about 100,000 psi. It will also be noted that the increase of modulus with pressure appears to take place in three separate stages. The anomaly near 80,000 psi is associated with a pressure-induced phase change (46) and that near 30,000 psi with a pressure-induced shifting of a low-temperature relaxation process.

From the experimental data now available on a number of both crystalline and amorphous polymers, it appears that the modulus–pressure relationship

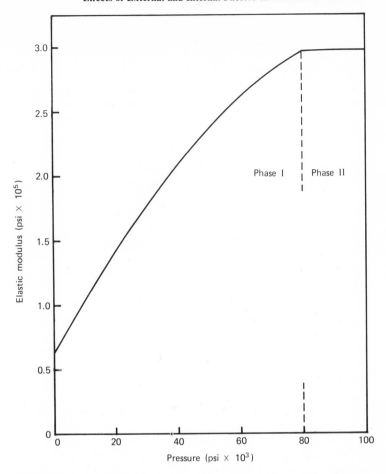

Figure 32   Elastic modulus versus pressure for polytetrafluoroethylene.

can frequently be represented, over a considerable pressure range, by the simple equation

$$E(p) = E_0 + mp \tag{162}$$

Here the parameter m may be treated as an experimental parameter whose value can be determined from the slope of the modulus pressure curve. However, it is also possible, as Sauer, Pae, and co-workers have shown (46, 55), to derive an equation of the form given above by consideration of nonlinear elasticity and finite strain effects. To a first approximation, the modulus is found to increase linearly with pressure and the coefficient m of Eq. 162 is given by

$$m = 2(5 - 4v_0)(1 - v_0) \tag{163}$$

where v_0 is Poisson's ratio, determined either at atmospheric pressure or at some modest pressure to overcome effects of possible voids in the sample. This continuum finite-strain theory appears to fit the data for different polymers quite well. This is shown by Table 8 where, for a number of different amorphous and crystalline polymers, a comparison is given of experimental and theoretical values of m. The theoretical values are computed from Eq. 163 using known values of Poisson's ratio, and the experimental values are determined from the observed slopes of the modulus pressure relation.

Table 8   Comparison of predicted and observed slopes of modulus vs. pressure curves (55)

Material	Poisson's Ratio	Theoretical Value of m	Experimental Value of m
Polytetrafluoroethylene	0.45	3.5	3.3
Polycarbonate	0.38	4.3	4.1
Polyvinyl chloride	0.40	4.0	3.2
Cellulose acetate	0.33	4.9	5.1
Polyethylene	0.34	4.8	4.1[a]
Polypropylene	0.32	5.1	4.1[a]

[a] Values of observed slopes in pressure range 35 to 100 ksi.

The agreement between the experimental and theoretical values is generally good. However, for polymers like PE and PP where T_g is near, but below, room temperature, the observed slope values are taken as those above the critical pressure required to shift T_g to room temperature. In fact, from the value of the critical pressure at which a break in the E versus P curve is found, one can estimate the pressure shift of the glass transition. Values obtained are about 20°C/kbar, in good agreement with values obtained by other methods.

4.   Influence of Radiation

High-energy radiation such as γ rays, high-speed electrons, or fast neutrons from a pile all have energies much greater than the activation energies for chemical reactions, and hence chemical changes and therefore changes in mechanical behavior are to be expected when radiation impinges on polymeric materials (27, 28).

The initial effects of radiation are to produce excitation and ionization, but subsequent effects depend on the chemical structure of the polymer. Insofar as mechanical properties are concerned, the important effects of radiation appear to be (a) breakage of C—C bonds and hence chain scission; (b) cross-linking of polymer chains by combining of adjacent free radicals; (c) changes of crystallinity in partially crystalline polymers.

The first of these effects tends to decrease the molecular weight and degrade the polymer, whereas the second effect, cross-linking, increases the molecular weight. The third effect is usually a loss in crystallinity as, with increasing radiation dose, crystal defects are produced and the melting temperature is lowered.

An example of the effects of relatively large doses of pile irradiation on a typical polymer, polyethylene, has been given in Fig. 5, where modulus–temperature data were presented for various pile dosages. From these data alone one can draw the following conclusions:

1. The irradiated polyethylene shows rubberlike behavior above its melting point, and hence radiation-induced cross-linking has occurred. Also, the degree of cross-linking is essentially proportional to the dose.

2. The melting point decreases with increasing radiation, owing to the pressure of radiation-induced defects in the crystallites.

3. The elastic modulus, over most of the temperature range, decreases with increasing dose, indicating that the degree of crystallinity, and hence the polymer stiffness, is being continuously reduced.

4. At very large doses the above trend reverses and the modulus value begins to increase with increasing radiation dose. This is an indication that the density of cross-linking has increased sufficiently to overcome the effects of reduced crystallinity.

Effects rather similar to those noted above have also been observed in the irradiation of other thermoplastic materials, such as nylon-6,6 and polypropylene. However, when polytetrafluoroethylene is irradiated, the situation is quite different. As Bernier et al. (81) have shown, for samples subjected to a ^{60}Co source, increasing the γ radiation from 0 to 200 mrad causes a significant increase in modulus over a very wide temperature range extending from 150 to 550°K. Accompanying this modulus increase are concomitant increases in density (from 2.146 to 2.235) and in degree of crystallinity (from 58 to 88%). These effects were attributed by the authors to a predominance in PTFE of chain scission versus cross-linking. It is well-known that in PTFE the molecular weight is unusually high, and evidently it is so high that it interferes with development of crystallinity. Therefore as scission occurs the average molecular weight is reduced, the local chain flexibility increases sufficiently to permit better packing of neighboring chains and higher crystallinity.

The type of radiation and the dose rate do not appear to be important variables insofar as mechanical properties are concerned. For example, in the study cited above one sample was given a dose of 78 mrad in the National Bureau of Standards ^{60}Co source and another sample was given a similar dose in the Penn State University reactor. The dynamic mechanical properties of these two samples were virtually identical over the whole temperature range studied, from 80 to 550°K.

It appears that polymers that contain aromatic rings in their chain structure are capable of receiving higher radiation doses without significant change of their mechanical properties than polymers that are not aromatic by nature. For example, both polystyrene and polyimide are known to possess good radiation resistance. On the other hand, PMMA and poly-isobutylene (PIB) are rapidly degraded by radiation. Another fact that shows the importance of chemical structure is that whereas the methacrylates are degraded the acrylates are cross-linked (82).

5.   Influence of Chemical Structure

In general much pertinent information concerning the influence of chemical structure on the resulting mechanical behavior can be obtained from knowledge of the glass-transition temperature and the melting temperature of the polymer. As discussed in Section A, if the ambient temperature is below T_g we expect most polymers to exhibit a modulus value in the 10^{10} to 10^{11} dyne/cm^2 region; if the temperature is above T_g and the polymer is amorphous, the material behaves like a rubber, with modulus of 10^6 to 10^7 dyne/cm^2. For partially crystalline polymers, where the ambient temperature lies between T_g and T_m, the polymer shows an intermediate degree of stiffness. An example of this for PP, PB, and PPen was given in Fig. 2.

The data of Fig. 2 also enable us to assess the influence of increasing the size of the side branch. From both Fig. 2 and Table 2 we note that, as the length of the side branch increases from methyl to ethyl to propyl to octyl, both T_m and T_g reduce, and hence the modulus value at any ambient temperature between T_g and T_m also reduces. Thus with increase of length of non-polar side branches there is greater interchain separation so that molecular mobility, and hence flexibility, increases. However, when side branches become very large it is possible to have an increase in stiffness or a rise in temperature location of relaxation or melting transitions. For example, in polyoctadecane T_g is at -30°C and T_m at 70°C as compared to -65°C and −38°C for the corresponding temperatures in polyoctane. These observed stiffening effects are due to onset of side chain crystallization.

Increasing the stiffness of side chains also tends to increase the elastic modulus and to raise transition temperatures. As an example of this effect we note that the T_g temperatures of poly(3-methylbutene) and poly(4-methylpentene) are 60 to 80°C higher than for PB and PPen, and lie above room temperature rather than below.

As another example of the influence of chemical structure on mechanical behavior, we give in Fig. 33 some results of Baumann (83) on measurements of shear modulus of various types of polymers over the temperature range from −200 to 400°F. One of the materials tested was a thermosetting polymer. This polymer shows very little variation of modulus with temperature and the modulus remains at a high value of about 500,000–800,000 psi over most of the temperature range covered.

Figure 33 also includes data on a rubberlike material, namely, a styrene–butadiene elastomer. For this material, we see that a sharp drop in modulus occurs in the T_g region of about −100°F and that above this temperature

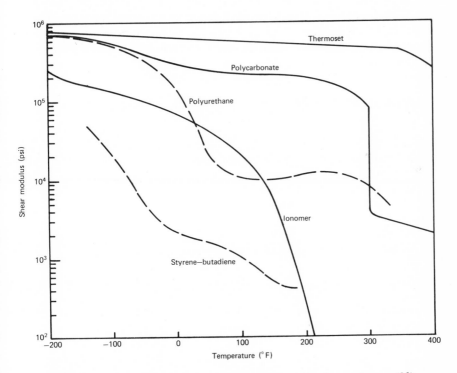

Figure 33 Effect of temperature on shear modulus of various polymers (83).

the modulus value is in the usual 10^2-10^3 psi range characteristic of a rubber. Results are also presented in Fig. 33 for a typical amorphous thermoplastic, namely, polycarbonate. Here we see that the modulus falls somewhat at $-100°F$, owing to a secondary relaxation, but that it remains comparatively high until elevated temperatures in the vicinity of its T_g (near 290°F) are reached. Here it falls rapidly with increasing temperature by a decade or more.

The influence of other chemical features on mechanical behavior is also evident from Fig. 33. Note that both the ionomer and the polyurethane show in the room-temperature range modulus values that fall between those of the rubbers and the thermosetting polymers. The reason for this behavior is that both have some of the characteristics of each type. In the ionomer there is a network structure, similar to that of inorganic crystals, in which intermolecular ionic forces arise owing to attraction between positive Na^+ ions and negative ions attached to the side branches of the polymer chains. In the polyurethane, intermolecular H bonds give rise to increased chain-to-chain cohesion and, in addition, it is believed certain primary bonds are reversibly disassociated at high temperatures. However, neither the ionic nor the H bond forces give as strong a degree of intermolecular cohesion as that arising from primary bonds in thermosetting materials; hence the moduli of these two polymers are much more affected by temperature than those of the thermosets.

6. Influence of Crystallinity

As the degree of crystallinity of a polymer increases, the modulus, the yield strength, and the hardness all rise. This effect can be seen by comparing the stress–strain properties of PE of various densities. This comparison is made in Table 9 for three different types of polyethylene as classified in ASTM Standard D1248-65T (84). Note that the modulus increases by more than 200% and the tensile strength almost doubles as the density, which is

Table 9   Mechanical properties of PE of varying crystallinity[a]

Property	Type 1	Type 2	Type 3
Density	0.910–0.925	0.926–0.940	0.941–0.965
Tensile strength (psi)	600–2300	1200–3500	3100–5500
Flexural modulus (10^3 psi)	8–60	60–115	100–260
Hardness, Rockwell D	41–48	50–60	60–70

[a] See Refs. 19 and 84.

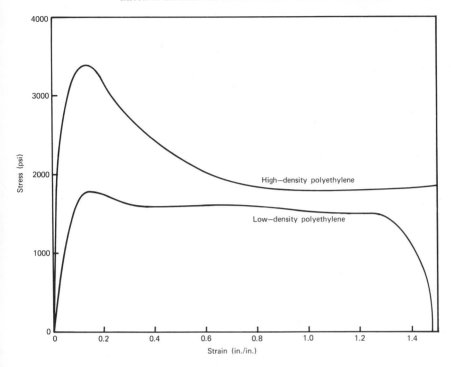

Figure 34   Tensile stress–strain curves of low-density and high-density polyethylene.

roughly proportional to the degree of crystallinity, rises from about 0.91 to about 0.96.

The influence of density, and hence crystallinity, on the stress–strain curves of commercial samples of polyethylene is shown in Fig. 34. The high-density sample has much higher modulus and yield stress, but both have good ductility.

For polypropylene as well as for polyethylene it has been shown (85) that the elastic modulus increases linearly with density. Hence any preparation or post-treatment procedures, such as slower cooling or subsequent annealing, that increase density and crystallinity also increase the modulus and stiffness of the polymer.

7.   Influence of Polar Substituents

Any substituents that give an increased degree of polarity to the polymer chains increase intermolecular forces and raise the glass-transition temperature appreciably. For example, as Tables 1 and 2 show, the T_g of atactic PP

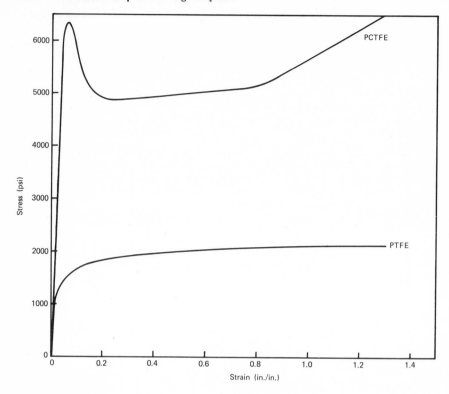

Figure 35 Tensile stress–strain curves of polytetrafluoroethylene and polychloro-trifluoroethylene.

is only $-20°C$, whereas that of PVC, in which one of the four H atoms in each monomer unit has been replaced by a strongly polar Cl atom, is 80°C. In view of this large change in T_g, it would be expected that similar large changes would occur in mechanical behavior upon introducing polar substituents in a polymer chain.

As another example of this effect, Fig. 35 shows the stress–strain curves of Teflon, PTFE, and polychlorotrifluoroethylene, PCTFE. Note the significant increase in yield stress and modulus that has taken place as one of the F atoms on each monomer unit has been replaced by the more polar Cl atom.

8.   Influence of Molecular Weight

In general, as the molecular weight of a polymer increases, T_g rises rather rapidly for low molecular weights and then gradually approaches a constant value as the molecular weight reaches some critical value. For example, in the case of PS, this occurs at a molecular weight of about 30,000–60,000.

In the usual molecular weight range of many thermoplastic materials, molecular weight does not have an appreciable effect on yield stress or modulus. This has been shown, for example, by tests carried out on PE fractions in which the molecular weight varied from 3500 to 165,000 (86). However, it is to be expected that rupture properties such as ultimate strength, ultimate elongation, and impact strength will be affected by molecular weight. Increasing molecular weight means more tie molecules between crystallities, more flexibility, and therefore greater toughness. In fact, when PE polymers are prepared from very high molecular weights ($\geq 10^6$) both the density and the degree of crystallinity decrease and hence the modulus is reduced and the elongation increased.

Extensive tests made on polypropylene fractions by van Schooten et al. (87) confirm the results obtained on PE. Yield stress and hardness were found to increase linearly with increase in degree of crystallinity but to be unaffected by molecular weight, whereas elongation to fracture and impact strength were both raised with increasing molecular weight.

9. Influence of Water and Monomer

In many polymers, such as polyethylene, polystyrene, and polytetrafluoro-ethylene, all of which absorb very little water in a 24-hr immersion test, water content and atmosphere, that is, humidity, are not important variables where mechanical properties are concerned. However, for those polymers, such as nylon, polyurethane, and cellulosic plastics, in which hydrogen bond forces contribute to the intermolecular cohesion, water in the polymer can have a large effect on polymer properties.

To illustrate the possible importance of this variable, values of various mechanical properties of nylon-6,6 are given in Table 10. In the first column

Table 10   Effect of moisture on mechanical properties of nylon-6,6[a]

Property	0.2% Water	2.5% Water
Tensile strength (psi)	11,800	11,200
Elongation at break (%)	60	300
Yield strength (psi)	11,800	8,500
Elongation at yield (%)	5	25
Flexural modulus (psi)	410,000	175,000

[a] See Ref. 35.

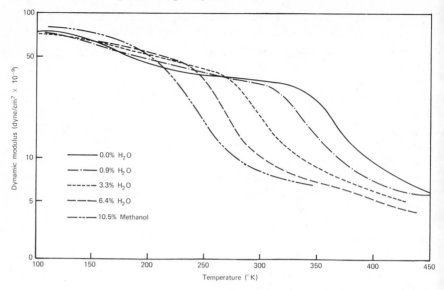

Figure 36   Effect of water content on modulus–temperature relaxations in nylon-6,6 (88).

are values for samples containing 0.2% water and in the second column values for samples containing 2.5% water. For the higher water content, strength and modulus values are lower and ductilities higher. Water acts as a plasticizer for nylon. It tends to disrupt the H bonds that are present between polymer chains and thus reduces the intermolecular forces.

As greater amounts of water are added the effects become much more pronounced. This is shown in Fig. 36, which gives the modulus–temperature variation of nylon-6,6 specimens having varying water contents (12, 88). Note that as water is added the modulus transition, centered near 350°K for the dry specimen, shifts progressively to lower temperatures and at the highest water content it falls well below room temperature. Methanol has a similar effect.

Comparable plasticizing effects can also be produced by methylating the nitrogen, that is, by replacing the amine hydrogen by a CH_3 group. This substitution also disrupts the intermolecular H bonding and hence one would expect that with increase of the amount of N methylation the modulus transition would drift successively to lower temperatures, as was noted above for the addition of water or methanol. This effect has been studied and the predicted behavior was obtained.

With regard to monomer, a situation prevails that is somewhat similar to that with regard to water. The monomer also acts as a plasticizer, but in the nylons it is not as effective as water; that is, for a given amount of additive the effects produced are not so dramatic. It has been observed, for example, that the modulus transition in nylon-6 shifts some 20°C to lower temperatures for each 1 % of water added, but it takes about 10 % of monomer to produce the same shift (89).

10. Influence of Plasticizer

In many polymers plasticizers are added in order to reduce viscosity and to improve processibility. They have a marked effect on mechanical behavior.

To illustrate the effects of plasticizer, modulus versus temperature values for PVC having plasticizer contents of 0, 10, 30, and 50 % are shown in Fig. 37 (90). The plasticizer in this instance was dioctyl phthalate (DOP). From Fig. 37 it is noted that at a temperature of 20°C the modulus drops with increase of plasticizer from a value of several hundred thousand psi for the

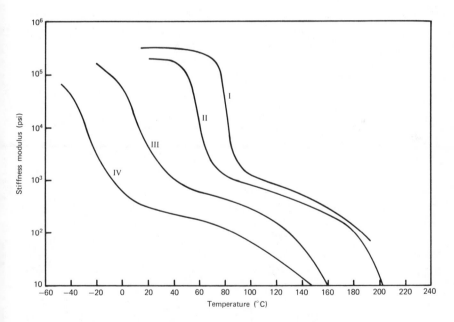

Figure 37 Modulus versus temperature for plasticized polyvinyl chloride (90). I, No plasticizer; II, 10 % dioctyl phthalate (DOP); III, 30 % DOP; IV, 50 % DOP.

unplasticized polymer to a value of several hundred psi for the 50% plasticized polymer. In terms of the temperature location of the modulus transition, the effect of plasticizer is to shift this transition to lower temperatures. The transition is at 80°C for the unplasticized sample, at 60°C for 10%, at about 10°C for 30%, and about −30°C for 50%. It is clear from data of this type why, in such an easily plasticized material as PVC, it is possible to have a whole spectrum of mechanical properties available with but one base resin.

Plasticizers need not be externally added ingredients or components. As an example of internal plasticization, we refer to the drop in T_g values that occurs on increasing the length of a side branch. For example, in PMA, T_g is at 10°C, but as the side branch increases in size T_g moves to lower temperatures. Also it occurs at −40°C in poly(ethyl vinyl ether), with two C atoms in the side chain, but in poly(n-butyl vinyl ether), with four carbons on the side chain, it is at −55°C. Other examples of this effect are evident in the data of Tables 1 and 2.

11. Influence of Copolymerization

In general, one can assess the general nature of the mechanical behavior of random copolymers by noting that to a first approximation the T_g of the copolymer will be a linear function of the weight percent of the added ingredient. Thus as vinyl acetate is copolymerized with increasing amounts of vinyl chloride, the T_g rises from about 30°C for the pure PVA to about 80°C for pure PVC (see Tables 1 and 2). Similarly, if one copolymerizes PVC with polyacrylonitrile, which has a T_g of about 107°C, then this T_g decreases as the percentage of vinyl chloride increases. These effects are shown graphically in Fig. 38, taken from the data of Reding, Faucher, and Whitman (91). The comonomer having the lowest T_g value may be thought of as essentially an internal plasticizer for the comonomer with the highest T_g value.

When physical blends of two polymers are prepared, the relaxation transitions that are present in each phase generally also show in the blend. For example, Hammer (92) has demonstrated that a 50-50 weight % blend of PVC and an ethylene–vinyl acetate copolymer (PEVA), shows the −150°C relaxation and the −20°C relaxation that are found in the pure copolymer and, at the same time, shows the 90°C transition that arises from the T_g of PVC. Sometimes, however, depending upon the particular polymers and compositions used, the compound material acts not as a two-phase blend but rather as a single-phase blend. Hammer noted this effect when he studied blends of PVC with a PEVA copolymer having 35% ethylene and 65% vinyl acetate. The copolymer, by itself, shows the same mechanical loss temperature spectra as another copolymer of 60% ethylene and 40% vinyl acetate.

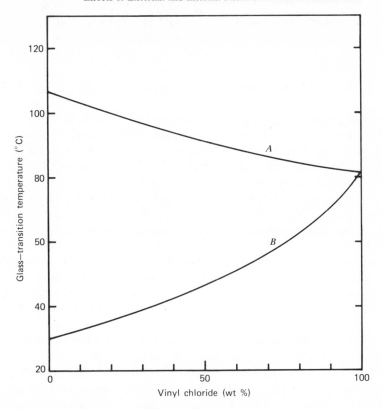

Figure 38 Glass-transition temperatures of copolymers of vinyl chloride–acrylonitrile (*A*) and vinyl chloride–vinyl acetate (*B*) (91).

However, when the latter copolymer was combined on a 50–50 basis with PVC, it no longer gave the $-20°C$ and the $90°C$ transitions that might have been expected based on each of the components added, but only a single relaxation or loss peak at about $25°C$. This new peak is interpreted as involving simultaneous and cooperative motions of both EVA chain segments and VC chain segments. In fact it was noted that if the proportions of PVC to PEVA were varied from 0 to 100, the transition temperature varied with the percent of PVC added, changing from $-23°C$ for 0% PVC to $90°C$ for 100% PVC. Furthermore, the T_g of the blend varied in the same fashion as would be expected when a plasticizer is added to a polymer, namely,

$$\frac{1}{T_g} = \frac{w_1}{T_{g_1}} + \frac{w_2}{T_{g_2}} \tag{164}$$

where w_1 and w_2 refer to the fraction of each blend present and T_{g_1} and T_{g_2} to the respective T_g's of each component by itself.

G. FRACTURE PHENOMENA IN POLYMERS

Although polymers are, in general, noted for their toughness, flexibility, and deformability, under appropriate conditions of temperature, strain rate, and state of stress, they can undergo brittle fracture, that is, localized fracture that occurs at some critical stress location without significant plastic deformation or dimensional changes. The principal conditions that give rise to brittle fracture are low temperatures, a high rate of loading (as may occur under shock or impact conditions), and application of alternating applied stresses for long periods of time, as can occur in components or parts subjected to dynamic and pulsating loads. In brittle fracture the strain at fracture is very low (usually $< 2\%$), even though the polymer may be capable of withstanding hundreds of percent of deformation under slowly applied static tensile loading. Under alternating conditions of loading, the polymer may fail after 10^6–10^7 cycles of loading at a stress level that is much lower than the yield or ultimate stress that it could withstand under static conditions. Similarly, even under a static load maintained for a long period of time, the polymer may rupture at a low stress value. Hence it is essential, if polymers are to be utilized effectively and efficiently in design applications, to be alert to the possibility of brittle fracture and to be aware of the possible effects of temperature, time, alternating stresses, high strain rates, and state of stress on the mechanical properties and behavior.

1.   Theories of Rupture and Brittle Fracture

The question of the expected breaking strength of materials has been approached from both an atomic or molecular point of view and from a continuum mechanics point of view. In the atomic or molecular method, one direct approach is to calculate a theoretical breaking or fracture stress by simply multiplying the number of atomic bonds or of polymer chains per unit area by the expected value of the force needed to break one atomic or molecular chain bond. Such calculations have been made by Frenkel, Polanyi, Orowan, and others, and the theoretical values are always much higher than observed ones (29). Theoretical calculations have also been made by Mark for highly oriented polymeric materials such as cellulose, but again the actual breaking strength was found to be much less than the expected theoretical strength (93). Further refinements have been made by Bueche (14). He was able to show that for a plastic subjected to constant load the time to break should depend both on the temperature and on the value of the imposed stress acting on the sample. For a given temperature his theory predicts a linear relation between breaking stress and log time.

The concept of a thermally activated molecular fracture process has also been developed by Zhurkov (94) and applied by him to a variety of materials, including metals and alloys, inorganic and organic glasses, plastics, and rubbers. According to Zhurkov, the time to fracture of a given material under a tensile load is an exponential function of the imposed stress, and at any given stress, σ, varies also with temperature according to the equation

$$\tau = \tau_0 \exp\left(\frac{U_0 - \gamma\sigma}{kT}\right) \tag{165}$$

Here τ_0 is a parameter comparable to the period of atomic vibrations, with a value of 10^{-12} to 10^{-13} sec, γ is a material constant to be determined, and U_0 is the activation energy for the breakdown process. It has a value comparable to the bonding energy of the chemical bonds. According to Zhurkov's equation, as the stress is increased the effective activation energy barrier $(U_0 - \gamma\sigma)$ is lowered in value and the time to fracture is thereby reduced. Equation 165 has been found to apply to a wide variety of materials over wide ranges of stress and many decades (8 to 10) of time. As the temperature of polymers is raised one expects more molecular uncoiling and extension for a given stress and a higher probability of bond rupture. The activation energy U_0 required to rupture chain bonds can be estimated for various polymers from stress–time tensile data taken at various stresses. Its value is found to fall in the range from 35 to 75 kcal/mole for most thermoplastics (30), whereas for metals the values obtained (namely, 25, 54, and 87 kcal/mole for zinc, aluminum, and nickel, respectively) are very close to tabulated sublimation energies. Also for polymers the U_0 values obtained from mechanical data are in good agreement with values obtained from thermal degradation data (16).

The principal continuum theory of brittle fracture is that first proposed by Griffith (95). In his phenomenological theory of strength Griffith pointed out that one could account for the large difference between experimentally observed rupture strengths in solids and theoretically expected values by presupposing that stress and associated strain energy are not uniformly distributed in a test specimen but are concentrated in the neighborhood of preexisting microcracks. Rupture or brittle fracture then occur by propagation of these preexisting microcracks—which presumably develop as a result of the preparation procedures—under the action of the high localized stresses at the crack tips. According to Griffith, a crack would propagate and a sample would fail in brittle fashion when the work required to extend the crack and create new surfaces was just balanced by an equal reduction in the strain energy of the specimen associated with the applied stress. Under these conditions no increase in total potential energy of the system occurs and the energy to make the crack self-propagating comes from the elastic energy stored near the crack tips. In this theory Griffith neglected the possible energy dissipation due to plastic deformation at the tips of the cracks, and assumed that the cracks had an elliptical shape.

Griffith used the Ingles (96) solution for the local stress, σ_{max}, occurring at the edge of a narrow elliptical crack in a uniformly stressed thin plate, namely,

$$\sigma_{max} = 2\sigma_0 \sqrt{\frac{c}{\rho}}$$

(166)

where σ_0 is the applied stress, $2c$ is the length of the elliptical crack, and ρ is the radius of curvature. Griffith was able to show that the critical applied tensile stress, σ_c, required to cause such a transversely oriented crack to spread was given by

$$\sigma_c = \sqrt{\frac{2E\gamma}{\pi c}}$$

(167)

where E is Young's modulus of elasticity, and γ is the specific surface energy, that is, the energy required to form a unit area of fracture surface. Thus the stress to propagate the crack and produce brittle fracture decreases with increase of preexisting crack size, but increases with increase of modulus or with the surface energy. Griffith investigated the validity of Eq. 167 by carrying out experiments on glass tubes and hemispheres in which he purposely introduced cracks of various sizes by use of a diamond cutter. He also determined E and γ from separate experiments. He was able to show that, for these materials, his theoretical equation was approximately satisfied and that σ_c did vary as the crack size to the negative one-half power. He also estimated that the radius of curvature need not be much larger than atomic dimensions, say, 5 Å, and under these conditions the estimated local stress at the crack tip does become comparable to expected theoretical stresses based on the magnitude of interatomic forces.

The Griffith theory was further developed and generalized by Irwin (97) and others, but the general form of the rupture stress–crack length relation remained the same. The principal difficulty in applying the Griffith criterion to polymers, or even to specific glassy and relatively brittle polymers such as polystyrene and polymethyl methacrylate, is that a considerable amount of energy can be dissipated in these materials by localized molecular orientation and plastic deformation processes that occur in the vicinity of the crack, and in the craze regions preceding the crack tips. The propagation of the

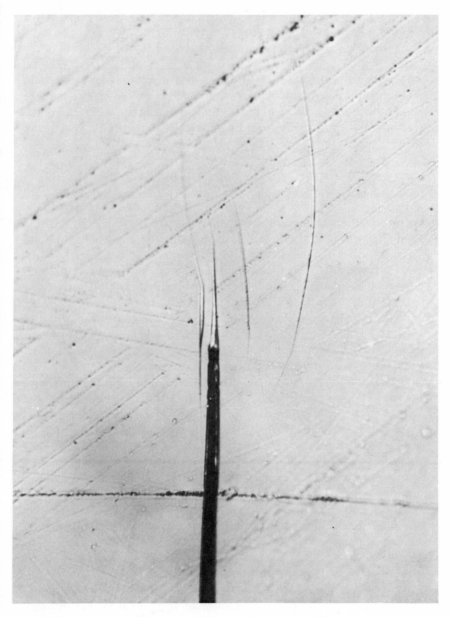

Figure 39 Micrograph showing crazed regions in a polymethyl methacrylate sample in vicinity of crack (98).

crack thus involves not only breaking of the interatomic bonds in the already crazed region but conversion of mechanical energy to heat energy by additional viscous and plastic deformation accompanying extension of the crazes.

The extent of crazing, and of local regions of plastic flow and chain orientation in the neighborhood of crack tips, has been studied by Kambour (98) and others. The presence of crazed regions extending well ahead of the crack tip is illustrated in Fig. 39, obtained by Kambour on specimens of PMMA.

2. Fracture under Steady Loading

In an investigation of rupture in polymeric solids, Berry (29) studied the applicability of the Griffith criterion by measuring the strength of PMMA and PS specimens in which cracks of varying size had been deliberately introduced. He found that the experimental dependence of the tensile strength on crack size was in accord with that predicted by Eq. 167, but that the value of γ determined from these experiments (namely, 2.1×10^5 and 17×10^5 erg/cm^2 for PMMA and PS, respectively) was approximately two orders of magnitude higher than expected values based on bond disassociation energies. This effect arises because much of the strain energy is used not just to create new surfaces but also in viscous dissipation processes accompanying chain deformation and orientation in the crazed regions near the crack tips. This energy is also proportional to the fracture surface area and hence it contributes to the observed high value of γ. By extrapolation of his results, Berry estimated that in PMMA the "natural" crack size was about 0.002 in.

Investigations similar to those mentioned above, in which tests were made on specimens with cracks of varying sizes, have been carried out on rubbery polymers as well as on thermoplastic ones (99). In this case the Griffith criterion was not satisfied and the dependence of tensile strength on crack length was found to be linear. It was concluded that for rubbers a rupture criterion based on critical stress is more in accord with the data. However, Rivlin and Thomas (100) have shown that a modified Griffith criterion, in which allowance is made for irreversible work based on viscous and plastic flow, is useful for study of the tearing characteristics of rubbers.

A given polymer may undergo a ductile or brittle type of fracture depending on test conditions. PS, for example, is normally considered a brittle polymer at room temperature because it fractures in a tensile test, run at normal speeds, after an elongation of only 1.5% or so, and has low impact strength. However, if similar test samples are tested at a temperature above the T_g value of 100°C, they show ductile behavior and have high impact strength and high extensibility. Similarly, PE and PP are normally considered ductile polymers. They exhibit a yield point in a tensile stress–strain test, they can be cold-drawn for 200% elongation or more without fracture, and they possess relatively high impact strength values. However, the processes of yielding, plastic flow, and cold-drawing all require time for chain segments to reorient under action of the applied stress. Hence if the strain rate is greatly increased or if the temperature is greatly lowered, the chain segments are not able to reorient in the time of the experiment. Therefore the stress developed cannot be relieved by plastic flow, and chain rupture and brittle fracture may result. Similarly, under high strain rates or low temperatures internal molecular motions are inhibited, and the polymer shows low impact strength and a brittle type of fracture. Imposition of high pressure also restricts molecular motion and increases segmental barriers to rotation; hence, in many polymers, like PTFE and PP, increase of pressure is somewhat analogous to increase of strain rate or to lowering of temperature, in that marked increases occur in modulus and yield stress, but ductility is reduced.

Under appropriate values of strain rate, temperature, and pressure all polymers exhibit a brittle–ductile transition. Higher temperatures and lower strain rates favor the ductile state, and lower temperatures and higher strain rates favor the brittle state. It has been reported (101) that PVC pipes fail in brittle fashion in short-time burst tests but show ductile types of failures when exposed to lower internal pressures for long periods of time. Even in short-time burst tests, a change to ductile behavior is found upon suitable increase of temperature. Brittle fracture may also result, even in otherwise ductile polymers, and at stresses far below those needed for fracture in air if the polymer samples are exposed to certain chemical environments. For example, kerosene is a known stress-cracking agent for polycarbonate, and usually any organic liquid that has a solubility parameter comparable to that of the polymer induces stress cracking and crazing. However, even low molecular weight substances such as ethylene glycol or fuel oil, which show no absorption in a simple immersion test, can embrittle materials such as PVC and ABS and cause marked reductions in time to failure for a given applied stress (102).

Various methods have been tried for reducing the tendency of polymers toward crazing, stress cracking, and brittle fracture. One method is to increase the molecular weight. Another is to improve molding procedures or to add an annealing operation so as to reduce the magnitude of residual stresses. A third is use of copolymerization. Notable increases in impact strength of so-called brittle types of polymers, such as PS, have also been achieved by blending or copolymerizing with a rubbery polymer such as butadiene.

Another factor that bears on the question of ductile versus brittle behavior of a polymer is the state of stress to which it is exposed. Stress conditions conductive to plastic deformation and viscous flow permit conversion of the applied energy into heat and thereby reduce the energy available for fracture. Hence a given material, as was demonstrated for a cast phenolic in Fig. 14, may be ductile under pure shear loading or under compressive loading but appear to be brittle when subjected to tension, as in this case there is a dilational component to the strain. It is known, for example, that crazing cracks originate at points of high local tensile stress and that crazing simply does not occur under either uniaxial compression or hydrostatic pressure. In fact the crazing cracks, or potential brittle fracture planes, are almost always propagated in a direction transverse to the maximum principal stress, and crazing occurs only if this stress is a tensile stress.

Various types of test measurements have been proposed for assessing the ability of a polymeric material to withstand impact loads. These include measurements of the area under a stress–strain curve run at rapid rates of straining, measurement of resistance to falling-ball impact (ASTM D1709-59T), and measurement of energy to cause failure under a rapidly applied blow by a heavy object. Perhaps the most widely accepted method is the Izod impact strength test. The method and procedure is fully described in ASTM Standard D256-56. It consists, in essence, of measuring the energy required to break a $\frac{1}{2} \times \frac{1}{2}$ in. notched cantilever-type specimen which is rigidly clamped at one end and then struck at the other by a pendulum weight. The test has the disadvantage that the actual rate of straining is not known (time to failure is of the order of 1 msec), and stresses are nonuniform throughout the specimen. This latter objection is overcome in the tensile type impact test described in ASTM D1822.

When polymers are studied by any of the means suggested above it is found that, under normal temperature and pressure conditions, they show a wide variation in impact resistance. This is not surprising in view of the wide range of T_g and T_m values shown in Tables 1 and 2 and in view of the marked effect of polymer structure and preparation conditions on the viscoelastic properties, as discussed in Section F.

Some represetative data on reported impact strengths of different polymers, taken from various literature sources, are presented in Table 11. From the table, and from other results reported by various investigators, one may draw the following conclusions:

1. Room-temperature impact energies or impact strengths vary widely from one polymer to another, as the data of the table indicate. For the more brittle polymers, such as cast phenolic or unmodified PS and PMMA, the Izod impact strengths fall well below 1 ft-lb/in. of notch. For the highly

Table 11   Izod impact energies of various polymers

Polymer	Grade	Value	Grade	Value
PS	General purpose	0.25–0.40	Impact	0.5–8
PMMA	Molding	0.3–0.5	High impact	1.4
Epoxy	Cast	0.2–1	Glass fiber filled	10–30
Phenolic	Cast	0.25–4	Glass fiber filled ($\sim 30\%$)	0.3–18
Nylon-6,6	Unmodified	1–2.5	Glass fiber filled (30%)	2.5
Nylon-6	Unmodified	1–3	Glass fiber filled (30%)	3.0
PP	Unmodified	0.5–2	Rubber modified	1–15
PVC	Rigid	0.4–20	Flexible	Varies
POM	Homopolymer	1.4–2.3	Glass fiber filled (25%)	1.5
ABS	Medium impact	3–6	Glass fiber filled (20–40%)	1–2.4
PTFE	Molding	3	FEP fluoroplastic	No break
PPO	Nonreinforced	5	Glass fiber filled (20–30%)	1.5
Ionomer		6–15		
PE	Low density	No break	High density	0.8–20
PB		No break		
PC	Unfilled	12–18	Glass fiber filled ($< 17\%$)	4–5

ductile polymers, like polyethylene, polybutene, and polycarbonate, the impact strengths are high, generally above 5 ft-lb/in., and sometimes so high there is no break.

2.   For polymers with impact strengths below 1 ft-lb/in. one can expect a brittle type of fracture if specimens are tested under tensile conditions at usual test speeds. Behavior of a brittle type generally occurs in polymers like PS, which have relatively stiff backbone chains and which show no significant internal relaxation processes, associated with segmental mobility, below room temperature.

3.   Even polymers of a brittle type can be given greatly increased impact strength by blending or copolymerizing with a second ingredient, such as butadiene, that itself has low-temperature relaxation modes and good impact strength and deformability at room temperature. Some examples of the significant improvement that can be obtained by material modification are given in Table 11, and many other examples of correlation between relaxation behavior and impact strength are given in the literature (103).

4.   Polymers having impact strength values above 1 ft-lb/in. are usually tough and ductile if subjected to simple tension conditions at normal test speeds. Many of the common polymers that are used in a variety of engineering applications fall in this class. These include, for example, PE, PP, and PTFE.

5.   Some polymers, such as POM, PC, and the polyamides, have an unusually good combination of both mechanical strength and toughness. Hence these materials are frequently used in applications where structural strength and rigidity, as well as ductility, are required.

6.   It is not possible to give an unequivocal statement regarding the effect of additives on impact strength. In general, addition of plasticizers decreases tensile strength but raises impact strength. Compare, for example, the data in the table for rigid and flexible PVC. Addition of glass filaments may raise the impact strength, as for epoxy, may lower the impact strength, as for polycarbonate, or affect it only slightly, as for PPO. The principal effect of addition of glass fibers is to raise the modulus and rigidity of the material.

3.   Fracture under Alternating Loading—Endurance Strength

Although reliable fatigue data on polymers are still relatively scanty in the engineering and scientific literature, it has been observed that under repeated loading polymers fail at stress considerably less than the static value and that the fracture itself is of the brittle type, with little indication of dimensional changes or marked plastic deformation. Failure generally occurs at some particular section of the specimen where localized stresses are high either because of the presence of surface imperfections or of other types of defects or impurities.

Fatigue consists of the progressive propagation of a crack, which may have been initially present or which was initiated by the applied stress. Final rupture probably occurs when the cross-sectional area of the specimen has been so reduced by the growth of the crack that the average stress intensity rises to that required for static rupture. Some attempts have been made to apply the Griffith criterion of an energy balance to fatigue as well as to static fracture, and it has been suggested (30) that the incremental growth of a fatigue crack of length c is given by

$$\frac{dc}{dN} = k\left(\frac{-\partial U}{\partial A}\right)^{n} \tag{169}$$

where N is the number of cycles, U the stored elastic strain energy, and A the interfacial area of the crack. To represent best the experimental data, the exponent n must take on a value that varies somewhat with N, but for PE the value of n is close to 2.

Another approach to fatigue phenomena is to consider statistical types of theories. If one assumes a distribution of preexisting crack sizes, then brittle fracture should occur, as in the Griffith theory, by crack propagation from that defect that has the largest local stress concentration. For a specimen of larger volume one would encounter more cracks, and the probability of a larger crack would be greater, and hence the fatigue strength less, than for a smaller specimen. In fact, data of this type have been reported for some materials and cited as evidence for a statistical theory of brittle fracture based on defects. Alternatively, one can develop a statistical theory based not on a distribution of microcracks of various sizes, but rather on a distribution of separation strengths of atomic or molecular bonds (104). This may well arise as a result of thermal energy fluctuations and Freudenthal has shown that these concepts, when applied to the problem of fatigue under alternating stress, can lead to an equation of the form

$$S_a = S_0 - k \log N \tag{169}$$

where S_a is the fatigue strength after N cycles of loading and S_0 is the "static" brittle strength. Equation 169 predicts that the fatigue stress sould be a linear function of the log of the number of cycles, and the data available are in approximate accord with this behavior. Note, for example, the S–log N plots given in Fig. 40, obtained from the fatigue studies on PS fractions reported by Foden et al. (105). However, in many polymers it appears that at some value of N a so-called endureance limit S_e is reached, such that for stresses $S_a < S_e$ the specimens essentially have infinite life. As the data of Fig. 40 show, the endurance limit for the PS fraction having an average molecular weight of 860,000 is about 1500 psi.

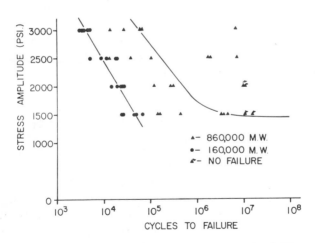

Figure 40 S–log N curves for two sets of polystyrene samples of different molecular weight; all samples subjected to alternating loads at zero mean stress (105).

One interesting feature of Fig. 40 is the obvious improvement in fatigue life that has occurred with increase of molecular weight of the polymer. The data indicate an approximate tenfold increase in fatigue life—for a given applied stress—for an increase in molecular weight from 160,000 to 860,000. This improvement is believed to be due to a reduction in chain ends, which are a potential source of voids and microcracks, to an increase in intermolecular interaction, and to better and more uniform stress distribution across the sample. Also, the greater deformability and orientation that is possible with high molecular weight material effectively lessens the severity of the stress concentration factor that arises at crack tips.

The fatigue properties of plastics have not been as widely studied as their static properties. One reason is that fatigue tests are time-consuming. Another is that difficulties arise owing to self-heating. In fact, in some polymers, severe temperature rises can take place in the specimen under alternating load conditions. The extent to which this occurs depends on the particular polymer and on its relaxation or viscoelastic behavior. It is not severe in PS tested at 1700 rpm, but if PE specimens are tested at the same speed they quickly soften and melt. Even at low testing speeds of only 100 rpm, temperature rises of 5°C or more may occur in PE specimens tested at nominal stress values.

From fatigue data that are available in the literature it appears that the ratio of endurance strength to static ultimate strength falls generally in the range about 0.20 to 0.35, with higher values associated with laminates and composites.

Table 12 gives literature values of static strength and endurance strength under alternating load conditions for a number of different polymers. These values should be viewed with some caution. One reason is that seldom is the specimen temperature during the test recorded or reported by the investigator, and hence the data may actually reflect behavior at some higher temperature, rather than indicate expected room-temperature performance. Also, wide variation in properties can occur, even for a given type of polymer, by variation of specific composition and preparation procedures.

With due recognition of these possible unknown variables, the data given in Table 12, together with other reported findings, enable us to draw some tentative conclusions relative to the performance to be expected of polymers subjected to alternating loading conditions. First, it must clearly be recognized that significant temperature rises can occur and these must be considered. Second, for many thermoplastic materials, the endurance limit appears to be about one-fifth of the static ultimate strength. Third, the ratio of endurance strength to static ultimate strength appears somewhat higher for reinforced polymers. Fourth, for some specific polymers, such as POM or PTFE, the ratio is unusually high, of the order of 0.4 or 0.5. Factors that

Table 12   Reported fatigue strength values of various polymers

Type	Polymer	Static Tensile Strength σ_u (psi)	Fatigue Strength at 10^7 Cycles σ_e (psi)	Ratio (σ_e/σ_u)
Thermoplastic				
	CA	5,000	1,000	0.19
	PS	5,800	1,250	0.21
	PC	10,000	2,000	0.20
	PPO	10,500	2,000	0.19
	PMMA	10,500	2,000	0.19
	Nylon-6,6 (25% H_2O)	11,200	3,400	0.30
	POM	10,000	5,000	0.50
	PTFE			
	40 cps	3,000	600	0.20
	30 cps	3,000	900	0.30
	20 cps	3,000	1,400	0.47
Thermosetting				
	Laminated phenolic			
	Longitudinal	21,400	5,000	0.23
	Transverse	18,200	4,000	0.22
	Epoxy, glass fiber reinforced			
	Unidirectional	158,000	25,000	0.16
	Cross-ply	75,000	22,000	0.29
	Random	27,000	12,000	0.45
	Laminated polyester			
	Unidirectional weave	78,000	22,000	0.30
	Square weave	43,000	12,000	0.28

would be expected to lead to higher values of this ratio are increased molecular weight, higher retention of static strength values with temperature rise, and less sensitivity to notch effects.

Some fatigue tests on polymers have been conducted by using the so-called Prot (106) method, that is, allowing the magnitude of the alternating stress to increase linearly with time. This type of test allows data to be assembled more rapidly and there are no runouts. However, the analysis of the data is more involved and heating effects may become a significant factor. It has been reported that, for annealed samples of PMMA, comparable values of endurance limit are found by both the progressive-stress method and the steady-stress method (107).

ACKNOWLEDGMENTS

We wish to acknowledge the valuable assistance given by S. K. Bhateja, A. A. Silano, and H. N. Yoon in carrying out many of the stress–strain investigations reported herein. Appreciation is expressed also to the Army Research Office (Durham), and the National Science Foundation, both of which provided financial support for the high-pressure studies. Some of the data have been assembled from various books and technical articles; these are acknowledged in the list of references.

We wish also to record our special thanks to R. Kambour, C. C. Hsiao, E. Foden, D. R. Morrow, and the editors of the *Modern Plastics Encyclopedia*, all of whom kindly supplied data or prints.

GENERAL REFERENCES

1. P. H. Geil, *Polymer Single Crystals*, Wiley-Interscience, New York, 1963.

2. H. D. Kieth, *Physics and Chemistry of the Organic Solid State*, Part I, Chap. 8, D. Fox, M. M. Labes, and A. Weissberger, eds., Wiley-Interscience, 1963.

3. P. J. Flory, *Principles of Polymer Chemistry*, Cornell University Press, Ithaca, N.Y., 1953.

4. L. R. G. Treloar, *The Physics of Rubber Elasticity*, 2nd ed., Oxford University Press, Oxford, 1958.

5. N. G. McCrum, B. E. Read, and G. Williams, *Anelastic and Dielectric Effects in Polymeric Solids*, Wiley, New York, 1967.

6. I. M. Ward and P. R. Pinnock, "The Mechanical Properties of Solid Polymers," *Br. J. Appl. Phys.*, **17**, 3 (1966).

7. G. M. Bartenev and Y. V. Zelenev, "Viscoelastic Behavior and Structure of Polymers," *Mater. Sci. Eng.*, **2**, 136 (1967).

8. A. E. Woodward and J. A. Sauer, "Mechanical Relaxation Phenomena," in *Physics and Chemistry of the Organic Solid State*, D. Fox, M. M. Labes, and A. Weissberger, (Eds.), Wiley-Interscience, New York, 1965, Chap. 7, Part II.

9. P. I. Vincent, "Mechanical Properties of High Polymers: Deformation," in *Physics of High Polymers*, P. D. Ritchie (Ed.), Iliffe, London, 1965, Chap. 2.

10. J. D. Ferry, *Viscoelastic Properties of Polymers*, Wiley, New York, 1961; 2nd ed., 1970.

11. L. E. Nielsen, *The Mechanical Properties of Polymers*, Reinhold, New York, 1962.

12. J. A. Sauer and A. E. Woodward, "Stress-Strain Temperature Relations in High Polymers," in *Polymer Thermal Analysis*, Dekker, New York, 1970, Chap. 3, Part II.

13. A. P. Boresi, *Elasticity in Engineering Mechanics*, Prentice-Hall, Englewood Cliffs, N.J., 1965.

14. F. Bueche, *Physical Properties of High Polymers*, Wiley-Interscience, New York, 1962.

15. A. V. Tobolsky, *Properties and Structure of Polymers*, Wiley, New York, 1960.

16. V. A. Kargin and G. L. Slonimsky, *Encyclopedia of Polymer Science and Technology*, Vol. 8, Wiley-Interscience, New York, 1968, p. 445.

17. A. X. Schmidt and C. A. Marlies, *Principles of High Polymer Theory and Practice*, McGraw-Hill, New York, 1948, p. 537.

18. E. R. Parker, *Materials Data Book*, McGraw-Hill, New York, 1967.

19. *Modern Plastics Encyclopedia*, Vol. 47, McGraw-Hill, New York, 1970–1971; Vol. 48, 1971–1972.

20. *Technical Data on Plastics*, Manufacturing Chemists' Association, 1957.

21. R. Hill, *Mathematical Theory of Plasticity*, Oxford University Press, Oxford, 1950.

22. W. Prager and P. G. Hodge, *Theory of Perfectly Plastic Solids*, Wiley, New York, 1951.

23. C. Ehringer, *Mechanics of Continua*, Wiley, New York, 1967.

24. P. W. Bridgman, *Studies in Large Plastic Flow and Fracture*, McGraw-Hill, New York, 1952.

25. T. Alfrey, Jr., *Mechanical Behavior of Polymers*, Interscience, New York, 1948.

26. B. Gross, *Mathematical Structure of the Theories of Viscoelasticity*, Hermann, Paris, 1953.

27. A. Charlesby, *Atomic Radiation and Polymers*, Pergamon, 1960.

28. A. Chapiro, *Radiation Chemistry of Polymeric Systems*, Wiley-Interscience, New York, 1962.

29. J. P. Berry, in *Fracture Processes in Polymeric Solids*, B. Rosen (Ed.), Wiley-Interscience, 1964, Chaps. IIA, IIB.

30. E. H. Andrews, *Physical Basis of Yield and Fracture*, A. C. Strickland (Ed.), Institute of Physics & Physical Society, London, 1966, p. 127.

SPECIFIC REFERENCES

31. K. D. Pae, J. A. Sauer, and D. R. Morrow, *Nature*, **211**, 514 (1966).

32. J. A. Sauer, *Annals N.Y. Acad. Sci.*, **155**, 517 (1969).

33. J. A. Sauer, R. A. Wall, N. Fuschillo, and A. E. Woodward, *Proceedings of 5th International Conference on Low Temperature Physics*, University of Wisconsin Press, Madison, 1958, p. 608.

34. A. E. Woodward, J. A. Sauer, and R. A. Wall, *J. Polym. Sci.*, **50**, 117 (1961).

35. *Zytel Nylon Resins*, E. I. du Pont de Nemours & Co., Inc., Wilmington, Del., 1962.

36. C. W. Deeley, J. A. Sauer, and A. E. Woodward, *J. Appl. Phys.*, **29**, 1415 (1958).

37. K. Schmeider and K. Wolf, *Kolloid Z.*, **134**, 157 (1953).

38. G. Murphy, *Properties of Engineering Materials*, 2nd ed., International Textbook Co., Scranton, Pa., 1947, p. 402.

39. C. W. Weaver and M. S. Paterson, *J. Polym. Sci. A2*, **7**, 587 (1969).

40. H. Alexander, Int. *J. Eng. Sci.*, **6**, 549 (1968).

41. R. S. Rivlin, *Phil. Trans. Roy. Soc. A241*, 379 (1948).

42. C. R. G. Treloar, *Trans. Faraday Soc.*, **40**, 59 (1944).

43. W. Whitney and R. D. Andrews, *J. Polym. Sci.*, **C16**, 2981 (1967).

44. P. B. Bowden and J. A. Jukes, *J. Mat. Sci.*, **3**, 183 (1968).

45. D. R. Mears, K. D. Pae, and J. A. Sauer, *J. Appl. Phys.*, **40**, 4229 (1969).

46. J. A. Sauer, D. R. Mears, and K. D. Pae, *Eur. Polym. J.*, **6**, 1015 (1970).

47. G. Biglioni, E. Baer, and S. V. Radcliffe, "Fracture-1969" Chapman & Hall, London (1969); A. W. Christiansen, E. Baer, and S. V. Radcliffe, *Phil. Mag.*, **24**, 188 (1971).

48. D. Sardar, S. V. Radcliffe, and E. Baer, *Polymer Eng. Sci.*, **8**, 290 (1968).

49. S. Rabinowitz, I. M. Ward, and J. S. C. Parry, *J. Mater. Sci.*, **5**, 29 (1970).

50. B. Crossland and W. M. Deardon, *Proc. IME*, **172**, 805 (1958).

52. H. Tresca, *Mém. présentées par divers savants*, **18**, 733 (1868).

52. R. V. Mises, *Z. Angew. Math. Mech.*, **8**, 161 (1928).

53. H. Hencky, *Z. Angew. Math. Mech.*, **5**, 116 (1925).

54. A. Nadai and E. A. Davis, *J. Appl. Phys.*, **8**, 205 (1937).

55. J. A. Sauer, K. D. Pae, and S. K. Bhateja, *J. Macromol. Sci.—Phys.*, **B8** (3–4), 631 (1973).

56. L. W. Hu and K. D. Pae, *J. Franklin Inst.*, 491 (June 1963).

57. J. A. Sauer and C. C. Hsiao, *Trans. ASME*, **1953**, p. 895.

58. D. K. Spurr and W. D. Niegisch, *J. Appl. Polym. Sci.*, **6**, 585 (1962).

59. R. P. Kambour, *Polymer*, **5**, 143 (1964).

60. J. A. Sauer and C. C. Hsiao, *J. Appl. Phys.*, **21**, 1071 (1950).

61. S. Sternstein, L. Ongchin, and A. Silverman, *Applied Polymer Symposium I*, Wiley-Interscience, 1968, p. 175.

62. J. A. Sauer, J. Marin, and C. C. Hsiao, *J. Appl. Phys.*, **20**, 507 (1949).

63. C. Chung and J. A. Sauer, *Polymer*, **11**, 455 (1970).

64. J. D. Ferry and K. Ninomiya, in *Viscoelasticity*, J. T. Bergen (Ed.), Academic Press, New York, 1960, p. 55.

65. A. P. Molotkov, Yu V. Zelenev, and G. M. Bartenev, *Mekhanika Polyimerov* (Polymer Mechanics), **4**, 445 (1968).

66. P. E. Rouse, *J. Chem. Phys.*, **21**, 1272 (1953).

67. M. G. Sharma, Chap. 4, *Testing of Polymers*, Vol. 1, J. Schmitz (Ed.), Wiley-Interscience, New York, 1965, Chap. 4.

68. H. Leaderman, *Elastic and Creep Properties of Filamentous Materials and Other High Polymers*, Textile Foundation, Washington, D.C., 1943.

69. M. L. Williams, R. F. Landel, and J. D. Ferry, *J. Am. Chem. Soc.*, **77**, 3701 (1955).

70. M. Gordon, *Physics of Plastics*, P. D. Ritchie (Ed.), Iliffe, London, 1965, Chap. 4.

71. A. K. Doolittle, *J. Appl. Phys.*, **22**, 1471 (1951).

72. G. M. Bartenev, I. V. Razumovskaya, D. S. Sanditov, and F. A. Lukyanev, *J. Polym. Sci. A-1*, **7**, 2147 (1969).

73. S. Saito, *Kolloid Z.*, **189**, 117 (1963).

74. A. E. Eisenberg, *Polym. Lett.*, **1**, 273 (1963).

75. Yu S. Urzhumtsev and R. D. Maksimov, *Polym. Mech.*, **4**, 318 (1968).

76. T. R. Smith, *J. Polym. Sci.*, **32**, 99 (1958).

77. I. H. Hall, *J. Appl. Polym. Sci.*, **12**, 731 (1968).

78. C. C. Hsiao and J. A. Sauer, *ASTM Bull.*, **172**, 29 (1951).

79. M. Nakamura and S. M. Skinner, *J. Polym. Sci.*, **18**, 423 (1955).

80. J. A. Sauer and R. G. Saba, *J. Macromol. Sci.—Chem.*, **A3** (7), 1217 (1969).

81. G. A. Bernier, D. G. Kline, and J. A. Sauer, *J. Macromol. Sci.—Phys.*, **B1**, 335 (1967).

82. J. C. Bevington and D. Charlesby, *International Symposium on Macromolecules*, suppl. to *La Ricerca Scientifica*, 1955.

83. G. F. Bauman, *Advances in Chemistry Series No. 49*, American Chemical Society, Washington, D.C., 1964, p. 96.

84. *Guide Book on Designing with du Pont Engineering Plastics*, E. I. du Pont de Nemours & Co., June 1965.

85. J. A.Faucher and F. P. Reding, *High Polymers*, Vol. 20, Wiley-Interscience, New York, 1965, Chap. 13.

86. L. H. Tung, *SPE J.*, **14**, 25 (1959).

87. J. van Schooten, H. van Hoorn, and J. Boerma, *Polymer*, **2**, 161 (1961).

88. J. A. Sauer, *SPE Trans.*, 57 (January 1962).

89. T. Lim, *Effect of Low Molecular Weight Substances on the Dynamic Mechanical Properties of High Polymers*, Ph.D. Thesis, Rutgers University, 1967; J. A. Sauer, *Polymer Science Symposium No. 32*, Wiley, New York, 1971, p. 69.

90. F. P. Reding, E. R. Walter, and F. J. Welch, *J. Polym. Sci.*, **56**, 225 (1962).

91. F. P. Reding, J. A. Faucher, and R. D. Whitman, *J. Polym. Sci.*, **57**, 484 (1962).

92. C. F. Hammer, *Macromolecules*, **4**, 69 (1971).

93. H. Mark, Melliand. *Textilber.*, **10**, 695 (1929); K. H. Meyer, *Natural and Synthetic High Polymers*, Interscience, New York, 1942, p. 328.

94. S. N. Zhurkov, *J. Tech. Phys. USSR*, **28**, 1719 (1958); *Int. J. Fracture Mech.*, **1**, 311 (1965).

95. A. A. Griffith, *Phil. Trans. Roy. Soc. A*, **221**, 163 (1921); *Proceedings International Congress Applied Mechanics*, Delft, 1924, p. 24.

96. C. E. Inglis, *Trans. Roy. Inst. Nav. Archit.*, **60**, 219 (1913).

97. G. R. Irwin, *J. Appl. Mech.*, **61**, 449 (1939); *Handbuch der Physik*, Vol. VI, Springer, Berlin, 1958, p. 551.

98. R. P. Kambour and R. E. Robertson, in *Polymer Thermal Analysis*, A. D. Jenkins (Ed.), North-Holland Publishing Co., Amsterdam, 1972, Chap. 11.

99. A. M. Bueche and J. P. Berry, in *Fracture*, B. C. Averbach et al. (Eds.), Wiley, New York, 1959.

100. R. S. Rivlin and A. G. Thomas, *J. Polym. Sci.*, **10**, 291 (1953).

101. L. S. Sansome, *SPE J.*, **15**, 418 (1959).

102. H. W. Kuhlmann, R. I. Leininger, and F. Wolter, *Gas*, 102 (October 1966).

103. J. A. Sauer, *Polymer Science Symposium No. 32*, Wiley, New York, 1971, p. 69.

104. A. M. Freudenthal, *Proc. Roy. Soc.*, **A187**, 416 (1946); *The Inelastic Behavior of Engineering Materials and Structures*, Wiley, New York, 1950.

105. E. Foden, D. R. Morrow, and J. A. Sauer, *J. Appl. Polym. Sci.*, **16**, 519 (1972).

106. E. M. Prot, *Tech. Rept. 52-148*, Wright Air Development Center, 1948.

107. E. L. Thorkildsen, *Engineering Design for Plastics*, E. Baer (Ed.), Reinhold, New York, 1964, Chap. 5.

DISCUSSION QUESTIONS AND PROBLEMS

Section A

1. Distinguish among rubbers, thermoplastics, and thermosetting polymers. Cite two examples of each type and show their chemical structure. What part do barriers to rotation play in their respective mechanical behavior?

2. Indicate how glass-transition temperatures are determined. Discuss the influence on T_g of increasing side branch length. What new factor arises and influences the location of T_g when the number of atoms, n, in an aliphatic side chain becomes large ($n \geq 8$)?

3. Discuss the morphology of polymer crystals and its dependence on preparation conditions. How is the morphology affected by annealing below T_m? Can polymer crystals undergo phase changes as temperature or pressure is varied? Cite examples. See Refs. 1 and 2.

4. Dislocations have played a prominent part in the development of theories of plastic flow in metals. Do dislocations occur also in polymers? What is the evidence? See, for example, M. L. Williams, "Polymer Science: Current Concepts and Civil Applications," *Annals N.Y. Acad. Sci.*, **155**, 539 (1969).

5. Name five different ways in which components or articles of polymers may fail in service. Indicate in each case the important mechanical parameters that measure the resistance of polymers to that type of failure.

Section B

1. Using the theory of rubber elasticity, compute the values of the tensile modulus and the shear modulus of an elastomer in which the molecular weight between cross-links is 5000 and the density at 23°C is 0.925 g/cm³.

2. For the elastomer discussed above, by how much would the expected modulus value change if one takes into account the correction due to finite molecular weight? Treat two cases, one in which M_n is 20,000 and one in which M_n is 100,000.

3. Give examples of some types of elastomers in which internal energy makes a significant contribution to rubber elasticity. What magnitude of error is introduced when the energy contribution is ignored compared to the entropy contribution? See, for example, M. Shen, *Macromolecules*, **2**, 358 (1969), and M. Shen and P. J. Blatz, *J. Appl. Phys.*, **39**, 4937 (1968).

4. When an elastomer is quickly stretched its temperature rises appreciably; one can easily note this effect by simply touching a rubber band to one's lips before and after stretching. Account for this effect. However, metals subjected to elastic strain cool in stretching. Why?

Section C

1. The general stress–strain relations for an isotropic material are given by Eq. 59. Apply these to consideration of simple stress states such as

simple shear, simple tension and simple hydrostatic pressure. Using your results, derive Eqs. 60, 61, and 62.

2. For an isotropic material prepare a table showing how the bulk modulus, K, and the shear modulus, G, are related to the tensile modulus, E, for various values of Poisson's ratio, v. Use values of v of 0.25, 0.30, 0.33, 0.40, 0.45, 0.48, and 0.50.

3. Give examples of polymeric materials that exhibit a brittle type of behavior in a simple tension test but give ductile behavior in a simple compression test run at the same test speed. Account for the difference in behavior in the two cases.

4. For an isotropic polymer subjected to tension, show that the true stress, σ' is related to the nominal stress σ_0, by the equation

$$\sigma' = \frac{\sigma_0}{1 - 2v\varepsilon}$$

where ε is the strain in the direction of the applied stress. If the applied stress is a compressive stress rather than a tensile stress, how should the formula given above be changed? Draw the true stress–strain curve of some given polymer and show how it differs from the nominal stress–strain curve. To what extent is the maximum stress value varied by consideration of true stresses? See Ref. 9.

Section D

1. Show that the von Mises shear energy theory of yielding for the two-dimensional stress case is given by the ellipse of Fig. 17, and that the Tresca simple shear theory is given by the inscribed polygon.

2. Using the Coulomb criterion of yielding, derive approximate expressions for the case of the following stress states and relate these to the appropriate expressions for simple tension: (a) simple shear; (b) simple compression; (c) axial tension superimposed on hydrostatic pressure.

3. Describe the phenomenon of crazing. Discuss both macroscopic and molecular aspects. State experimental conditions conducive to development of crazing and also indicate ways to suppress or retard crazing. See Refs. 58, 59, and 60.

4. How do single crystals of polymers react to applied deformation? Discuss each of the modes of plastic deformation that have been observed. See Refs. 1 and 2.

Section E

1. Give the general form of the stress–strain relation for Hookean elasticity, for Newtonian viscosity, for linear viscoelasticity, and for nonlinear

viscoelasticity. Describe three experimental manifestations of linear viscoelasticity. Are these observed in real polymers? If differences arise, what is their cause? See Refs. 10 and 11.

2. Describe briefly the various relaxation processes usually designated by the letters α, β, γ, and δ. Give approximate ranges of activation energy for each of these and approximate temperature locations. Can the α relaxation be adequately described by an Arrhenius process? See Refs. 5 and 8.

3. Describe three different experimental methods for obtaining values of dynamic mechanical properties and indicate the approximate frequency range covered in each case. What types of polymers show dielectric as well as mechanical relaxations and what are the advantages of using both mechanical and dielectric methods? Cite two polymers that have been studied by both techniques and compare the results obtained. See Refs. 5 and 73.

4. Derive the appropriate expressions for deformation of a Maxwell model subjected to constant stress and for stress relaxation in a Voight model subjected to constant strain. Contrast the predicted behavior with the observed creep and deformation behavior of real polymers.

5. Using the data given in Table 6, and a factor of safety of 4, state the design tensile stress value you would recommend for a component made of Delrin 500 so that the tensile strain, after 10^5 hr of loading, would not exceed the initial strain that you would obtain at an applied stress of 1500 psi.

Section F

1. Discuss the influence of pressure on the temperature location of the primary or of secondary relaxations in polymers. What is the magnitude of the observed effect and what is its molecular interpretation? Will pressure also affect the location of the melting temperature and, if so, give a thermodynamic equation that can be used to determine the magnitude of the effect.

2. Give an example of a polymer that exhibits relaxation behavior above T_g, at T_g, and below T_g. Discuss the nature of each of these relaxation processes.

3. Discuss some of the expected changes that may occur in low molecular weight materials when they are subjected to radiation. Can these materials be polymerized by use of radiation? If so, give examples. See Refs. 27 and 28.

4. Indicate ways in which a given relaxation process can be shifted to higher temperatures both by change of external variables and by controlled changes of structure. Do polymer single crystals show relaxation behavior? What is the effect of annealing at successively higher temperatures on this behavior?

5. Cite significant uses and applications of polymers in the communications industry, the automotive industry, the construction industry, and the pollution control field. See, for example, Part IV of "Polymer Science: Current Concepts and Civil Applications," *Annals N.Y. Acad. Sci.*, **155**, Art. 2 (January 1969).

Section G

1. The Griffith criterion for fracture has been applied to thermoplastics. From the breaking stress values obtained on samples in which cracks of various sizes were deliberately produced, one can estimate the size of the "intrinsic" flaw in such materials as PS and PMMA. What are these estimates and how reasonable do they appear to you? See Ref. 29.

2. Describe three different ways in which resistance of materials to dynamic or impact loads is measured. Make a plot of impact energy values versus percent ductility in tension of a series of plastics and discuss the degree of correlation that exists between these variables.

3. What is meant by such concepts as fracture toughness and tearing energy? How can these be determined for a given polymer or rubber? See Ref. 100.

4. Describe the various phases that occur in the development of fracture of polymer specimens owing to application of alternating loads. Can crack extensions and velocities be measured? How do these depend on stress magnitude? See, for example, V. Havlicek and V. Zilvar, *J. Macromol. Sci.—Phys.*, **85** (2), 317 (1971).

5. What is the degree of correlation, if any, between toughness of polymers and the presence of molecular relaxation processes? Is the location of reorienting groups in the polymer chain of significance in this regard? See, for example, I. J. Heijboer, *J. Polym. Sci. C*, **16**, 3755 (1968).

CHAPTER 8

Rheology

JAMES F. CARLEY
University of Colorado
Boulder, Colorado

A. INTRODUCTION

"Rheology" is a word made by the fusion of two Greek stems and literally means "the study of flow." In the development of this branch of science, the flow of all kinds of materials in which flow is the least bit complicated by unusual features—solidlike behavior, suspended solids, extremely high or low flow, nonuniform response to stress—has been the province of rheologists.

The flow of polymeric materials is of interest because most processing techniques involve some form of directed flow of polymer solids or melts, and many mechanical properties are affected by the relatively small amount of flow or creep that may occur at ambient temperatures in the solid-state polymers. The flow of solutions of polymers plays an important part in the identification and characterization of their molecular structure. Because solution properties are the subject of Chapter 4, little is said about rheology of solutions here.

Fluid flow is characterized by a number of descriptive terms: dimensionality, time dependence, turbulence, and non-Newtonian. If the distribution of velocities in a flowing system can be completely described with only one spatial variable, the flow is said to be one-dimensional; if two are required, it is two-dimensional; etc. The flow of water through a long capillary tube is one-dimensional, whereas if the tube cross section were rectangular the flow would be two-dimensional. If the rectangular tube section were changing along its length, either in size or shape, the flow would be three-dimensional. Three-dimensional and two-dimensional flow are mathematically complicated, but sometimes they may be treated approximately as one-dimensional situations.

If the velocities everywhere in the system are not changing with time, the flow is said to be "steady" and the system is time-independent. If each "particle" in the stream traces out a smooth line that is more or less parallel to the bounding surfaces, the flow is called *streamline* or *laminar*. If the particles follow jiggly lines and circulate through small local eddies crosswise to the main stream, the flow is *turbulent*. For turbulence to exist in liquids, the velocity must be high, the viscosity must be low, and/or the channel must be of large cross section. The criterion of turbulence is that the Reynolds number, a dimensionless group of key variables, must exceed about 3000. The Reynolds number, *Re*, is defined by

$$Re = \frac{DV\rho}{\mu} \tag{1}$$

where D = channel diameter (actual or, for noncircular channels, an equivalent)

V = average velocity of the stream

ρ = fluid density

μ = fluid viscosity

The units of the quantities making up the Reynolds number must be consistent to make the number itself dimensionless. In the cgs system, for example, D would be in centimeters, V in centimeters per second, ρ in grams per cubic centimeter, and μ in grams per centimeter-second. This metric unit of viscosity is known as the *poise*, in honor of the French scientist Poiseuille, who experimentally investigated flow through capillary tubes more than 100 years ago. In polymer melts, because of their extremely high viscosities, turbulent flow is never observed. In the dilute solutions, however, it is quite feasible to have turbulence.

Non-Newtonian flow includes a great many different kinds of deviations from Newton's law of flow, which states that the rate of change of fluid velocity with position is always proportional to the applied shear stress (Fig. 1). The deviations include such behavior as variation of viscosity with shear stress, mixed elastic and viscous behavior, viscosity changing with duration of flow, and failure of flow to occur until some critical stress has been exceeded. It has long been recognized, even for fluids that do obey Newton's law, that viscosity—which can be thought of as the internal resistance of the fluid to being made to flow, or as the energy dissipated in causing

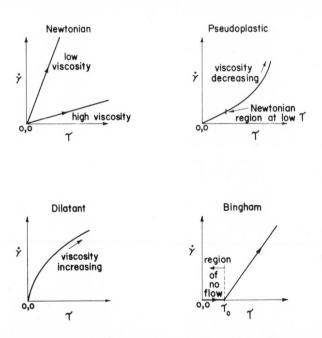

Figure 1   Steady-state flow behavior of Newtonian and non-Newtonian fluids at constant applied stress. In each diagram, shear rate ($\dot{\gamma}$) is the ordinate, shear stress (τ) the abscissa. If viscosity is defined as the ratio of shear stress to shear rate (see Eq. 2), then a Newtonian material, for which viscosity is independent of the shear state, gives a linear plot. For pseudoplastics, viscosity decreases as shear stress (or rate) increases, whereas for dilatant materials the reverse is true. Bingham plastics behave like Newtonian materials provided the stress exceeds a certain "yield" value, τ_0, below which there is no flow at all.

a unit cube of the fluid to flow with a unit velocity gradient—changes with temperature and pressure in the fluid. Most studies of flow situations have been simplified by assuming that temperature and pressure remain constant, even when it is known that they do not. With the exception of the silicone family and a few others, the viscosities of most liquids drop rapidly with rising temperature.* The viscosity of molten polyethylene, for instance, drops about 30% with a 10°C rise in temperature because molecular motion is increased and there is more chain slippage. Most other molten polymers exhibit even sharper viscosity–temperature coefficients. Rising pressure, on the other hand, *increases* melt viscosities, and much less sharply. Only in injection and compression molding, where pressures can sometimes be 20,000 psi and higher, is the pressure effect on viscosity of practical importance.

Generally speaking, polymers are viscoelastic materials. That is, they exhibit a flow behavior that is a combination of irreversible viscous flow due to chain slippage as well as a reversible elastic deformation which is time dependent and due to molecular entanglements that constrain large-scale molecular motion. The exact behavior under stress of a given polymer depends very much on temperature. Below the so-called glass-transition temperature (see Chapter 7) they are essentially elastic solids but can be made to flow a little; above the melting range they are mostly like liquids, but do exhibit evidences of elasticity. Between the glass-transition temperature and the melting range, they gradually become less and less like elastic solids and more and more like viscous fluids. However, at all temperatures they retain some of the behavior typical of both ideal classes, so the term "viscoelastic" was coined to describe them. The study of viscoelastic behavior, the region in which both the fluid and solid aspects of polymers are important, has been most successfully developed in connection with the creep and stress–relaxation behavior of polymeric objects under stress (see Chapter 7). In molten polymers, elastic behavior is seen in the swelling of extruded products as they emerge from dies, and in the resistance to tearing and to severe sagging of sheets that have been heated prior to forming. The relative importance of the elastic response of a polymer depends on the time scale of the operation, response being much faster at higher temperatures. The popular "Silly Putty," a silicone material, is interesting in that at room temperature its elastic response time is such that if it is stressed very rapidly, it behaves like a brittle solid, whereas if stressed very slowly it behaves like an inelastic liquid. You can shatter it, bounce it, pull strings from it, and pour it.

In some fluids, notably suspensions of solids, the viscosity changes with time even though temperature, pressure, and velocity gradient are un-

---

* In gases, viscosity *increases* with rising temperature, but relatively slowly.

changing. For example, in the behavior sequence of chilled heavy cream during whipping, the changes are due partly to alteration of the cream structure caused by the mechanical action and partly to the stable inclusion of air bubbles. If whipping is continued further, the butterfat globules coagulate into lumps and the air bubbles collapse. This is an example of irreversible time-dependent effects. Suspensions of potato starch seem fluid when at rest, but become "thicker," that is, more viscous, when shaken, then revert to their "thin" state shortly after stirring ceases. Fluids that become more viscous upon application of stress are called *rheopectic*; those that thin on stirring are known as *thixotropic*. Such modes of behavior may be encountered in suspensions of plastics in fluids, but have not been observed in molten plastics. They may be reversible or/and irreversible.

Some materials won't flow at all unless applied stress exceeds a minimum critical value called the *yield stress*, whereas at stresses above the yield stress the flow rate is in proportion to the difference between the applied stress and the yield stress. Such materials are called *Bingham plastics*, in honor of one of the founders of modern rheology. Most toothpastes behave in this fashion: unscrew the tube cap and invert the tube without squeezing, and you can wait indefinitely, and vainly, for any paste to emerge; squeeze firmly and out it flows (see Fig. 1).

Having viewed this opening panorama of rheology, we can now "zero in" on those parts of it that will be discussed further. First we must understand what viscosity is and some simple ways of measuring it. We then discuss in detail the special type of non-Newtonian behavior typical of plastic melts, and its applications to some simple flow geometries. A brief introduction to pressure–temperature–volume relationships in polymers is presented.

B. VISCOSITY AND NEWTONIAN FLOW

Consider the liquid-filled, infinite parallel plates in Fig. 2. The bottom plate is still, the top one moving at a constant velocity V in the x direction. The space between is completely filled with a liquid at constant temperature, that is, constant viscosity, and the liquid wets both plates. The liquid molecules touching the lower plate are not moving, and those touching the top plate are moving along with velocity V. The liquid between the plates, though it is actually made up of molecules of small but definite size, is for practical purposes a continuum. It may be thought of as a great many layers, all parallel to the plates, each separated by a distance dy from its adjacent layers.

If in fact such an ideal setup could be made available for experimentation, we would find that a definite force F must be exerted to keep the top plate

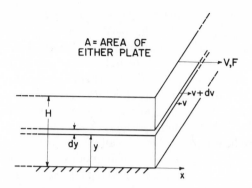

Figure 2   An idealized parallel-plate situation, in which the space between the plates is filled with fluid and the top plate moves with steady velocity V in the x direction. Fluid wets, that is, adheres to, both plates.

moving steadily, and this force would be in proportion to V. Since there is no acceleration, there can be no unbalanced forces on the top plate, so the force F must be balanced exactly by an equal and opposite force. This force is the drag of the liquid on the plate and is due to the resistance of the liquid molecules to being sheared apart. If we now consider the liquid layer just below the top one, we realize that it is being dragged forward by the same force with which it is dragging backward on the top layer. The same situation prevails with respect to all the layers of liquid, each one transmitting the drag force to the next, the bottom layer being dragged on by the bottom plate. Thus every layer is in shear, and subject to the same shear stress, F/A. If we now were to reduce the plate separation H to half its value, we would observe that the force F would have to be doubled to maintain the velocity V. In other words, the applied shear stress, F/A, varies directly with V and inversely with H. On a one-layer basis the ratio V/H reduces to the *velocity gradient*, dv_x/dy. Newton's law is usually stated in terms of the local gradient, which in this simple case is constant over y, as follows:

$$\frac{F}{A} = -\tau_{yx} = +\mu \frac{dv_x}{dy} \left(= \mu \frac{V}{H} \text{ here} \right) \tag{2}$$

where v_x = velocity in the x direction
   τ_{yx} = shear stress on the lower side of the liquid layer, x-directed and acting on a surface perpendicular to the y direction
   μ = liquid viscosity

The negative sign of the equation signifies that τ_{yx} at y is directed to the left, that is, in the negative x direction, whereas v_x is increasing with increasing y.

This equation defines the fluid property, viscosity; that is, μ = stress/rate of shear.

In addition to the velocity distribution, shear rate, and shear stress, the rate of doing work in flowing systems is also of interest. In the situation of Fig. 2, there is no rise in the fluid pressure, nor, because the flow is steady, is there any change in the kinetic energy of the stream as a whole. This means that the entire power consumption, $F \times V$, is used to overcome the viscous resistance of the liquid and is converted to heat. The volume in which this heat is being generated is just $A \times H$, so the power expenditure per unit volume is $FV/AH = (F/A)(V/H)$, that is, the product of shear stress and shear rate. For more complex situations, in which these quantities are varying, the rate of viscous dissipation becomes a point function:

$$p = \tau_{yx} \frac{dv_x}{dy} = \tau \dot{\gamma} \tag{3}$$

The symbol $\dot{\gamma}$, widely used for its conciseness, is the time derivative of shear deformation, that is, the shear rate or velocity gradient. Note that, by Newton's equation, $p = \mu \dot{\gamma}^2$.

The simplest concept of shear stress is that of a force acting over a unit area, requiring units such as dynes per square centimeter and pounds (lb_f) per square inch, or psi. If the same length unit is used for both the velocity and the y coordinate, shear rate (dv_x/dy) has the unit reciprocal time, usually reciprocal seconds (sec^{-1} or recsec). For Eq. 2 to be dimensionally consistent, viscosity must have the units of (force)(time)/(area). Typical are the dyne-second per square centimeter, or poise, and the pound force-second per square inch, or psisec. The poise and centipoise are conveniently sized for ordinary liquids and dilute polymer solutions, but the psisec, which equals 6.89×10^4 P, and the kilopoise, which equals 0.01 N-sec/cm^2, are much more appropriate for the very viscous polymer melts.

Some authors, following the unified transport ideas, prefer to think of τ_{yx} as a component of momentum transport. Because each layer in Fig. 2 moves at a different speed than the adjacent layers, the upper layers have more momentum per unit volume than the lower ones. Momentum is therefore being transferred in the y direction from layer to layer. From this viewpoint, the proper dimensions for τ_{yx} are those of momentum flux per unit volume, or (mass $\times$ velocity)/(area crosswise to flux direction)(time), which reduce to (mass)/(length)(time)^2. Typical are the gram per square centimeter-second and the pound mass per square inch-second. The corresponding viscosity units are the gram per centimeter-second ($\equiv 1$ P) and the pound mass per inch-second ($= 178$ P). Because many flows of interest are caused by applied pressures, the units of which are compatible with those of shear stress, we use the force-based units exclusively in this chapter. The reader should be aware, however, that many different viscosity units are current in the scientific and engineering literature and he should keep fully alert in using viscosity data. Table 1 contains some conversion factors for the commonly used viscosity units.

The reciprocal of the viscosity is known as the *fluidity* of the material.

Equation 2 may be written in a slightly different, more general way, as follows:

$$\tau_{yx} = -\frac{\mu}{\rho}\frac{d(\rho v_x)}{dy} \tag{4}$$

where ρ = the material density.

The quotient μ/ρ is usually symbolized by ν and is called the *kinematic viscosity*. Note that in this form the derivative on the right side is the rate of change of momentum concentration in the y direction, so that the equation takes on the form of the diffusion equation, the momentum flux being proportional to the momentum concentration gradient. Thus the kinematic viscosity may be thought of as the momentum diffusivity of the material, with the usual diffusivity units.

If the fluid is Newtonian, the viscosity is constant over all values of shear stress and velocity gradient at any given temperature and pressure. Polymers, both the melts and the solutions, are non-Newtonian in that, when the shear stress is greater than about 1 psi, the viscosity as defined by Equation 2 begins to drop as shear rate and shear stress increase. Such behavior and materials are described as *pseudoplastic*. This behavior is dealt with in detail later.

Equation 2 applies only to one-dimensional flow. If the fluid velocity is changing in such a way that more than one space coordinate is needed to describe it, or is changing with time, then the shear stress τ_{yx} becomes one of many changing components of fluid stress and more complicated momentum equations must be invoked. In the most general case of three-dimensional, time-varying flow, the stress tensor τ has nine changing components and the momentum–conservation equations are so complicated that, to date, no such flow situation has been successfully solved. The general derivations are available in Ref. 12. We deal here only with a few of the simplest, one-dimensional cases.

Although the question of how liquid molecules flow has been under study for more than 70 years, it is still not understood clearly enough to allow us to predict viscosity accurately from molecular structure, nor even from other properties of the liquid, such as density, "free volume" (the fraction of the molar volume not filled by the actual molecules, based on known molecular dimensions), and polarity. The most successful approach to date has been that of Eyring and co-workers (14–16), in which the liquid is visualized as

Table 1 Viscosity units and conversion factors

Units

A	poise (P) $\equiv$ 1 dyne-sec/cm^2 $\equiv$ 1 g/cm-sec
B	centipoise (cP)
C	kilopoise (kP)
D	Newton–second per square meter (N-sec/m^2)
E	Newton–second per square centimeter (N-sec/cm^2)
F	Kilogram(force)–second per square centimeter (kg$_f$-sec/cm^2)
G	Pound(force)–second per square inch (psisec)
H	Pound(force)–second per square foot (psfsec)
I	Pound (mass)/foot-second (lb$_m$/ft-sec)
J	Pound (mass)/foot-hour (lb$_m$/ft-hr)

Conversion Factors

To Convert to → from ↓	A	B	C	D	E
A	1	100	0.001	0.1	10^{-5}
B	0.01	1	10^{-5}	0.001	10^{-7}
C	1000	10^5	1	100	0.01
D	10	1000	0.01	1	0.0001
E	10^5	10^7	100	10^4	1
F	9.81×10^5	9.81×10^7	981	98100	9.81
G	6.89×10^4	6.89×10^6	68.9	6890	0.689
H	478	47800	0.478	47.8	0.00478
I	14.88	1488	0.01488	1.488	0.0001488
J	0.004134	0.4134	4.134×10^{-6}	0.0004134	4.134×10^{-8}

To Convert to → from ↓	F	G	H	I	J
A	1.020×10^{-6}	1.451×10^{-5}	0.002090	0.0672	241.9
B	1.020×10^{-8}	1.451×10^{-7}	2.090×10^{-5}	0.000672	2.419
C	0.001020	0.01451	2.090	67.2	2.419×10^5
D	1.020×10^{-5}	0.0001451	0.02090	0.672	2419
E	0.1020	1.451	209.0	6720	2.419×10^7
F	1	14.23	2050	65900	2.372×10^8
G	0.0703	1	144	4630	1.667×10^7
H	0.000488	0.00694	1	32.12	1.156×10^5
I	1.518×10^{-5}	0.0002159	0.03110	1	3600
J	4.217×10^{-9}	6.00×10^{-8}	8.64×10^{-6}	0.0002778	1

being made up of molecules interlaced with "holes," or molecular-sized bits of free volume. By acquiring, in a statistically distributed manner, some extra energy, a molecule may shift suddenly from its present position to a nearby hole, thus creating a new hole. Flow is pictured as being made up of a multitude of such shifts. The energy is known as the *activation energy of flow* and is usually given on a molar, rather than molecular, basis. For simple liquids it has been shown experimentally that this energy averages about 0.4 times the heat of vaporization at the normal boiling point, but the factor varies from liquid to liquid.

For polymer molecules, with their long chains, the mechanism is pictured a little differently, because the holes are small relative to the whole polymer molecule. Flow is thought to take place by jumps of chain segments, called *flow units*, about 20–30 chain atoms in size, so the activation energy is based on a mole of such flow units rather than a mole of assorted whole polymer molecules.

The Eyring model has passed through some permutations of form during the years since Ref. 14 was published, but in one form, useful to us here, it was as follows:

$$\mu = 7.71 \times 10^{-4} \frac{\sqrt{M}\, T^{2/3}}{(P_i + P)V^{5/3}} \exp\left[\frac{(P_i + P)V}{nRT} - \frac{\Delta S}{R}\right] \qquad (5)$$

where μ = fluid viscosity (P)

M = mass per mole (g)

T = absolute temperature (°K)

P_i = internal pressure of the fluid (on the order of hundreds or thousands of atmospheres) (atm)

P = external pressure

V = molar volume (cm^3/g-mole)

R = molar gas constant = 82.06 cm^3-atm/g-mole-°K

ΔS = "entropy of activation" for flow (cm^3-atm/g-mole-°K)

n = an empirical factor, between 2 and 3 and about 2.45 for many fluids, which is the ratio of the latent heat of vaporization of the fluid to its observed "activation free energy" (ΔF) of flow.

The group $(P_i + P)V/n$ is the activation energy of flow.

The form of this equation (16) has proved useful in interpreting the actual flow behavior of liquids, including polymer solutions and melts. It shows that the temperature dependence of viscosity is essentially exponential, although the fixed ΔS term in the exponent and the $T^{2/3}$ multiplier give warning that there should be some curvature in plots of $\ln \mu$ versus $1/T$, and in fact there is. Over reasonably small temperature ranges, however, say, 50 to 100°C, the deviations from linearity are often not serious enough to warrant taking into account in practical calculations. Equation 5 also indicates that there should be an exponential pressure effect which, however, won't become important until $P \gtrsim 0.1P_i$, a relatively high pressure. The available viscosity–pressure isotherms are in fact reasonably well fitted by this exponential form.

It must be recognized that the special flow characteristics of polymers are due to the long-chain nature of the molecules and their associated interactions and entanglements. The longer the molecules (i.e., the larger the molecular weight), the greater the level of entanglement and the greater the resistance to flow, that is, the higher the melt viscosity. This in fact leads to the relationship

$$\mu(\text{melt}) = kM^{3.4} \tag{6}$$

where k = temperature-dependent constant
M = average molecular weight
$\mu(\text{melt})$ = melt viscosity

This equation generally holds for molecular weights corresponding to degrees of polymerization greater than several hundred. This further points up the special importance of chain entanglement and its effect on polymer melt flow.

In most polymer rheology to date, the working model for temperature dependence has been simplified to the convenient form:

$$\mu = \mu_0 \exp\left[\frac{E}{R}\left(\frac{1}{T} - \frac{1}{T_0}\right)\right] = \mu_0 \exp\left[\frac{E}{R}\left(\frac{T_0 - T}{T_0 T}\right)\right] \tag{7}$$

where E = "activation energy of flow"
R = molar energy constant
μ_0 = viscosity at a convenient base temperature, T_0

This activation energy, rather than the slightly different one of Eq. 5, is the one nearly always meant in literature references to "activation energy of flow." The "mole" on which this energy is based is presumably a gram-molecular weight of flow units of the polymer. However, since flow units vary in size, even in a particular material, the definition of E on a molar basis in polymers is mostly a carryover from the similar definition by Andrade and by Eyring and co-workers for simpler fluids of definite molecular weight.

Some typical activation energies of flow and the conditions over which they apply are listed in Table 2, along with some typical viscosity values for familiar materials. To translate these activation energies into practical terms, the increase factor for viscosity over a 10°C temperature drop is also tabulated.

Table 2 Temperature Dependence of Viscosity in Liquids

Liquid	Base Temperature, t_0 (°C)	Apparent Viscosity $(\pi R^4 \Delta P/8QL)$ at 100 recsec $(4Q/\pi R^3)$ and t_0 (poise)	$E_\tau{}^a$ at t_0 (kcal/g-mole)	Factor of Approximate Increase in Viscosity with 10°C Drop in Temperature at t_0
Gasoline (sp gr = 0.68)	20	0.0028	3.40	1.12
Water	20	0.01005	7.50	1.33
Mercury	20	0.0158	0.88	1.03
SAE 10 motor oil	20	0.73	21.4	2.00
Glycerol (USP)	20	10–12	32.4	2.86
Corn syrup (dextrose equiv. = 64; sp gr = 1.685; 16.2% H_2O)	38	180	42	3.88
Nylon-6,6 (Zytel 101)	290	1170	42	1.45
Polymethyl methacrylate (Lucite 140)	260	4100	64	1.88
Polystyrene (Styron 666)	240	4400	42	1.55
Linear polyethylene (Marlex 50/15)	230	7900	12.5	1.15
Rigid polyvinyl chloride (Geon 8750)	177	54,000	38	1.69

[a] The subscript τ indicates that the activation energy was computed from the change in shear rate with temperature at constant stress, an important qualification for non-Newtonian fluids such as the polymer melts. E_τ for pseudoplastic fluids is greater than E_γ; whereas for Newtonian fluids they are identical. The fluids gasoline through corn syrup are Newtonian, so for them the viscosities reported are applicable to all shear rates, not just 100 recsec.

C. VISCOSITY MEASUREMENT

1. Instruments and Equations

A fascinating variety of methods and instruments has been devised to measure viscosity and other flow properties of liquids of all classes. No attempt is made here to survey all these methods; we merely explain a few of those that are most widely used. (An excellent survey of methods and equipment is provided in Ref. 21.) Nearly all viscometric systems fall into two basic classes: those in which flow is caused by a difference in pressure from one part of the liquid to another, and those in which flow is caused by controlled relative motion of the confining solid boundaries of the liquid. In the first class are the capillary types; in the second are the rotational, sliding-plate, falling-ball, and vibrating-reed types. The simple capillary tube is certainly the granddaddy of them all and probably is still, in its many varieties, the most widely used type of viscometer. It consists of a reservoir for the liquid charge, a small-diameter tube, or capillary, of known dimensions, a method of applying a controlled pressure to the liquid in the reservoir so that it is forced to run through the capillary at a steady rate, and methods for measuring the quantity of the liquid that passes through the capillary and the time of passing. This sort of viscometer is represented diagrammatically in Fig. 3. The information it yields is the pressure drop across the capillary, the capillary dimensions, and the volume or mass flow rate through the capillary. In some models, these quantities are measured by elaborate auxiliary equipment, not shown in Fig. 3. We will see shortly how viscosity may be calculated from such information.

In the other class the most popular type is the rotational instrument, in which two cylindrical surfaces confine the liquid. Typically, one of the surfaces is held still while the other is rotated at one of many constant speeds. The torque required to hold the stationary surface at rest is measured, also the rotational speed of the other cylinder, along with the dimensions of the space filled by the liquid. If the clearance between the cylinders is small relative to their radii, and to the length of the cylinder, then the geometry closely approximates the infinite parallel plates of Fig. 2, and the velocity gradient and shear stress are very nearly constant throughout the liquid sample. In practice there are a number of corrections that must usually be made to get accurate results, but it would be out of place to dwell on such details here.

Refer now to Figs. 3 and 4. Figure 4 is a differential shell of fluid at any radius r within the stream moving through the capillary tube of the rheometer in Fig. 3. The high pressure P acts on the upstream face of the shell, with area $2\pi r\, dr$, and the pressure $P + dP$ acts on the same area at the downstream face. The net pressure force on the two end faces is $dP(2\pi r\, dr)$. with area $2\pi r\, dr$, and the pressure $P + dP$ acts on the same area at the down-stream face. The net pressure force on the two end faces is $dP(2\pi r\, dr)$.

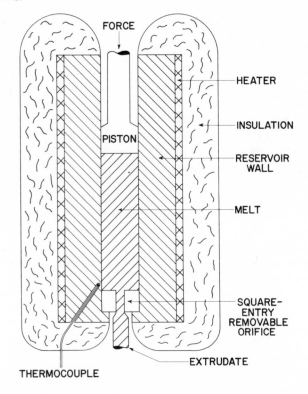

Figure 3 Schematic drawing of simple piston-type, capillary rheometer. The widely used Melt Indexer is an instrument of this type.

However, in steady flow, which we deliberately set up in the rheometer, there are no velocities that are changing with time and therefore no un-balanced forces, so this net pressure force must be exactly opposed by the shear forces acting on the cylindrical surfaces of the element. At radius r, the shear stress is τ_{rz}, and the shear force is $(2\pi r\, dz)\tau_{rz}$. On the inside surface of the shell, the shear stress is directed downstream, since the interior is moving faster than the shell, whereas on the exterior it is directed upstream. The change in the shear force must be exactly equal and opposite to the net pressure force, giving us

$$2\pi\, dz\, d(r\tau_{rz}) = -2\pi r\, dP \tag{8}$$

Dividing through by $2\pi\, dz$, integrating, and dividing by r, we obtain

$$\tau_{rz} = -\frac{r}{2}\frac{dP}{dz} + \frac{C_1}{r} \tag{9}$$

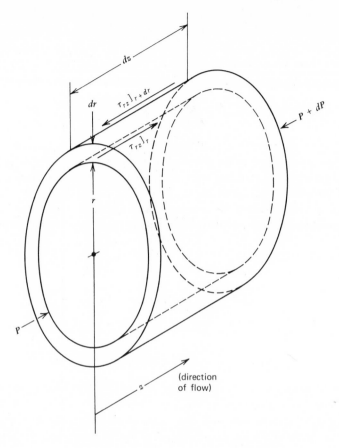

Figure 4 Differential shell of melt stream inside the capillary tube of a capillary rheometer, showing mathematical quantities used in deriving Eqs. 8–17.

The stress must be finite throughout the stream; therefore, since $r = 0$ at the center, C_1 must be zero. For a Newtonian fluid in laminar flow, by Eq. 2,

$$\tau_{rz} = -\frac{r}{2}\frac{dP}{dz} = -\mu\frac{dv_z}{dr} \tag{10}$$

Solving Eq. 10 for the velocity gradient and integrating again, we obtain

$$v_z = +\frac{1}{2\mu}\frac{dP}{dz}\frac{r^2}{2} + C_2 \tag{11}$$

We need a boundary condition and the usual assumption is that the liquid in contact with the bounding surfaces has the velocity of those surfaces. This boundary condition has been generally verified experimentally, though exceptions have been reported. For our case, this condition is $v_z]_{r=R} = 0$. Applying this, we get

$$v_z = -\frac{1}{4\mu}\left(\frac{dP}{dz}\right)(R^2 - r^2) \tag{12}$$

which shows that the velocity distribution of a Newtonian fluid in laminar flow in a circular tube is parabolic in r, with the greatest velocity at the center of the tube. The minus sign results from the fact that flow is in the direction in which P is decreasing.

We are also interested in the flow rate, since this is much easier to measure than local velocities. The contribution to the total flow by the element of Fig. 4 is its velocity multiplied by its cross-sectional area perpendicular to the flow direction, that is,

$$(2\pi r \, dr)v_z = -\frac{\pi}{2\mu}\frac{dP}{dz}(R^2 r - r^3)\, dr.$$

Integrating this flow element between the limits $r = 0$ to $r = R$, we obtain the total volume flow rate:

$$Q = \frac{\pi R^4}{8\mu}\left(-\frac{dP}{dz}\right) \tag{13}$$

For a channel of constant cross section, the pressure gradient $(-dP/dz)$ remains constant over the entire length L of the channel and may be replaced by $\Delta P/L$. This leads to a more familiar form of Eq. 13, known as the Poiseuille equation.

$$Q = \frac{\pi R^4 \Delta P}{8\mu L} \tag{14}$$

This is the form of the equation used in capillary rheometry. If we solve it for the viscosity, we see that the rheologist can express the viscosity of the working fluid in terms of his primary measured variables, Q, ΔP, R, and L.

$$\mu = \frac{\pi R^4 \Delta P}{8QL} \tag{15}$$

Returning to Eq. 10, and substituting $r = R$, we find that the shear stress at the tube wall is

$$\tau_{Rz} = \frac{\Delta P R}{2L} \tag{16}$$

Equation 10 also tells us that the shear rate at the wall should be given by the quotient of the stress at the wall by the viscosity. If this ratio is extracted from Eq. 14, we obtain

$$\frac{\Delta PR}{2\mu L} = \frac{4Q}{\pi R^3} \qquad (17)$$

which expresses the shear rate at the tube wall in terms of primary variables. Note that the viscosity in Eq. 15 is just the ratio of the shear stress at the wall to the shear rate at the wall. For a Newtonian fluid, this ratio should remain constant, at given temperature and pressure, regardless of the tube dimensions. This constancy does not hold up for non-Newtonian fluids, but it can be shown that where there is no time dependency of viscosity, and if the assumption of no slip at the wall is valid, there is a single-valued functional dependence of the quantities $4Q/\pi R^3$ and $\Delta PR/2L$ upon each other. For this reason, and because it is convenient, most flow measurements on polymer melts have been reported by giving the "apparent" shear rate at the tube wall, $4Q/\pi R^3$, versus the applied nominal shear stress at the wall, $\Delta PR/2L$. If we review the derivation of Eq. 9, we see that the expression for shear *stress* anywhere in the fluid in no way depends on the fluid properties. The quantity $4Q/\pi R^3$, however, *does* depend on an assumption of Newtonian behavior, and therefore cannot be correct for pseudoplastics such as polymer melts. Correcting this apparent shear rate to a true value is not difficult, as we shall see.

Equation 15 has been widely used in rheological circles to define for non-Newtonian fluids a quantity known as the *apparent viscosity*, the ratio of the nominal shear stress at the capillary wall to the apparent shear rate at the wall. This apparent viscosity, though superficially convenient, is not the same as the ratio of corrected shear stress to true shear rate for the given material under the test conditions; the true value depends on the L/D ratio of the orifice and, to some extent, on the shape of the orifice entry. Apparent viscosities obtained in two different orifices for a given material at a given temperature and shear level are in general not equal. Even greater discrepancies arise between apparent viscosities obtained in capillary instruments and those obtained on rotational instruments. However, compatible true values can be obtained from both types (17, 21).

By a similar process of setting up a shell balance in the fluid as was used in analyzing tube flow, the bob-and-cup rheometer may be analyzed. If the clearance between the two containing surfaces is small relative to their radii, say, $0.1\ R_b$ or less, then the situation idealizes accurately as the parallel-plate model. If the clearance is large, vortices that lead to substantial inaccuracy in measured viscosity may be established, even though the calculated

Reynolds number is well below the critical value of 3000. The working equation for this type of rheometer is

$$\mu = \frac{T}{8\pi^2 LN}\left[\frac{1}{R_b^2} - \frac{1}{R_c^2}\right] \tag{18}$$

where R_b and R_c = radii of the inner (bob) and outer (cup) cylindrical surfaces in contact with the fluid

L = immersed depth of the bob and cup

N = constant angular velocity (rev/sec) of the cup relative to the bob

T = observed torque on either member at the angular velocity set

If torque is measured in dyne-centimeters and the lengths in centimeters, the viscosity is in poises.

It is difficult to construct a bob-and-cup rheometer that has negligible end effects, and some adjustment of the data for the drag at the bottom end and meniscus deformations at the top end must usually be made. As the ratio $L/(R_c - R_b)$ increases from one design to the next, end effects become less important. The usual way of dealing with them is to add a fictitious extra length L_e to L; once this length has been accurately determined by calibration with fluids of known viscosity, or by making runs with one fluid at several depths of fill, the corrections are thenceforth made automatically by using $(L + L_e)$ in place of L in Eq. 18.

Equation 18 applies only to Newtonian fluids at constant temperature. The condition of constant fluid temperature is worth verifying in a situation where (a) viscosity and/or rotational speed are high, and (b) where the clearance between bob and cup is small. From Eq. 3, and with the assumption that the liquid heat capacity $\cong 0.5$ cal/g-°C, it can be shown that if a liquid is subjected to a viscous dissipation rate of 0.02 J/(cm^3)(sec) for 1 min, the adiabatic temperature rise would be close to 1°C. It is clear from Table 2 that viscosity changes associated with such a temperature rise can be large. The dissipation rate of 0.02 J/(cm^3)(sec) corresponds to a shear rate of 450 recsec in a material whose viscosity is 1 P, or 4.5 recsec if the viscosity is 10^4 P.

For the bob-and-cup rheometer, the Newtonian shear rate at the cup surface is

$$(-\dot{\gamma})_{r=R_c} = -\frac{dv_\theta}{dr}\bigg]_{r=R_c} = -\frac{4\pi NR_b^2}{R_c^2 - R_b^2} \tag{19}$$

and at the bob surface it is

$$(-\dot{\gamma})_{r=R_b} = -\frac{4\pi NR_c^2}{R_c^2 - R_b^2} \tag{20}$$

If the clearance h $(= R_c - R_b)$ is equal or less than $R_b/10$, the shear rate is almost uniform throughout the fluid and the average value may be used.

$$\dot{\gamma}_{av} = \frac{4\pi N}{R_c^2 - R_b^2}\left(\frac{R_c + R_b}{2}\right)^2 = \frac{\pi N(R_c + R_b)}{R_c - R_b} = \frac{\pi \bar{D} N}{h} = \frac{\bar{v}}{h} \qquad (21)$$

Comparing this with Eq. 2, we see that for small h, the concentric cylinders closely approach the parallel-plate idealization.

The corresponding Newtonian shear stresses (for zero end correction) are

$$\tau_{r\theta}]_{r = R_c} = \frac{T}{2\pi R_c^2 L} \qquad (22)$$

$$\tau_{r\theta}]_{r = R_b} = \frac{T}{2\pi R_b^2 L} \qquad (23)$$

$$\tau_{av} = \frac{2T}{\pi \bar{D}^2 L} \qquad (24)$$

One or another of these shear rates, and corresponding shear stresses, are the values usually reported by users of rotational rheometers, even when dealing with non-Newtonian fluids. In this geometry, it is the shear rates that are independent of material behavior* and are essentially true values for pseudoplastic materials, whereas the stresses depend on flow behavior and are "apparent" values.

2. Dependence of Density on Temperature and Pressure

It is clear from Eq. 15 that measurement of viscosity (or apparent viscosity) by the capillary method requires that we measure the volume rate of flow through the capillary. Many types of capillary rheometer are designed to do this directly, but many others, e.g., the widely used "Melt Indexer" (9), find the flow rate by weighing the extrudate collected in a given time. To obtain the volume flow rate from such data, we need information on the density of the melt in the rheometer.

There are other good reasons for studying the density behavior of polymers. As every molder knows, there is a distressingly large difference between the dimensions of a molded piece of plastic and the dimensions of the mold cavity in which it was molded. These differences arise from two fundamental properties of materials: expansion and contraction under the influence of pressure and temperature changes. The compressibilities of polymer melts are comparable to those of simpler organic liquids, but the

---

* These shear-rate expressions, however, can break down if the material has a definite yield stress for flow.

processing range of pressures is much higher than that ordinarily encountered in processing simple compounds. In injection molding, polymers may be subjected to pressure changes of from 5000 to 40,000 psi. At 300°F poly-methyl methacrylate undergoes a volume compression of about 3.6% as the pressure applied on it rises from atmospheric to 10,000 psi. For low-density polyethylene, the figure is 5.5% at 300°F; for nylon-6,6 it is about 3.5% at 560°F, and for polystyrene (Styron 666) it is 5.1% at 300°F. As a rough bench mark, the compressibility of polymer melts can be taken to be about $\frac{1}{2}$%/1000 psi of pressure change.

Temperature changes effect even larger, and more complicated, changes in polymer density. For the polymers just mentioned, a drop in temperature of 100°F causes volume contractions of about 3.3, 3.5, 2.8 and 2.4%, respectively, at pressures near 2000 psi. These changes are for the *molten* polymers. For the crystalline polymers—polyolefins, nylons, fluorocarbons, polyethylene terephthalate—there are additional large, abrupt volume changes associated with freezing and melting. There are several polymers for which detailed thermal expansion and compressibility data have been published; some of these data are collected in Ref. 2.

Some fairly successful attempts have been made to express the volume of a plastic as a function of the temperature and pressure, by analogy with the equations of state used for gases. Spencer and Gilmore (19) proposed a modified van der Waals equation containing three parameters characteristic of the polymer which are evaluated from measurements of volume at various temperatures and pressures. Their equation, which fits existing P–V–T data reasonably well, is

$$(P + \pi)(V - b) = \frac{RT}{M} \tag{25}$$

where V = specific volume of the polymer
P = applied pressure
R = universal gas constant, in compatible units
T = absolute temperature
M = molecular weight of an "interaction unit" of the polymer
π = internal pressure of the polymer
b = an empirical constant with units of specific volume

The inventors of this equation fitted it to dilatometric data and reported values of constants M, π, and b for five polymers: polystyrene, polymethyl methacrylate, ethyl cellulose, cellulose acetate–butyrate, and low-density polyethylene. By differentiating V in Eq. 25 with respect to P and T, one can obtain relationships giving the compressibility and thermal-expansion coefficient in terms of the equation-of-state constants. From measurements of these two properties alone, however, it is *not* possible to evaluate the equation-of-state constants.

The volumes calculated from Eq. 25 and, for that matter, volume changes calculated from coefficient of expansion and compressibility, assume that the polymer sample has come to volume equilibrium at the given conditions. Reaching this equilibrium can take many hours near room temperatures, less at higher temperatures. Thus the *rate* of cooling or heating and the *rate* of applying or relaxing pressure have a definite influence on polymer volume and therefore on piece dimensions. In injection molding, the plastic is subjected to a very complex system of shear stresses, pressure, and severe cooling in the mold, so the dimensions of a freshly molded piece most likely will not be able to be reliably computed from the cavity dimensions and the equilibrium $P-V-T$ relationship, assuming such a relationship were available for the solid polymer. The solution of *additional* equations involving *rate* relationships will be necessary. On the other hand, if the molded piece is rewarmed to a temperature at which the frozen-in flow stresses can relax in a reasonable time, and crystallization, if any, is allowed to reach equilibrium, then the final dimensions can be more reliably controlled. Such annealing is a common practice where dimensional requirements are tight.

Table 3 lists approximate values of the equation-of-state constants for some polymer melts for which they have been determined. With pressure in atmospheres and temperature in degrees Kelvin, these constants will give the specific volume in cubic centimeters per gram, correct within about 1 %.

Equation 25 is also satisfactory for solid polymers that do not form crystals, using the same constants. However, for crystal-forming polymers, such as the polyolefins and nylons, the constants are different for the crystalline phases and, unless the degree of crystallinity is known, specific volumes of such solids cannot be reliably calculated from Equation 25.

Table 3

Polymer	π(atm)	b(cm^3/g)	M(g/mole)
Crystal polystyrene	1840	0.822	104
Polymethyl methacrylate	2130	0.734	100
Ethyl cellulose	2370	0.720	60.5
Cellulose acetate butyrate	2810	0.688	54.4
Polyethylene (RT density = 0.92 g/cm)	3240	0.875	28.1
Polyhexamethylene adipamide (nylon-6,6)			
Polypropylene (melt only)	3045	0.914	30.0

D. THE POWER-LAW MODEL FOR PSEUDOPLASTIC FLOW

In this section we show how a simple and often adequate model for pseudo-plastic flow can be used to calculate flow rates and pressure drops in simple channels. The model is generally referred to as the "power law" for flow, but is also known as the Ostwald–deWaele model. For one-dimensional flow it takes either of two equivalent forms:

$$\eta = \eta^0 \left| \frac{\tau}{\tau^0} \right|^{(n-1)/n} \tag{26a}$$

$$\eta = \eta^0 \left| \frac{\dot{\gamma}}{\dot{\gamma}^0} \right|^{n-1} \tag{26b}$$

where η = ratio of true shear stress to true shear rate at any point in the moving stream, that is, the prevailing viscosity

η^0 = the value of this viscosity at a specified standard state, which may be either a standard shear stress, τ^0, or a standard shear rate, $\dot{\gamma}^0$. The standard state is arbitrarily chosen to be convenient. The values of η^0 for the two forms of the power law are the same *only* if we define $\eta^0 \equiv \tau^0/\dot{\gamma}^0$

n = an exponent, usually between 0.25 and 0.9 for molten polymers, which gives this "law" its name

The two forms of the law are completely equivalent, but normally Eq. 26b is used when the flow is caused by relative motion of the bounding surfaces and Eq. 26a is preferable when flow is caused by a pressure gradient.

The viscosity as defined by this model is used in Newton's equation (Eq. 2) instead of the constant Newtonian viscosity. If that substitution is made, we get two equivalent equations relating shear rate and shear stress in the lower-law material.

$$\dot{\gamma} = \frac{1}{\eta^0} \left| \frac{\tau}{\tau^0} \right|^{(1-n)/n} \tau \tag{27a}$$

$$\tau = \eta^0 \left| \frac{\dot{\gamma}}{\dot{\gamma}^0} \right|^{n-1} \dot{\gamma} \tag{27b}$$

Although other forms of the power law are extant in the literature, the form offered here has some good features: (1) the prevailing viscosity has the dimensions of viscosity no matter what the value of n is; (2) the quantities in the absolute-value brackets are dimensionless; and (3) enclosing of the non-integral parts of $\dot{\gamma}$ and τ within absolute-value brackets avoids some of the very troublesome problems with negative signs that arise in using simpler-looking forms of the law.

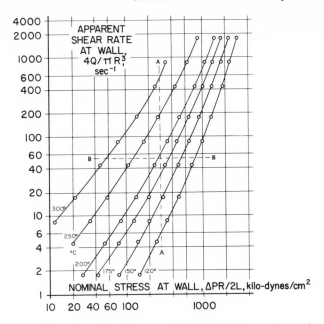

Figure 5   Flow curves of a low-density polyethylene resin at various temperatures.

Logarithmic plots of shear rate versus shear stress for a number of molten commercial thermoplastics (13) show a definite though gentle curvature which approximates linear behavior. Thus the power law can provide a useful approximation of the actual behavior, particularly if the range of interest is not too wide and is in the high-shear range. Where higher fidelity to the data is needed, more complex mathematical models may be used, and some of these are presented and discussed in Chapter 2 of Ref. 6.

Consider now the flow curves plotted in Fig. 5 for a low-density polyethylene resin (18) at many different temperatures. Clearly we have a choice in estimating the effect of temperature. We can follow a line of constant stress (line $A-A$), observing the increase in shear rate as the temperature rises; or we can follow a line of constant shear rate, observing the decrease in stress from one temperature to the next, along $B-B$. We see that for equal temperature changes, the increments of shear rate at constant stress are bigger than the stress increments at constant shear rate. Referring now to Eq. 7 and thinking in terms of prevailing viscosities rather than Newtonian viscosities, we see that it is possible to define the activation energy E in terms of the changes with temperature in *either* shear rate or shear stress while holding the other constant, and that two different values of E result. These are denoted E_τ and $E_{\dot\gamma}$. For power-law materials it can readily be shown that $E_{\dot\gamma} = nE_\tau$.

However, real materials rarely obey the power law perfectly; furthermore, E varies with the state of shear, decreasing as shear stress and shear rate rise. Analysis of the data of Fig. 5, for example, and other similar plots for other resins, shows that if apparent viscosity is calculated at various temperatures for given levels of stress and shear rate, and log viscosity is plotted against reciprocal absolute temperature, the viscosities calculated for constant shear stresses give much straighter lines, with more nearly constant slopes, than do the viscosities calculated for constant shear rate. For these particular data, E_τ ranged from 12.5 kcal/g-mole at 0.800 N/cm^2 down to 10.3 kcal/gmole at 10.00 N/cm^2. Over the corresponding range of shear rate, from 1.74 to 1740 recsec, $E_{\dot\gamma}$ dropped from 9.66 to 3.59, and with estimated errors about twice those of the E_τ values.

The practical significance of these findings is that E_τ will be a more reliable, more shear-stable value than $E_{\dot\gamma}$ and is definitely preferred in interpolating and extrapolating melt-flow data for polymers.

E. SPECIAL FLOW CHARACTERISTICS OF MOLTEN POLYMERS

When liquids of low viscosity flow from a reservoir or channel of large cross section into a channel of small cross section, the velocity in the small channel is often so much higher than that in the larger channel or reservoir that a considerable part of the liquid pressure is converted to kinetic energy of motion. An additional part is lost because of turbulence at the channel contraction. These pressure reductions are commonly referred to as entrance losses by hydraulic engineers, although that part of the pressure converted to velocity energy is more or less recoverable and so is not really "lost." When molten polymers flow, even the highest attainable velocities are low, because of the very high viscosities involved (about a million times that of water). For the 0.060-in.-diameter die used in this work the average velocity was only 13.0 in./sec, which corresponds to a kinetic energy of about 0.006 in.-lb$_f$/in.^{3}. At this same shear rate and 250°C the shear stress at the wall was 8.8 $\times$ 10^5 dyne/cm^2, or 12.8 psi. This required an applied pressure of about 3400 psi, corresponding to a pressure energy of 3400 in.-lb$_f$/in.^{3}. Therefore, less than 0.0002% of the pressure energy was converted to kinetic energy. Even at the much higher shear rates attained in wire-coating operations (involving thin walls and small diameters) kinetic energy is still negligible, and turbulence, in the Reynolds sense, doesn't exist.

Rheologists have found, however, that if a polymer melt is forced to flow at the same volumetric rate through shorter and shorter dies of a given radius, the observed pressure required to cause the flow does not diminish to zero as Eq. 8 suggests it should. Instead, there is clear evidence that, even if the tube length could be reduced to zero, so that the melt would be flowing through a sharp-edged, lengthless orifice, the pressure drop would still be very substantial. There must be some energy-transforming mechanism operating that consumes the pressure energy. Cinematographic studies of transparent melts containing tracer particles and flowing through transparent channels indicate that most of this energy seems to be transformed into elastic strain energy related to reforming the melt from a thick short stream into a long narrow one. In these viscoelastic melts, if the orifice tube is short, some of this strain energy may be recovered and is seen as crosswise swelling of the extrudate as it emerges from the outlet end of the tube. The melt to some degree "remembers" the shape it had in the reservoir and tries to return to that shape. However, its memory is rather short-lived, being characterized by a *relaxation time*, or by a family of such times, and if the orifice tube is very long, the recovery will be less. This extrudate-swelling behavior is different for each different resin family and there will be variations even within families. It also depends on extrusion conditions, and generally increases with nominal shear stress and entering melt temperature.

The longer the flow tube, at a given flow rate, the longer time the stressed melt has to relax and the larger the proportion of the entrance pressure drop that will be converted into heat by lateral viscous dissipation. Because in most extrusion situations, the melt emerges at atmospheric pressure and, as we have seen, with negligible kinetic energy, all of the pressure drop through the tube and most of the entrance pressure drop are finally converted into heat through viscous dissipation. The resulting average rise in melt temperature may vary from 2 to 6°C/1000 psi of pressure drop, depending on the volumetric specific heat of the resin. Polyethylene is at the low end of this range, polyvinylidene chloride at the high end. In capillary flow, because the shear rate and shear stress are much higher near the tube wall than at the stream center, the local rise in temperature may be much more than 2 to 6°C. This temperature rise owing to viscous dissipation is important in the processing of heat-sensitive resins. Since it also affects the fluid viscosity as it flows through the tube, we should, in theory, correct capillary-flow data for this temperature rise. Because the correction is enormously complicated, it is not normally made and the error has the effect of making the pseudoplasticity seem more severe than it really is at high shear rates, that is, of lowering the values of η and the η^0 in Eqs. 26. This can lead to discrepancies between calculated and observed results when rheological data from one geometry are used to estimate performance values in different geometries.

If one forces a polymer melt through a capillary at a very high rate, a new phenomenon is observed: the emerging melt not only swells, but it may exhibit a number of other distortions. As shear stress is increased, an alert observer will notice a fine roughness of the extrudate surface. Examined closely, this roughness appears to have a regular pattern like that of inter-secting wavelets. If the stress is increased some more, the extrudate begins to twitch and wiggle and will take on a spiral shape, still remarkably regular. As stress is further increased, the surface becomes very rough and torn look-ing and the spiral becomes erratic. At still higher rates, the extrudate will be torn into small, shaggy fragments. This phenomenon was named *melt fracture* by one of its early investigators, J. P. Tordella (20).

What seems to be happening is that the melt, as it enters the capillary and is distorted very suddenly from its thick short shape to the long thin shape, is stressed beyond its relatively low strength, even though the short-time mod-ulus is low at these flow temperatures; and it is stressed so fast that the visco-elastic stress–relaxation mechanism cannot react appreciably to reduce stress, so that the material is actually ruptured. The ruptures cannot be seen in the flowing melt as it continues through a transparent capillary because the high hydrostatic pressure prevents the creation and maintenance of finite void volumes; but before these rips can "heal," the melt is out of the capillary and the fragments spring apart. The earlier stages reveal the less catastrophic failure of the severely compressed melt along the natural 45° shear planes of the material. This failure naturally has its start at the surface, where shear stress due to flow, combined with the compression-caused stresses, is highest.

Rheologists have also noted that under conditions in the melt–fracture range, sudden jumps in extrusion rate can take place with only slight increases in applied stress. This phenomenon has been named *melt–flow instability* and appears to be caused by slip at the tube wall, that is, failure of the boundary condition of Eq. 12 and its power-law analogue, namely, that the fluid wets the wall and has the wall's velocity at that point. The phenomenon holds out some promise of obtaining much higher processing rates at small cost in pressure and power if it can be exploited, and is now under intense investiga-tion.

F.   TRUE FLOW CURVES FOR PSEUDOPLASTIC·FLUIDS

As we have seen, neither the shear stress at the wall nor the associated ap-parent shear rate at the wall, as given by Eqs. 16 and 17, is accurate for molten polymers. The stress is inaccurate because it includes the entrance loss, so the actual stress in the capillary tube itself is less than this value. The apparent shear rate is inaccurate because it presumes Newtonian behavior when in fact we have pseudoplastic behavior. We now present, without derivation, methods for obtaining from the observed Q and ΔP the true shear stress and shear rate at the tube wall for a pseudoplastic melt with viscoelastic character-istics.

The stress may be corrected in either of two simple ways. If data are available giving Q versus ΔP at a specified temperature for two or more tubes of equal radius and substantially differing lengths, we begin by plotting ΔP versus Q for each tube. Following any constant-Q line, we can pick off the several corresponding ΔP values. These are then replotted with tube length L as the abscissa and a straight line connecting the points is extrapolated to zero L to get the value of ΔP corresponding to no tube length. This ΔP is the entrance loss for the particular entrance geometry, material, and temperature involved. The procedure should be repeated at several well-spread Q values and the resulting entrance losses averaged. This average value may be subtracted from all the measured ΔP values to get ΔP_{ec}, the entrance-corrected pressure drop. The true shear stress at the wall is

$$\tau_w = \frac{\Delta P_{ec} R}{2L} \tag{28}$$

A second, somewhat similar procedure that leads to a more broadly useful result, has been proposed by Bagley (10), who thoroughly investigated the entrance loss for low-density polyethylene. He suggested that the entrance loss could be conveniently handled by imagining it to be due to a fictitious extra length of the capillary tube, expressed as a number of tube radii or diameters. To evaluate this fictitious length, of course, we must have data giving nominal stress versus apparent shear rate at the wall for capillaries of at least two different L/R ratios, at a given temperature. Nominal stress (i.e., $\Delta PR/2L$) is plotted against L/R for given values of $4Q/\pi R^3$, and a line is drawn. The line is extended leftward until it intersects the L/R axis. The negative L/R value thus found is the equivalent fictitious entrance length in tube diameters. If we designate this ratio as m, the entrance-corrected pressure drop is $\Delta P_{ec} = \Delta PL/(L + mR)$ and the true stress at the wall is given by

$$\tau_w = \frac{\Delta PR}{2(L + mR)} = \frac{\Delta P}{2[(L/R) + m]} \tag{29}$$

This is a form convenient for design purposes, since it permits us to use all the regular flow equations by replacing L by $L + mR$. However, a particular m value applies to a particular entrance geometry and should not be used carelessly with another. Also, Bagley, and others since, found that m depends mildly on the shear rate. So, although either method makes it possible to correct the stresses for a particular group of flow measurements, we must be cautious in using those curves to judge the corrections for new geometries.

Correction of the shear rate is more tedious. Rabinowitsch and Mooney* independently showed that, if the liquid wets the wall so that our basic

---

* Ref. 16, p. 73, contains the derivation of the Rabinowitsch–Mooney equation.

boundary condition is valid, and flow is steady, then no matter what the nature of the deviations from Newtonian behavior in a fluid, there is an equation relating the true shear rate at the wall to the apparent shear rate, $4Q/\pi R^3$, and its derivative with respect to the true shear stress at the wall. We present here a convenient modification of their result.

$$-\dot{\gamma}_w = \frac{4Q}{\pi R^3}\left[\frac{3}{4} + \frac{1}{4}\left(\frac{d[\log(4Q/\pi R^3)]}{d[\log(\Delta P_{ec}R/2L)]}\right)\right] \qquad (30)$$

The derivative in parentheses is just the slope of the plot of $\log(4Q/\pi R^3)$ versus $\log(\Delta P_{ec}R/2L)$.* For a power-law material this slope is constant with the value $1/n$, and Eq. 30 reduces to

$$-\dot{\gamma}_w = \frac{4Q}{\pi R^3}\left[\frac{3}{4} + \frac{1}{4}\frac{1}{n}\right] = \frac{4Q}{\pi R^3}\left[\frac{3n+1}{4n}\right] \qquad (31)$$

If the log–log plot of apparent shear rate versus corrected shear stress is curved, then the fraction $1/n$ may be interpreted as the slope of the curve at any specific value of $\Delta P_{ec}/2L$. To obtain a corrected flow curve, then, we first correct the nominal stress values for entrance loss, then plot $4Q/\pi R^3$ versus $\Delta P_{ec}/2L$. For various values of stress, the slope of the curve, $1/n$, is determined at that stress. Then Eq. 31 is applied to get the true shear rate at the wall for each stress and a new, true flow curve is plotted. For a power-law material, this is simply a parallel straight line at a distance $\log[(3n + 1)/4n]$ above the uncorrected line.

This may seem to the student to be rather laborious, and in truth, it is sometimes unnecessary to develop the corrected flow curve. However, if the data are to be used for calculations involving a geometry different from the one in which they were obtained, then the true flow curve usually must be established. Such a curve should apply to data obtained on both capillary and rotational instruments, or any other type that may be devised. Correction procedures will be different for the rotational data, of course, and those procedures are discussed in Refs. 17 and 21.

The corrections can be rather large. For example, in a short orifice, say, $L/D = 4$, the entrance-corrected stress is about half of the nominal. If $n = 0.4$, a fairly typical value, the true shear rate is 37% more than $4Q/\pi R^3$. Taking the quotient of the stress and rate, the viscosity, η, at the wall would be only 36% of the apparent viscosity computed from $(\Delta PR/2L)/(4Q/\pi R^3)$.

---

* When Rabinowitsch and Mooney (independently) derived their equations in 1929 and 1930, they were unaware of the large entrance loss in polymer melts and assumed that pressure drop was proportional to tube length. Thus their equations, and all reproductions of them to date, have contained ΔP, the nominal pressure drop, rather than ΔP_{ec}.

G. INTRODUCTION TO POLYMER PROCESSING

Polymer rheology is a consideration in any operation that involves flow of polymers or their solutions: pumping, extrusion, molding, forming, spraying, and others. The field of "polymer processing" has been defined (11) as "an engineering specialty concerned with the operations carried out on polymeric materials or systems to increase their utility. These operations produce one or more of the following effects: chemical reaction, flow, or a permanent change in a physical property." Combined operations involving chemical reaction, for example, the curing of thermosetting resins during molding or casting, are so complex that they have so far defied complete analysis, though much is known about the chemical reactions involved and their kinetics. In addition to having all the ramifications already discussed, the flow of hot thermosetting resins is further complicated by the fact that the resin is cross-linking as it flows, sometimes also envoling gas, and usually getting hotter because of the exothermic heat of reaction. On the other hand, the principles so far discussed have been very useful in analyzing the operations involved in processing thermoplastic materials.

Much of thermoplastic processing can be divided into two main types: (1) molding, in its many variations, and (2) extrusion. In *extrusion*, the plastic is softened by heating, then forced through a shaped hole and chilled. The *extrudate* retains a cross-sectional shape that is related to, but usually not identical to, that of the die opening. By this process such consumer and industrial products as pipe, film and sheet, wire coatings, gasketing, monofilaments and staple fibers, building siding, and structural shapes are produced from plastics (as well as from rubbers and metals).

Injection molding, probably the most important of all the plastics-conversion processes, consist of softening the plastic, as in extrusion, then ramming it very quickly into a closed chilled mold, where it is held under high pressure until partially or wholly frozen. Typical products are trash cans, automobile parts, appliance handles, tool handles, toys, artificial flowers and plants, parts for business machines and other industrial products, and bottle caps.

In addition there are several other processing techniques that depend upon flow of polymer melts. These are all considered in subsequent chapters.

GENERAL REFERENCES

1. T. Alfrey, Jr., *Mechanical Behavior of High Polymers*, Wiley-Interscience, New York, 1948. Though not new, this book is still a monumental treatment of its subject, including sections on rheology, plastoelasticity, cross-linked polymer crystallization, solutions, and strength properties. Familiarity with graduate-level mathematics is presumed, but many parts of the book should be useful to college seniors with the usual mathematics training.

2. E. C. Bernhardt (Ed.), *Processing of Thermoplastic Materials*, Reinhold, New York, 1959.

3. F. W. Billmeyer, Jr., *Textbook of Polymer Science*, 2nd ed., Wiley-Interscience, 1972. An excellent survey of polymer structure and properties, polymerization reactions, and kinetics, with chapters on processing and fiber and elastomer technology.

4. F. R. Eirich (Ed.), *Rheology—Theory and Applications*, Academic, New York. This is the most complete treatise on the subject, covering most of its great range. The first of the five volumes appeared in 1956, the second in 1958, and others have followed. Many of the articles that make up this encyclopedia of rheology have been contributed by the leaders in all parts of the field. Mathematical level ranges from undergraduate to very advanced.

5. A. G. Frederickson, *Principles and Applications of Rheology*, Prentice-Hall, Englewood Cliffs, N.J., 1964. An advanced treatment of rheology, attempting to be broad in scope and general in its mathematical modeling of complex flow behavior. The author takes the continuum-mechanics viewpoint and relies heavily on tensor mathematics.

6. J. M. McKelvey, *Polymer Processing*, Wiley, New York, 1962.

7. E. T. Severs, *Rheology of Polymers*, Reinhold, New York, 1962. A readable, nonmathematical survey of flow of polymer melts, solutions, and suspensions with an introduction to processing.

8. A. H. P. Skelland, *Non-Newtonian Flow and Heat Transfer*, Wiley, New York, 1967.

SPECIFIC REFERENCES

9. ASTM Test Method D1238-57T, *ASTM Standards for Plastics*, American Society for Testing Materials, Philadelphia, 1958, p. 398.

10. E. B. Bagley, *J. Appl. Phys.*, **28**, 624 (1957).

11. E. G. Bernhardt and J. M. McKelvey, *Mod. Plast.*, **35**, 154 (July 1958).

12. R. B. Bird, W. E. Stewart, and E. N. Lightfoot, *Transport Phenomena*, Wiley, New York, 1960.

13. J. F. Carley, et al., *How to Choose an Extruder*, Prodex Corporation, Fords, N.J., 1963.

14. R. H. Ewell and H. Eyring, *J. Chem. Phys.*, **5**, 726 (1937).

15. S. Glasstone, K. J. Laidler, and H. Eyring, *The Theory of Rate Processes*, McGraw-Hill, New York, 1941, Chap. 9.

16. J. F. Kincaid, H. Eyring, and A. E. Stearn, *Chem. Rev.*, **28**, 301 (1941).

17. J. A. Lescarboura, F. J. Eichstadt, and G. W. Swift, *A.I.Ch.E. J.*, **13**, 169 (1967).

18. R. A. Mendelson, *Polym. Eng. Sci.* (formerly *SPE Trans.*), **5**, 34 (1965).

19. R. S. Spencer and G. D. Gilmore, *J. Appl. Phys.*, **21**, 523 (1950).

20. J. P. Tordella, *J. Appl. Phys.*, **27**, 454 (1956), and subsequent papers in rheological journals.

21. J. R. Van Wazer et al., *Viscosity and Flow Measurement*, Wiley-Interscience, New York, 1963.

DISCUSSION QUESTIONS AND PROBLEMS

1. Look up the origin of the word "viscosity." Does its modern meaning correspond with the meaning of the words from which it came? Do those words relate to the boundary condition for Eq. 11?

2. In one kind of viscosity-measuring device for pastes and softened polymers, the specimen, preformed into a stubby, right-cylindrical shape, is

pressed between two parallel plates with the upper plate descending toward the lower plate at a known, constant speed. The specimen is thus flattened out into a disk. What is the dimensionality of this flow? Is it steady?

3. A 60% solution of cane sugar (sucrose) in water is being pumped iso-thermally through a 1-in. pipe (ID = 1.049 in.) at the rate of 0.35 gal/min. At the prevailing 20° temperature, the density and viscosity of this solution are 1.287 g/cc and 56.5 cP. Is the flow laminar or turbulent?

4. The viscosity of the Newtonian sucrose solution of Problem 3 was determined by allowing it to flow from a pressurized reservoir through a horizontal capillary tube. The tube was 2.000 mm in diameter (preci-sion-bored capillary) and 6.00 in. long between the manometer taps. The manometer was filled to half its depth with mercury over which rested the sucrose solution being tested. The flow rate was determined by collecting and weighing the solution emerging from the capillary over a 5.000-min interval. If the manometer differential reading during this test was 39.22 cm, what must the flow rate, in g/min, have been? What was the pressure drop between the manometer tapes in N/m^2?

5. A stream of molten resin undergoes a pressure drop of 2500 psi as it flows through a tube. Its properties are as follows: density = 67.5 lb/ft^3; thermal conductivity = 0.155 $Btu/(hr)(ft^2)(°F/ft)$; specific heat = 0.52 $cal/(g)(°C)$; zero-shear (Newtonian) viscosity = 3.7 psisec. Find the average adiabatic temperature rise during this decompression. (Ignore Joule–Thomson effect, since the coefficient is not known, and neglect kinetic-energy changes.)

6. (a) Compare the values of M in Table 3 with the "mer" weights of the same polymers. (b) How does Eq. 25 differ from van der Waals' equation of state for gases? Why is the simplification made by Spencer and Gilmore permissible for amorphous polymers? (c) Interpret the constant b in physical terms.

7. Show that for a material obeying the power law, $E_{\dot\gamma} = nE_r$. How will a power-law fluid appear on a plot like Figure 5?

8. Equation 3 gives the local rate of viscous dissipation in a liquid under-going laminar Newtonian flow. What is the corresponding equation for a power-law fluid? For a Bingham plastic?

9. A geometry of special interest to wire-coaters and makers of pipe and tubing is the channel whose section is a circular annulus. The shell balances (Eqs. 8 and 9) are applicable unchanged to the annulus. (Why?) Obtain the velocity distribution and flow-rate/pressure-drop equation for the Newtonian fluid flowing isothermally through such an annulus with stationary walls. Do your equations reduce to Eqs. 12 and 13 when the inside radius goes to zero?

CHAPTER 9 _____

Extrusion

IMRICH KLEIN
Scientific Process & Research, Inc.
Highland Park, New Jersey

JULES W. LINDAU
Lindau Chemicals, Inc.
Columbia, South Carolina

457

A. INTRODUCTION

Extrusion is one of the most important plastics processing methods in use today. Most plastic materials in the world are processed in extruders. A substantial percentage of all plastics products pass through two or more extruders on their way from the chemical reactor to the finished product.

Among the products manufactured in extruders are pipe, rod, film, sheet, fiber, and an unlimited number of shapes or profiles. However, the extruder is also used in compounding and in the production of plastic raw materials such as pellets (also for reclamation of scrap plastics).

In addition to the continuous products mentioned above, for which the extruder is particularly suited, it is also used to manufacture noncontinuous products, either by using special molds, as in the case of blow-molding machines, or by reciprocating the screw, as in the case of reciprocating-screw injection molding.

Because the extruder is the basis of all these processing techniques, understanding its operation is extremely important.

The heart of the extruder is the Archimedean screw rotating in a barrel. It is capable of pumping a material under a set of operating conditions at a specific rate. This pumping rate depends on the resistance at the delivery end against which the extruder is required to pump. The performance characteristics of an extruder, therefore, more closely resemble those of a centrifugal pump than those of a positive-displacement pump. The latter maintains a nearly constant pumping rate regardless of the pressure it is required to develop.

After the pumping action and the shaping of the extrudate by dies the finished product is taken up by one of several take-up devices (which are not covered here).

B.   TERMINOLOGY AND SCREW GEOMETRY

The basic terminology associated with the extruder can be seen in Fig. 1; the screw is a very important part of the extruder and its details are given in Fig. 2. The screw rotates in a barrel the inner diameter of which is D_b. A radial flight clearance exists between the barrel and flight δ_f. The flight has a width e; the width of the channel formed between two flights is W and varies with radial location. The helix angle θ is that formed by the flight with the direction normal (perpendicular) to the screw axis. It is also a function of the radial location. The channel depth, H, is the distance between the inner wall of the barrel and the root of the screw.

The channel depth of an extruder screw is rarely kept constant over its whole length. Figure 3 shows two so-called "metering type" screws in an 8-in.- or larger-diameter extruder. The three geometrical sections in a simple screw of this type are called feed, compression, and metering sections. The significance of these names is mostly historical and they do *not* define the part of the screw where solid material is being fed or where melting or melt pumping may take place. The various functional zones of an extruder are shown in a schematic drawing in Fig. 4; they are the hopper, solids conveying zone, delay (in melting) zone, melting zone, and melt pumping zone.

C.   THE EXTRUDER AS A MELT PUMP

In a melt pump the extruder channel, defined by two flights, the root of the screw and the barrel, is filled with melt. For ease of explanation, the barrel in this instance is considered to be rotating and the screw stationary. The melt adheres to the walls of the channel and its advancement in the down-channel direction is caused by the down-channel component of the velocity of the barrel. Expressing the melt velocity at every point across the channel mathematically and summing these velocities (integrating) in both the cross-channel and radial directions yield the volumetric flow rate passing the particular point of the channel. Because this flow rate is caused by the melt being dragged by the barrel, it is referred to as *drag flow*; for a simplified case of a Newtonian melt under isothermal conditions it can be expressed as

$$Q_d = \frac{n}{2} W H V_{bz} \tag{1}$$

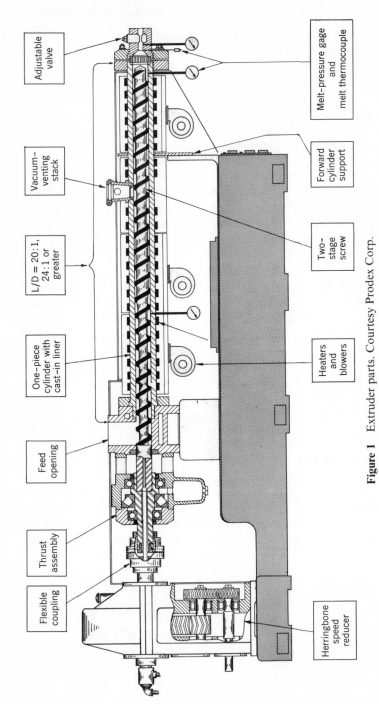

Figure 1 Extruder parts. Courtesy Prodex Corp.

Adjustable valve

Melt–pressure gage and melt thermocouple

Forward cylinder support

Vacuum– venting stack

Two– stage screw

L/D = 20:1, 24:1 or greater

One–piece cylinder with cast–in liner

Heaters and blowers

Feed opening

Thrust assembly

Flexible coupling

Herringbone speed reducer

460

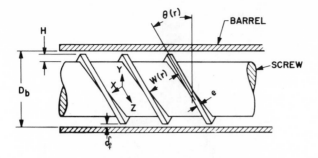

Figure 2   Schematic drawing of extruder screw.

where Q_d = volumetric drag flow rate in extruder channel (cm^3/sec)
n = number of parallel channels, dimensionless
W = width of screw channel (cm)
H = channel depth (distance between screw root and barrel surface), (cm)
V_{bz} = component of barrel velocity in down channel direction (cm/sec)

The output from an extruder seldom equals the drag flow, and as a result pressure must build up in the extruder channel. The portion of the flow rate created by the pressure gradient in the down-channel direction is called pressure flow and can be expressed as

$$Q_p = n \frac{WH^3}{12\mu} \left(-\frac{\partial P}{\partial z} \right)$$
(2)

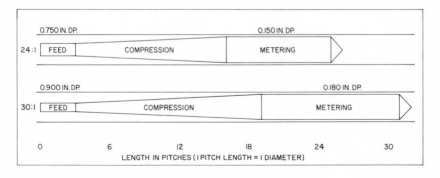

Figure 3   Extruder screws of the metering type. Channel depth values are representative for 8 in.- or larger-diameter machines. Courtesy *Plastics Technology*.

462

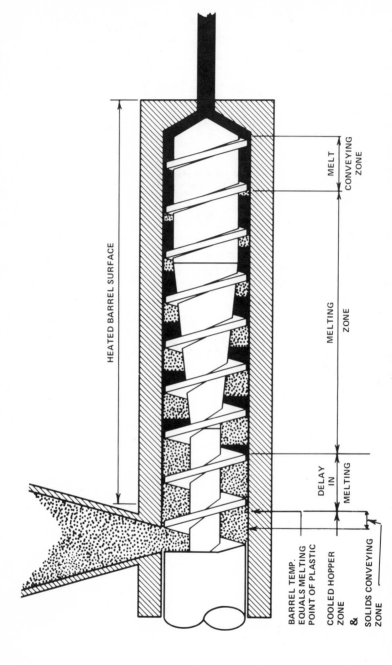

Figure 4 Schematic drawing of an extruder showing the various functional zones: hopper, solids-conveying zone, delay in melting start, melting zone, and melt-pumping zone.

HEATED BARREL SURFACE

MELT
CONVEYING
ZONE

MELTING
ZONE

DELAY
IN
MELTING

BARREL TEMP.
EQUALS MELTING
POINT OF PLASTIC

COOLED HOPPER
ZONE
&

SOLIDS CONVEYING
ZONE

where the terms not previously defined are as follows:

Q_p = volumetric pressure flow in the extruder channel (cm^3/sec)
μ = viscosity (dyne-sec/cm^2)

$$-\frac{\partial P}{\partial z} = -\frac{P_2 - P_1}{l}\sin\theta = \text{pressure gradient in the helical direction}$$

P = pressure (dyne/cm^2)
z = rectangular coordinate in down-channel direction (cm)
l = axial distance (cm)
θ = average helix angle between surface of screw and barrel (rad)

The output or melt flow of an extruder pumping melt can be expressed as

$$Q = Q_d - Q_p \tag{3}$$

Equation 3 is a result of combining (superimposing) at each point of the channel cross-section the down-channel velocity of the melt created by the relative velocity of barrel and screw with the velocity created by the pressure gradient, as can be seen in Fig. 5.

If the channel depth and fluid density are constant along the whole screw, the output of the extruder cannot exceed the drag flow term. This is so because the pressure must monotonically increase from zero at the entrance to the extruder to the final pressure value at the die end. As pointed out earlier, the channel depth usually varies and consequently the value of the drag flow as calculated by Eq. 1 also varies from point to point in an extruder. The net flow in an extruder must, however, be constant at all axial locations, and as a result the pressure flow term (Eq. 2) must also change. This latter requirement means that the pressure gradient or slope of the axial pressure profile must continuously change in a tapered section of an extruder. At a certain channel depth the computed drag flow exactly equals the net flow in the extruder channel. At this point pressure flow is zero. No pressure flow is synonymous with a zero pressure gradient and represents a maximum in the axial pressure profile.

It can be concluded, therefore, that a screw which consists of several geometrical sections, but operating at atmospheric pressure at the die end, a condition usually referred to as "open discharge," must contain a pressure maximum at some axial location of the compression section. At this axial location the drag-flow term computed from the dimensions of the channel equals the net flow or the actually measured extruder output.

It is also customary to express the extruder output in Eq. 3 as

$$Q = \alpha N - \frac{\beta}{\mu}\cdot\frac{\partial P}{\partial l} \tag{4}$$

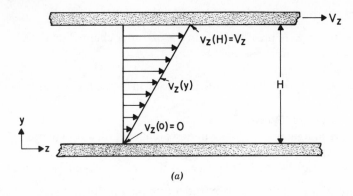

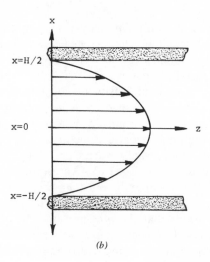

Figure 5 Down-channel velocity distribution created by the movement of the barrel relative to the screw (*a*) and that created by the down-channel pressure gradient (*b*).

where N = screw speed (rpm), α and β are parameters of screw geometry, which for cases when flight clearance and flight width can be neglected are expressed as

$$\alpha = \frac{\pi^2 D_b^2 H}{2} \sin \theta \cdot \cos \theta \qquad (5)$$

$$\beta = \frac{\pi D_b H^3}{12} \sin^2 \theta \qquad (6)$$

The above simple equations are the result of gross oversimplification of the actual conditions in an extruder. Among these are the following assumptions:

1. There is only one flight, that is, $n = 1$.
2. Conditions in an extruder are isothermal. This is invalid as temperature changes in both the radial and axial directions.
3. Plastic melts are so-called Newtonian fluids; that is, viscosity depends on temperature only, and is therefore constant at each temperature. This assumption is not valid because viscosity even at constant temperature greatly depends on shear rate.
4. The superposition principle applies. This principle applies only for Newtonian fluids, which plastic melts very rarely are, and velocities due to the barrel and those due to an axial pressure drop are not additive. As a result net flow (extruder output) cannot be expressed in two simple terms as drag and pressure flow.
5. The depth, H, is small in comparison with both the diameter, D_b, and the lead, t.
6. There is no leakage over the flight between one turn of the channel and the adjacent ones.
7. The thickness of the flight is negligible.
8. Physical properties of the melt are constant.

Despite the invalidity of the assumptions, the simplified equations are valuable in gaining a semiquantitative insight into conditions of an extruder. This is particularly so because the error in using the superposition principle is zero when pressure flow is absent, that is, at the point where the axial pressure profile passes through a maximum. The simplified equations are at least valuable in indicating the increasing or decreasing nature of the pressure profile in a particular screw geometry under a particular set of operating conditions.

D. EXTRUDER OPERATING AT MAXIMUM CAPACITY

From Eqs. 1–6 it can be seen that at a specified set of operating conditions— screw speed, barrel temperature profile, screw temperature profile, etc.—a screw having a certain geometrical configuration will pump a particular melt in a clearly defined manner. What this means is that there is a single relationship between extruder output and extruder end pressure (head pressure). For each end pressure there is one corresponding output, or conversely, for a particular output the extruder pumps at a specific pressure.

A series of pressure profiles can be developed in an extruder at one screw speed, each end pressure corresponding to a different output. The maximum possible output corresponds to a zero (atmospheric) pressure at the discharge end of the extruder or open discharge.

E. STARVE FEEDING OF EXTRUDERS

The clearly defined relationship between output and end pressure is valid only if the extruder is free to seek its own output level to match either an end-pressure requirement or a die resistance that has a certain relationship between output and pressure drop across the die, commonly referred to as the *die characteristic*.

When exit pressure of an extruder is lowered, the output of an extruder immediately increases. If, however, the material fed to the extruder is metered and kept below the level corresponding to the specific output pressure, the extruder is starve-fed. In this case the output cannot increase to satisfy the lower end-pressure requirement and must remain at the level at which the melt is being supplied to the extruder.

At first this seems to violate the relationship between output and extruder head pressure, which is defined by screw geometry, operating conditions, and physical properties of the melt. The extruder therefore takes internal evasive action, by filling the first portion of the screw channel only partially, causing the pressure to rise only further down-channel. A semifilled channel causes the plastic to lose contact with part of the barrel, resulting in a decrease in the drag flow term, which is normally computed under the assumption of a completely filled channel.

F. TWO-STAGE VENTED SCREWS

In some applications it is found that volatiles such as monomers, moisture, or even air must be removed from the melt at an intermediate point in the extruder. This is done by venting. A vent must be located at a point where the channel is not completely filled and the melt is at or below atmospheric pressure. If this is not the case, plastic tends to invade the vent and may clog it completely or overflow.

Figure 6 represents three typical two-stage screws with channel depth data for 6-in.- and larger-diameter extruders. As can be readily observed from the figure, a two-stage screw is essentially equivalent to two single-stage metering-type screws mounted on a single shaft and driven by one motor and drive. Thus it can almost be referred to as two single-stage screws rotating at the same speed. About single-stage screws we now know that the first portion of the extruder channel—part or all of the feed section and sometimes also the compression or metering section—can be semifilled by starve-feeding.

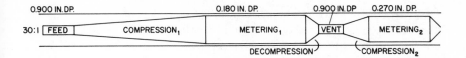

Figure 6 Extruder screw of the two-stage or vented type. Channel depth values are representative for 6-in.- and larger-diameter machines. Courtesy *Plastics Technology*.

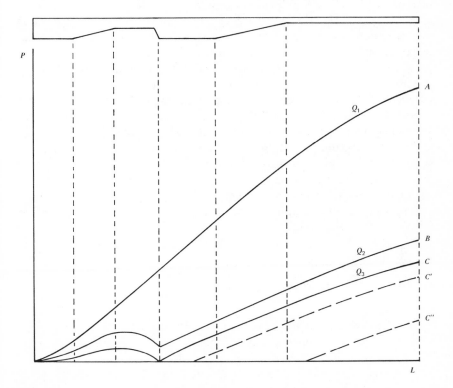

Figure 7 Axial pressure profiles in two-stage screws. Curve A at output Q_1 and curve B at a higher output Q_2 represent conditions under which the extruder can *not* be vented. Curves C, C', and C'' are all at the maximum output Q_3 the extruder is capable of delivering with the particular screw under identical operating conditions (screw speed, etc.); venting is possible. L = the distance along the length of the screw.

Once the channel fills up at any point in the extruder, the pressure can not drop to zero before the end of the extruder. In the case of two-stage screws this limitation applies to the first stage, and pressure cannot drop back to zero before the end of this stage.

Figure 7 represents the performance of a two-stage screw of any geometry. Curve A shows the pressure profile in a two-stage screw with a relatively low output, Q_1. If without changing screw speed we lower the die resistance to flow, or lower the end pressure in the extruder, the result is a higher output (curve B). Lowering the end pressure further, without increasing screw speed, permits a maximum output Q_3 with a pressure profile shown in curve C. It was pointed out that the first stage of the screw is to be regarded as a self-contained unit. Because of this, its maximum output is determined by the pressure profile along the channel in this stage, as is the case in any metering-type screw. Lowering the pressure at the end of the screw further calls for an increased output from the second stage. This call, however, cannot be satisfied by the first stage, which already pumps at its maximum capacity, its end pressure dropping all the way to zero. The end result is a starve-fed second stage with a semifilled vent zone, as can be seen in curves C' and C''. In the latter case the level of the end pressure determines the length of the semifilled portion of the extruder channel, which may even extend to the tip of the screw if operating at open discharge. In the last three cases, namely, C, C', and C'', the extruder is pumping at its maximum capacity for the particular screw speed. The output from the extruder under these conditions is constant and the head pressure or pumping pressure can vary from zero to the one where the channel becomes completely filled, as shown in curve C. The system acquires an additional degree of freedom, namely, the length of the filled portion of the channel.

G.   DESIGN OF EXTRUDER SCREWS

As described above, the axial pressure profile in an extruder at each output for a particular set of operating conditions depends on the screw geometry. The screw geometry has to be carefully selected, therefore. To understand the difference in performance it must be realized that the so-called screw characteristics, that is, the line describing the output of the extruder as a function of the pressure rise along the extruder, is a straight line under isothermal conditions, provided that the plastic melt also behaves as a Newtonian fluid. Figure 8 shows that if ΔP is zero, the melt flow is the highest and equals the drag flow. As the pressure rise along the extruder grows, so does the pressure flow, which being in the opposite direction, lowers the net flow

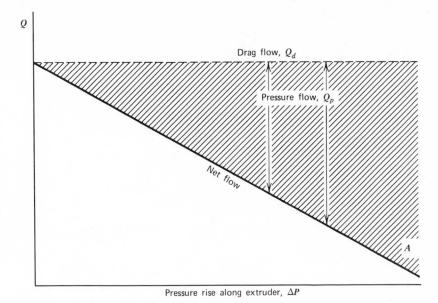

Q

Drag flow, Q_d

Pressure flow, Q_p

Net flow

A

Pressure rise along extruder, ΔP

Figure 8 Screw characteristic (output vs. pressure rise) of a melt extruder pumping a Newtonian fluid under isothermal conditions (Curve A). The channel depth is constant along the extruder. The dotted line represents the drag flow (independent of pressure) and the area between drag and net flow represents the pressure flow.

(curve A). By consulting Eq. 4 it becomes clear that the value of drag flow depends only on screw geometry and screw speed and not on the change in pressure. Drag flow can accordingly be represented in Fig. 8 as a line parallel to the ΔP axis. Pressure flow, on the other hand, increases with ΔP and is represented by the crosshatched area. Because the pressure flow has a negative sign, its value is negative when pressure rises along the extruder (ΔP positive) and therefore has to be subtracted from the drag flow to yield the net flow (curve A).

It is important, however, to understand the effect of screw geometry on the performance of the extruder in order to be able to improve the performance of existing equipment or to select new screws.

Figure 9 shows that the drag flow (value of Q when $\Delta P = 0$) for the deeper channel is higher than that of the shallow channel. This is so because, as can be seen in Eq. 5, α is proportional to the channel depth H. On the other hand, Eq. 6 shows that β is proportional to H^3 and therefore the slope of the Q versus ΔP line is bigger for the deeper channel. As a result the two lines

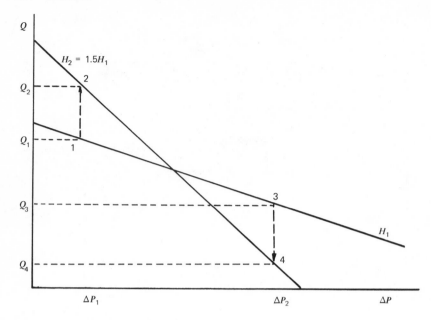

Figure 9   Screw characteristic curves of a shallow and a deep screw show that deepening the screw increases output when operating at ΔP_1 pressure level, but lowers output when operating at ΔP_2 level.

intersect. This causes a behavior which greatly depends on the actual location of the screw characteristic at which we are operating. For example, deepening the screw from H_1 to H_2 results in a change in performance from point 1 to point 2 if operating at the ΔP_1 level. As can be observed in Fig. 9, the corresponding output for the same pressure rise (ΔP_1) increases from Q_1 to Q_2. Operating the same two screws at the ΔP_2 level shows that deepening the screw means change in performance from point 3 to 4 or a decrease in output from Q_3 to Q_4. The effect of deepening the screw depends on whether we operate to the left or the right of the crossover point of the two screw characteristic lines.

Furthermore, the location of the crossover point also depends on the operating conditions. Figure 10 shows the effect of channel depth on the screw characteristic line at two levels of screw speed. As can be seen, the shapes and location of the screw characteristic lines change with screw speed, and so does the crossover point between the two lines representing two channel depths.

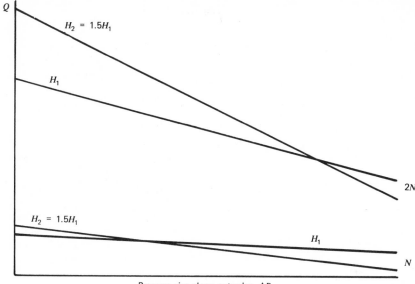

Figure 10   Screw characteristic curves of two screws with different channel depths at two screw speed levels show the effect of screw speed on the crossover point of screw characteristic curves.

H.   EFFECT OF DIE CHARACTERISTICS ON EXTRUDER PERFORMANCE

Up to this point the operation of an extruder has been described at specified outputs or alternately pumping against a specific pressure. The pressure against which the extruder is to pump depends on the die characteristics or the relationship between output and pressure drop along the die. This relationship can be expressed as

$$Q = k_d \frac{\Delta P}{\mu_d} \qquad (7)$$

where k_d is the die constant (conductance), which depends on the die geometry and μ_d is the average viscosity in the die. The graphic representation of the die characteristics in Q versus ΔP coordinates similar to those used to represent screw characteristics in Figs. 8 to 10 is a straight line through the origin. Dies with a large die constant (low resistance to flow) at a certain pressure drop have a higher flow rate than dies with a small die constant at

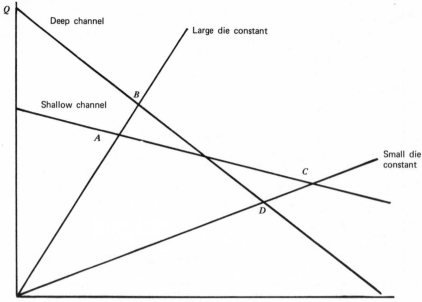

Figure 11 Effect of screw geometry on the pumping and pressure raising capacity of an extruder operating with two different dies. Deepening the channel with a large die characteristic increases both output and head pressure from *A* to *B*. The same deepening of the channel with a small die characteristic lowers both output and head pressure from *C* to *D*.

the same pressure drop. This together with the screw characteristics can be seen in Fig. 11, which reveals that deepening the screw channel, using a die having a large die constant, increases output from *A* to *B*. On the other hand, a die having a small die constant shows a lowering of output from *C* to *D* when the screw channel is deepened.

I.   EFFECT OF TEMPERATURE ON EXTRUDER PERFORMANCE

The above description of the extruder as a melt pump refers to operation under isothermal conditions. These, however, can seldom be maintained in actual operation. The major effect of temperature is on viscosity, and this appears in the denominators of both the pressure-flow term (Eq. 2) and of the die characteristics. Because increase in temperature means a lower viscosity value, the slope of the die characteristics as well as of the screw characteristics increases with temperature. The extruder output is seriously affected by temperature changes because its performance is very sensitive to viscosity variation. For this reason we cannot overemphasize the need for very precise temperature control in all extrusion operations.

J. NON-NEWTONIAN MELTS

Very few plastic melts behave as Newtonian fluids. As pointed out earlier, the equations, which were derived for Newtonian materials, do not really apply. The size of the error depends on how non-Newtonian the melt really is. For a Newtonian material

$$\tau = \mu \dot{\gamma} \tag{8}$$

where τ = shear stress (dyne/cm^2)
μ = viscosity (dyne-sec/cm^2)
$\dot{\gamma}$ = shear rate (sec^{-1})

A non-Newtonian material is often described by the power law,

$$\tau = \eta^0 \left| \frac{\dot{\gamma}}{\dot{\gamma}^0} \right|^{n-1} \dot{\gamma} \tag{9}$$

where η^0 and n are characteristic for the material and $\dot{\gamma}^0$ is the standard state. The degree of non-Newtonian character is expressed in n. For a Newtonian fluid $n = 1$, whereas for a non-Newtonian fluid n is between 0 and 1. The lower the value of n, the more non-Newtonian the fluid. The non-Newtonian character of the plastic melt causes both the screw characteristics and die characteristics to curve.

K. THE PLASTICATING EXTRUDER

Understanding the performance of a melt pump with a Newtonian fluid is not sufficient to grasp the reality in any actual extruder. Most extruders are not melt pumps. They are fed with an unmelted plastic, and the most important task of the extruder is to melt the solid plastic, fed into the extruder as pellets, flakes, powder, etc., at a desired rate and to generate an extrudate of satisfactory quality. Because the most important objective of these extruders is melting or plasticating, they are called plasticating extruders.

The actual melting mechanism and progress of melting in a plasticating extruder can be observed in the following experiment. The extruder is stopped and the barrel and possibly also the screw are cooled. The screw is pushed out of the barrel and the solidified plastic is taken off the screw.

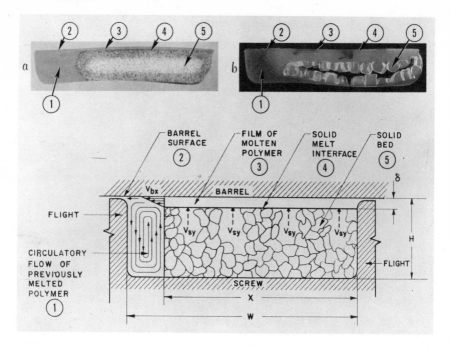

Figure 12 Experimentally obtained channel cross section for (*a*) rigid PVC powder, and (*b*) low-density polyethylene and their idealized counterparts (1). Courtesy Van Nostrand-Reinhold Company.

Mixing some colored plastic into the feed aids in visualizing the melting mechanism in a screw extruder. Cutting sections perpendicular to the down-channel or flow direction of the helical ribbon peeled off the screw, one finds that the cross section of the channel contains both melted and unmelted plastic (Fig. 12). Figure 12*a* is a cross section of the extruder channel extruding a rigid PVC powder, whereas Fig. 12*b* is that of a pelletized low-density polyethylene. The schematic drawing of the channel cross section in the melting zone reveals that three areas can be distinguished: the melt film on the barrel surface, the solid bed, and the melt pool next to the solid bed.

L.   MELTING IN A CONSTANT DEPTH CHANNEL

The melting zone starts at the point where the film of melt formed on the barrel surface increases in thickness to such a degree that melt starts to accumulate in front of the advancing flight, not only in the radial direction (melt film) but also in the cross-channel direction, and forms the melt pool. Melting is caused by heat being conducted from the barrel into this film, as well as by the viscous heat dissipation induced by constantly shearing this thin film. The melt film thickness is always several times the flight clearance when the melt pool first appears. The viscous heat dissipation is

$$E = \dot{\gamma}\tau \tag{10}$$

or

$$E = \mu\dot{\gamma}^2 \tag{11}$$

and roughly depends on the square of the shear rate, which in itself linearly decreases with film thickness.

$$\dot{\gamma} = \frac{\Delta V_b}{\delta_f} \tag{12}$$

$$E = \mu\left(\frac{\Delta V_b}{\delta_f}\right)^2 \tag{13}$$

where ΔV_b = velocity difference between barrel and solid bed (cm/sec)
δ_f = film thickness (cm)

The thickness of the film is always larger than the flight clearance and depends on the flight clearance. An increase in flight clearance can therefore greatly reduce the melting capacity of an extruder.

At the point where the steady-state melting mechanism starts to operate the advancing flight scrapes off a portion of the melt film and the melt is forced to accumulate in front of the advancing flight. A melt pool is formed at the side of the solid. Moving in the down-channel direction the melt pool increases in width, exerting a pressure on the solid bed and forcing it to continuously deform and supply solids to contact the melt film and barrel surface and to replace the newly melted plastic in the interface between the solid bed and the melt film. The solid bed is continuously "flowing" into the melt film, in addition to its movement in the down-channel direction. The width of the solid bed is, therefore, decreasing continuously from the hopper to the die.

The melting mechanism becomes considerably more complicated when the channel geometry changes with axial location. In a compression section, where the cross section of the channel decreases, the average down-channel velocity of the melt and solid material must increase (provided density changes do not compensate for the decrease in channel cross section). However, it was experimentally found that the solid bed tends to resist acceleration. Since the "bottom" of the channel rises in the tapered section, it causes the solid bed to move toward the melt film. Experimental evidence shows that, despite the decrease in channel cross section, the solid bed retains its low velocity as it proceeds down channel; thus part of the solid melting in the film is being resupplied by this decrease in channel depth. Therefore, less solid has to be pushed from the side of the solid bed to replace the melting solid. As a result, in a compression section, the width of the solid bed (X) decreases at a lower rate than it does in a constant-depth channel. However, this does not reduce the rate at which melting takes place. Quite the contrary is true. The melting rate increases with the area of the solid bed exposed to the melt film on the barrel surface and melting ends at an earlier point in the extruder.

This situation causes the melt in the melt pool, which is now generated at the higher rate, to flow at even higher velocities in the down-channel direction than would be the case if the whole cross section were to be accelerated. This is so because the cross-sectional area available to the melt flow is now decreased even more than the total cross section. Higher melt velocities cause more heat generation in the melt pool and consequently result in higher power consumption and higher melt temperature. Not all plastic materials can withstand these high temperatures, and caution has to be exercised in design of the melting zone. Rigid polyvinyl chloride is a good example of a sensitive plastic material in this respect. (Owing to this heat sensitivity the barrel of modern extruders as well as their screws are equipped to cool and also to heat the material when necessary to achieve an acceptable melt temperature.)

If the compression section is short and the compression ratio relatively high, more solid is supplied to the melt film than the extruder is able to melt. In such a case, the width of the solid bed increases with the down-channel direction and can even cause the whole channel cross section to be occupied by the solid bed. The slow moving "solid plug" formed toward the downstream end of the compression section prevents the melt pool from flowing in the down-channel direction. A pressure drop is therefore generated across the plugged portion of the channel. This pressure drop increases with time, as melting continues upstream of the solid plug. As soon as the pressure drop overcomes the strength of the plug, the plug is blown out in the down-channel direction, causing large variation in output commonly referred to as *surging*.

Another explanation of the same phenomenon is that as soon as the channel plugs with unmelted plastic, the output suddenly drops. This is so because of the considerably lower velocity of the solid bed than of the melt pool. The melt formed upstream of the plug is now prevented from reaching any point downstream of the plug. The sudden decrease in extruder output, at otherwise identical operating conditions, such as screw speed and barrel temperature, results in a sudden improvement in melting without plugging the channel in the compression section. As soon as the channel unplugs, output increases, leading back to the conditions that caused plugging originally. The end result is a cyclic variation in output, that is, surging. Pressure, temperature, and quality fluctuations of the extrudate are experienced at the discharge end of the extruder. Such a situation, which is an extreme condition, is detrimental to most operations and interferes with product uniformity and extrudate quality.

M. MELTING IN A PLUGGED CHANNEL

The length of the channel which is packed with solid does not permit a melt pool to exist. In the absence of the melt pool in the solid-filled portion of the melting zone, the advancing flight is unable to scrape off the melt film from the barrel surface, because no room is left for the melt pool to accumulate. The newly formed melt must remain in the melt film, which becomes thicker with time. The thickness of the melt film also lowers the shear rate in the melt film, as was shown earlier. It was also pointed out that viscous heat dissipation in the melt film is roughly proportional to the square of the shear rate. As a result, the rate of melting rapidly diminishes in the solid-filled portion of the melting zone. In certain cases, when the solid bed is soft and resistance to acceleration not too high, the pressure developed by the melt pool may be high enough to force the solid bed to accelerate even prior to its filling up the whole channel cross section; therefore surging does not develop. It must be pointed out that the rate of melting increases with an increase of the taper angle in the compression section of an extruder. However, exceeding the optimum taper angle even slightly results in an unfavorable plugging of the channel with solids, which can often result in a detrimental phenomenon of surging. In such a case, if sudden pressure surges also develop, the extruder head may blow, or damage to the barrel may occur. The geometry of the melting zone of a screw extruder has to be carefully designed, therefore, and the barrel temperature precisely controlled.

N. ALTERNATE MELTING MECHANISM

Several authors report that in certain cases the melting mechanism observed is different from the one described above. This is particularly evident in extruding certain polyvinyl chloride formulations. The most striking characteristic of this alternative melting mechanism is that the melt pool does not form at the pushing side of the flight, but rather at the trailing side. The reason for this phenomenon is the rather small cross-channel pressure gradient existing in these cases, which is inadequate to deform the solid bed and cause the melt to penetrate between the pushing side of the flight and the solid bed to form the melt pool. The result of this is that the melt remains in the melt film, except the portion leaking through the flight clearance, which then accumulates behind the trailing side of the flight. In effect this is *not* an alternate melting mechanism, but the extension of the *delay zone* to the whole length of the extruder.

O. SOLIDS CONVEYING

The degree to which solids are conveyed in the first portion of the extruder channel depends on the geometry of this zone, as well as on the friction coefficients on the barrel and screw surfaces, on bulk density, and on screw speed. The friction coefficients depend on the metal, its roughness, and its temperature and the pressure developed at each point in the channel. The capacity of a plasticating extruder or a molding machine can be limited by the efficiency of the solids-conveying zone. The importance of the proper design for a solids-conveying zone, therefore, cannot be stressed sufficiently. Only if pressure rises in the solids-conveying zone is it able to pump solids at the particular rate. If, on the other hand, pressure along this zone rises only to a very small degree, the pressure developed is insufficient to compact the solid bed. In such a case, as soon as melting starts, the channel becomes semifilled, causing the pressure profile across the rest of the extruder to be significantly lowered, and also lowering the pressure drop along the die. The lower pressure drop along the die also results in lowering the output of an extruder–die combination.

The solids-conveying zone starts at the downstream end of the hopper and continues to the point where the surface of the solid bed in contact with the barrel heats up to the melting point of the plastic. This can be caused by heating the barrel surface or by heat generation as a result of friction on the cold barrel surface. The ratio of the pressure at the end of the solids-conveying zone to the pressure at its beginning depends exponentially on the length of the solids-conveying zone.

The performance of the solids-conveying zone can be improved by increasing the coefficient of friction on the barrel surface, and by lowering it on the screw surface. This is often accomplished by controlling the temperature of these surfaces rather accurately. Various methods have been devised, such as cutting grooves in the barrel in the axial direction, maintaining the barrel diameter larger around the hopper end, and conically decreasing it along the solids-conveying zone.

Increasing the bulk density of the solid also improves the solids-conveying capacity. This is often accomplished by forced feeding, which unfortunately creates a side effect by preventing the air from between the solid pellets or powder to escape through the hopper. In many cases this results in the air being released in the product and destroying its appearance. This problem is often eliminated by using two-stage vented screws when forced feeding is applied to enable the air to escape through the vacuum vent.

The length of the cooled but unheated barrel from the downstream end of the hopper (throat area) also determines the performance of the solids-conveying zone. When it is too short, it is incapable of developing sufficient pressure to compact the solid bed; on the other hand, if it is too long, it may cause the plastic to melt owing to friction before reaching the heated portion of the barrel. In the latter case power consumption may become excessive and may even exceed the power actually available for the extruder from the electric motor. As a result the screw speed will have to be lowered, which will also greatly reduce the melting and pumping capacity of the extruder.

1. Hopper

Because of the exponential dependence of the pressure ratio on the length of the solids-conveying zone, the actual value of the pressure at the input end of the solids-conveying zone, namely, at the bottom of the hopper, acquires special importance. Small fluctuations in this initial pressure can in certain cases cause enormous fluctuations at the end of the solids-conveying zone, which are transmitted all the way to the die at the end of the extruder. The hopper has to be designed in such a way as to minimize these pressure fluctuations. Also, the level of solid feed, whether pellets, flakes, powder, etc., has to be maintained above a predetermined level so that small fluctuations in the height of the solid plastic column in the hopper do not affect the pressure below this column to any significant degree.

2. Delay Zone

As soon as the temperature of the solid plastic plug in contact with the barrel reaches the melting point of the plastic, a melt film is formed on the barrel surface. However, the so-called melting zone does not start at this point, but considerably farther down the channel. The reason for this is that the thickness of the film on the barrel surface must exceed the flight clearance before the advancing flight is capable of scraping part of the melt film off the barrel surface. This zone forms a natural transition between the solids-conveying zone, where the solid bed advances as a plug and slips on the surface of both the *screw* and *barrel*, and the melting zone, where the melt coats both surfaces. In the delay zone the solid plug usually rubs against the screw, yet on the barrel surface the melt film adheres to the metal exhibiting a viscous drag instead of friction. The delay zone is several diameters long and can now be accurately computed for any plastic in any size of extruder.

P. MIXING DEVICES

It is often found that complete melting cannot be obtained under any circumstances in a screw extruder. Various devices have to be incorporated into a screw design to permit rapid heat transfer from high-temperature melt to the relatively low-temperature solid. To accomplish this the continuity of the solid bed must be terminated and the heat transfer area between solid and melt or the solid and the barrel surface dramatically increased.

Figure 13 shows typical mixing rings installed perpendicular to the screw axis. The flights are interrupted and the number of "holes" usually varies according to the need to keep the pressure drop along the ring low. Mixing devices affect the rate at which melting occurs. They also affect the temperature of the melt in the melt pool and the extrudate temperature. The reason for this is that mixing devices break up the continuity of the solid bed, cause the solids to be mixed in with the melt, and cause melting due to conduction from the melt to occur at considerably higher rate than in normal extrusion operation. Because melting occurs in this case, owing to conduction from the

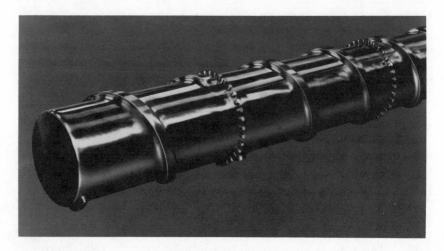

Figure 13   Mixing ring.

melt pool, it also results in lowering the temperature of the melt pool. The degree of this temperature lowering depends on the latent heat of fusion, heat capacities of polymer melt and solid, solids content at the particular axial location, and melting efficiency of the mixing device. The location and geometry of a mixing device are therefore very important; inserting it at a location where very little melt exists does little good. Indeed, in this case more harm than good can result from a mixing device. The same may apply to mixing devices located in the melt pumping zone where no unmelted plastic is present.

Mixing devices can be and often are designed to create the necessary pressure drop needed in two-stage screws of vented extruders where atmospheric pressure must be ensured around the vent in order to prevent extrusion of plastic melt into the vent. The plastic might solidify in the vent and cause plugging of the vent and would not permit a vented operation.

Q. SURGING IN EXTRUDERS

Cyclic variations in extruder head pressure and, as a result, also in production rate and extrudate quality, are often experienced. There are two reasons for surging in extruders: the first is the plugging of the channel by unmelted plastic; this has been described. The second reason for surging is solids conveying. When conveying of solids controls the operation, and the pressure buildup in the solids-conveying zone is inadequate to compact the solid bed, the melt film formed penetrates the voids in the loosely packed solid bed, thereby greatly decreasing its volume. As a result, a portion of the melting zone is starved. This starved or semifilled channel starts at the point where the melt film starts to form on the barrel surface and may continue into the compression section to the point where channel cross section becomes filled with plastic and pressure starts to rise in the channel. The length of the semifilled channel depends on the head pressure determined by the die. Small cyclic changes in barrel temperature can make solids conveying erratic and the length of the semifilled portion of the channel can greatly vary with time. This situation can be further aggravated by changes in the level of solids in the hopper. Since this phenomenon causes the length of the filled channel to vary, it also causes severe variations in the pressure level at any point in the extruder and therefore in the extruder head.

R. MELTING INSTABILITY

In many cases the extrudate quality varies with time without actually causing pressure changes. This phenomenon originates in the melting zone where the

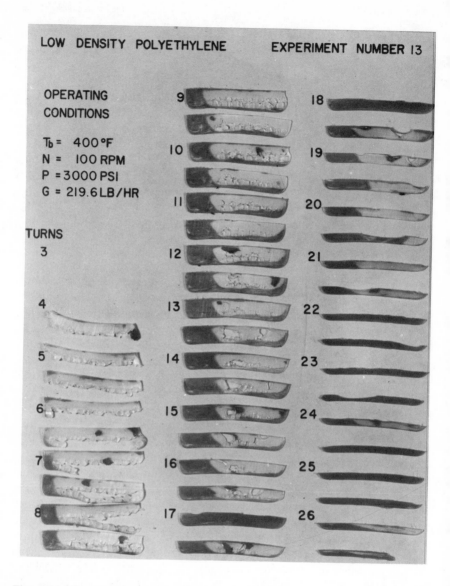

Figure 14 Cross sections obtained from a cooling experiment (1). Courtesy Van Nostand-Reinhold Company.

482

solid bed breaks up and becomes discontinuous. This can be clearly seen in cooling experiments, where cross sections of the solidified plastic at regular intervals show the channel to be completely filled with melt, yet farther down channel the solid bed reappears. Figure 14 shows mounted cross sections for a low-density polyethylene assembled from a cooling experiment. The channel is melt-filled at 17 turns (diameters), 18 turns, 20.5 turns, from 22 through 23.5 turns, and from 24.5 through 25.5 turns, whereas the solid bed reappears in between.

Melting instability can be remedied by strengthening the solid bed, for example, cooling the screw, increasing compression ratio, or permitting the melt in the melt pool to move at a lower velocity, thereby decreasing the force acting on the solid bed. Various screw modifications have been suggested to eliminate melting instability or the breaking up of the solid bed. Maillefer, for example, installs an auxiliary flight totally blocking the channel and forcing the plastic melt to pass over the flight clearance.

It has been found, however, that when the solid bed breaks up, the pressure-building capacity of the extruder is considerably higher than when it remains continuous. As a result it seems more beneficial to encourage the breakup of the solid bed than to prevent it. Various devices installed close to the end of the extruder to prevent unmelted solid particles from reaching the die are therefore often more desirable than those that prevent melting instability.

S. COMPUTERIZATION OF THE SINGLE-SCREW EXTRUDER

It can be seen that, contrary to popular belief, the plasticating extruder is not merely a simple pump, but a complex system comprised of a slow moving solid in a hopper, a solids-conveying zone, delay in melting zone, melting zone, melt-pumping zone, and often also a vent zone. Every portion of the extruder contributes very significantly to the pumping and pressure-raising capacity of the extruder. However, they also all contribute to plasticating, mixing, heat transfer, and viscous heat dissipation. The interaction of these portions is therefore so complex that no manual computation is really capable of describing what is going on in an extruder. A computerized physical model Extrud* had to be constructed, therefore, to show the intricacies of the whole process and the effect of each of its components on the overall performance of an extruder. This computerized model of the extruder permits us to computer simulate the operation of any size of extruder under any set of operating conditions, extruding any resin. The simulator is also applicable to blow molding and injection molding.

* Trademark of Scientific Process & Research, Inc.

T. TWIN-SCREW EXTRUDERS

In certain special applications, such as rigid PVC, it was found that single-screw extruders heat the plastic to a higher temperature than it is capable of withstanding. However, certain twin-screw extruders do not overheat the plastic. Several types of twin-screw extruders are on the market.

1. Nonintermeshing Screws

In this type of extruder the pressure versus output relationship is similar to that in single-screw extruders and resembles, therefore, the operating characteristics of a centrifugal pump.

2. Intermeshing Co-Rotating Screws

In this type of extruder the screw profile is often conjugated; that is, no gap is left between the screws. As a result tolerances between screws, as well as between barrel and screw, can be smaller, reducing the residence time of the plastic and reducing thermal degradation of particularly sensitive grades. The material does not pass between the screws, but as soon as it reaches the other screw, continues to travel down the channel with it.

3. Intermeshing, Counter-Rotating Screws

Here the two screws penetrate each other, but their centers are slightly more than a screw diameter apart. Since the plastic material passes the clearance between the screws it becomes less of a positive-displacement pump and its pumping capacity depends somewhat on the pressure. In addition the existence of the clearance also prevents one screw from wiping the other clean.

Twin-screw extruders combine heat generation by viscous heat dissipation with mixing in the high-temperatue melt with the solid and probably using up this sensible heat to melt some of the remaining solid plastic. Thus they tend toward more uniform temperatures with greater control. Their disadvantage, however, is their considerably higher cost, compared to single-screw extruders, and the lack of a thorough understanding of what is really occurring inside the extruder during processing.

The availability of a good theoretical model for the single-screw extruder enables us to locate mixing devices strategically and predict their effects. As a result, in many cases the design of screws using modern computerized methods can eliminate the need for expensive twin-screw extruders.

U. EXTRUSION DIES

The extruder is the heart of most plastics processing lines, but it is by no means the only component. Although the extruder melts the plastic and raises its pressure to required level, the die ensures that the plastic product acquires the desired shape.

Although no melting should take place in a die, nor does the die pump plastic melt, the flow pattern in a die may be quite complex and its mathematical analysis presents a formidable problem. The reason for this is, as in the case of extruders, that plastic melts are non-Newtonian fluids and their viscosity is not constant at a constant temperature. Even the assumption of a power-law fluid, namely, that the function of logarithm of viscosity against the logarithm of shear rate can be represented as a straight line with a negative slope, is valid over only a limited (one to three decades) range of shear rates. Combining the complex flow behavior of the melt with the complex shape makes a simple analytic solution to a die problem impossible. Luckily the high-speed digital computer comes to the rescue, for it permits a numerical solution to flow problems in dies. This introduces another problem, namely, that of computer programming, which in some cases may be very complex.

To illustrate some of the problems associated with die design Fig. 15a shows a sheet die and Fig. 15b shows the die halves. This die consists of a manifold that carries the plastic melt from the extruder to the distributing section. To obtain a sheet or film of constant thickness across the sheet, the melt must be uniformly distributed. This implies that the pressure drop through the lips at the center of the die equals that at the edges. The manifold, which is at an angle, must change diameters from the center of the die to its edge. Naturally, one such design would only ensure a completely uniform distribution of a single melt, at one particular production rate and at a single temperature. To make the die somewhat more versatile, it is usually equipped with a dam or choker bar which compensates, to a certain degree, for the drop in pressure along the manifold.

Some applications of extrusion using particular die design, that is, tubing and profiles, are shown in schematic drawings of the entire system in Fig. 16.

Film extrusion may be carried out using a flat die similar to that shown in Fig. 15 or through a tubing die with a blowing arrangement which causes the soft, semimolten tube to balloon out as shown in Fig. 17. This tube of plastic is oriented by the blowing action and is taken up on rolls as a flat collapsed thin-walled tube.

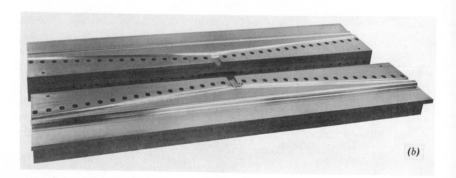

Figure 15 (a) 130-in. sheet die. (b) Inside of 130-in. sheet die. Courtesy Sterling Extruder Corporation.

Tubing Extrusion Line Using Vacuum Sizing

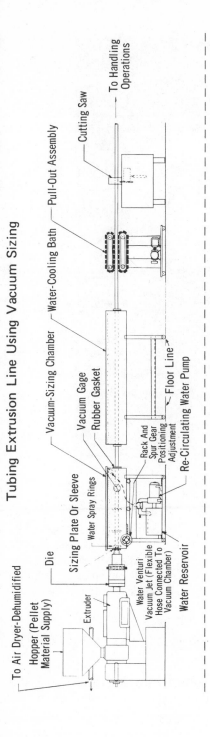

Profile Extrusion Line

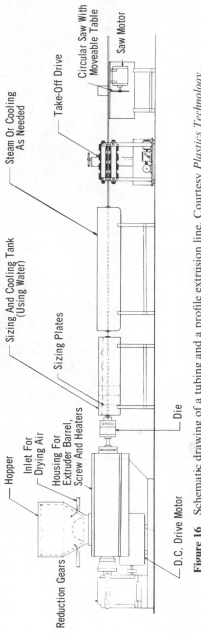

Figure 16 Schematic drawing of a tubing and a profile extrusion line. Courtesy *Plastics Technology*.

487

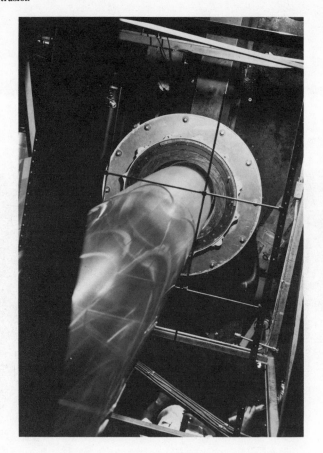

Figure 17   Polyethylene blown film extrusion. Courtesy Union Carbide.

Fiber extrusion is a major application of the extrusion technique. The general concept is essentially the same as previously discussed. The dies are very complex and the heating and cooling steps are very critical because they affect the degree of orientation and strength as well as the hand of the resulting fibers. This application is not discussed further here because of its complexity.

V. SOME MOLECULAR ASPECTS OF EXTRUSION

From a molecular point of view the process of extrusion involves the conversion of a solid mass of entangled and intertwined molecules, which may or may not be crystalline, into a viscous, formable liquid mass which emerges from the die. This is accomplished by external heat input along the barrel of the extruder and by development of frictional heat as the screw covers the material with shearing forces.

The nature of the process requires that the hot melt have good form and strength stability as it emerges from the die. Without this the product would not retain its desired shape as it is taken up and cooled. This requires relatively high molecular weight, long-chain polymeric materials (in accordance with the melt viscosity–molecular weight relationship in Section E of Chapter 8). In general, therefore, extrusion operations require higher molecular weight polymers than are used in injection or compression molding, where the melt is contained in a closed mold until it is cooled.

As the extrudate emerges from the die it is usually hauled off with some degree of tension applied by the take-up equipment. At the same time it is cooled by a water bath, chill rolls, or cold air jets. The nature of the cooling curve is important, especially for crystalline polymers, in that more or less molecular order or crystallinity can be developed in the part. Slow cooling gives higher crystallinity; rapid cooling tends toward lower crystallinity.

The degree of tension on the take-up equipment may also result in orientation of the extrudate beyond what may be already present owing to the shear forces, which tend to align the molecules in the direction of flow.

Accompanying the plasticization (liquefaction or melting) of the polymer mass by head and shear is the ever-present possibility of molecular degradation. The net effect may then be to produce a product that has a lower molecular weight than desirable for optimum properties, and a higher degree of orientation (because shorter chains are easier to orient) along the axial or extrusion direction, which may be deleterious to the product properties.

An example of this phenomenon is the extrusion of pressure pipe. With excessive degradation and shear the molecular chain length is too low to assure maximum entanglements, which is necessary for toughness, and axial orientation leads to weakness in a radial direction so that the pipe will have a low burst strength.

REFERENCES

1. Z. Tadmor and I. Klein, *Engineering Principles of Plasticating Extrusion*, Van Nostrand-Reinhold, New York, 1970.

2. A. L. Griff, *Plastics Extrusion Technology*, Van Nostrand-Reinhold, New York, 1968.

3. W. Mink, *Grundzuge der Extruder-technik*, Rudolf Zechner Verlag, Speyer am Rhein, 1966.

4. E. G. Fischer, *Extrusion of Plastics*, Iliffe, London, 1964.

5. G. Schenkel, *Kunststoff—Extrudertechnik*, Carl Hauser, Munich, 1963.

6. J. M. McKelvey, *Polymer Processing*, Wiley, New York, 1962.
7. H. R. Jacobi, *Grundlagen der Extrudertechnik*, Carl Hauser, Munich, 1960.
8. E. C. Bernhardt (Ed.), *Processing of Thermoplastic Materials*, Van Nostrand-Reinhold, New York, 1959.

DISCUSSION QUESTIONS AND PROBLEMS

1. Review the various sections of an extruder screw and indicate the function of each.
2. Problems associated with single-screw extrusion are:
   a. Shear or high temperature degradation of the polymer.
   b. Orientation of the extrudate along the extrusion axis.
   Can you suggest ways of minimizing these effects? (Include die design and extruder modifications in the consideration.)
3. What is the purpose of a vent on an extruder?
4. Review the problems associated with conveying of solids in an extruder.
5. Discuss some of the considerations associated with die design. Include orientation and output considerations.

CHAPTER 10

Injection Molding*

IRVIN I. RUBIN
Robinson Plastics
Jersey City, New Jersey

* Material in this chapter is from Irvin I. Rubin, *Injection Molding—Theory and Practice*, Wiley, New York, 1972.

Injection molding is a major processing technique for converting thermoplastics, and now thermosetting materials, into all types of products. Approximately 25 % of the 13 billion lb of thermoplastics sold in the United States in 1971 were injection molded, and about 36 % (4320) of the 12,000 processing plants in the United States injection molded (1). By 1976 the amount of thermoplastic sold almost doubled to 24 billion lb. Furthermore, in 1970 about 5000 injection molding machines were purchased in this country, which brought the total of injection machines in-place to about 58,000. Since there were 130,000 processing machines, injection machines represent close to 45% of all processing units.

Sixty percent of the machines use a reciprocating screw, 35% a plunger (concentrated in the smaller machine sizes), and 5% a screw pot.

Injection molding requires four things, material, mold, machine, and management. Some aspects of the last three of these are covered in this chapter.

A. INJECTION MACHINE—DESCRIPTION AND OPERATION

Figure 1 shows a reciprocating screw injection molding machine with a clamping capacity of 250 tons, using a 2-in. reciprocating screw which delivers a maximum of 13 oz of polystyrene per shot. The machine basically is a tool for the following.

1. Raising the temperature of the plastic to a point where it will flow under pressure.
2. Allowing the plastic to cool and solidify in the mold, which the machine keeps closed.
3. Opening the mold to eject the solid plastic.

The injection molding machine is an instrument for performing the above functions automatically under controllable conditions of temperature, time, and pressure so that quality parts may be manufactured on an economical basis. The nature of the process is such that the molecular structure, the molecular weight, and the molecular weight distribution (all of which control melt viscosity), orientation of the polymer molecules, and crystallizability play an important role and must be considered for their effect on product properties.

Figure 2 is a schematic representation of the clamping end of an hydraulic machine, and Fig. 7 shows the injection end of an in-line reciprocating screw plasticizing unit. The injection side of the mold is clamped to the stationary platen, and the ejection side of the mold is clamped to the moving platen. The mold has an empty space in the configuration of the part to be molded. This empty space is filled with melted plastic under high pressure.

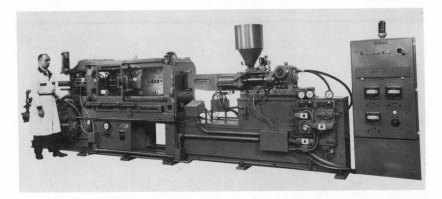

Figure 1   2-in. reciprocating-screw, 250-ton injection molding machine. Courtesy Lester Engineering Co.

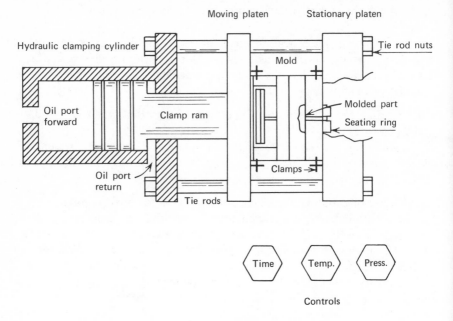

Figure 2   Schematic drawing of hydraulic clamping system.

The moving platen rides on four steel bars called tie rods (or tie bars). The clamping force is generated by the hydraulic mechanism pushing against the moving platen and stretching the tie rods.

The steps of the molding process for a reciprocating screw machine with hydraulic clamp follow.

1.   Plastic material is put into the hopper. (The virgin powder is normally granulated to $\frac{1}{8}$–$\frac{3}{16}$ in. spheres or cubes.)

2.   Oil behind the clamp ram moves the moving platen, closing the mold. The pressure behind the clamp ram builds up, developing enough force to keep the mold closed during the injection cycle. If the force of the injecting plastic material is greater than the clamp force, the mold opens, an unacceptable condition causing plastic to flow past the parting line on the surface of the mold, producing "flash," which has to be removed or the piece has to be rejected and reground.

3.   The two hydraulic injection cylinders now bring the screw forward, injecting previously melted material into the mold cavity. The injection pressure is maintained for a predetermined length of time. Most of the time there is a valve at the tip of the screw that prevents material from leaking

into the flights of the screw during injection. It opens when the screw is turning, permitting the plastic to flow in front of it.

4. The oil velocity and pressure in the two injection cylinders develop enough speed to fill the mold as quickly as needed and maintain sufficient pressure to mold a part free from sink marks, flow marks, welds, and other defects.

5. As the material cools, it becomes more viscous and solidifies to the point where maintaining injection pressure is no longer of value.

6. The material is melted primarily by the turning of the screw which converts mechanical energy into heat. It also picks up some heat from the heating bands on the plasticizing cylinder (extruder barrel). As the material melts, it moves forward along the screw flights to the front end of the screw. The pressure generated by the screw on the material forces the screw, screw drive system, and the hydraulic motor back, leaving a reservoir of plasticized material in front of the screw. The screw continues to turn until the rearward motion of the injection assembly hits a limit switch, which stops the rotation. This limit switch is adjustable, and its location determines the amount of material that remains in front of the screw (the size of the "shot").

The pumping action of the screw also forces the hydraulic injection cylinders (one on each side of the screw) back. This return flow of oil from the hydraulic cylinders can be adjusted by the appropriate valve. This is called "back pressure," which is adjustable from 0 to about 400 psi.

7. Most machines retract the screw slightly at this point to decompress the material so that it does not "drool" out of the nozzle. This is called the "suck back" and is usually controlled by a timer.

8. Heat is continually removed from the mold by circulating cooling medium (usually water) through drilled holes in the mold. The amount of time needed for the part to solidify so that it might be ejected from the mold is set on the clamp timer. When it times out, the movable platen returns to its original position, opening the mold.

9. An ejection mechanism separates the molded plastic part from the mold and the machine is ready for its next cycle.

The process is not new. John and Isiah Hyatt received a patent in 1872 for an injection molding machine, which they used to mold camphor-plasticized cellulose nitrate (celluloid). In 1878 John Hyatt introduced the first multicavity mold. In 1909 Leo H. Baekeland introduced phenol-formaldehyde resins which are now injection moldable with the screw molding machine.

The experimental and theoretical works of Wallace H. Carothers led to a general theory of condensation–polymerization that provided the impetus for the production of many polymers, including nylon. At the end of the 1930s modern technology began to develop and great improvements in materials permitted injection molding to become economically viable. A similar advance in machine technology is developing now.

There are both advantages and disadvantages to injection molding. Advantages are as follows:

1. Parts can be produced at high production rates.
2. Large volume production is possible.
3. Relatively low labor cost per unit is obtainable.
4. The process is highly susceptible to automation.
5. Parts require little or no finishing.
6. Many different surfaces, colors, and finishes are available.
7. Good decoration is possible.
8. For many shapes this process is the most economical way to fabricate.
9. Process permits the manufacture of very small parts which are almost impossible to fabricate in quantity by other methods.
10. Minimal scrap loss results as runners, gates, and rejects can be reground and reused.
11. Same item can be molded in different materials, without changing the machine or mold in some cases.
12. Close dimensional tolerances can be maintained.
13. Parts can be molded with metallic and nonmetallic inserts.
14. Parts can be molded in a combination of plastic and such fillers as glass, asbestos, talc, and carbon.
15. The inherent properties of the material give many advantages such as high strength-weight ratios, corrosion resistance, strength, and clarity.

Some disadvantages and problems of injection molding are the following:

1. Intense industry competition often results in low profit margins.
2. Three shift operations are often necessary to compete.
3. Mold costs are high.
4. Molding machinery and auxiliary equipment costs are high.
5. Process control may be poor.
6. Machinery that is not consistent in operation, and whose controls are not directly related to the end product.
7. Susceptibility to poor workmanship.
8. Quality is often difficult to determine immediately.
9. Lack of knowledge concerning the fundamentals of the process causes problems.
10. Lack of knowledge about the long-term properties of the materials may result in long-term failures.

Figure 3 shows an operator removing a molded piece from a molding machine. Note also the four tie bars upon which the movable platen rides. The rubber hoses circulate fluid for mold temperature control. Each cavity and

Figure 3   Operator removing a molded part.

core has its own set, so that the temperature on each can be adjusted independently. The mold is held on the platen by clamps.

Figure 4 shows a cutaway view of a $2\frac{1}{2}$-in. reciprocating-screw, 300-ton machine. The injection end is always to the right of the operator. This permits him to open the safety gate with his left hand and remove the molded piece with his right.

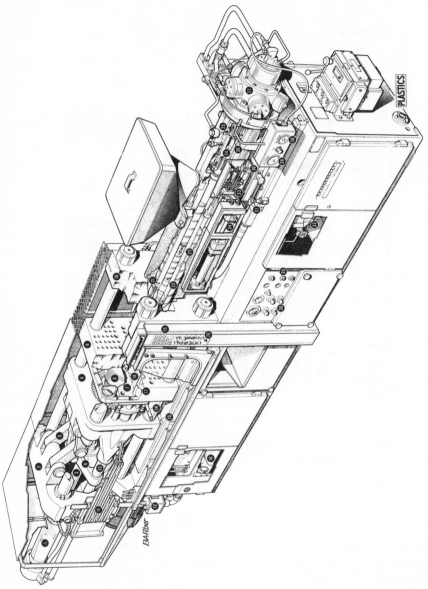

Figure 4 Cutaway view of a $2\frac{1}{2}$-in. reciprocating-screw, 300-ton machine (Peco molding machine). Courtesy *British Plastics*.

498

KEY

1. Hydraulic cylinder
2. Tail stock plate
3. Hydraulic piston extension
4. Toggle crosshead
5. Toggle link
6. Moving back plate
7. Ejector plate
8. Mold height adjustment screw
9. Moving platen
10. Fixed platen
11. Linear limit switch stops
12. Lubrication pump
13. Toggle cross head guide bar
14. Mould height adjustment
    mechanism
15. Moving plate support pad
16. Hydraulic tank
17. Ejector bar
18. Hydraulic ejector

19. Solenoid indicator lights
20. Manual control panel
21. Cylinder
22. Screw
23. Air tube and bore
24. Screw coupling
25. Bearing
26. Motor drive shaft
27. Motor
28. Screw speed indicator
29. Injection pressure gauge
30. Shot volume control mechanism
31. Retraction stroke limit switch
32. Screw speed control
33. Injection follow-up pressure
    control
34. Injection unit retraction cylinder
35. Water on/off cocks
36. Hydraulic controls

1. Injection End

The injection end has a hopper for holding the molding material. Some materials such as nylons, polycarbonates, acrylics, acrylonitriles, and acetates are hygroscopic and require drying before molding. A magnet is placed in the hopper throat to catch any iron accidentally introduced into the hopper or material.

In all machines the control of the amount of material is critical. If too little material is injected into the mold, the parts do not fill out, and voids, sink marks (depressions usually opposite heavy sections), poor surface, poor physical properties, and other faults occur. If too much material is injected the shot is called "packed." This results in difficult mold release, overweight, flash (the mold is forced open by the material and it spills over the edges of the cavity), and probable poor quality.

2. Controls

The electrical controls can be mounted in a separate enclosure (Fig. 1) or on the machine (Fig. 4). The manual control panel (Fig. 4, *20*) is mounted on the machine for the operator's convenience. Hydraulic gauges and controls (Fig. 4, *28, 36*) are mounted for convenience in piping. The location of the hydraulic components and the electric motors depend on the individual machine. Most oil reservoirs (Fig. 4, *16*) are in the base of the machine. Fully hydraulic clamps can have the tanks above the clamping end. The machine should be designed for easy maintenance, with as many components as possible readily accessible. Unfortunately, this is not always the case.

The hydraulic system generates heat which is removed with a water-cooled heat exchanger (cooler). Provision for bringing water to and from the machine is required.

3. Safety

The satefy gate is part of the safety system. The molding machine is dangerous, having caused lost limbs and fingers and fatalities. The gate should be large enough and high enough so that it is impossible for anyone to get any part of one's body in the platen area when the platen is closing. On smaller machines a cover is essential. There must be electrical and hydraulic safety systems. Mechanical devices are also necessary. The rear of the machine should be guarded with a gate electrically interlocked so that the machine cannot operate without it. At no time should any repairs between the platens be made with the motor running. Moving parts, such as toggles, should be guarded. The heating cylinder should have a cover to prevent direct contact with the heating bands. The purging controls should be located so that the operator is protected from spattering by overheated material. Electrical locks should be provided so that no one can start the machine when mold or machine repairs are being made.

The requirements of the Occupational Safety and Health Act of 1970 (OSHA) must be met.

4. Knockout System

The knockout (KO) system causes a special knockout plate in the mold to change its location relative to the rest of the mold. Attached to this plate are knockout (ejector) pins or other devices which push against the molded parts as the mold completes its opening. This ejects the parts from the mold. On most machines there is an adjustable stationary ejector plate (Fig. 4, 7) to which are attached the ejector bars (Fig. 4, 17), which go through the movable platen. As the mold returns, the ejector bars go through holes in the back plate of the mold and contact and stop the knockout plate. The rest of the mold continues, moving a predetermined distance, and stops. Pins attached to the knockout plate and projecting to the mold parting surface are pushed by the injection side of the mold as the mold closes. This pushes the knockout plate back to its normal position. Such pins are called push-back pins. A much superior but more costly KO system uses hydraulic cylinders to activate the KO system. The advantages are activating and retracting the knockout system at any time and with any speed or pressure.

5. Clamping Systems

The two major methods of clamping the machine are an hydraulically operated toggle system and a fully hydraulic one. In both instances a hydraulic cylinder provides the force to stretch the tie bars, which causes the clamping action. In a straight hydraulic system, a large diameter hydraulic cylinder is attached directly to the movable platen. The clamping force is rated in tons. Force should not be confused with pressure. They are related by the following equation:

$$F = P \times A \qquad (1)$$

where F = force (lb)

P = pressure (lb/in.^{2}, psi)

A = area (in.^{2})

For example, a press with an hydraulic clamp has a 20-in.-diameter clamping cylinder. The maximum working line pressure is 2000 psi. The clamping force is

$$F = 2000 \text{ psi} \times 314.2 \text{ in.}^2 = 628{,}400 \text{ lb}$$
$$= 314.2 \text{ tons}$$

This press would be called a 300-ton press. Clamping force is one of the main machine specifications. Machines available today range from 5.5 tons for a machine whose maximum shot is 10 g to 5500 tons for a giant (2) that can mold 54 lb of polystyrene per shot. Hydraulic clamping (3) permits unlimited pressure selection, which can be continually monitored with a pressure gauge.

As molds get bigger, longer strokes are required for the movable platen. Additionally, the larger molds require larger clamp forces and larger-diameter cylinders. Cylinders of such size are excessively costly. This is overcome by using a much smaller diameter cylinder for rapid movement and the large diameter cylinder for generating the clamping force.

There are many different systems of toggles (4). Basically the hydraulic cylinder (Fig. 4, 1) is attached to a stationary plate (Fig. 4, 2). It moves forward, eventually spreading the toggle links (Fig. 4, 5) so that they are in a straight line and holds them there. The mechanical advantage is from about 20/1 to 30/1. The toggle system is less expensive to build. It requires good maintenance, because wear reduces the clamping tonnage. Pressure adjustment is possible, but not as easily or accurately controllable as with an hydraulic system.

The clamp stroke is the maximum distance the clamp can move. The maximum daylight is the farthest distance the platens can separate from each other. The difference is the minimum die thickness that can be put into the press and still maintain clamping pressure. This minimum distance can always be decreased by adding a bolster plate in front of or behind the movable platen. These are important specifications. They tell how deep a piece may be molded and whether a mold of given depth will fit in the machine.

The clearance between the tie rods (inches) is the determining factor whether a mold of a given length or width will fit. For example, a press has a 20-in. clearance vertically and an 18-in. clearance horizontally. Therefore a mold less than 20 in. wide but over 20 in. long will fit vertically; a mold less than 18 in. high but over 18 in. long will fit horizontally. The length and width dimensions of a mold are often determined by the side parallel to the knockout plate.

6. Injection End Specifications

Injection end specifications vary for the different methods of injection. They all include the following.

The maximum weight of material that can be injected in one shot (machine capacity) is given in ounces and based upon molding general-purpose polystyrene. This is what is meant when one speaks of a "16-oz machine." The more accurate specification is the volume of material per shot (in.^{3}). This is a measure of the geometry of the cylinder and feed system. The injection stroke is the maximum distance that the injection plunger can travel. The injection speed (in./min) is the speed of the injection plunger, usually with no material in the cylinder. The maximum injection rate (in.^{3}/min, oz/min) is the rate at which the injection cylinder can eject fully plasticized material into the air. This is somewhat different from the speed achieved during molding.

An important quality of the machine is the amount of pressure that is placed directly upon the material. In a two-stage machine or in-line screw, this is easy to determine. It depends upon the diameter of the screw or plunger, the piston diameter of the hydraulic injection cylinder, and the oil pressure. In straight plunger machines, the injection pressure on the material is sometimes given based upon the same factors. This is erroneous for there is at least a 30% pressure loss upon the material from the feed end to the nozzle. Such pressure loss is related to the granular condition of the cold pellets in the back of the cylinder. Material pressures of 20,000 psi are required to mold some of the new materials.

The plasticizing capacity is the most important injection end specification and the most difficult to verify. The Society of the Plastics Industry (SPI) has adopted a specification developed by a technical committee of the Society of Plastics Engineers (SPE (5)).

The rate of recharging (oz/sec) tells how much material the screw can produce running continuously. In most machines the screw runs intermittently and the output is estimated based on the type of molding. A good approximation for the average custom molder is 50%.

B. TYPES OF INJECTION ENDS

There are several types of injection ends in use today.

1. Single-Plunger System

The single-state plunger (Fig. 5), or plunger machine, uses a plunger to force the material over a spreader or torpedo. The heat is supplied by resistance heaters, and the material is melted by conduction and convection.

The channels between the torpedo and the cylinder body are a compromise between wide passages to minimize pressure loss and small passages for better heat transfer. The flow in the cylinder is basically laminar, giving the least homogeneous melt of all plasticizing systems.

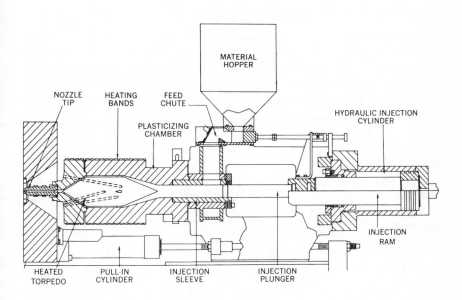

Figure 5 Schematic drawing of injection end of a single-stage plunger machine. Courtesy HPM Division of Koehring Co.

There is a pressure loss of at least 30% between the injection ram and the material in the mold. This is because of the cold granules in the back of the plasticizing chamber. It has the result of maintaining a springlike pressure on the material. Until 1960 almost all molding machines in the United States were of this type. Virtually all new equipment today uses a screw to plasticize the polymer.

2.   Two-Stage Plunger System

The single stage plunger machine has some severe limitations that are overcome by using a "preplasticizing" system (6). The material is melted in one chamber and transferred into a second chamber, where it is forced into the mold by the direct action of another plunger on the melted material. This gives rise to a number of significant advantages.

1.   The melt is more homogeneous because it mixes as it passes through the small opening connecting the two chambers.
2.   There is direct pressure on the material by the injection plunger.
3.   Faster injection is possible.
4.   Better injection pressure control is possible.
5.   Better shot weight control is possible.
6.   Because plasticizing can take place throughout most of the cycle, more pounds per hour per dollar of machine can be had.

The two-stage plunger (or plunger into a "pot") type of machine is illustrated in Fig. 6. The plasticizing chamber is that of Fig. 5. It is connected by a three-way valve shown in the plasticizing position. As the stuffer plunger reciprocates plasticized material is forced in front of the injection plunger. The injection plunger moves back a predetermined distance until it hits a limit switch (not shown) which stops the preplasticizer. At the appropriate time the rotary valve is turned, and the injection plunger comes forward, filling the mold. While this is happening the plasticizing chamber is melting the next shot.

A two-stage plunger machine always gives better results than a single-stage. A superior way to plasticize is to use a screw. For this reason two-stage plunger machines are obsolete, a screw being used in place of the plasticizing plunger.

3.   Injection Screw

Screws have been used to extrude thermoplastics for many years. Their superiority in the quality of the melt and in the speed of plasticizing has led to their almost universal adoption on new equipment. Existing equipment can

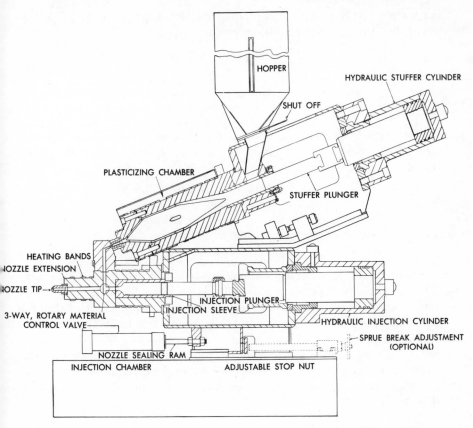

Figure 6 Schematic drawing of injection end of two-stage plunger machine. Courtesy HPM Division of Koehring Co.

be readily converted (7). A plunger machine has four main variables, a screw machine is more complicated. It is necessary to understand the theory of screw plasticization to properly operate this equipment.

Torque

The work done in a screw machine (melting the material) is done by rotating a screw in a stationary barrel. The rotational force is called torque. It is the product of the tangential force and the distance from the center of the rotating member. For example, if a 1-lb weight were placed at the end of a 1-ft bar attached to the center of the screw, the torque would be 1 ft × 1 lb or 1 ft-lb. Torque is related to horsepower:

$$hp = \frac{torque\ (ft\text{-}lb) \cdot rpm}{5252} \tag{2}$$

$$hp = \frac{torque\ (in.\text{-}lb) \cdot rpm}{63{,}024} \tag{3}$$

It is clear that the torque output of an electric motor of given horsepower depends on its speed. A 30-hp motor has the following torque at various speeds:

Speed (rpm)	Torque (ft-lb)
1800	$87\frac{1}{2}$
1200	133
900	175

The speed of a given horsepower motor is built into that motor. Obviously, a higher torque unit (lower speed) motor has a larger frame than the lower torque unit. The change in speed and torque can also be accomplished by changing the output speed of the motor by using a gear train. The change in torque varies inversely with the speed.

Most screws are driven by a hydraulic motor. The motor can be attached to the screw through a gear reduction unit. It can also drive the screw directly, in which instance it can be in the enclosure containing the injection unit (Fig. 4, 27).

Screw Strength and Speed

The strength of the screw limits the input horsepower. As input horsepower is increased, a point is reached where the torque that can be provided is above the yield strength of the metal screw. The strength of the screw varies with the cube of the root diameter. For a $2\frac{1}{2}$-in. screw at 200 rpm, the maximum permissible drive input is about 40 hp. For a $3\frac{1}{2}$-in. screw at 200 rpm it is 75 hp, and for a $4\frac{1}{2}$-in. screw at 150 rpm it is 180 hp. For a given horsepower, the slower the speed, the higher the torque (Eq. 2). Too rapid a shear rate degrades the material. The shear rate is highest at the barrel wall. The maximum surface speed with present screw technology is about 150 ft/min. Some of the more shear sensitive materials limit the surface speed to 100 ft/min.

Horsepower Requirements

The work done in a screw plastication causes the material temperature to rise from room temperature to the molding temperature. Assuming that all the energy comes from turning the screw and that the mechanical efficiency is 100%, the work done would be the product of the average specific heat and the temperature rise. Neither of these assumptions is correct. A small amount of heat is supplied by the heating bands, and corrections for machine efficiency must be made. Since the screw acts as a pump, energy is also required when pressure is generated. This is a relatively small and is disregarded in subsequent calculations. Therefore

$$HP = C \cdot (T_p - T_f) \cdot Q + \Delta P \cdot Q \qquad (4)$$

where HP = horsepower
C = average specific heat (btu/lb-°F)
T_p = temperature plasticized material (°F)
T_f = temperature feed material (°F)
Q = throughput lb/hr
ΔP = back pressure (psi)

Converting into consistent units, we get

$$HP = 0.00039 \cdot C \cdot (T_p - T_f) \cdot Q \qquad (5)$$

and for 70% efficiency

$$HP = 0.00056 \cdot C \cdot (T_p - T_f) \cdot Q \qquad (6)$$

For example, high-impact polystyrene has an average specific heat of 0.42. How many horsepower are required to plasticize 1 lb/hr ($Q = 1$) when the room temperature is 70°F and the material temperature 400°F?

$$HP = 0.00056 \cdot 0.42 \cdot (400 - 70) \cdot 1 = 0.078 \text{ hp}$$

This is equivalent to 13 lb/hr for each 1-hp input. Molding materials range from 6 to 14 lb/hr for each 1-hp input.

If a 30-hp motor is used at a room temperature of 80°F, what is the maximum output of low-density polyethylene ($C = 0.8$) at 380°F?

$$30 = 0.00056 \cdot 0.8 \cdot (380 - 80) \cdot Q$$
$$Q = 223 \text{ lb/hr}$$

Supposing the material temperature were raised to 450°F. Would the output increase?

$$30 = 0.00056 \cdot 0.8 \cdot (450 - 80) \cdot Q$$
$$Q = 181 \text{ lb/hr}$$

Thus raising the temperature *lowers* the maximum output. The molder therefore molds at the lowest possible melt temperature. This gives maximum screw output and reduces the time needed for reversing the process, that is, polymer cooling in the mold.

It is interesting to note that the output of the machine as defined in Eq. 5 is completely independent of the screw diameter. If, for example, $2\frac{1}{2}$- and $3\frac{1}{2}$-in. screws, each having the same length/diameter (L/D) ratio and the same input drive horsepower, are operated at their maximum capacity, they both deliver the same output (lb/hr). The plastic remains longer in the $3\frac{1}{2}$-in. screw. Why, then, have large screws? The answer is found in Eq. 2. Screw speeds must be kept low to prevent degradation. With a constant horsepower, the slower speeds can develop a torque high enough to shear the screw. For a $2\frac{1}{2}$-in. screw at 200 rpm, the maximum permissible drive input is 40 hp. Therefore, the higher horsepowers required for higher output need larger-diameter screws to prevent breaking the screw. The torque a screw can safely carry is proportional to the cube of its diameter. At 200 rpm, a $3\frac{1}{2}$-in. screw can handle 120 hp and a $4\frac{1}{2}$-in. screw, 240 hp.

It is obvious, then, that in a screw machine the horsepower rating available for screw rotation is a very important specification. Assuming similar efficiency for different screw designs, the maximum output, which is a primary concern of injection molders, is largely determined by the horsepower rating of the screw. With this criterion, analyzing machine specifications of different manufacturers becomes more productive and interesting.

Chapter 9 on extrusion, explains the typical screw, how it melts plastic materials, drag and pressure flow, and other parameters that determine the quality and quantity of the plastic output.

Advantages of Screw Plasticizing

In a screw the melting of the plastic is caused by the shearing action of the screw. As the polymer molecules slide over each other they convert the mechanical energy of the screw drive into heat energy. The heat is applied directly to the material. This and the mixing action of the screw contribute to its major advantages as a plasticizing method:

1. High shearing rates lower the viscosity, making the material flow easier.

2. Good mixing results in a homogeneous melt.

3. The flow is nonlaminar.

4. The residence time in the cylinder is approximately three shots compared to the 8 to 10 shots of a plunger machine.

5. Most of the heat is supplied directly to the material.

6. Because little heat is supplied from the heating bands the cycle can be delayed for a longer period before purging.

7. The method can be used with heat-sensitive materials, such as PVC.

8. The action of the screw reduces chances of material holdup and subsequent degradation.

9. The screw is easier to purge and clean than the plunger.

There is considerable literature comparing screw versus plunger plasticizing and describing the operation of the in-line screw (8–16).

Screw Drive

It is desirable to plasticize at the lowest possible screw speeds. The drive must supply enough torque to accomplish this, but not enough to mechanically shear the screw. Changes in torque are needed because of the different processing characteristics of plastics. Much higher torque is required to plasticize polycarbonate than polystyrene. The speed of the screw regulates the quality and output rate of the polymer. It is desirable to have two, and preferably three, torque speed ranges for handling the various materials. If a material is being molded with a minimum torque, at a given speed, increasing the speed increases the horsepower requirements.

The ability to control torque and speed is very important. One method of applying the driving force to the screw is to attach it to an electric motor through a speed-reducing, gear train coupling with various speed ranges (17). The other is to connect an hydraulic motor driven by an hydraulic pump either directly to the screw or through a speed-reducing coupling (18, 19). The advantages, disadvantages, and operating characteristics of each are shown in Table 1.

Reciprocating Screw System

A reciprocating screw (in-line screw) (Fig. 7) is the most popular injection end in the United States. The material is fed from the hopper, plasticized in the screw, and forced past a one-way valve at the injection end of the screw. The material accumulates in front of the screw, forcing back the screw, its drive and motor. When the screw reaches a certain position a limit switch is contacted, stopping the screw rotation. Two hydraulic injection cylinders, one on each side of the carriage, bring the screw assembly forward, and use the screw as an injection ram. The one-way valve prevents the material from going back over the flights. If this valve is not functioning correctly the screw rotates during injection, preventing controlled shot size (20–22).

Screw-Plunger System

A screw-pot (screw plunger) is shown in Fig. 8. It is essentially similar to the plunger-plunger machine (Fig. 6) except that a fixed screw is used for plasticizing. A reciprocating screw could be used which permits continual operation of the screw throughout the whole cycle. It has significant production advantages but is not commonly used because of the initial cost.

The reciprocating screw and screw-pot are both preplasticizing systems. The difference is the location of the pot, which is in front of the reciprocating screw and is a separate cylinder in the two-stage machine. Most machines

Table 1   Hydraulic versus electric screw drive

Characteristics	Hydraulic	Electrical
Efficiency	Low, 60–75%	High, 95%
Screw safety	Relief valve prevents screw damage	Overload protection difficult, particularly with small-diameter screws
Size	Small and light-weight; good shot weight control	Heavy; poorer shot weight control; more stresses in the system
Torque	Constant output; infinitely adjust-able; lower starting torque requires larger input horse-power; poorer at lower screw speeds	Varies with screw speed; excellent starting character-istics and at low screw speeds
Speed	Stepless control easily adjusted; gives best control of molding con-ditions	Limited number of speeds; more difficult to adjust and maintain
Melt quality	Best	Acceptable

sold are of the reciprocating type, but many molders have found significant advantages in screw-pot equipment. Some of them follow:

1.   Because the screw does not act as the injection ram, lighter bearings can be used. There is no need for the heavy thrust assemblies found on reciprocating screws. This reduces maintenance cost.

2.   The extruder barrel need only be strong enough to maintain the pressure of the material during plasticization which is rarely over 5000 psi. In contrast the barrel for the reciprocating screw must contain the 20,000 psi used.

3.   There is less wear because the screw does not move.

4.   The connection between the two stages can be a ball check valve, which is troublefree and easy to maintain. It presents minimum flow re-sistance.

5.   The nonreturn valves at the tip of a reciprocating screw wear rapidly, sometimes do not seat properly (preventing consistent molding), can cause

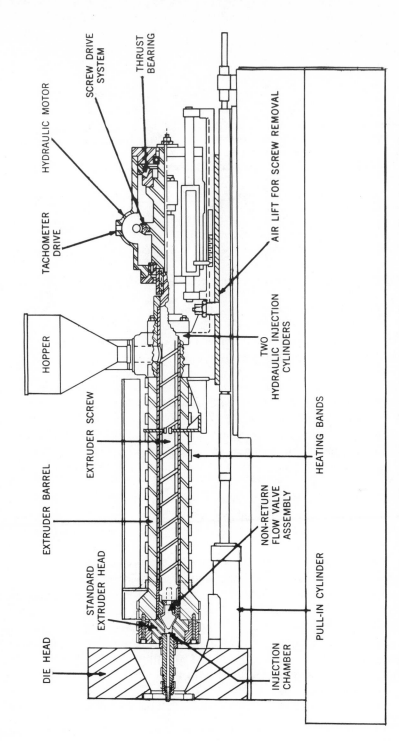

DIE HEAD

EXTRUDER BARREL

EXTRUDER SCREW

STANDARD
EXTRUDER HEAD

HOPPER

TACHOMETER
DRIVE

HYDRAULIC MOTOR

SCREW DRIVE
SYSTEM

THRUST
BEARING

AIR LIFT FOR SCREW REMOVAL

TWO
HYDRAULIC INJECTION
CYLINDERS

NON-RETURN
FLOW VALVE
ASSEMBLY

HEATING BANDS

INJECTION
CHAMBER

PULL-IN CYLINDER

Figure 7 Schematic drawing of injection end of reciprocating-screw machine. Courtesy HPM Division of Koehring Co.

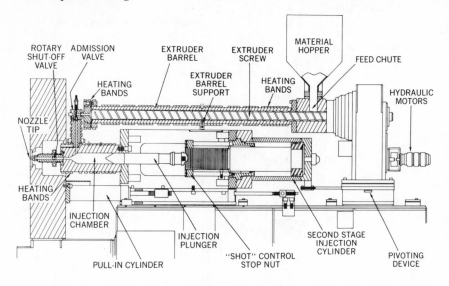

Figure 8   Schematic drawing of injection end of two-stage screw-plunger machine. Courtesy HPM Division of Koehring Co.

wear in the barrel, are a place for material to hang up and degrade, and are much more expensive than a ball check valve.

6.   The small clearances between the plunger and barrel of the second stage help in degassing the material.

7.   The connection between the two stages result in better mixing of the melt.

8.   A very important advantage of a two-stage machine is that all the material goes over the full flights of the screw, receiving the same heat history. In a reciprocating screw only the first material in goes over the full length of the screw.

9.   In a two-stage machine, the screw pumps only against the injection ram which is floating in oil in the hydraulic cylinder. The reciprocating screw must push back the whole weight of the carriage and all the equipment on it. For this reason the shot size control is considerably more accurate in the two-stage machine.

10.   Because part of the energy input is used to push back the heavy carriage, an in-line machine gives slightly less output per unit input.

11.   There is better injection speed control.

12.   There is better injection pressure control.

13.   Because of the preceding two factors a larger projected area can be molded.

14. Extremely high injection pressures are available.
15. The size of a pot in front of a reciprocating screw is limited by the length of the feed section. If the screw goes too far back the material does not plasticize correctly. In a two-stage machine there is no theoretical limitation. Thus a 2-in. reciprocating screw normally has a maximum shooting capacity of 13 oz, whereas the same diameter screw can be readily used to shoot 60 oz in a two-stage machine.

There are a number of disadvantages of the two-stage machine.

1. It requires two cylinders and two sets of heat controls.
2. It is slightly more difficult to clean.
3. It is slightly more difficult to set up.
4. It does not process materials sensitive to high heat as well as a reciprocating screw does.
5. Cylinders for molding thermosets and rubber are designed only for reciprocating screws.
6. It takes up more space than a reciprocating screw.
7. It costs more.

In molding heavy sections or when the shooting capacity of the machine is not adequate, intrusion molding can be used. In this method the screw continually turns, filling the cavity directly. When the cavity is filled a cushion is extruded in front of the plunger, which then comes forward to supply the needed injection pressure.

Thermosetting materials can be molded on reciprocating screw machines. The screw is designed differently and the barrel is usually heated with hot water. The material cannot be permitted to cool in the cylinder. If it does the screw has to be removed and the thermosetting material chipped out.

C. MACHINE SPECIFICATIONS

A machine is classified by its clamping and injection end. The clamping end is specified by its type (hydraulic, toggle) and its clamping tonnage, the injection end by its type (plunger, two-stage plunger, screw, two-stage screw) and its capacity. The capacity is given in oz/shot and lb/hr of plasticizing ability. Space does not permit a discussion of hydraulic and electrical components and the corresponding machine control circuits. There are many simple books on hydraulics for those who care to pursue the subject (23–26).

Electrical machine circuits are a little more difficult to follow than hydraulic circuits because of the many interactions of relay contacts. A study of electrical components, particularly timers, and a knowledge of the hydraulic circuit will make it considerably easier.

Almost all machines built before 1968 use simple hydraulic components without any servomechanisms and use regular electric components such as timers, relays, and limit switches. In essence the circuits are the same as those of the earlier machines built in the late 1930s (27). The mechanical unreliability and the lack of feedback cause serious molding problems and have prevented further development of the state of the art. Recently machinery manufacturers have finally begun to develop solid-state-circuitry and servomechanisms in the hydraulic systems so that comparison and feedback techniques can be used.

D.   DESIRABLE CHARACTERISTICS OF MOLDING MACHINES

There are a number of accessories that are standard or optional on some equipment which are very useful in molding:

1.   Hydraulic ejection on the movable side.
2.   Hydraulic ejection on the stationary side.
3.   The ability to stop and start the movement of the platen in either direction at any location.
4.   Two injection pressures.
5.   Two injection speeds.
6.   Two clamping pressures.
7.   Two clamping speeds.
8.   Safety interlocks for the mold so that the machine cannot close unless a switch is contracted.
9.   Hydraulic pressure testing points.
10.   Hydraulic takeoff for operating auxiliary mechanisms.
11.   Melt temperature recorder for nozzle.
12.   Pressure transducer in nozzle.
13.   Extra timers and relay.
14.   110-V outlets on the machine.

E.   MOLDS (28)

The injection mold is the mechanism into which the hot plasticized material is injected and maintained under pressure. When the plastic material has sufficiently solidified, the machine opens, separating the mold. The plastic pieces are ejected. The quality of the part and its cost of manufacture are strongly influenced by mold design, construction, and excellence of workmanship.

As machine capacity increases, the molds become larger and more expensive. It is not uncommon, for example, for a mold for a garbage can and cover to cost over $35,000. Molds for 12-oz machines usually range from $3,000 to $9,000. This is only a small part of the investment. The original idea, market testing, samples, prototypes, advertising, selling, and the commitment for the initial order cost many times the cost of the tool.

The two most critical steps in the production of a plastic part are the piece part design and the mold design. A failure of the first naturally results in an unacceptable part. Mold design failure does not necessarily result in piece part failure, though it may well do so. It does cause low productivity, high mold maintenance, and the probability of reduced part quality. An improperly functioning mold requires excessive supervisory time in the plant and tool maintenance department. This can cause neglect of other operations, compounding the normal problems of running a molding plant.

It is for this reason that the molder, mold maker, and on occasion the purchaser, should give maximum attention to the design. Unfortunately this is not always done. The last opportunity to change the part easily is when the mold is designed.

The first step in designing a mold is to have an accurate fully dimensioned drawing, noting tapers and where they start, tolerances, shrinkage specifications, surface finish specifications, material in which it is to be molded, part identification, and any other pertinent information. It is usually desirable to have a model of the part to be molded, particularly if it or the drawing is complicated. Mold design is significantly more reliable and molding problems are more readily anticipated. At this time the metal for the cavity and the core is selected. The cavity and core are those parts of the mold that, when held together by the closing of the mold, provide the air space into which the molten plastic is injected. The following are now decided: the number of cavities; the parting line of the piece (where the faces of the cavities and cores touch); the type and location of the gates (the entry point of the hot plastic into the cavity); the gating system (which brings the hot plastic into the cavity); the runner system (which brings the hot plastic from the plasticizing chamber to the gate); the method of ejection (which removes the molded plastic parts from the mold); the location of the ejecting devices; location and size of the temperature control channels; and the type and location of the venting system (which removes the air that is displaced by the incoming plastic).

Any additional information required by the molder or moldmaker is discussed and drawings of the mold are submitted. These must be reviewed very carefully.

1. Mold Base

The steel parts that enclose the cavities and cores are called the mold base, mold frame, mold set, die base, die set, or shoe. Figure 9 shows an exploded view of the parts of a standard mold base. The locating or seating ring centers the sprue bushing on the stationary or injection platen directly in line with the nozzle of the injection cylinder. The sprue has an opening in the center of a concave spherical surface, whose counterpart is an equivalent convex surface on the nozzle of the injection heating cylinder. The opening

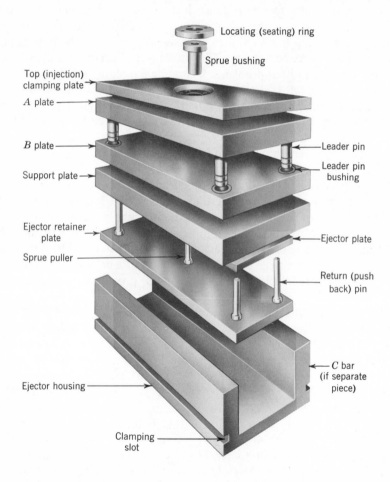

Figure 9   Exploded view of standard mold base. Courtesy D-M-E Corp.

of the sprue bushing must always be larger than the opening in the nozzle, so that when the plastic hardens it does not form an obstruction larger than the sprue opening and cause the sprue to stick. The sprue has a generous taper to facilitate easy removal of the plastic. The standard radii for sprue bushings and nozzles are $\frac{1}{2}$ and $\frac{3}{4}$ in.

The injection or top clamping or backup plate supports the cavity or A plate. The B plate has the cores attached to it, and these match the cavities in the A plate.

Ejector or knockout (KO) pins are mechanically attached to the ejector plate assembly so that the molded parts can be knocked out and removed from the mold.

A number of companies manufacture standard mold bases and parts. Because of their high volume they have equipment which usually makes their mold bases less expensive and superior to those manufactured by the moldmaker. In addition replacement parts are standard and readily available to the molder at a low cost.

There are cavities and cores of such size or shape that a mold base is best built around them. There are several types of steel available for the mold base. It is strongly urged that the best quality be used for any mold that requires high-quality parts or a long production run (28a).

2. Mold Types

The injection mold is identified descriptively by a combination of some of the following terms. They are described in the text.

Parting line
    Regular
    Irregular
    Two-plate mold
    Three-plate mold
Material
    Steel
    Stainless steel
    Prehardened steel
    Hardened steel
    Beryllium–copper
    Chrome plated
    Aluminum
    Brass
    Epoxy

Number of cavities
Methods of manufacture
   Machined
   Hobbed
   Cast
   Pressure cast
   Electroplated
   EDM (spark erosion)
Runner system
   Hot runner
   Insulated runner
Gating

Edge	Diaphragm
Restricted (pin pointed)	Tab
Submarine	Flash
Sprue	Fan
Ring	Multiple

Ejection

Stripper ring	Removable insert
Stripper plate	Hydraulic core pull
Unscrewing	Pneumatic core pull
Cam	Knockout pins

3.   Two-Plate Molds

Figure 10 shows a schematic representation of the cross section of part of a regular two-plate injection mold. The part being molded is a shallow dish. The six cavities are gated on the edge. There are temperature control channels in both backup plates and in the cores and cavities. Because there is significant insulation between two pieces of metal, the use of channels in cores and cavities gives better and more efficient temperature control.

Note the support pillars which are anchored to the back plate and support the backup plates underneath the cores in the B plate. A machine knockout bar is shown. As mentioned, they remain stationary and as the moving platen returns, they stop the ejector plate. The mold opens on the parting line and the undercut, which in this instance is shaped like a "Z," pulls the molded sprue and runner with it. The part design and molding condition keep the plastic on the core. As the ejector mechanism works, the parts are pushed off the core and the sprue puller moves out of its hole, allowing the parts to be removed.

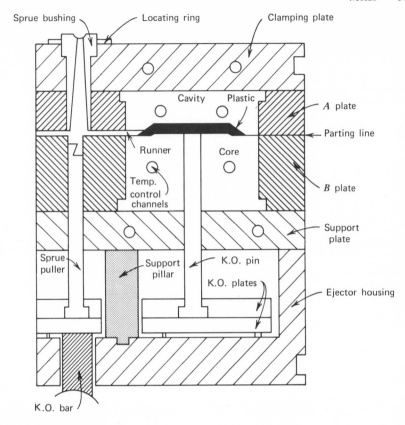

Sprue bushing   Locating ring   Clamping plate

Cavity   Plastic   A plate

Parting line

Runner   Core

Temp.
control   B plate
channels

Support
plate

Sprue
puller   Support   K.O. pin
pillar

K.O. plates   Ejector housing

K.O. bar

Figure 10   Schematic drawing of two-plate mold. Courtesy Robinson Plastics Corp.

4.   Three-Plate Molds

Suppose the dish were deeper and could be gated only in the top center section. The mold could be constructed as a one-cavity mold feeding directly from the sprue. If a multiple cavity mold is needed there are a number of alternatives, one of which is shown in Fig. 11. Six cavities could be located in two parallel rows of three. One cavity and core are shown with other significant parts of the mold.

The difference between this type mold and the one illustrated in Fig. 12 is that it separates between the *A* plate and the clamping plate as well as at the parting line. This is called a three-plate mold even though a third plate is not always used. The plastic is injected through the sprue bushing into the

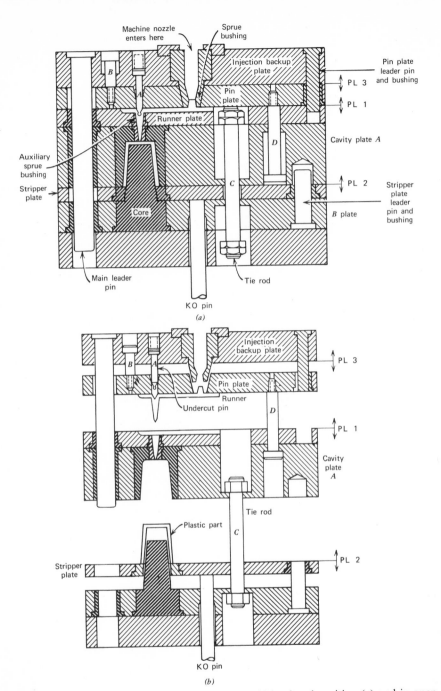

Figure 11 Schematic drawing of three-plate mold in closed position (*a*) and in open position (*b*).

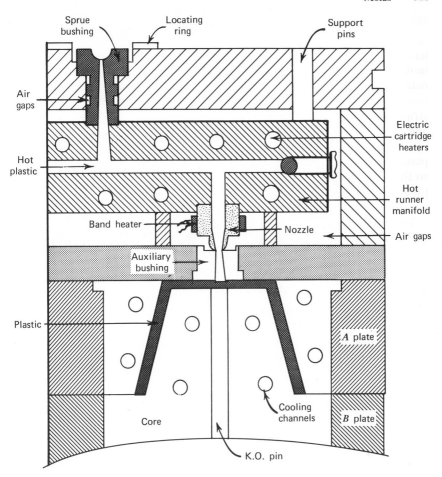

Figure 12   Schematic drawing of hot runner mold.

runner, which is cut into the plate with a trapezoidal cross section. The flat back is on the pin plate. The plastic flows into the part through an auxiliary sprue bushing. Although this can be machined directly into the *A* plate and cavity, it is good practice to have a separate bushing so that it can be replaced or changed.

When the mold opens, the *A* and *B* plates move together. Sometimes this occurs normally. Other times latching or spring mechanisms are needed. The mold opens initially on Parting Line 1. This breaks the gate and leaves the runner attached to the pin plate because of the undercut pins attached to the injection backup plate and extending into the runner. After the separation has occurred at PL 1, the mold continues to open, separating at PL 2. The molded pieces are on the core. They are then ejected in a conventional manner, in this instance by a stripper plate. The pin plate is limited in its travel by stripper bolts *B*. When it is moved forward (by latches, chains, stripper bolts, or ejector bars), the runner is stripped off the undercut pins, and the plastic sprue is moved forward out of the sprue bushing. The runner can be removed by hand, an air blast, or mechanical wiper. The runner plate and *A* plate always stay on the leader pins. The pins must be long enough so that the plates can separate far enough to remove the runner. It is good practice to support the pin plate on its own leader pins, attached to the backup plate. This prevents it from binding on the main leader pins.

This type of mold works very well provided the workmanship is of good quality and the components fully sized and adequately designed. If not, cocking and binding occur relatively quickly on heavy molds. It is sometimes necessary to put an extra set of leader pins and bushings to support the *A* plate. These should not be used to line up the *A* and *B* plates. In other instances small leader pins and bushings are put into the *A* and *B* plates to assure good line-up and compensate for wear on the longer leader pins.

5. Hot Runner Molds

The plastic runners must be reground and reused, if permissible. A logical extension of the three-plate mold overcomes this and is called a hot runner mold (Fig. 12). This mold has a hot runner plate, which is a block of steel heated with electric cartridges, usually thermostatically controlled. This keeps the plastic fluid. The material is received from the injection cylinder and is forced through the hot runner blocks into the cavities. It is a more difficult mold to build and operate than a three-plate mold, but produces parts less expensively on long-running jobs (29, 30).

6. Insulated Runner Molds

A cross between a hot runner mold and a three-plate mold is called an insulated runner mold. The gating system is very similar to that of a three-plate mold except that the runners are very thick, at least $\frac{3}{4}$ in. in diameter. There is no runner plate, and the backup plate and *A* plate are held together by latches. The material, usually an olefin, is injected. The outside of the runner freezes but insulates the center, permitting the core to remain fluid. This acts as a hot runner. If the runner freezes during startup the two plates are separated and the runner system is removed. As soon as the runner reaches equilibrium, the latch is closed and the mold is operated that way. These are more difficult to start and operate than a three-plate mold and are usually restricted to the olefins (31).

7. Materials for Cavities and Cores

Steel is the most often used for injection mold sets, cavities, and cores. A discussion of steels is not in order here and is available in many books and in the literature of steel manufacturers.

The type of steel selected depends on the end use, the size of the part, and the method of fabrication. It should be free from defects, have minimal distortion during heat treatment, be easily machinable, polish well, and weld readily.

To last under the stresses of injection molding and give reasonable protection against damage during production, steel has to have a minimum hardness. It is usually designated on the Rockwell C scale in America. Foreign sources may use the Brinell system. If the steel is too hard it becomes brittle. If too soft, it does not provide enough protection against damage and wear. A Rockwell C (R_c) of 50 to 55 gives good results. Steel this hard is difficult to machine even with carbides. It is easily worked by grinding and electrical or chemical removal equipment. The cavity or core is machined in the soft condition as it comes from the steel mill. The important alloy in iron relating to hardening is carbon: iron with carbon is called steel. When steel is heated to its critical temperature it changes its structure. If this hot steel is quenched or quickly cooled, a hardened structure occurs. It can be cooled in air, oil, or water and is correspondingly called air hardening, oil hardening, or water hardening steel. After these steels have been quenched they are hard and brittle and have to be tempered or drawn. The steel is heated to a temperature below its critical temperature and cooled slowly under controlled conditions, which determine its final hardness. To anneal or soften hardened steel, it is heated to a temperature just below its critical temperature and slowly cooled under controlled conditions. Hardening and annealing specifications are provided by steel manufacturers.

Iron is case-hardened by a process called carburizing. It is heated in contact with carbon, and absorbs the carbon on the surface or "case." The depth of the case depends on the time and temperature of the heating. The iron in the case combines with carbon and is now steel and susceptible to hardening. It has a hard outside and a soft inside. The parts are then annealed to produce the required surface hardness. Mold parts are sometimes nitrided, which means they are subjected to ammonia gas at temperatures up to 1200°F for 50 to 90 hr. There is practically no distortion and a very hard, though thin, case is produced.

The drawbacks to hardening are distortion and the ever-present, but remote, possibility of cracking during heat treating. Corrections are difficult to make. To overcome this, a series of prehardened steels were developed. These steels are in the R_c 35 to 42 range and are readily machinable, though not as quickly or easily as soft steel. They are hard enough to give long satisfactory service. Most large size molds are made in prehardened steel.

8. Surface Finish

The surface finish of a mold is an important specification. It affects appearance, ejectability, and cost. Molds are polished by stoning with stones of different grades, starting with the coarsest and ending up with a number 400 or 600. The steel is then polished with diamond compounds, carborundum, or special "soaps" that contain polishing material.

Steel is highly susceptible to rust, especially after polishing, and should be protected by a suitable rust preventive when not in use. Molds that are run cold should be brought to room temperature before being coated with a rust preventive. If they are not, water condenses on the mold and causes rust damage, particularly in humid air. Using stainless steel or chrome-plating the mold prevents water damage. Maintaining a proper polish is the responsibility of the molder.

The second major material used in cavities is beryllium–copper. The material is an alloy of copper containing approximately $2\frac{3}{4}\%$ beryllium and $\frac{1}{2}\%$ cobalt.

An advantage of beryllium–copper is its high thermal conductivity (32). It adds to or removes heat from a mold several times faster than steel. Since the time required for cooling a plastic in the mold is a function of heat removal, a beryllium cavity should give faster cycles. If cooling is the limiting factor in the molding cycle, this is true. The costs of beryllium and steel cavities are very close, and the material selection depends on the mold properties desired. Beryllium–copper takes a very high polish, is not affected by water, and when flash chrome-plated, gives an excellent, durable, molding surface.

Other cavity and core materials used for sampling or very low production molds are brass, aluminum, and steel-filled epoxy. These are not materials of choice for production runs or quality parts.

9. Methods of Fabricating Cavities and Cores

Most fabrication of metals for molds is done with toolroom equipment. They are used for machining mold bases, cores, cavities, pins, blocks, etc. More advanced equipment permits running of toolroom equipment by electronically punched tape. This tape can be computor generated. For those not familiar with these methods, a very brief description follows.

A drill press is a tool that has a stationary table above which is a rotating motor-driven shaft. The shaft contains a chuck to which the drill or other tool is attached. The drill moves up and down. It can be hand or automatically fed.

A miller is a drill press with a table that can be moved left and right, backward and forward, and up and down. The rotating shaft or spindle in the head moves up and down. This is an indispensable tool. These movements can be automatically or manually controlled. A separate attachment has a stylus which moves to trace a three-dimensional replica of the part to be cut in steel. As the tracer (stylus) moves in one direction, the cutting tool moves in the same direction with a proportional movement. This is now called a duplicator.

A lathe has a rotating head to which is attached the material to be cut. The material rotates and the carriage, which contains the cutting tools, moves along the length of the bed or across it. The tail stock is equipped with a chuck for drilling and reaming.

A grinder has a rotating head to which is attached a grinding wheel. The table reciprocates left and right and can move in and out at a predetermined distance per reciprocation of the table. The height of the grinding wheel above the work is accurately adjusted. The grinder is used to obtain accurate dimensions and a good surface finish. It readily grinds hardened steel.

Band saws consist of two wheels around which rotate an endless belt of saw blades. Cut-off saws have straight blades which reciprocate.

Hobbing

Hobbing is the cold forming of metal. The term is used in plastics to designate the cold displacement of one material by another, caused by high pressures. For example, if a piece of plastic is left on a mold and the clamping pressures of the machine forces the plastic into the steel, the plastic is said to have hobbed itself into the mold. The term is used in mold making for the process that takes a hardened steel replica (hob) of the plastic part and, by means of high pressure, forces it into a soft iron block. Iron is very ductile. It flows around the hob, giving an identical but reversed impression. This is much the same as forcing a coin into a piece of clay.

Hobbing is a fast economical way to produce multiple cavities. All the cavities are the same size compared to each other and the hob. A high polish on the hob is transferred to the cavity. Since the cavities are iron they must be carburized after they are machined to size. Figure 13 shows the hob for a plastic column. Beneath the hob (a) is the molded plastic part (d), which is identical in shape. The size of the molded parts is smaller than the hob because the plastic shrinks on cooling (33).

Figure 13 (*a*) Hob; (*b*) hobbed cavity; (*c*) finished cavity; (*d*) molded part. Courtesy Robinson Plastics Corp.

Pressure Castings

Beryllium–copper cavities and cores are fabricated by pressure casting or, more accurately, hot hobbing. A hob is made proportional in size but larger than the finished plastic part. After hobbing, the beryllium shrinks as it cools just as plastic shrinks as it cools after molding. The hob is made of a good hot working die steel which will not deform under the temperatures and pressures of casting. For one or two cavities a different alloy of beryllium can be used for the hob.

The hob is placed in the bottom of an insulated cylindrical container. The melted beryllium–copper is poured over it and a plunger comes down, exerting pressure on the beryllium and hob. When the beryllium is cool the hob is separated and can be used again. The cavity has a surface finish depending on the quality of the hob. Because the beryllium is poured on in a liquid state, delicate fins and parts can be hobbed; this would be impossible with cold steel hobbing. There high pressure would cause the hob to snap. The dimensions of the cavities are not as accurate as those from cold hobbing because of shrinkage factors. For most purposes this is no problem.

When the parting line is uneven, hot hobbing can significantly reduce costs. The parting line can be cast so that a minimum of fitting is required. The cost of beryllium and steel cavities are similar, and the choice is made by engineering consideration (34).

Casting

Recent technique for casting is so improved that this method is readily adaptable for injection molds. Any metal can be successfully cast, particularly with the Shaw process. In this patented process a sample (or a plaster reversal) of the part in plastic, wood, metal, or other material is cast against a ceramic slurry. The slurry is fired and gives a reverse ceramic reproduction with a micrograin structure filled with small air gaps. The gaps act as vents so that the molten metal can achieve a good reproduction of the surface. The resultant cavity is not as dense as those produced by other methods, and there is the possibility of small pits. The appropriate shrinkage factors for the slurry, metal, and molding must be calculated. A new slurry casting must be made for each new cavity. The major advantages of casting are its speed and cost. A cavity can be made in less than a week. The economics depend on the size and nature of the part (35).

Electroforming

Electroplating is old. It was tried unsuccessfully for mold cavities and failed primarily because the stress in the plating caused a deformation during molding. This has now been overcome. A master, sometimes called a mandrel, is an exact reverse of the cavity. On it is plated approximately 0.15 in. of a nickel–cobalt compound. Behind that is electroplated copper, which is harder than mold steel. The finished cavity has good dimensional stability, is rustproof, has high thermal conductivity, and is very precise, within 0.0001 in.

These properties indicate the main reasons for using electroformed cavities. They are primarily used where accuracy is required, such as in gears, and where fine reproduction is needed. Irregular parting lines can be made with a near-perfect match. Irregular shapes that would be difficult to machine can be easily electroformed. Since the cavity is in the finished condition there is no distortion from hardening. The plating solution can be flushed into deep crevices, forming very narrow and thin slots.

Duplicating

Duplicating is mechanical reproduction by cutting tools which are guided by a master, proportional in size to the desired finished parts. Duplicating is mostly used for large parts, for hobbing and casting will usually reproduce a smaller one more economically. Large automatic duplicators are basically powerful horizontal millers with hydraulically controlled feeds. With feedback and electronic techniques maximum cutting speed is obtained. Such processes as producing mirror images are easy. They can be automatically run and tape controlled. Small duplicators are often used in making hobs or engraving small designs, letters, and numerals on cavities and making carbon electrodes for EDM'ing. A major disadvantage of duplicating is its poor surface finish (36).

Erosion (EDM)

An extremely useful method for removing steel is by electrical discharge machine (EDM). An electrode, usually made of carbon, though it can be made of any conducting material, is made in the reverse shape of the part to be produced. The steel and electrode are immersed in a circulating solution, which serves to flush away the eroded material and cool the work. Ac power is rectified and charged into a capacitor system. This discharge between the electrode and the cavity creates a spark which erodes the steel. The electrode is eroded about one-eighth as fast as the steel. Roughing electrodes are used to bring the cavity to its approximate shape and a finishing electrode brings it to size.

The process is accurate, produces good detail, can be used with hardened steel so that no heat distortion takes place, and can be used for cutting thin slots. By eroding on one plate the distance between cavities can be reduced. Cutting is relatively slow. The preparation of the electrodes and the operation of the equipment require an excellent toolmaker. Spark erosion is widely used in changing and correcting hardened steel activities (37).

Chemical removal of steel is slowly being accepted in the mold making industry (38).

10. Tolerance

A tolerance (39) is the total permissible variation of size, form, or location. It is indicated as a unilateral tolerance, in which the variation is from a given dimension in one direction only, or a bilateral tolerance, where the variation is permitted in both directions. Tolerances are important because they prescribe the limits for the part. Ideally, these limits should be controlled by function and aesthetics.

Suggested tolerances on molded parts in different plastics have been published by the Society of the Plastics Industry, Inc. (SPI). Figure 14 shows tolerances for polystyrene. If finer tolerances are required, the rejection rate rises significantly and the part should be priced accordingly.

Tolerances in mold making depend on the need for fits and the plastic dimensions.

11. Parting Line

When a mold closes, the core and cavity meet, producing an air space into which the plastic is injected. If one were inside of this air space and looking toward the outside, this mating junction would appear as a line. It appears so on the molded piece and is called the parting line. A piece might have several parting lines if it has cam or side actions. The expression "parting line" is usually restricted to that line which is related to the primary opening of the mold.

The selection of the parting line is largely influenced by the shape of the piece, method of fabrication, tapers, method of ejection, type of mold, aesthetic considerations, post-molding operations, venting, wall thickness, the number of cavities, and the location and type of gating.

12. Venting

When the hot plastic is injected into the mold it displaces air. In a well built, properly clamped mold without vents, the injecting material may compress the air to such an extent as to prevent proper molding. The heat of compression might burn the material. The force of the compression might open the mold, causing flash. The resistance of the compressed air might prevent the mold from completely filling. Vents are usually ground on the parting line. Their size depends on the nature of the material and the size of the cavity.

NOTE: The Commercial values shown below represent common production tolerances at the most economical level. The Fine values represent closer tolerances that can be held but at a greater cost.

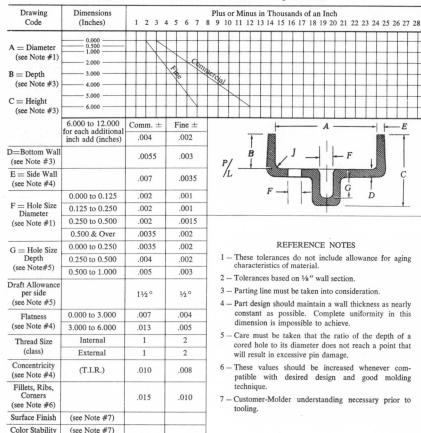

Drawing Code	Dimensions (Inches)		Plus or Minus in Thousands of an Inch
A = Diameter (see Note #1)	0.000 0.500 1.000 2.000		
B = Depth (see Note #3)	3.000 4.000		
C = Height (see Note #3)	5.000 6.000		
	6.000 to 12.000 for each additional inch add (inches)	Comm. ± .004	Fine ± .002
D=Bottom Wall (see Note #3)		.0055	.003
E = Side Wall (see Note #4)		.007	.0035
F = Hole Size Diameter (see Note #1)	0.000 to 0.125	.002	.001
	0.125 to 0.250	.002	.001
	0.250 to 0.500	.002	.0015
	0.500 & Over	.0035	.002
G = Hole Size Depth (see Note#5)	0.000 to 0.250	.0035	.002
	0.250 to 0.500	.004	.002
	0.500 to 1.000	.005	.003
Draft Allowance per side (see Note #5)		1½°	½°
Flatness (see Note #4)	0.000 to 3.000	.007	.004
	3.000 to 6.000	.013	.005
Thread Size (class)	Internal	1	2
	External	1	2
Concentricity (see Note #4)	(T.I.R.)	.010	.008
Fillets, Ribs, Corners (see Note #6)		.015	.010
Surface Finish	(see Note #7)		
Color Stability	(see Note #7)		

REFERENCE NOTES

1 – These tolerances do not include allowance for aging characteristics of material.

2 – Tolerances based on ⅛″ wall section.

3 – Parting line must be taken into consideration.

4 – Part design should maintain a wall thickness as nearly constant as possible. Complete uniformity in this dimension is impossible to achieve.

5 – Care must be taken that the ratio of the depth of a cored hole to its diameter does not reach a point that will result in excessive pin damage.

6 – These values should be increased whenever compatible with desired design and good molding technique.

7 – Customer-Molder understanding necessary prior to tooling.

Figure 14 Molding tolerances for polystyrene. Courtesy The Society of the Plastics Industry Inc.

A typical vent would be 0.001 in. deep and 1 in. wide. After the vent extended for $\frac{1}{2}$ in. the depth would be increased to 0.005 in. Clearance between knockout pins and their holes provide venting. Sometimes special pins are placed in the mold just for venting purposes.

The gate location has a lot to do with venting, and one is often restricted in gating because of the inability to completely vent the mold. Rapid injection is desirable in many moldings. Inadequate venting materially slows down the injection. Some authors have even gone so far as to recommend enclosing the mold and drawing a vacuum in it before molding. The location and size of vents are still governed mainly be experience. Vents are put in the obvious places before testing the mold. Additional vents are added as required.

13.   Ejection

After the part is molded it must be ejected from the mold. Parts are ejected by KO (knockout) pins, KO sleeves, stripper rings or stripper plates, either singly or in combination. The considerations for ejection are similar to those for parting lines. Additionally, the quality of the molded piece is affected. We do not consider undercuts at this point, except to say that an undercut is an interference by the mold to delay or prevent mechanical ejection of the plastic parts.

The geometry of the parts and the plastic material are the major factors in selecting the knockout system. Most parts eject readily with a taper of $1°$ per side. They can be ejected with smaller tapers, but this should be done only if required. A high polish is not always required for easy ejection. The direction of the polish is more important. Draw polishing (stoning and polishing in the direction of ejection rather than randomly) is important in difficult cases. With some materials, such as the olefins and nylon, fine sand blasting may help. Normally a moderately polished surface does not present ejection problems.

The cross-sectional area of the knockout pins or rings must be large enough so that the knockout does not damage the piece. Aside from the obvious, when the knockout pins go through the molded parts, serious stressing can be caused in the knockout area. Birefringence studies of transparent molded parts show this clearly. It is desirable, though not always possible, to use large-diameter knockout pins.

Figure 15 shows a stripper plate ejection system; the core pins are stationary. Around them are hardened stripper bushings which are mounted in the stripper plate or plates. There is clearance in the lower part of the stripper bushings to minimize wear. The knockout pins cause the stripper plate to move in relation to the core pin, leaving the part either on the plate or free to fall off.

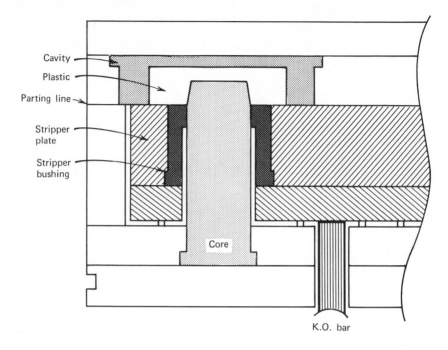

Figure 15   Stripper plate ejection system.

Sometimes it is necessary to eject in two stages. Figure 16 shows a double acting KO system. When the KO bar is activated the main KO plate moves forward ejecting by means of ejector pins A. As the ejector plate continues to move it hits the second ejector plate, which actuates ejector pins B. A separate push-back system is needed for returning the second KO plate.

Figure 17 illustrates a condition known as entrapped material. The item molded is a cup with an inner circular compartment. When the mold opens the molded piece remains on the core. The annular plastic ring A is entrapped between the core C and the annular steel ring B. This can be very difficult if not impossible to eject, particularly if A is thin. The proper way to build the mold is to have C retract in relation to the rest of the core before ejection.

In molding deep parts it is sometimes necessary to vent the core. If not, the vacuum prevents ejection.

Cam acting molds are common in the injection field. The cams are primarily used for molding parts with undercuts (40) and holes that if left in place would prevent ejection in the direction of the machine movement. They

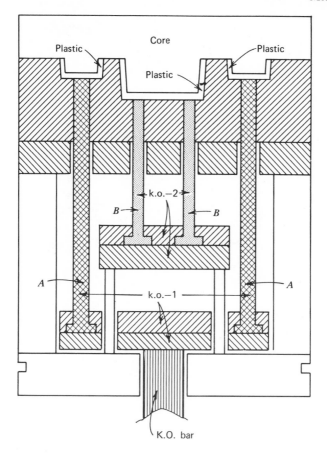

Figure 16   Double-acting ejection system.

are also used for engineering consideration relating to ejection, venting, and gate locations.

Internal threads can be molded automatically by using a "collapsible core" (41). Automatic unscrewing driving mechanisms include racks and pinions, gears, sprockets, electric motors, and hydraulic motors. Automatic unscrewing molds are considerably more expensive to build and maintain. One can also use inserts in the mold which are removed with the piece and unscrewed on the bench. Extra inserts are used to save machine time. Thought should be given to using threaded inserts in the mold, adding metal inserts in a post-molding operation, or tapping the hole in the plastic.

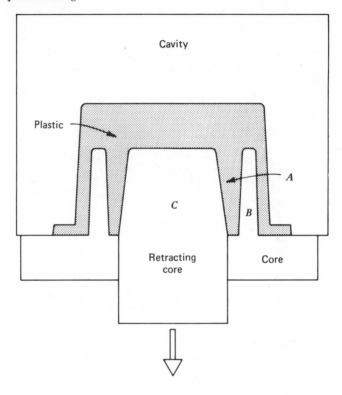

Figure 17   Use of retractable core to prevent ejection difficulties caused by entrapped material.

F.   RUNNERS

The runner is the connection between the sprue and the gate. It is a necessary evil. It should be large enough to allow rapid filling and minimum pressure loss, but not so large as to require the cooling cycle to be extended for the runner to harden enough for ejection. Most jobs permit the runner to be reground and reused. Regrinding is expensive, wastes material, is a source of contamination, a place for foreign material such as screw drivers and other metal parts to enter, and causes probable lowering of the physical specifications of the plastic.

The full round runner is preferable because, for a given cross section, it permits the greatest flow. It has the highest ratio of cross-sectional area to circumference, minimizing the cooling effect. The material feeds from the center which has the hottest material. When a runner has to be on one side only, the best compromise is a trapezoidal shape. Half round and rectangular runners should not be used. The runner should be polished. This gives less turbulence in the flow and slightly faster filling rates.

G. GATES

The gate is the connection between the runner and the molded part. It must permit enough material to enter and fill the cavity, plus the extra amount required to prevent excess shrinkage. The literature is full of articles relating to the size, type, and location of gates and their effect on the molding process and the physical properties of the molded part.

Gates can be classified as large or restricted (pin pointed). Restricted gates are circular in cross section and for most materials do not exceed 0.060 in. in diameter. The more viscous materials may have restricted gates as large as 0.115 in. in diameter. An example of a large gate, which is usually square or rectangular, is $\frac{1}{4}$ in. wide by $\frac{3}{16}$ in. high. They are used for molding heavy sections and where the restricted gates give a surface blemish problem.

The restricted gate is successful because the viscosity of the plastic is sensitive to the shear rate. The faster it moves, the less viscous it becomes. As the material is forced through the small opening its velocity increases. The velocity is directly related to the shear rate. In addition, some of the kinetic energy is transformed into heat, raising the local gate area temperature. Once the gate is opened to the extent where it loses this shear rate–viscosity improvement, a much larger opening is required to get any reasonable flow. This is why there is a jump in size from a restricted to a large gate.

Gates are also described by location, such as edge gated, back gated, submarine gated, tab gated, and nozzle gated. Figure 18 shows examples of various types of gates. A sprue gate feeds directly into the piece from the nozzle of the machine or a runner. It has the advantage of a short direct flow, with minimal pressure loss. Its disadvantages include the lack of a cold slug, the possibilities of sinking around the gate, the high stress concentration around the gate, and the need for gate removal. Most single-cavity molds of any size are gated this way.

Edge gating is the most common. It can be the large type or restricted. If the edge gate is spread out, it is called a fan gate. If the gate is extended for a considerable length of the piece and connected by a thin section of plastic, it is called a flash gate. Sometimes it is necessary to have the gate impinge upon a wall. This distributes the material more evenly and gives improved surface conditions. Walls are not always available. To overcome this a rectangular tab is milled into the piece and the gate is attached there. This is called a tab gate.

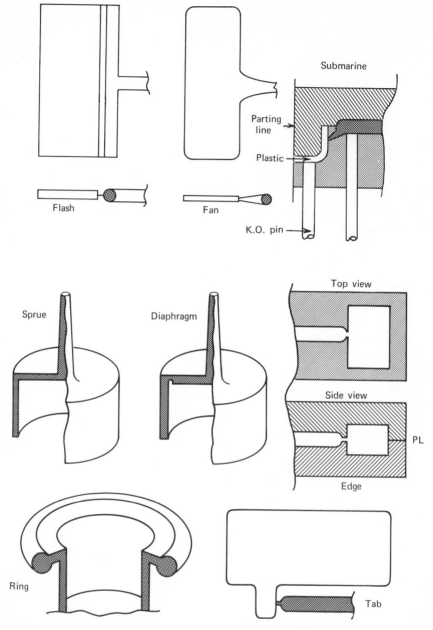

Figure 18 Various gating designs. Courtesy Robinson Plastics Corp.

In gating into hollow tubes, flow consideration can require an even injection flow pattern. A single gate is not sufficient. Four gates 90° from each other often give four flow lines down the side of the piece, which can be objectionable. To overcome this a diaphragm gate is used. The inside of the hole is filled with plastic directly from the sprue and acts as a gate. It must be machined out later. A ring gate accomplishes the same thing from the outside.

A submarine gate is one that goes through the steel of the cavity. When the mold opens the part sticks to the core and shears the piece at the gate. A properly placed knockout pin, using the flexibility of the plastic, ejects the runner and pulls out the gate. This type of gate is usually used in automatic molds.

Restricted gates have the benefit of better mixing. It is virtually impossible to mold a good variegated pattern (mottle) without using a large gate. Dispersion or mixing nozzles on the machine use the principle of the restricted gate. Many small restrictions are placed on a plate inserted between the nozzle and the cylinder. This restricts the flow because the effective cross section is much smaller than in an unrestricted nozzle (42).

In multiple-cavity molds, the gate size must be adjusted so that all the cavities fill at the same rate. If not, severe molding and dimensional problems may develop. Such parts are more subject to long-term failure.

H. TEMPERATURE CONTROL

Accurate control of the mold temperature is required for consistent molding. Refrigerated water, hot water, and heating media above 200°F are required for each machine. It is not always possible or necessary to predict the best temperature for a given mold and material. With thermostatically controlled temperatures, trial and error is not difficult. In many instances there are several different temperatures maintained for different parts of the mold. Equipment is available to provide refrigerated water either from one central unit or from smaller portable chillers. Separate portable heaters are employed where the circulating fluid is above room temperature. Heat can also be supplied electrically with bands, cartridges, and strip heaters. They are best controlled with pyrometers rather than proportioning controls or auto transformers.

Even though the expression "heating a mold" is used, the purpose of the temperature control system is to remove heat from the plastic part at a controlled rate. Obviously the lower the temperature of the controlling fluid, the quicker the heat is removed. The amount of heat removed depends on the material, the metal in which it is contained, the size of the cooling channels, their locations in relation to the molded parts, the cleanliness of the channels, the rate of flow of the heat exchange fluid, and its temperature. Air is an effective insulator and it is desirable to locate the cooling channels in the cavity and core themselves. The minimum size of cooling channels should by $\frac{1}{4}$ in. pipe, though $\frac{3}{8}$ and $\frac{1}{2}$ in. are preferable.

Mold temperature control is so important that molds are built at considerable extra cost to achieve greater cooling (43, 44).

I. AUTOMATION

Many times one hears the expression "automatic machines" when automatic molding is discussed. This is redundant for all machines today are automatic. What makes automatic molding automatic is the mold. There are a number of requirements for automatic molding. The machine must be capable of consistent, repetitive action. The mold must clear itself automatically. This means that all the parts have to be ejected using a runnerless mold, or that the gate and the parts have to be ejected in a conventional mold. There usually is some method for assisting in the removal of the pieces, usually in the form of a wiper mechanism or an air blast. Some systems weigh the shot after ejection and stop the machine or sound an alarm if the shot is too light. All machines used automatically must have a low-pressure closing system which prevents the machine from closing under full pressure if there is any obstruction between the dies. The machine is shut off and/or an alarm is sounded.

Automatic molding does not necessarily eliminate the operator. Many times an operator is present to pack the parts and perform secondary operations. Automatic molding gives a better quality piece and a more rapid cycle. Usually in automatic molding an experienced person attends several machines. Unless the powder feed and part removal are automated this person takes care of them.

Automation is expensive to obtain. It requires excellent machinery, controls, molds, trained employees, and managerial skill. When the quantity of a part permits, it is a very satisfactory and economical operation.

Now that we have an overall view of the injection molding process and equipment, let us examine the behavior of the plastic in molecular terms.

J. MOLECULAR ASPECTS OF INJECTION MOLDING AND PHYSICAL PROPERTIES OF MOLDED PARTS

The polymeric materials used in the molding process may be all amorphous or part amorphous and part crystalline and require varying amounts of energy input to provide the plastic flow condition that is necessary to obtain good moldings. The energy is made up of frictional or shear energy and the thermal energy from the heating units on the machine. In this condition of plastic flow the viscosity of the melt is sufficiently low so that the injection pressures available from the machine can force the plastic into the relatively cold molds.

On a molecular basis, as the temperature increases the energy absorbed by the polymer chains causes vibrational, rotational, and translational or segmental motion of the polymer molecules. This is essentially Brownian motion, and tends toward a random molecular arrangement in the viscous liquid mass of molecules.

If a unidirectional force is applied to a polymer above its glass-transition temperature, it begins to move in the direction away from the force. If the force is applied very slowly, so that the Brownian movement can overcome the orienting forces caused by the flow, the mass of the polymer moves with a rate proportional to the applied stress. This is Newtonian flow, as discussed in Chapter 8. As the molecules move more rapidly (under the influence of higher pressures in the injection process), the chains tend to orient in the direction of flow. Chains are moving so rapidly that there is not sufficient time for the Brownian motion to have an appreciable effect. In addition the molecules, being oriented in one direction and being less entangled, tend to slide over each other more easily. Therefore, the increased shear rate is no longer proportional to the shear stress. The unit increase in shear stress gives a larger increase in shear rate than would occur with a Newtonian liquid. This is characteristic of plastic or polymer flow; that is, shear rate is no longer linearly proportional to shear stress. This means that generally one may assume that injection molding (or extrusion) causes a degree of orientation of the molecules as they are transported to the sprue and gate and are about to be injected into the mold.

As the flow rate increases it reaches the final stage where all the polymer molecules have oriented to their maximum level. There is no further untangling. Therefore any increase in the shearing stress gives a proportional increase in the shearing rate and the material acts as a Newtonian fluid.

As one would expect, the viscosity or resistance to flow is temperature dependent. The relationship is expressed by the Arrhenius equation

$$\eta = A \cdot \exp\!\left(\frac{E}{RT}\right) \qquad (9)$$

where η = viscosity
A = constant depending upon the material
R = gas constant
T = temperature ($^\circ$K)
E = activation energy for viscous flow

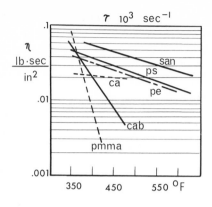

Figure 19 Effect of temperature upon polymer viscosity. san, Styrene–acrylonitrile; ps, polystyrene; pe, polyethylene; ca, cellulose acetate; cab, cellulose acetate butyrate; pmma, polymethyl methacrylate methacrylate (acrylic).

Plots of the logarithm of the viscosity versus any linear function of the temperature are a straight line over narrow temperature ranges and describe viscous flow fairly accurately (Fig. 19); the slope of the line is E. It is a measure of the temperature dependency of the viscosity. The higher the activation energy of viscous flow, the more highly dependent is the viscosity on the temperature.

A graph of this kind has practical value in molding. For example, if a part molded in cellulose acetate were not filling out, raising the temperature would not help significantly. On the contrary, an acrylic part would be very sensitive to any increase in melt temperature. Accurate cylinder temperature control is much more important in molding acrylic or butyrate than in molding acetate, styrene, polyethylene, or acrylonitrile.

Orientation

Let us now consider what happens when the material enters the cold mold. As the material hits the cold wall it freezes. Regardless of the orientation caused by the gate, the turbulence in the flow is enough to randomize the outside molecular layer. This outside frozen layer is therefore relatively unoriented. Consider the next layers of polymer. Part of the polymer is frozen in the outside layers. The flow of the material is pulling the remainder of the molecules in that direction. We would expect these layers to be the most highly directional or oriented. For the same reason it would be the most highly shear stressed. Nearer the center of the part the molecules have less orientation caused by flow. Because the outer layers thermally insulate the inner portion, it remains warmer longer, allowing more time for Brownian movement and disorientation.

If a clear injection molded part is placed between two polarizing filters and one is rotated, a characteristic series of colored bands appear. They are

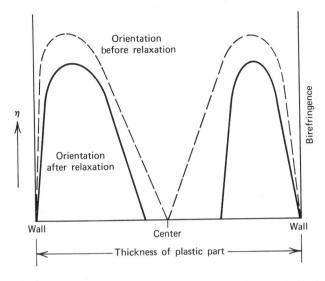

Figure 20   Amount of birefringence versus distance from wall. (Ref. 45a.)

primarily related to orientation stress in the part. This is called birefringence.

Figure 20 shows a schematic representation of the amount of birefringence (in the molded part) versus the distance from the mold wall. The dotted line indicates the initial condition and the solid line the final condition when the part cools. Measurements of various moldings were taken with the aid of a polarizing microscope (45). The peak orientation was between 0.025 and 0.030 in. from each side of a 0.100-in. thick slab. The two peaks were not identical in each specimen, reflecting the difference in temperature of each side of the mold.

If our interpretations are correct, milling off about one-third of either side produces a part that has molecules randomly placed on one side and highly oriented on the other side. If the specimen is heated above its T_g, the oriented molecules should assume a more random structure, thus causing a shrinkage of the oriented portion. The specimen should act like a bimetallic unit and bend in the direction of the oriented layer. This is what occurs (Fig. 21) (46).

What is the effect of orientation on the physical properties of the parts? Consider regular polystyrene sheet, which has a tensile strength of 6000 to 7000 psi and is quite brittle. Heat the sheet slightly above its T_g and stretch it. Chill it while it is under tension, to retain its orientation. The tensile strength now is 9000 to 12,000 psi, depending on the percent elongation and processing temperature. The brittleness disappears. If the material were allowed to cool slowly its orientation would disappear and the properties would be similar to the starting sheet (47).

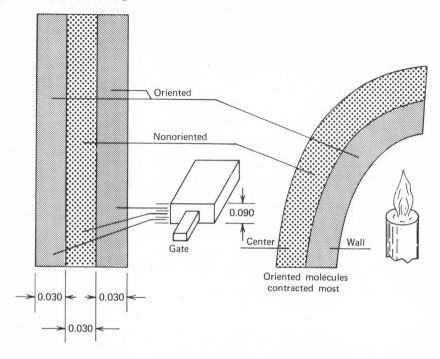

Figure 21 Demonstration of the effect of orientation—nonuniform shrinkage. (Ref. 46)

The effect of orientation, as measured by birefringence, upon the tensile strength, elongation at failure, and notched Izod impact test is shown in Fig. 22.

There are two forces that hold the polymer together. C–C linkages have a disassociation energy of 83 kcal/mole, and van der Waal's forces have a disassociation energy of 3 to 5 kcal/mole. The latter are electrostatic in nature and decrease exponentially with the sixth power of the distance.

One would expect, therefore, increased tensile strength in the oriented direction of flow because there are more C–C linkages to be broken. These are much stronger than the van der Waal's forces, which are the major forces holding the polymer together perpendicular to flow. This is also true for the percent elongation at failure and the notched Izod impact.

How Molding Conditions Affect Orientation

Molding conditions, the thickness of the part, and gate size affect orientation. The net orientation is the difference between the orientation caused by flow and that lost by relaxation. One would expect that a higher stock (or material) temperature would give less orientation because hotter material relaxes more. This is ordinarily the case. However, in thin sectioned parts the hotter material may permit faster flow. With large gates hotter material keeps the gate open longer, increasing the flow time and hence the birefringence.

The higher the mold temperature, the more time for relaxation and the lower the orientation. Increasing cavity thickness has a dramatic effect on decreasing overall orientation. Because of the low thermal conductivity of plastic the interior remains hotter longer, increasing relaxation.

Increasing the ram forward time materially increases orientation by permitting significantly longer flow time. This effect stops when the gate seals off. Higher injection pressures on the material increase orientation. They keep the gate open longer, permitting longer flow time. In addition, the pressure compresses the molecules inhibiting Brownian movement. The

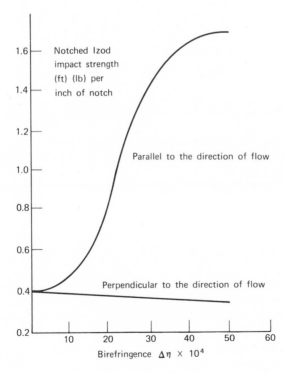

Figure 22 (*a*) Effect of orientation on impact strength. (Ref. 45)

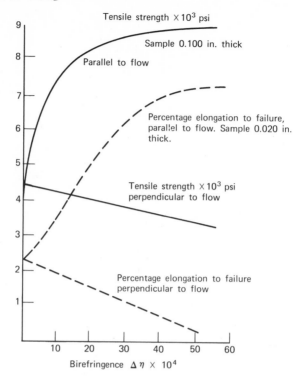

Figure 22 (*b*) Effect of direction of flow upon physical properties of polystyrene as measured by birefringence. (Ref. 45)

larger the gate size, the longer the seal off time, assuming the appropriate molding conditions.

We have seen how some physical properties vary depending on the direction of flow as evidenced by orientation. From our concept of molecular structure we would expect that there would be more shrinkage in the direction of flow. This is because the C–C linkages stretch, whereas the other molecular forces, once decreased by distance, have no mechanism to force them back.

The effect of different shrinkage parallel and perpendicular to flow is considerable in injection molding. Let us consider the molding of a center-gated 4-in.-diameter 0.060-in.-thick polypropylene disk, (Fig. 23). Consider a segment encompassing a 60° angle. When the material flows in hot, it has a 2 in. dimension on each side of the triangle. The polypropylene typically shrinks 0.020 in./in. in the direction of flow and 0.012 in./in. perpendicular to flow. When the material cools the radius is 1.960 in. and the chord 1.976 in.

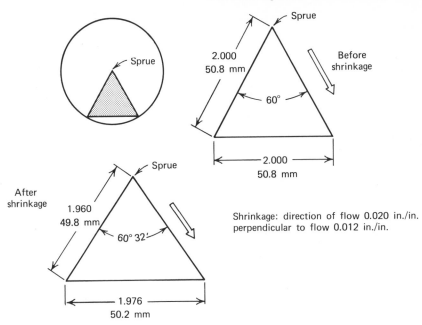

Figure 23 Warping of center-gated polyolefin part caused by different shrinkage parallel and perpendicular to flow.

The new angle is 60° 32′. For the whole 360° circle, there is an increase of 3.20°. Unless the material is rigid enough to overcome this stress, there is warping to allow for the extra 3.2° of material. If this were a thin-walled cover it would mean that regardless of what was done during molding, a warpfree part could not be produced.

If a rectangular tray of the same material were molded (Fig. 24), a center gate would give a distorted tray unless the walls were heavy enough to overcome the stress. If the part were gated in diagonal corners there would be twisting. If the part were edge gated at one end it would be warpfree but the material would flow around the rim on the parting line and trap air, giving either burns or poor welds. The best way to gate the piece would be to place two gates at one end. This would give maximum linear flow without air entrapment and produce a warp-free part.

One well-known cause for stress is packing at the gate. As the polymer cools there is a limited volume into which can flow the extra material forced in by the injection ram. The extra material is required to minimize shrinkage. If too much material is forced in it will be overstressed and fail immediately or

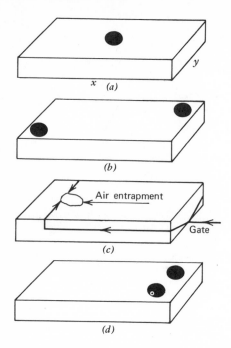

Figure 24   Effect of gate location on polyethylene tray.

at a later date; failure may also occur because of special environmental conditions.

The main parameter is the ram forward time. In molding a center-gated polyethylene tumbler with a ram forward time of 10 sec, there were no failures the first day and 7% 14 days later. Increasing the ram forward time to 25 sec gave a 1 day failure of 70% and a 14 day rate of 88%. Injection pressure and gate size also affect packing. Although packing is concentrated at the gate it can extend throughout the parts. It is a prime cause for sticking in a mold. This is the result of better adhesion to the mold surface, deflecting the mold and possibly distorting the core. Obviously the amount of feed and the temperature affect the amount of plastic that flows into a cavity and hence the packing.

All materials shrink when they cool; however, crystalline materials shrink to a greater extent than do amorphous materials. The degree of crystallization depends on the rate of quenching or temperature drop to the mold temperature. Thus this temperature is a critical variable in influencing finished part properties. Thus a relatively cool mold tends to freeze in the amorphous arrangement of the melt with lower levels of crystallinity. On the other hand, warmer molds permit the polymer to cool more slowly and thus permit greater levels of crystallization. The increased crystallinity results in greater shrinkage, higher tensile strength, lower elongation, and greater hardness.

When a part is molded a number of complex and opposing forces come into play. The hot material is injected into the cavity, where it shrinks as it cools. The material is compressed by the injection pressure, permitting more material to flow into the given volume, to compensate for the shrinkage. The part is cooling during the injection process, continually reducing its volume. The amount of additional material forced in depends on the gate size, injection rate, temperature conditions, ram forward time, and pressure. This is why it is difficult to maintain really fine tolerances for injection molded parts. The goal of proper injection molding is to so balance the machine and mold conditions that acceptable parts are produced.

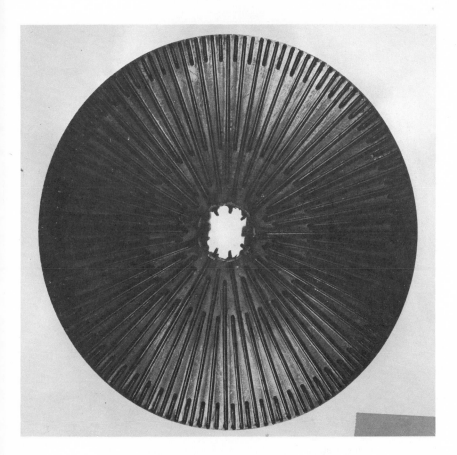

Figure 25   12-in.-diameter filter disk, poorly designed.

K. EXAMPLES OF INJECTION MOLDED PARTS

A number of injection molded parts have been selected to illustrate some of the principles discussed. They were chosen by carefully considering the needs of the end user, molder, and mold maker.

Figure 25 shows a 12-in.-diameter filter grid molded in high-impact polystyrene, over which is sewn a polypropylene fiber bag. A number of these are stacked on a center tube, the unit enclosed in a container, and filtering medium put in the container outside the bags. Dirty water flows through the filtering medium on to the grids, and the filtered water is carried off through

Figure 26   12-in.-diameter filter disk, overdesigned. Courtesy American Machine Products Inc.

the center hollow tube. A significant number broke radially while in operation: more broke when the user dismantled and cleaned the unit.

The original end user and molder had decided to economize. The thickness was minimal. Because the part was center gated, the strength was much stronger in the direction of flow. The hoop or circumferential strength was minimal. Because the ribs (which were used to keep the polypropylene bag off the surface of the filter grid) were in the direction of flow, they could not contribute to the strength of the piece in the circumferential direction.

Another supplier decided to build a disk engineered to overcome these faults. Figure 26 shows the results. The part is thicker and reinforced circumferentially where it is weakest. The other side of the piece has identical ribs

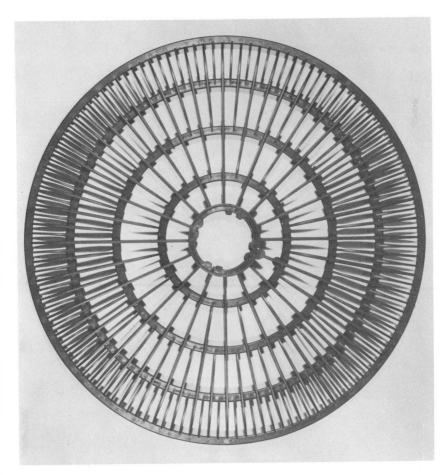

Figure 27   12-in.-diameter filter disk, properly designed.

but displaced one-half the angular distance of the segment, so that there was always at least one rib supporting the flat part of the filter disk. The part was overdesigned because of consumer resistance caused by the "cheaper" unit. After consumer acceptance was established another unit was made (Fig. 27) which was sufficiently strong and considerably less expensive.

One of the advantages of injection molding is that parts can be produced which require little or no finishing. An advantage of plastic is its corrosion resistance, particularly for water applications. Figure 28 shows the parts of a four-way valve for swimming pool applications. It was originally made in

Figure 28 Polycarbonate replacement of aluminum die-cast valve which overcame corrosion and cost problems. Courtesy American Machine Products Inc.

aluminum die casting. The rough casting had to be machined and the valve assembled. After 1 year many were returned because corrosion caused the valve to jam. When leverage was used on the handle, the "freeze" was so severe that the metal handle bent or broke.

The part was converted into polycarbonate. The design was chosen to be very close to the die-cast part for two reasons. Die casting is very much like injection molding. There were a large number of valves in the field and the outside dimensions had to be similar. There was also a large stock of internal parts for the metal valve.

Figure 29   Lamp shade, 8 in. high and 4 in. in diameter, molded with a four-cam mold. Courtesy Robinson Plastics Corp.

The only machining required was to tap the eight holes that hold the cover and one body hole for a pipe connection; converting to polycarbonate reduced the cost by two-thirds. The tooling cost for the plastic part was approximately double. Rejection in the field because of corrosion was eliminated.

A lamp part 8 in. high and 5 in. in diameter molded of general-purpose, heat-resistant, antistatic polystyrene is shown in Fig. 29. The bottom cap is solvent cemented after molding. The material was selected for its low cost, beauty, and ease of molding. The main problem was the tool design. In order to achieve aesthetically pleasing triangular cutouts, four cams were required. The cutout effect could have been achieved with two cams, but the triangles close to the parting line would be severely distorted. The cam blocks were massive and held together by a large ring which acted as a cam lock. To prevent tearing the cams move at right angles to the core. The cams are on the ejection side and are activated by the pins that are attached to the injection side. The cores must be maintained accurately in their position when the machine closes. If not, the core activating pins will not locate. If one visualizes the mold in a horizontal press one can see that the top cam must be supported against the pull of gravity. This is done with springs and detents. The cams and knockout plate are electrically interlocked. If the knockout plate is not completely back, the cams hit the knockout pins, damaging the mold.

These few illustrations serve to show some of the factors that are considered in designing and producing a plastic part. It would be good practice to purchase a number of inexpensive parts and try to determine what the engineers did in the design and production of each item.

REFERENCES

1. L. I. Naturman, "Machinery in Place," *Plast. Technol.*, 37 (Feb. 1970).
2. "Giant Machines for the Big Jobs," *Mod. Plast.*, 101 (Jan. 1969).
3. T. Erwin, "Why Choose a Hydraulic Clamp," *Plast. Technol.*, 35 (Jan. 1968).
4. T. Debreccени, "Why Choose a Toggle Clamp," *Plast. Technol.*, 39 (Jan. 1968).
5. "New Standards for Plasticating Performance of Screw Injection Machines," *Mod. Plast.*, 107 (March 1968).
6. E. G. Fisher and W. A. Maslen, "Preplasticizing in Injection Moulding," Part 1, *Br. Plast.*, 417 (Sept. 1959); R. Wood, Part 2, 468 (Oct. 1959); E. G. Fisher and W. A. Maslen, Part 3, 516 (Nov. 1959).
7. C. J. Waechter and L. J. Kovach, "Converting to Screw Plastication," *Mod. Plast.*, 125 (March 1963).
8. R. B. Staub, "How to Get Best Performance in Screw Injection Molding," *SPE J.*, 1182 (Nov. 1963).
9. "Screw Injection Molding Analyzed," *Plast. Des. Process.*, whole issue (Oct. 1964).

10. L. J. Kovach, "Injection Molding with the In-Line Single Plasticizer Screw," *Plast. Des. Process*, 23 (Dec. 1962).

11. H. Frimberger and J. G. Fuller, "Single Screw Injection Molding," *Plast. Technol.*, 53 (May 1961).

12. W. C. Filbert, "Screw Plastication—A Status Report," *Mod. Plast.*, 123 (July 1964).

13. T. C. Bishop, "Screw Machine Operating Characteristics," *SPE J.*, 459 (May 1963).

14. W. G. Kriner, "Graphic Comparison of Screw and Plunger Machine Performance," *Mod. Plast.*, 121 (May 1962).

15. C. L. Weir and P. T. Zimmermann, "Ram vs. Screw Injection," Part 1, *Mod. Plast.*, 122 (Nov. 1962); Part 2, 125 (Dec. 1962).

16. L. W. Meyer and J. W. Mighton, "Ram vs. Plunger Screw in Injection Molding," *Plast. Technol.*, 39 (July 1962).

17. R. M. Norman and R. J. Lindsey, "Electric or Hydraulic Screw Drives for Injection Machines," *Plast. Technol.*, 31 (April 1966).

18. J. Newlove, "Hydraulic Injection Screw Drives," *Mod. Plast.*, 237 (May 1966).

19. M. Jury, "Screw Injection Machine Design," *Mod. Plast.*, 119 (April 1965).

20. E. Bauer, "Non-return Valves in Screw Injection Moulding Machines," *Br. Plast.*, 83 (Feb. 1969).

21. A. R. Morse, "Reciprocating Screw Tip Designs," *Plast. Des. Process.*, 18 (Feb. 1969).

22. A. R. Morse, "Reciprocating Screw-Tip Shutoffs," Part 1, *Plast. Technol.*, 46 (July 1967); Part 2, 46 (Aug. 1967).

23. *Industrial Hydraulics Manual 935100*, Vickers, Inc., Troy, Mich., 1965.

24. *Basic Course in Hydraulic Systems: Machine Designs*, Penton Publishing Co., Cleveland, Ohio. Excellent.

25. *Fluid Power Directory*, Penton Publishing Co., Cleveland Ohio. Contains circuits and sources.

26. Pippenger and Koff, *Fluid Power Controls*, McGraw-Hill, New York. Good on components and circuits.

27. J. Newlove, "How to Trouble-Shoot Your Injection Molding Machine," *Plast. Technol.*, 39 (Nov. 1966).

28. R. G. W. Pye, *Injection Mould Design*, Illiffe, London. Strongly recommended.

28a. J. R. Schettig, "How to Select Mold Steels," *Plast. Technol.*, 36 (Nov. 1964).

29. "Hot Runner Systems," *Plastics*, 1311 (Nov. 1967).

30. G. B. Thayer, "New Standards of Hot-Runner Molds," *Mod. Plast.*, 92 (March 1969).

31. J. N. Scott, D. L. Peters, and P. J. Boeke, "Insulated Runner Systems for Injection Molding," *SPE J.* (Sept. 1959).

32. W. J. B. Stokes, "Thermal Consideration in Mold Design," *SPE J.*, 417 (April 1960).

33. I. Thomas and E. W. Spitzig, "How and When to Hob," *Mod. Plast.*, 115 (Feb. 1955).

34. I. Thomas, "How to Hot Hob Beryllium Copper," *Mod. Plast.*, 101 (July 1961).

35. I. Lubalin, "Cast Mold Cavities," *Mod. Plast.*, 147 (Oct. 1957).

36. I. Thomas, "The Art of Engraving," *Int. Plast. Eng.*, 496 (Nov. 1961); I. Thomas and R. Koegl, *Mod. Plast.* (May 1945).

37. P. J. C. Gough, "Tool Making by Spark Erosion," *Int. Plast. Eng.*, 399 (Sept. 1961).

38. R. Wolosewicz, "Electrochemical Machining," *Mach. Des.*, 160 (Dec. 11, 1969).

39. "Dimensioning and Tolerancing," *MIL-STD-8B 16*, Nov. 1959.

40. G. Ward, "Undercuts on Injection Mouldings," *Int. Plast. Eng.*, 274 (July 1961).

41. J. Andras, "Collapsible Core," *SPE J.*, 35 (May 1967).

42. D. G. Briers, "Feeding (Gating) Techniques for Injection Moulds," *Int. Plast. Eng.*, 102 (April 1961); 166 (May 1961).

43. H. A. Meyrick, "What You Should Know About Mold Cooling," *Mod. Plast.*, 219 (Oct. 1963).

44. L. Temesvary, "Mold Cooling; Key to Fast Molding," *Mod. Plast.*, 125 (Dec. 1966).

45. G. B. Jackson and R. L. Ballman, "The Effect of Orientation on the Physical Properties of Injection Moldings," *SPE J.*, 1147 (Oct. 1960).

45a. R. L. Ballman and H. L. Tour, "Orientation in Injection Molding," *Mod. Plast.*, 113 (Oct. 1960).

46. W. Woebcken, "Effect of Processing Techniques or Structure of Molded Parts," *Mod. Plast.*, 146 (Dec. 1962).

47. C. T. Hathaway, "Orientation Characteristics in Plastic Film and Sheet," *SPE J.*, 567 (June 1961).

DISCUSSION QUESTIONS AND PROBLEMS

1. What are the advantages and disadvantages of injection molding?

2. What are the advantages of (a) preplasticizing, (b) screw plasticizing?

3. Describe the flow pattern of plastic in the injection process.

4. How does orientation occur and why is it important?

5. Obtain at least 5 different samples of injection molded parts and answer the following on each:

    a. Why was the particular material used?

    b. What criteria might have influenced the design of the part?

    c. Was orientation considered?

    d. Is the part satisfactory?

    e. Can you improve it?

CHAPTER 11 ⎯⎯⎯⎯⎯⎯⎯⎯

Polymer Fabrication Processes

JOHN M. McDONAGH
Stauffer Chemical Co.
Westport, Connecticut

A. INTRODUCTION

Processing of plastics covers two broad areas. One is the production of the polymer into some form, for example, pellets and powder; the second is the fabrication of the polymer into end use items. As each year passes, new fabrication processes come into existence and new modifications arise for the older methods.

The two most common fabrication methods, extrusion and injection molding, are covered elsewhere in this book. The other primary processes available for fabrication of polymers are as follows:

Blow molding
Rotational molding
Thermoforming
Thermoset molding (compression and transfer molding)
Foamed plastic processing
Cold forming
Glass reinforcement

Other fabrication techniques for plastics are used but these are of minor importance compared to the above methods.

This chapter summarizes each of the above methods to give the reader an overall view of the process, its capabilities, and limitations. The general references listed amplify and expand on the discussion of various fabrication methods in the chapter.

The section on thermosets presents the basic molding techniques used for these materials. It should be mentioned that thermosets may be processed by other fabrication techniques including glass reinforcement methods, rotational molding, and potting.

A number of these processing techniques may be used to produce a given item. For example, a plastic pallet can be produced by blow molding, rotational molding, injection molding, and thermoforming. The choice of the best method is dictated by cost and the performance requirements of the pallet. This is true for all plastic applications—they must be evaluated on a cost performance basis. Though this seems obvious, numerous items have been produced that meet cost or economic requirements, but fall short on performance requirements. During the 1960s and early 70s the advent of engineering resins, improved fabrication techniques, and greater demand on engineering design of plastic items has improved the cost-performance of plastics so they now compete very effectively with metals.

B. BLOW MOLDING

Blow molding was practiced with glass in ancient times. In the early 1900s doll components were produced commercially from celluloid. Today countless objects are blow molded ranging in size from less than 1 oz to a few hundred gallons (Fig. 1). Polyethylene resins are used in the largest

Figure 1 Various bottles produced on blow molding machines. Courtesy Kautex Machines, Inc.

volume for blow molding. High-density polyethylene compromises 85% of the market. Other resins used are ABS, acrylics, cellulosics, polypropylene, plasticized and rigid PVC, nylons, ethylene copolymers, polystyrene, acrylonitrile, acetates, and polycarbonate.

1. Process Description

In blow molding a molten tube of thermoplastic (*parison*) is contained in a mold having the shape of the part to be produced. A gas, usually air, is introduced into the tube to expand it against the walls of the mold where it solidifies. From this process a hollow object is produced, as in rotational molding and "clamshell" thermoforming (to be described later). With the recent introduction of larger blow molding machines this process has become competitive with rotational molding in the production of large hollow objects.

Generally the process consists of three stages: melting resin, formation of the parison, and blowing the parison to produce the final shape. The main steps of a blow molding cycle are shown in Fig. 2. The blowing step may take from a few seconds to more than a minute for large parts. As in most processes the rate limiting step is cooling of the molded parts. Production rates for various size containers are given in Table 1.

2. Plasticizing the Resin

There are two techniques of plasticizing resin (making the material flow) and forming the parison:

1.   Extrusion, which produces a continuous parison that has to be cut: this is the most common method used.

2.   Injection molding: the parison is formed in one mold, then transferred into another mold for blowing.

Extrusion blow molding uses many arrangements for making and forming the parison. Figure 3 shows the cross-section of an extrusion blow molding machine. The function of the extruder is to melt, plasticize, and mix the resin, and then deliver it to the die(s) at the proper temperature and rate to be formed into the parison(s).

In the simplest method of extrusion blow molding, a mold is mounted under the die. The parison is extruded between the open halves of the mold. When the parison reaches proper length, the extruder is stopped and the mold closes around the parison. Air is injected through a blow pin to form the product. Sufficient time is allowed for the part to cool before the mold is opened and another cycle is started. To use the full capacity of the extruder numerous systems are used.

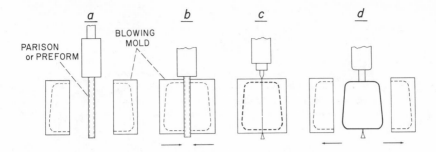

Figure 2 Schematic of the blowing stage, (*a*) The molten, hollow tube—The parison or preform—is placed between the halves of the mold. (*b*) The mold closes around the parison. (*c*) The parison, still molten, is pinched off and inflated by an air blast which forces its walls against the inside contours of the cooled mold. (*d*) When the piece has cooled enough to have become solid, the mold is opened and the finished piece is ejected.

Table 1   Blow molding performance data[a,b]

Type of Container	Capacity	Weight (g)	Average Production[c] (no./hr)	Die Heads
Milk	$\frac{1}{2}$ gal	50	2250	5
	1 gal	90	1200	4
Oil	1 qt	40	1600	5
Detergent	12 oz	30	1500	5
	22 oz	40	1390	5
	32 oz	50	1285	5
Bleach	1 qt	45	1500	5
	$\frac{1}{2}$ gal	75	1000	4
	1 gal	120	675	3
Dry cleanser	21 oz	23	2225	5
	35 oz	36	1800	5

[a] Continuously reciprocating screw, $3\frac{1}{2}$-in. extruder, 380 lb/hr HDPE.
[b] Data courtesy Uniloy Division, Hoover Ball and Bearing Co.
[c] Production varies according to design and unfinished weight.

559

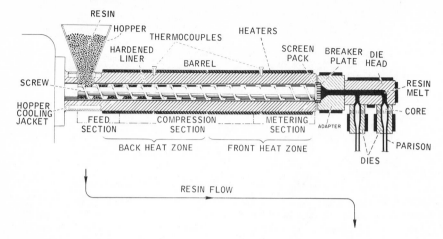

Figure 3   Cross-section of an extrusion blow molding machine. Courtesy USI Chemical Company.

The simplest is to have several molds nearby and when the parison reaches the proper length manually place it into a mold. That mold is closed and the blowing and cooling are accomplished while another parison is being extruded. The hand operating features of this system result in loss in quality control as well as requiring relatively greater manpower. Therefore, several systems have been devised to automate the blow molding operation.

In one system manifold extrusion dies are constructed so that the melt flow can be directed to one die or one set of dies. When the parisons are the proper length, the flow is redirected to another set of parison dies, while molds close around the first set. By the time the parisons in the second set of dies are the proper length, the products in the first set have been blown, cooled, and removed, so the manifold valve changes to redirect the melt flow back to the first set. This arrangement has many variations. It may consist of only two molds with the melt flow alternately directed, or it may consist of many dies and molds.

Full capacity of the extruder can also be utilized by using moving molds. The extruder extrudes parisons continuously while molds come into position around the parison, close, and at the same time cut the parison, then move out of the way for the completion of the cycle while another parison is being extruded. An accumulator may be used to store the molten plastic and a ram or plunger forces the melt through the die forming the parison. Multiple mold arrangements permit high-speed ram extrusion to compete with continuous extrusion in the bottle industry. A more recent version of the

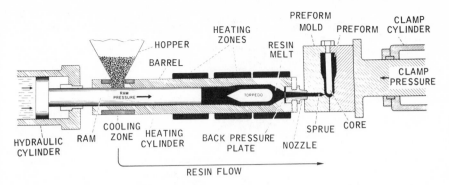

Figure 4 Cross-section of an injection blow molding machine. Courtesy USI Chemical Company.

ram accumulator system is the use of a reciprocating screw in which the whole screw is moved back and forth by a hydraulic ram.

Injection blow molding, in contrast to extrusion blow molding, is a non-continuous cyclic process. A schematic of an injection molding machine with a preform mold is shown in Fig. 4. The parison is formed in one operation and then transferred by hand or automatically to a second, blowing operation to form the final part. To operate the injection molding machine near capacity, several molds must be available to allow for core pin insertion, blowing, cooling, and part removal. The biggest advantage of injection molding over extrusion blow molding is that finished parts usually require no further trimming, reaming, or other finishing steps.

3. Process Parameters

Molds for blow molding require much lower clamping pressure (100–300 psi) than used in injection molding. There are three main methods of producing molds: machining, hobbing, and casting. Molds are usually cast from aluminum if more than one is to be produced. The main areas of attention to the mold designer are the "pinch-off" sections at the top and bottom of the mold and the provisions for cooling. The pinch-off areas pinch the ends of the parison and seal the edges together when the mold closes. These surfaces are usually made from steel inserts when using aluminum copper alloy or beryllium–copper, because of a higher wear rate. Another advantage of pinch-off inserts is that they can be replaced if damaged.

Cooling is particularly important in mold design because it bears heavily on cycle time and therefore on production economics. Coolants can be circulated through the molds in several ways. One of the most common is to

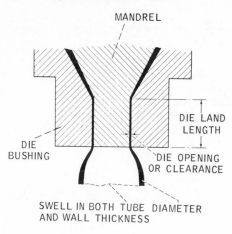

MANDREL

DIE LAND
LENGTH

DIE
BUSHING

DIE OPENING
OR CLEARANCE

SWELL IN BOTH TUBE DIAMETER
AND WALL THICKNESS

Figure 5   Swell in both tube diameter and wall thickness which takes place at the die face. Courtesy USI Chemical Company. Some blow molders recommend the ratio of the die land length to be roughly between 1.0:1 and 4.0:1.

drill holes through the mold block at various locations. Excess openings are plugged and hose connections are inserted for the entrance and exit cooling lines. *Weight and wall thickness* of the parison is determined by the die opening, that is, the distance between the usually cylindrical inside and outside parts of the die. Figure 5 is a schematic of a die head. The outside of the die is termed the die bushing and the inside core pin is termed the mandrel. Wall thickness of the end product—bottle, container, etc.—is controlled by the length and thickness of the extruded parison and the ratio at which the parison is blown out. The higher the blow-up ratio, that is, the maximum diameter of the blown piece divided by the parison diameter, the greater the danger of obtaining nonuniform wall thickness. For high-density polyethylenes the blow-up ratio should usually not exceed 5:1. The blowing process invariably produces molecular orientation in the finished part. This may be an advantage because it increases the stiffness; however, in some cases where the part is to be heated in use, shrinkage or distortion may result. Because many shapes must be produced in blow molding, special parison and wall thickness control techniques have been developed. The main technique used is programming. Programming primarily controls wall thickness distribution of the parison. The use of programming is largely determined by part size, geometry, melt strength, and the rate of parison extrusion. The most versatile programming method is the use of a variable orifice die. This consists of raising or lowering a conical shaped die bushing or mandrel to change the die opening. Another programming method utilizes the fact that parison weight changes with extrusion rate. Therefore, extrusion rate is changed during parison drop to vary the wall thickness as desired.

4.   Parison Parameters

Parison *swell* or *shrink-back* is an increase in both parison diameter and wall thickness after leaving the die opening (Fig. 5). The swell occurs because resin molecules oriented in the die direction (owing to shear) during extrusion tend to return to an unordered or isotropic orientation as they leave the confines of the die. Swell results in a retraction in parison length and requires that more resin be extruded to obtain the desired length. Swell can be reduced by increasing the die land length. Generally high melt index (short molecules) resins have lower swell characteristics because there is less intermolecular entanglement.

Sag (drawdown, neck-down) is the distortion of the parison due to its own weight. This results in excessive wall thickness at the bottom and thinning at the top. This sometimes can be an advantage in part design, but usually is not. Sag can be corrected by parison programming or by use of a lower melt index resin with a higher molecular weight and consequently higher melt strength.

Figure 6   Fully automated production of 1-gal dairy containers utilizing a 5-cavity mold. Courtesy Uniloy Division, Hoover Ball and Bearing Co.

5.   Future of Blow Molding

In the last few years the development of large blow molding machines has opened new horizons for the process. Better and more complete automation in various industries is inevitable. The dairy bottling industry is typical of fully automated facilities (Fig. 6). Packaging of granular foods on form, fill, and seal blow molding machines is indicated. Presently blow molded gas tanks are being used in limited production of automobiles and trucks.

One of the largest markets to be explored for blow molding is carbonated drinks. Special resins have been developed to overcome permeation and stress cracking problems.

C. ROTATIONAL MOLDING

Rotational molding or casting has become a processing technique competitive with injection molding, blow molding, and vacuum forming in certain areas. The first materials to be rotocast in significant volume were the vinyl plastisols. In the early 1960s high- and low-density polyethylenes began to be extensively used and today constitute the major volume of resins used in this market. During the last few years a number of the engineering thermoplastics have been used in various applications including ABS, acetal copolymers, nylon-6, nylon-11, and polycarbonate.

At this time the process is still far from mature. Theoretical knowledge of the intricacies of the process is not extensive. The future of this process looks fairly bright as larger and more automated equipment becomes available, along with resins tailored for rotational casting.

The prime advantages of this process are production of large parts, producing different parts on the same machine, and economic production in low quantity runs. Actual economics for producing parts depends primarily on the machine size used and mold depreciation per part. Today injection and blow molding, along with thermoforming (because of the larger machines now produced) can be competitive with rotational casting in various applications.

1.   Process Description

In rotational molding an enclosed hollow part is formed while applying heat to molds which contain polymer powder and which rotate biaxially. Rotation assures coverage of all parts of the mold by the polymer powder as it melts. There are four steps in a production cycle—loading and heating raw material, molding the part, cooling or curing, and unloading finished articles.

Loading

Raw material, as a powder (or liquid), is loaded into the mold or cavities and the mold halves are mechanically locked together. The amount of material loaded depends on the desired wall thickness.

Molding the Part

The loaded mold is next rotated into a closed chamber where it is subjected to heat while rotating biaxially. Rotation in each plane is generally in the range 0 to 40 rpm. Heat penetrates the mold walls, causing powdered raw materials to fuse and semiliquid or liquid materials to begin gelation. In each case polymer coats the inside of the mold. In the case of powdered materials particles fuse to each other during the heating cycle to build up the wall thickness. Heating must be sufficient to allow the particles to flow and eliminate grain boundaries between particles in order to produce a homogeneous cross section with minimum voids or porosity.

Cooling

After an appropriate heating time to allow for wall buildup the molds are cooled while still rotating biaxially. This causes the parts to cool evenly with even wall distribution and allows the molds to reach handling temperatures. Most cooling is done by rotating in air and then spraying the molds with water after a determined time.

Unloading

Like loading, this is usually accomplished manually by opening the mold halves and physically removing the parts. It is possible to utilize forced air to assist ejection; however, fully automatic unloading methods seldom warrant their cost.

There are numerous modifications to the above cycles. In some instances molds are purged and pressurized with inert gas (nitrogen or helium) during heating to retard degradation and discoloration of some resins. Cold gas is occasionally injected into molds to accelerate the cooling of exceptionally thick parts.

2. Process Variables

Heating Cycle

After the mold is charged with material, it is rotated in an oven to initiate the fusing process. The temperature of the heating medium is a function of a number of variables: (a) the material used; (b) the method of heating, that is, hot air, molten salts, or circulated oil through jacketed molds; and (c) wall thickness of finished part. The prime requirement of the heating system is that the mold be heated uniformly at a reasonable rate. The time–temperature relationship must be properly balanced so that the desired wall thickness can be obtained without resin degradation. Degradation occurs first at the mold and fused resin interface where the temperature is highest.

The molten liquid–salt system (owing to its higher specific heat) is reported to offer a shorter heating cycle than recirculated hot air. This difference has been reduced in the last few years by improved hot-air equipment which utilizes high-velocity blowers and air displacement rates (ft^3/min of air through oven). Molten salts, though, have some disadvantages associated with their use (such as freeze up or corrosion) which have to be balanced against the possible advantages of a shorter cycle.

Systems are presently available that recirculate hot oil or a heat transfer fluid through jacketed molds for heating and heat exchanged oil for cooling. This reduces the amount of equipment and floor space needed for processing (no oven or cooling station needed). Also, much better control of temperature is obtained compared to circulating air or molten salt heating systems. One drawback is that jacketed molds are more expensive than standard rotocasting molds.

Cooling Cycle

The cooling cycle has a pronounced effect on properties of the rotocast part. It has been reported that a rotocast part has virtually no "built in stresses" since it is cast at nearly atmospheric pressure. It is true that the heating cycle doesn't cause stress in the final part, but the cooling cycle does. Rate of cooling affects part shrinkage, final density of the material, and part brittleness because of its effect on crystallinity and frozen in stress The effects of *rapid* cooling are as follows: (a) reduced part shrinkage; (b) lower final density (owing to lower crystallinity); and (c) improvement of impact resistance owing to lower crystallinity developed but counteracting this, frozen-in stresses and shrinkage gradients might lower impact resistance. The rate of cooling must be controlled in order to balance part warpage against any physical property advantages gained from rapid cooling.

During the cooling cycle the mold is continually rotated. The prime methods of cooling are as follows: (a) rotation in air, (b) applying water spray to mold, and (c) a combination of these. In some systems the mold is removed from the oven and immersed directly in a water bath. Of course, warpage can be excessive using this method depending on the material used.

A method to control warpage during rapid cooling is to introduce chilled air, water vapor, or liquid carbon dioxide into the mold. Thus the part is cooled on the inside as well as on the outside and is held against the mold by the induced pressure. Parts evenly cooled on both sides using this method tend to have less warpage but may have frozen-in stress due to rapid cooling.

Mold Rotation

In the rotocasting process the mold rotates biaxially. Axes are at a 90° angle for simplification of mechanical design. For any part being cast there exists an optimum rotational speed for each axis. Axes are usually termed major and minor—the major axis having the higher rotational speed.

The factors in achieving uniform wall thickness are rotational speed ratio and the magnitude of rotational speeds. Usually a rotational speed ratio of major to minor axis is chosen that is greater than 1 but that does not give a whole number. This is done so that the melt does not continually flow over the same path during the heating cycle. It is also advisable to mount molds off-center to avoid possible dead spots.

During the rotocasting cycle the pressure in the mold is controlled by a number of methods:

1.   Most molds are vented during the heating and cooling cycles. Vents are usually packed Teflon tubes inserted far enough into the mold so that material does not fill and plug the vent. Proper venting reduces flash of material from pressure created during heating and reduces warpage that may be caused by vacuum created during cooling.

2.   Molds can be designed so that they may be pressurized during the cooling cycle to reduce warpage.

3.   To reduce oxidation of some polymers molds are purged before heating with an inert gas such as nitrogen or carbon dioxide. If the gas continues to flow into the mold during heating, there is the possibility that the material will be cooled and the cycle lengthened. An additional charge of gas as the mold enters the cooling chamber prevents air from entering the mold as the internal gases cool.

4.   Closed molds (no vent) may be run for some materials. On cooling, parts do not warp and excessive flash is prevented by proper mold design.

Vented molds cannot be used successfully with molten salt heating systems. There is the possibility that salt may enter the vent and contaminate the material or final part.

Whether to vent, purge, or pressurize molds is dictated by the material, labor involved, type of equipment used, and part design.

3.   Rotocasting Equipment

The main trend in development of rotational molding machinery has been the design of large machines. The largest commercial machine is capable of rotating a total weight of 5000 lb within a 17-ft sphere. Machinery is rated by the total weight that can be loaded on an arm (including resin, molds, and spider) and the spherical diameter of swing, which is somewhat misleading. Depending on the type of rotating arm utilized, the total volume of the sphere is not always available to rotate molds. Machines are classified mechanically into three basic types: batch, semiautomatic (shuttle), and continuous (rotary). Also machines are classified by the heating system used: circulated hot air, molten salt, and hot liquid (through jacketed molds).

Batch Type

Used in prototype of low volume production where a varied product mix is desired. These machines require the least capital expense but the most manual labor.

Semiautomatic

Semiautomatic or shuttle-type machines utilize a common oven with one or more cooling and loading–unloading stations (Fig. 7). These machines offer a number of advantages; modular design can be used, production can continue when one side of the machine is inoperational, and they are more efficient when different heating times are required for each arm or heating cycle.

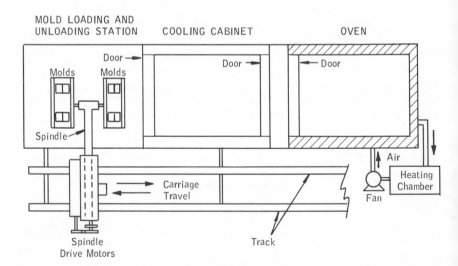

Figure 7   Schematic of a semiautomatic shuttle-type machine.

Continuous

These machines include three basic stations arranged 120° apart with arms attached to a central hub containing the drive mechanism. One type of continuous machine is shown in Fig. 8. These machines are most efficient when running parts of the same design on each arm. In practice different parts are sometimes molded on different arms to fully utilize the machine's capabilities.

Advantages of continuous machines are minimal labor and high production rates. Many features can be added to these machines. At present continuous machines are used predominantly by processors.

Features that are important in any machine are individually programmed arms and variable axial speeds, constant mold rotation from one station to the next, and drilled arms for inert gas introduction. Some machines produced have a fixed rotational ratio, usually 3.5:1 or 4:1. This limits flexibility in obtaining even wall thickness on some parts of irregular geometry.

Figure 8 A production rotational molding machine operating on a merry-go-round principle. One mold is entering the cooling chamber on the right while the oven is on the left. A finished part is being unloaded in the center and then the mold will be refilled with powdered resin. Courtesy McNeil Corp.

4. Heating Systems

Circulated Hot Air

Approximately 70–80% of machinery uses this type of heat. The efficiency of this type of heating has improved over the last few years so that cycles are close to hot salt systems. Maximum air temperatures of approximately 900°F are common.

Molten Salt

With this system molten hot salt is pumped through spray nozzles over the molds as they rotate in the oven. Molding temperatures are lower ($\sim 450°$F) and salts provide a more effective heat transfer medium compared to hot air systems. A number of disadvantages occur with molten salt including salt recovery, possible freeze up and need for more precisely made molds to prevent molten salt from leaking through a mold parting line.

Hot Liquid

Jacketed molds are used in this system and one station is used for the entire cycle. Judged on economics, this is probably the least economical heating system. Because the economics of producing cast parts are partly based on cycle time, and of course mold cost, production costs are probably higher than either of the above two methods because heating and cooling are done in the same station. Owing to the heating efficiency there is the possibility that the overall cycle in producing some parts might be equivalent to other heating methods. For resins requiring precise temperature control on heating and cooling, this heating method would offer obvious advantages.

5. Molds

The rotocasting process reproduces the mold surface very accurately. Grained, textured, and matte finish are reproduced according to mold finish. Three types of molds are in common use: cast aluminum, sheet metal, and electroformed. Cast aluminum molds (Fig. 9) are used when a number of molds are to be produced, when the part has intricate design, and when the parting line lies in more than one plane. Sheet metal molds are used for low-cost large molds of fairly simple design. Electroformed copper and copper–nickel molds are the most expensive but offer the advantage of a very smooth finish.

It is essential that molds used be nonporous. During cooling, water spray can be trapped in mold pores. During the next heating cycle this water will generate steam which can affect part surface. Porous molds may also trap polymer, which would be degraded to form impurities when subjected to repeated heat cycles.

Figure 9 Cast aluminum molds for rotational molding. Mold on left has spider frame for mounting to rotating arm. Courtesy McNeil Corp.

Mold Clamping

Owing to the constant thermal cycling of the rotocasting process, molds tend to warp with use if proper care isn't exercised. Clamping force should be kept to a minimum and shouldn't be much more than that needed to stop molten material from exuding or flashing at the parting line during heating.

Molds and spiders (devices for holding molds) should be stress relieved before final matching of parting lines, especially for production molds.

Clamping devices vary from the use of C-clamps and vise grips to spiders with built-in clamping devices.

6. Part Design

The rotocasting process gives wide flexibility in the shape or size of parts that can be produced. As long as the mold configuration allows contact between the material and the mold surface, the mold shape can be reproduced. Mold size must be contained and rotated within the limitations of the heating and cooling chambers. Generally wall thickness can be controlled within ±5%. Parts using various resins have been molded with walls as low as 0.015 in. or as thick as 0.500 in. Little or no draft angle is necessary for release because the part is not formed under high pressure and tends naturally to shrink from the mold.

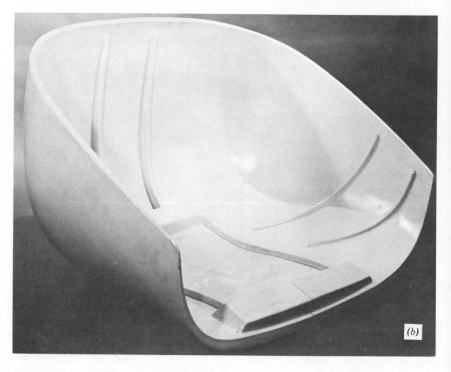

Figure 10 (*a*) 2000-l. rotationally molded tanks being inspected. Courtesy Krauss–Maffei Co. (*b*) Expandable ABS resin chair form. Courtesy McNeil Corp.

There are some design limitations in the process:

1.   Varying wall thicknesses cannot be easily cast in the same piece; varying rotational speeds (and ratio) and insulating mold sections accomplish this to a degree, but wall thickness cannot vary sharply in a short distance.

2.   *Sharp* threads are impractical because material tends to bridge in a confined area.

3.   As a general rule, to prevent bridging in ribs or bafflelike shapes, the distance between interior walls should be four times the wall thickness.

4.   As in injection molding, wall thickness transitions should be as gradual as possible to prevent built-in stress owing to differential cooling.

5.   Molds *must* be designed using proper shrinkage. Because only low pressures are used in rotocasting, shrinkage values for various resins might be somewhat greater than for injection molded parts. Shrinkage is a function of process conditions, primarily cooling rates and wall thickness. One should be reminded that crystallizable polymers have greater shrinkage than amorphous polymers.

7. Applications

As mentioned, a number of thermoplastic resins are being rotationally molded. Limited work is being conducted to mold thermosets by this process. Virtually any enclosed shape can be rotationally molded. Two applications are shown in Fig. 10.

Some applications involve the molding of two or more materials in a layer effect to take advantage of the properties of each material. This is usually done by casting each material in a separate step.

Large applications such as boat hulls, pallets, and chemical storage vessels are being produced. A number of these parts are foam filled in a separate step so as to combine a tough outer skin with a lightweight, insulating core.

D.   THERMOFORMING

Thermoforming is one of the oldest and simplest methods of fabricating plastic materials. Basically, a sheet of thermoplastic is heated to its softening temperature and by some pneumatic or mechanical means the sheet is forced against the mold contours. The material now in the shape of the mold is allowed to cool and a part is removed that retains the pattern and detail of the mold.

New forming equipment now available is capable of producing large and complicated parts including boat hulls, shower stalls, car hoods, refrigerator

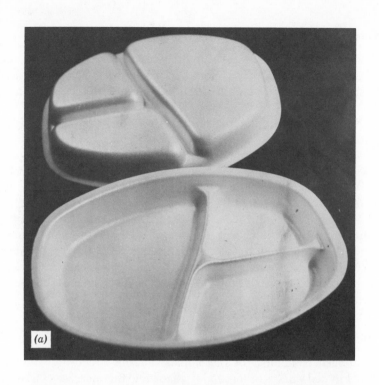

(a)

(b)

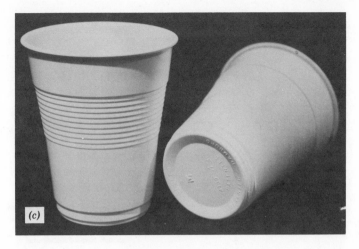

Figure 11 Thermoformed applications. (*a*) Food tray—expanded polystyrene foam, 0.060 in., 3-sec cycle. (*b*) ½-pt bottle—high-impact polystyrene, 0.037 in., 3-sec cycle. (*c*) 7-oz cup—high-impact polystyrene, 0.030 in., 3-sec cycle. Courtesy Brown Machine Co.

liners, and tote boxes. Clamshell and twin-shell forming techniques (11) allow hollow objects to be produced, thus allowing thermoforming to compete with blow molding and rotational molding. Examples of some thermoformed parts are shown in Fig. 11.

1. Process Description

There are a number of basic thermoforming techniques, a simple schematic is shown in Fig. 12. The best technique for producing any part combines the shortest production cycle, the best distribution of plastic material, and use of the thinnest gauge sheet possible.

All thermoforming techniques may be grouped into three basic methods: *vacuum forming* is the method predominantly used for forming (shaping) sheet. In the simplest technique, heated sheet is clamped into place and mechanically brought into contact with the mold. A vacuum is drawn between the hot, elastic sheet and the mold until the sheet is pressed (by atmospheric air pressure) against the contours of the mold. After a short cooling period, the sheet cools enough to take a set conforming to the mold surface and then the part can be removed.

Mechanical forming techniques stretch the hot sheet over a mold or use matched molds for more complex shapes. No vacuum or air pressure is used.

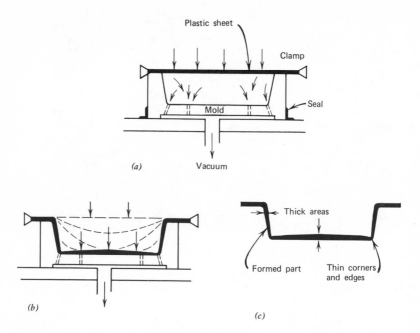

Figure 12 Vacuum forming—steps in the process (*a*) clamped heated sheet has vacuum applied (*b*) formed plastic contact mold and cools (*c*) areas of sheet reaching the mold last are most stretched and thinnest.

In the *air blowing* process, compressed air is used to form the sheet. Compressed air up to 500 psi is used to produce thin gauge containers. One variation of the air blowing technique is *sheet blow molding*. This technique is similar to regular blow molding. Two sheets are held vertically, heated, and then moved to a forming station. The mold closes over the sheets, and air is forced between the sheets to form a hollow part. One advantage of this technique over extrusion blow molding is that two different colored sheets can be used to form the part.

2. Sheet Variables

All thermoplastic sheet materials can be stretched when hot, but this property varies with different resins. The term used for stretching heated sheet is *hot elongation* and it is associated with *hot strength*. It is dependent on the resin and its molecular weight, sheet temperature, and forming speed. Materials with poor hot elongation cannot be formed into intricate or deep-drawn parts because the material tears and does not stretch evenly.

Hot elongation and hot strength are related to the viscosity and molecular structure of the sheet material. (The higher the molecular weight, the higher the hot strength.) Some materials become too fluid at forming temperatures and cannot be formed properly. This type of material has poor hot strength. The term is related to hot elongation, but differs slightly in that a material can have good hot elongation but poor hot strength.

Sheet Thickness

Nonuniform thickness causes uneven heating, resulting in variations of the formed parts. With deep draws thin walls can cause fracture or tearing of the wall. On deep draws, the maximum variation in sheet thickness is usually 2-3 %. The more uniform the sheet temperature before forming, the better are the chances of forming acceptable parts.

Sheet Orientation

Extrusion formed sheet may have considerable orientation (usually in the machine direction) and this can significantly affect the forming operation. In extruded sheet orientation in the machine direction is caused by stretching between the polished rolls and the take-up unit. In calendered sheet there may be varying degrees of orientation in the machine and transverse directions. The orientation causes different shrinkage in the two axes of the sheet and must be taken into account in mold design. If excessive shrinkage takes place in one direction there might be unbalanced pull from the clamping frames during forming, and possibly the sheet could pull lose. It is necessary to know shrinkage in both directions for the material and wall thickness of sheet being used to be able to design molds capable of producing the desired final part dimensions.

3. Process Variables

As we have already seen, the behavior of thermoplastics on being heated depends on molecular weight and whether the material is crystalline or amorphous. For amorphous materials the polymer softens over a rather broad range of temperature. In crystalline materials there is a more abrupt change from the solid to the soft phase but again long chains provide the hot melt strength necessary for forming. The degree of elasticity and melt strength depends upon the molecular weight. Since the viscosity and melt strength are highly dependent on temperature this variable emerges as a very important aspect of forming processes after the resin has been chosen. Excessive heat can cause easy tear (owing to lack of hot melt strength); on the other hand, inadequate heat can cause tear owing to lack of flow.

To produce consistent quality parts the sheet must be heated evenly throughout. Sheets that have uneven temperature distribution pull unevenly when formed, with the hotter material stretching more. Sandwich heating on sheet materials greater than 30 mils thick avoids overheating one side of the sheet. The heating rate on thick sheet must be balanced so that when the center of the sheet is at the forming temperature the outside surfaces are not overheated.

Since hot elongation and hot strength are a function of temperature it is very important that each sheet used has the same heat history. The heating method, handling the heated sheet, and subsequent operations until the formed part is cooled must be consistent from part to part.

The two primary methods of heating are forced air ovens, gas or electrically heated; and infrared radiant heat. The factors that influence the speed of heating are (1) sheet thickness, (2) thermal conductivity and specific heat of the material, (3) the absorption characteristics of the material in the wavelength of the source used for infrared heating, and (4) the possibility of surface degradation in thick sheet.

Sheet materials are heated on the thermoforming machine or they are pre-heated on auxiliary equipment to a certain temperature and then are transferred to the clamping station of the machine. Infrared heating is usually used on forming machines. Heaters are placed on one side of the sheet in some setups, but more commonly sandwich heaters are used so that the sheet is heated from both sides. These heaters can move up and down independently so that when the sheet sags during heating the bottom heater can be lowered. The advantages of heating on the machine are more precise temperature control, less manual handling, and shorter overall forming cycle. Convection ovens are used for materials that are heat sensitive and for thick gauge sheet where evenness of heating is critical.

4. Molds

Compared to molds for other processes, thermoforming molds are fairly inexpensive. Tooling can vary from prototype hardwood molds to polished steel molds for long-term production. The most common mold material is aluminum, because it provides a good combination of light weight, durability, castability, and thermal conductivity.

In design, a male primary mold allows deeper draw than a female mold because the plastic can be draped or prestretched over it. However, if a male plug assist is used to prestretch the material for a primary female mold, the advantage is nullified.

Female molds provide easier release, are less likely to be damaged, and provide good definition on the outside of parts; the converse is true for male molds. A disadvantage of female molds is that parts have thinner bottoms, but good design can greatly alleviate this problem. Male molds are usually cheaper.

All molds used in vacuum or air blowing techniques require channels for the evacuation of air or pressure buildup. The channels or slits are small, usually 10–25 mils to avoid marks on the formed part. Vacuum holes are kept to the minimum and are usually countersunk on the backside for more rapid air removal.

Temperature control of molds is essential. Production molds are temperature controlled, being cored for a temperature controlled liquid to pass through as in injection and blow molding. Cooling conditions significantly affect shrinkage, with approximately half the shrinkage taking place on the mold during cooling. Therefore *good* mold temperature control is needed to minimize part warpage and distortion.

5. Thermoforming Machinery

Machinery is usually classified according to its mechanical arrangement, as in rotational molding.

Single Station Machines

These machines perform one operation at a time and the total cycle is the sum of the times for loading, heating, forming, cooling, and unloading. Single station machines are quite versatile although the production cycle is relatively long. Machines vary from small laboratory type to large units with movable upper and lower platens, movable clamping frames, and sandwich heaters.

Most single station machines are fed with single sheets but some models have a roll-feed mechanism which indexes roll stock after each forming cycle. These roll-feed mechanisms eliminate some loading–unloading labor but are far from a continuous type of operation.

In-Line Machines

These machines are designed for large volume production of containers and closures from rolled sheet stock or, for in-line operation with an extruder, producing sheet. Heating, forming, cooling, and trimming operations are carried on at the same time at different positions inline; therefore the longest step in the process determines the overall cycle time. These machines are widely used in the packaging industry.

One of the newer concepts in thermoforming is a variation of in-line machines used in producing packages. This machine operates in a similar fashion to a four-station rotary machine in that four sheets can be processed at once. This type of machine reportedly outcycles a rotary machine and also requires less floor space. Model variations of this machine are being built to accommodate molds up to 120 × 300 in. Maximum draw will be 72 in.

Rotary Machines

These machines are built on a horizontal circular frame. There are three station machines consisting of loading–unloading, heating, and forming–cooling stations. The fourth stage in some machines is used for trimming formed parts. These machines approach continuous operation and are used for high volume production. Four-station machines reduce the forming cycle by one-third compared to the three-station rotary.

E. THERMOSET MOLDING

Thermosetting resins are plastics that in their formed or cured state are essentially infusible and insoluble owing to cross-linking. Resins are often liquids at some stage in manufacturing or processing, and are cured by heat, catalysts, or other chemical means. After being fully cured, thermosetting resins cannot be resoftened by heat. Therefore, processes for fabricating thermosets have to allow for the fact that sprues, runners, and reject parts cannot be reused, as they can with thermoplastic materials.

Owing to improvements in machine design and more sophisticated resin formulations the plastics industry is today witnessing a resurgence in the use of thermosets. This is mainly because of the use of screw injection molding techniques with thermosets. Prior to 1930, all thermosets were processed by compression molding.

During the 1930s, transfer molding techniques began to be used. Automation of compression and transfer molding equipment occurred in the 1950s and screw injection equipment began to be used in the late 1950s. Table 2 (13) shows that compression molding is the method used most commonly, and screw injection molding is expected to grow significantly.

Table 2   Thermoset Processing (13)

	Percent of Total	
Process	1966	1972
Compression molding	75	60
Transfer molding	18	15
Screw injection	7	25

1.   Compression Molding

In compression molding a predetermined amount of material, either at room temperature or preheated, is placed in the lower half of a heated mold (Figs. 13 and 14). The top of the mold, or force plug, is then advanced into the matching bottom cavity. Pressures in the mold during the curing stage range from 2000 to 8000 psi, depending on mold geometry. The mold temperature and pressure required to form a satisfactory part must be predetermined for every material/part combination to be produced. The weight of the resin charge must be greater than that of the final part. When the hold is closed, excess material is forced out at the parting line in the form of flash which must be trimmed.

Mold temperatures are maintained by using electric heaters, circulated oil, or high pressure steam. Maintaining mold temperatures within $\pm 5°F$ is fairly important. The mold temperature significantly affects material flow, cure time, and properties of the final part. In the last few years, radio frequency

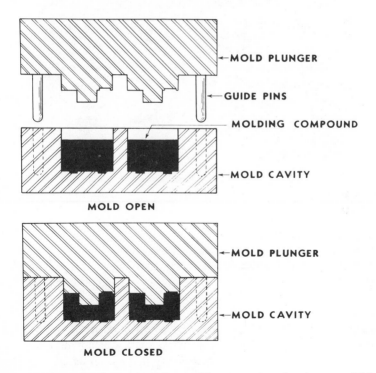

MOLD PLUNGER

GUIDE PINS

MOLDING COMPOUND

MOLD CAVITY

MOLD OPEN

MOLD PLUNGER

MOLD CAVITY

MOLD CLOSED

Figure 13   Schematic of a compression molding operation showing material before and after forming. Courtesy HPM Division of Koehring Co.

Figure 14 Thermosetting resin premix logs are compression molded into lamp housings for automobiles. Courtesy HPM Division of Koehring Co.

heating of preforms and powders before molding has lowered cycles significantly. When these preheated charges are used, lower mold temperatures are required to avoid premature curing of the resin before the mold has closed.

Gassing or breathing of molds may be necessary in compression molding some thermosets, for example, phenolic, melamine, or urea materials. These materials generate moisture during molding, and during compression air can be sealed in the mold. To release these volatiles, the molds are opened slightly to allow the gases to escape from the cavity. Certain materials require a timed "dwell" with the mold open before it is closed again. Gassing the mold results in denser parts, reduces internal voids, and reduces the molding cycle.

Most compression molding machines are semiautomatic, the cure and ejection cycles are performed automatically, and mold charging and cycle initiation are performed manually.

Compression molding offers the following advantages:

1. Waste of material is minimized—no gates, runners, or culls.
2. Molds are relatively inexpensive owing to absence of gates and runners.
3. A greater number of cavities can be used in a given mold base because there are no gates or runners.

Some disadvantages of compression molding are as follows:

1. Sections thicker than $\frac{3}{8}$ in. thick cannot usually be molded because of uneven cure. Outer surfaces cure fully, whereas the part interior is undercured.
2. Long cycle times are required—about 1 min for each $\frac{1}{8}$ in. of thickness.
3. Inserts are difficult to mold in because of the high pressures involved.

It should be noted, too, that compression molding may be used with thermoplastics. This provides a convenient laboratory technique for obtaining molded shapes for tests. Compression molding is often the method of choice in production of large cylindrical shapes or billets of thermoplastic, which are then machined to final shape. Finally, compression molding is the primary technique for fabrication of ultra high molecular weight polymers such as polytetrafluoroethylene (PTFE) or polyethylene. The fabrication of PTFE is interesting in that it involves room-temperature compression molding to "coin" or preform the part. This compacted preform is then sintered in an oven maintained above the melt temperature of the polymer (327°C or 520°F).

2. Transfer Molding

Transfer molding applies to the process of forming parts in a closed mold, by conveying the thermoset material under pressure in a molten state from an auxiliary heated chamber or "pot." The material is loaded into the pot and, when heated to a molten state, is injected through a runner system into a closed mold cavity.

Figure 15 is a schematic of a transfer molding operation. The press ram is used to hold the mold closed with sufficient clamping pressure. The plunger is used to force the molten material along the runners and into the two cavities shown.

Transfer molding of thermosets is somewhat analogous to injection molding of thermoplastics with certain important differences. Because of the

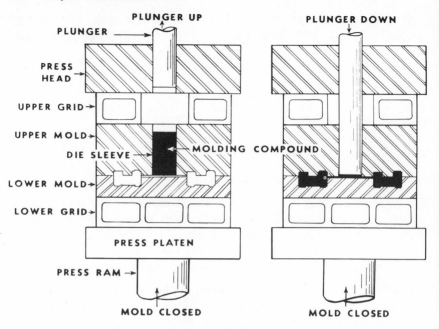

Figure 15 Schematic of a transfer molding operation. Courtesy HPM Division of Koehring Co.

curing characteristics of thermosets, it is not usually feasible to hold material at a high temperature any length of time before molding. The material could polymerize into a solid mass in the cylinder or melting chamber. Therefore, in transfer molding only enough material for a single shot is heated at one time. Also, the melting chamber must be thoroughly purged and cleaned between shots. In transfer molding, the sprue taper is the reverse of that used in injection molding, because it is desirable to keep the sprue attached to the cull (the excess of material remaining in the pot) rather than to the molded part.

Advantages of transfer molding include the following:

1. Loading time is shorter than in compression molding.
2. Thick sections can be obtained with uniform cure.
3. Undercuts can be molded.
4. Cycle time is shorter than in compression molding.
5. Gas and moisture inclusion is less than in compression molding, although molds are usually vented from the parting line to the outside surface for gas removal.

The principal disadvantages of transfer molding are the amounts of scrap material from sprues and runners, along with higher mold costs compared to compression molding.

Since 1963, automatic transfer molding using the reciprocating screw principle has gained wide acceptance in the molding of thermosets. In this system, resin is plasticized by a reciprocating screw in a cylinder as in normal injection molding. The reciprocating (forward) action of the screw forces molten resin into the pot. From the pot a plunger forces the material into the molds as in straight pot type transfer molding. Molds are usually mounted with parting lines horizontal so a mechanical device has to be used to comb or eject parts. Advantages of automatic transfer molding are shorter cures, shorter load and unload cycles, and usually less machine downtime.

3. Screw Injection

The newest innovation in thermoset processing is reciprocating screw injection molding (discussed in Chapter 10). This carries the automatic transfer process one step further and eliminates the pot-plunger system of transfer molding, as well as the mechanical systems used to "comb" finished parts from the runner system. Molds used on screw injection molding machines have vertical parting lines, so that parts and sprues gravity-drop, thereby reducing overall cycle time compared to automatic transfer molding.

Among the thermoset materials that have been screw injected successfully are phenolics (filled and unfilled), ureas, melamines, diallyl phthalate, and alkyds. "For optimum results, transfer grades of materials should be used, but a wide range of flows has been found suitable" (14).

Machines for screw injection of thermosets are similar to those used for thermoplastics with some essential differences:

1. Shear on material in the barrel is kept to a minimum to prevent precure of resin. Screw compression ratios are approximately 1:1 rather than the 2:1 to 4:1 used for thermoplastics. Screws are usually of constant depth for two-thirds of the length followed by a slight decompression zone (15).

2. Barrel and material temperatures must be carefully controlled to prevent resin precure. Barrels are usually water jacketed to provide temperatures up to 120°C (248°F).

3. The front section of the cylinder is easily removable (Fig. 16) in case resin has cured in the front section. This occurs because the barrel nozzle is in direct contact with the hot mold and curing can occur prematurely. Therefore, nozzle cooling is used with some machinery.

The main disadvantages of this process are higher mold costs and the waste material from sprues and runners.

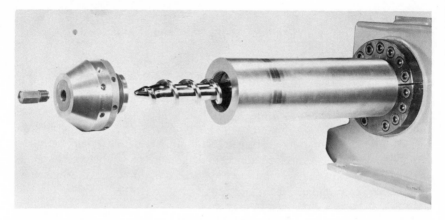

Figure 16   Barrel section of screw injection machine for thermosets (without insulation). Note that cylinder head can be removed easily for removal of precured material in nozzle section. Courtesy Van Dorn Plastic Machinery Co.

4.   Future of Thermoset Molding

Today with automatic transfer molding machines and direct screw injection machines available, the future of thermoset molding appears bright. As molding cycles become shorter, thermosets can and will become competitive with thermoplastics in certain application areas. Thermosets can offer advantages over thermoplastics such as higher temperature performance, weatherability, and rigidity. The main area of competition between thermoplastics and thermosets is in higher temperature performance markets. Thermoplastics such as thermoplastic polyester, polysulfone, polyphenylene oxide, polycarbonate, and the specialty polyimides as well as filled thermoplastics are currently in direct competition with a number of thermosets, for example, phenolics, alkyds, and diallyl-phthalates (DAP). The thermoplastics have replaced a number of thermosets, especially in electrical applications and even in some engineering applications.

Where higher performance is required, such as for high-temperature end use, thermosets such as DAP, epoxies, and modified alkyds will continue to be chosen over most engineering thermoplastics.

Major growth areas for thermosets in the last few years have been in existing applications as replacement for metal parts. These applications have been mainly in the electrical, automotive, and aircraft industries.

F.   FOAMED (CELLULAR) PLASTICS

Foamed or cellular polymers have been used in a variety of applications since the 1940s. The high strength to weight ratio, insulating properties,

Table 3   Common Foamable Plastics[a]

Polyethylene (low and high density)	Cross-linked PVC
Polypropylene	Ethylene copolymers
Polyurethane	Ionomers
Polystyrene	Silicones
Phenolics	Cellulose acetate
Epoxy	
Expanded vinyls	

[a] See *Modern Plastics Encyclopedia* for a discussion of each type of material.

and cushioning properties have provided incentive for a rapid growth for foamed polymers. The 1969 market for polyurethane foams alone has been estimated at 650 million lb and by now is probably nearing 1 billion lb.

Plastic foams are classified as either flexible or rigid, and may have open- or closed-cell structure.* Flexible urethane dominates the market at present with flexible polystyrene second. A number of polymers can be foamed, including thermoplastics and thermosets. Table 3 lists the more common foamed plastics.

1.   Preparation of Foams

Foams are prepared by a number of methods—mechanical, physical, and chemical. The most important process consists of expanding a fluid polymer phase to a low-density cellular structure and then maintaining this structure. Other methods of producing cellular structures are leaching out solids previously dispersed in a polymer, sintering of small particles, and dispersing small cellular particles in a polymer. These latter processes are of relatively minor importance.

The expansion process usually consists of three steps: initiating small cells in a fluid polymer, causing these cells to grow to a desired size, and stabilizing the desired structure by physical or chemical means. This is shown schematically in Fig. 17.

The formation of cells is caused or initiated by "foaming agents," which can be any material or combination of materials—solid, liquid, or gaseous—capable of producing a cellular structure.

Physical foaming agents are compounds that change their physical state during cellular growth. Compressed gases expand when pressure is released,

---

* In an open-cell structure the cells are interconnected, whereas in closed-cell structure most cells are closed and separate and may enclose gases that can modify the foam properties, for example, insulation.

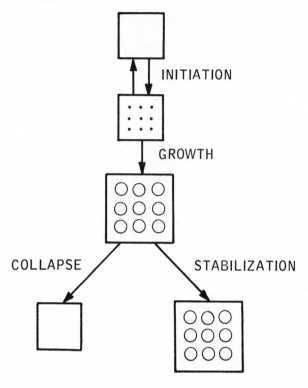

Figure 17   Steps in preparation of foamed polymers.

soluble solids leave pores when leached out, and volatile liquids create cells when they change from a liquid to a gas.

The volatile liquids are the most important physical foaming agents. Usually their boiling points are below 110°C at atmospheric pressure. The vaporization of these liquid foaming agents is a reversible endothermic reaction and can serve as a decent means of controlling the process temperature. Aliphatic and halogenated hydrocarbons are the most widely used agents, although low-boiling alcohols, ethers, ketones, and aromatic hydrocarbons have been used (16).

Chemical blowing agents, frequently called foaming agents in the plastics industry, are materials that decompose under heat to at least one gaseous decomposition product. The most important property of chemical agents is the decomposition temperature. This is so because foaming must be initiated for thermoplastics within a specified viscosity range, and for thermosets when the desired degree of cross-linking is obtained. Therefore, no one foaming agent is suitable for all resins. The most common chemical foaming agents are organic compounds that release nitrogen as the major component of the gaseous phase.

2. Foaming Techniques

There are numerous methods of producing foams and foamed parts. The more important are summarized here, along with the polymers most commonly used.

Structural foams are strong expanded materials having solid integral skins; this differentiates them from open-pore type foams. These foamed parts are usually formed by modified injection molding techniques. Parts produced from structural foams have up to three times the rigidity as those made from the equivalent weight of solid plastic. A number of methods exist for producing structural foams.

The Union Carbide process can be used to produce almost any thermoplastic structural foam, although polyethylene and polypropylene offer the best combination of properties. The process (Fig. 18) consists of three steps:

1.   An extruder feeds melt containing blowing agent to the accumulator. The temperature of the melt is above foaming temperature, but at a high enough pressure to prevent foaming.
2.   The melt is discharged from the accumulator into the mold.
3.   The mold opens to discharge the part while the extruder fills the accumulator.

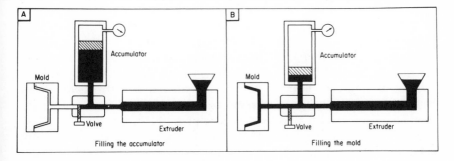

Figure 18   Union Carbide structural foam process. Extruder feeds accumulator until predetermined charge is reached (left). Accumulator plunger then rams melt past open valve into mold (right), where expansion takes place.

Because the mold is under low pressure, the melt automatically expands as it is forced by the piston from the accumulator into the mold. This type of process is termed "low pressure."

Parts with smooth glazed surfaces can be produced by using relatively high mold temperatures. Lower mold temperatures produce parts with a grainy or woodlike appearance. High L/D ratio extruders are preferred (24/1) in order to disperse the gaseous blowing agent in the melt during plasticizing. Molding pressures are low, 200 to 350 psi; therefore cast aluminum molds are usually used. Properties of the foamed parts are controlled by molding temperature, pressure, gate size, injection speed, and the type and amount of blowing agent.

Recent developments in technology allow a number of structural foams to be molded on modified screw injection molding machines. In one technique, a resin concentrate containing blowing agent is mixed with normal resin and pigment and then loaded into the molding machine hopper. The resin is then plasticized in the cylinder but with enough heat to decompose the blowing agent. The gases are held in the cylinder until the melt is injected into the mold at which time it foams in the mold cavity. The material entering the mold first forms a solid skin on the surface. Additional material expanding between the skins completes the filling of the mold. In one process (Fig. 19) the foaming is controlled by injecting a full shot and then controlling the opening of the mold. This is termed a high-pressure process and reportedly imparts a smoother surface with fewer flow marks compared to low-pressure processes that inject a partial shot into the cavity.

Urethane foams range from very flexible foam used in pillows to rigid materials used in the building industry. Urethanes are formed by a polyol and polyisocyanate in the presence of a blowing agent and a surfactant. The surfactant is used to stabilize the foam and to control cell size. With variations of ingredients urethane foams can be made with densities ranging from 1 to 60 lb/ft^3. Urethane foams have several key properties that make them the most versatile of the foamed polymers. These include insulating efficiency, good adhesion, buoyancy, ease of application, and good strength.

Production of rigid urethane foam involves metering the ingredients listed above into a high-speed mixer. Because of the rapid reaction, timing is critical and the mixing and metering equipment required is fairly sophisticated. Slab stock is produced by moving a mixing head back and forth over a moving belt. The rising foam is held between side plates to form a continuous bun 3 to 4 ft wide and 1 to 2 ft high. In 5 to 10 min the foam is firm enough to be cut into lengths for curing and cooling.

Urethane mixtures can also be poured directly into a cavity or mold. The liquid flows into recesses and cracks and forms a strong seamless part. Pouring is used for filling curtain wall sandwich panels and wall cavities primarily as insulation material.

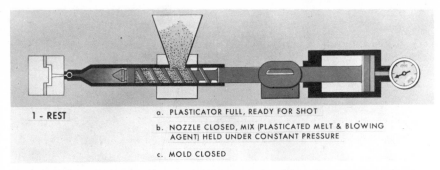

1 - REST   a. PLASTICATOR FULL, READY FOR SHOT

b. NOZZLE CLOSED, MIX (PLASTICATED MELT & BLOWING AGENT) HELD UNDER CONSTANT PRESSURE

c. MOLD CLOSED

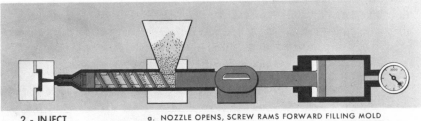

2 - INJECT   a. NOZZLE OPENS, SCREW RAMS FORWARD FILLING MOLD

b. AS MELT PASSES THROUGH NOZZLE TEMPERATURE RISES

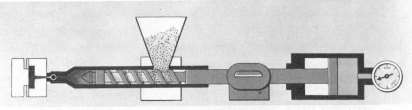

3 - MOLD EXPANSION   a. INJECTION COMPLETED, NOZZLE CLOSES, PLASTICATOR BEGINS REFILL AT CONSTANT BACK PRESSURE

b. PLATEN MOTION BEGINS AT COMPLETION OF MOLD FILL, EXPANDING CAVITY VOLUME AT CONTROLLED RATE

Figure 19 Cycle sequence for direct injection molding of foamed plastics. Courtesy USM Corporation.

Spray-up techniques are also used with urethane and other foams. Special guns mix and atomize the materials. Thin layers of urethane foam can be built up on large surface areas without the need for special forms owing to the adhesion of urethane. Fast reacting chemical systems are used to prevent sagging or running of the foam.

Rigid polystyrene foams are processed primarily by two methods, injection molding and extrusion. Expandable polystyrene beads containing foaming agent, when exposed to heat, expand to form a series of noninterconnecting cells. The degree of expansion can be controlled by various methods within

50 times the volume of the unexpanded beads. This characteristic is used to injection mold polystyrene foam into many items including foamed drinking cups. Basically the machines are similar to normal injection molding machines. The main differences are that some method must be used to convey beads to the mold cavities and heat must be applied to the filled cavities to expand the beads. Beads are usually conveyed pneumatically along channels into the cavities and the molds are designed to inject steam through separate nozzles to heat the beads.

Extruded polystyrene foam is produced by free expansion of hot polystyrene, blowing agents, and additives through a slit orifice to about 40 times the preextrusion volume. The cooling of the foam is controlled to minimize stress that might collapse the cell walls. Because of its high strength and rigidity polystyrene is easily machined and fabricated. Integral skin polystyrene foam is easily bonded and decorated and is seeing ever-increasing markets in the furniture industry.

3. Applications

Numerous applications are available for foamed plastics, because each material can be made with a wide range of density and properties. Foamed plastics have better acoustical, thermal, and electrical insulation than the solid polymers. They exhibit improved mechanical damping, dielectric properties, and usually higher flexibility. Last but not least, the lower specific gravity of foamed plastics can lead to considerable cost savings in many applications.

In the furniture and mattress markets, flexible urethanes continue to replace latex foams. Flexible foams are also used to a large extent in household furnishings and automotive cushioning.

In the near future, foams will be seen more and more in the building industry as they pass various building codes. Polystyrene foams are used principally in the construction industry for walls and cold storage insulation, as well as in many packaging applications. Flexible PVC foam is used in automobile upholstery and in the furniture industry. Rigid PVC foams are gaining acceptance in the boat industry, but the high cost limits their penetration into some industrial markets at present. Foamed engineering thermoplastics are currently being utilized in business machines, appliances, and automotive end uses.

G. COLD FORMING OF THERMOPLASTICS

During the last few years a number of metalworking techniques have been used to fabricate thermoplastics. Normally thermoplastics are formed from a

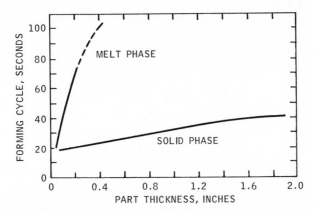

Figure 20   Effect of part thickness on forming cycle for polypropylene.

viscous melt in a molding or extrusion operation. The rate limiting step in most of these operations is cooling. In cold forming operations the finished part is produced from the material in the solid phase; although sometimes materials are preheated as in the forging process. Figure 20 shows a comparison of production cycles for producing polypropylene parts by melt forming and by solid phase forming (17). Solid phase forming is far less sensitive to part wall thickness.

Cold forming offers other advantages:

1.   Tooling costs average 20 to 35% of those for injection molding. Press costs are approximately equal.
2.   Solid phase forming usually improves clarity.
3.   Impact resistance is improved by cold forming processes.
4.   Cold forming methods are the only practical processing methods available for some materials, for example, ultra high molecular weight polyethylene.
5.   Parts require little or no trimming, and sprue marks and weld lines are eliminated.

There are four primary cold forming techniques being used today: cold stamping and drawing, forging, rubber pad forming, and cold heading. Before entering into a discussion of techniques, a definition of terms is in order.

Stamping: the rapid application of force to cut or form a material into a desired shape.
Braking: the application of a force to form an angle in flat sheet.

Blanking: the cutting of a predetermined shape (preform or blank) from sheet stock.

Drawing: the forming of a flat piece into a seamless hollow shape. The operation is performed by a punch that causes the material to flow into the die cavity.

Roll Forming: a continuous bending operation executed by a series of matched driven rolls. The sheet is progressively shaped until the final form is achieved.

1. Cold Stamping and Drawing

Fabricators of cold drawn steel, aluminum, or other metals can cold stamp or draw certain types of thermoplastic sheet. Both single and multiple cavity dies can be used with mechanical punch presses and fluid forming machines. Recently ABS sheet has been cold stamped successfully in production equipment. The sheet is registered and fed automatically from rolls by standard metal strip feeders, adapted to allow for the difference in rigidity between the plastic and metal sheets. Speed of operation, printing, ease of decoration and minimum metal usage are the main advantages of cold stamping. Table 4 compares the economics of cold drawing to other processes.

After the blanks or preforms are stamped they are usually cold drawn into the final shape. During the drawing operation, the surface area is maintained while being formed into the final shape. The blank is drawn radially; thus there are compression and tensile stresses being exerted during the operation. A drawpad is used to maintain control of the material as it flows under pressure into the die cavity. The drawpad is activated by an air cushion which provides enough pressure to prevent thickening and wrinkles from forming in the part. Thus the thickness of a section taken from a properly drawn part is close to the original wall thickness. This is a definite advantage over thermoforming processes. Thermoformed sheet must be heavier gauge in order to form parts with the same minimum wall thickness. Therefore, for the same weight cold drawn containers should be stronger than thermoformed containers.

A schematic of the cold drawing operation is shown in Fig. 21. With ABS sheet the first draw can reduce the diameter up to 35%. Further drawing operations reduce the form to its final dimensions.

Materials

A number of materials have been cold drawn including ABS, polycarbonate, rigid PVC, and cellulose acetate. Thermoplastics used in the drawing process require sufficient ductility and toughness to allow them to be drawn with up to 45% diameter reduction. One of the main problems in forming

Table 4 Economics of Cold Stamping ABS Sheet (18)

Comparison Factor	Injection Molded	Vacuum Formed	Cold Drawn
Cycle time (sec)			
Thin section, small part	25	3–4	0.4–0.5
Heavy section, large part	60	60–120	1–10
Number of cavities			
Thin section, small part	1–16	1–35	1–5
Heavy section, large part	1	1	1
Parts/hr			
Longest cycle, one cavity	60	30	360
Shortest cycle, max. number of cavities	2304	42,000	45,000
Shortest cycle, one cavity	144	1200	9000
Operating cost ($/hr)	5–30	15–25	15–25
Forming cost ($/1000 parts)			
Longest cycle, one cavity	500.00	833.00	69.50
Shortest cycle, max. number of cavities	3.25	0.595	0.555
Shortest cycle, one cavity	34.75	12.50	1.66
Printing and decoration cost ($/1000 parts)			
Round shapes, thin wall, limited decoration	2.50–3.50	2.50–3.50	0.50–1.00
Round shapes, thin wall, quality decoration	500 and up;	5.00 and up;	0.50–1.00
Non-round shapes, thin wall, limited decoration	may be prohibitive	may be prohibitive	0.50–1.00

most thermoplastics is springback. Springback is the elastic recovery that takes place when the force of the forming or drawing tool is released. This relaxation is also characteristic of the commonly formed metals. This problem is compensated for by overforming and allowing the part to relax to its final shape. Springback is very prevalent in PVC forming because of a high elastic memory and because the yield strength is not surpassed in forming (19). The yield and ultimate strengths are almost equivalent.

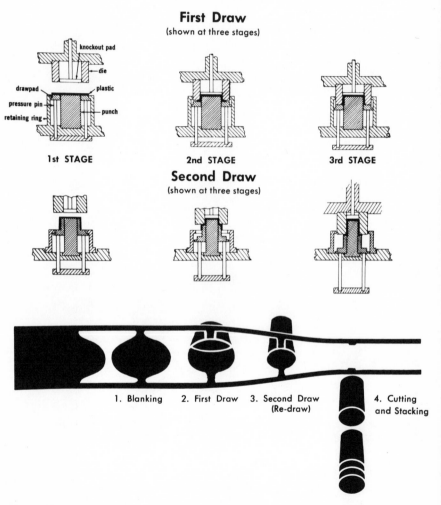

Figure 21   Representative cold drawing operation—web feeding. Courtesy Marbon Chemical.

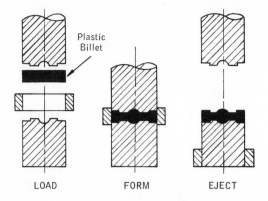

LOAD   FORM   EJECT

Figure 22   Forging process.

2. Forging

In this process a preheated plastic billet of predetermined weight and shape is loaded into a press. The opposing punches close and under fairly high pressure the billet is formed into the desired shape (Fig. 22). Billets are preheated to slightly under the melting point of the polymer; therefore, the process is sometimes referred to as "warm forging."

The main advantage of this process is that ultra high molecular weight materials (polyolefins) may be formed that cannot be formed by other processes. These materials offer superior impact properties compared to their lower molecular weight counterparts. Other advantages are the capability of forming thick walled parts and parts of varying cross section.

Equipment

The equipment required for forging polyolefins consists of a hydraulic press and some type of heating oven to heat billets at a reasonable rate. If precut billets are not available a die cutting press is needed. At temperatures near the melting point approximately 1 ton force/in.^{2} of finished part area is needed to form ultra high molecular weight polyethylene (MW > 1,000,000). Hydraulic presses are the most suitable for control of dwell time. Mechanical presses can be used but they don't often have suitable dwell time control.

Materials

Most forging done to date has been conducted with polyolefins. High molecular weight materials are best suited for the process and it has been found that polymers with poor melt strength don't keep their shape in billet form.

In one study (20) a number of materials were experimentally forged including acetal copolymers, polysulfone, and polyphenylene oxide (PPO). Rigid, noncrystalline polysulfones and PPO required billet temperatures above the second-order transition.

3. Rubber Pad Forming

In rubber pad forming, a solid block of rubber replaces one-half of the matched die set used in metal stamping (Fig. 23). The rubber acts as a hydraulic medium under pressure. Springback with rubber pad forming is much less than with conventional stamping operations, providing good dimensional stability of fabricated parts. When the press closes, the mold forces the sheet into the rubber pad. The pressure of the compressed rubber on the sheet results in good retention of mold detail because the pad forces the sheet into the mold recesses.

Parts made by pad forming are generally of large area and relatively shallow draw. The greatest advantage is in using sheet greater than 0.100 in. thick. For thinner sheet thermoforming may be a faster method. One important difference between rubber pad and thermoformed parts is the quality of corners. Thermoformed corners are thinned and can be susceptible to impact damage. For the same part weight, pad formed parts have thicker corners, which have higher impact resistance owing to the cold working.

4. Cold Heading

This is a high-speed forging method and requires the least revision to standard metal forming tools. Typical parts made by cold heading are rivets, nails, bolts, and similar shapes.

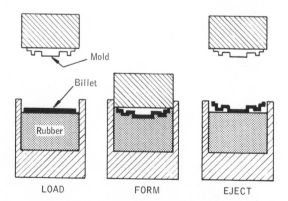

Figure 23   Rubber pad forming process.

In the process a continuous coil of material is used. The cold heading unit is a multiple station unit producing 50 to 600 parts/min. The coil is cut to length, hit on the head to form a preformed head, then transferred to a second cavity. In this cavity, the part is hit a second time to form the finished head.

Cold heading units can accept up to 1.5-in.-diameter rod, and thread rolling equipment is used for up to 6-in. pipe. The process is for very high volume production.

H. GLASS REINFORCEMENT

The most important reinforcing agent used currently with plastics is glass, in continuous roving or strand, chopped strand, milled glass, or hollow sphere forms.

Numerous other reinforcing agents are being investigated and used today, including the more exotic boron and graphite fibers and whiskers. Because of the scope of reinforced plastics processing, only the more important glass reinforcement techniques are summarized in this section. For further study reference is made to an excellent book on fiberglass and advanced plastics composites (21).

The field of reinforced plastics is approximately 25 years old, yet the growth rate with thermosets and thermoplastics has been spectacular. From virtually no consumption in 1940, well over a billion pounds per year of fiberglass reinforced plastics were used in the early 1970s.

1. Advantages of Glass Reinforcement

Various types of glass fiber are available and their properties are mainly a function of chemical structure and the method in which the fibers are drawn.

The main characteristics of glass fiber that makes it a very good reinforcement include the following:

1.   Very high tensile strength and modulus of elasticity to weight ratios.
2.   Excellent thermal properties coupled with a low coefficient of thermal expansion.
3.   Dimensional stability.
4.   High dielectric strengths and low dielectric constants.
5.   Relatively low cost compared to other reinforcing agents; especially high performance reinforcements.

The most important requirement of glass reinforced plastics is good adhesion between the fiber and the plastic matrix. If adhesion is weak, or weakened by

Table 5 Commercial Forms of Glass Fiber Reinforcements (22)

Nominal Form	General Description	Process	Nominal Glass Content of Typical Laminates (%)	Typical Application
Rovings	Continuous strands of glass fibers	Filament winding, continuous panel, preforming (matched die molding), spray-up pultrusion	25–80	Pipe, automobile bodies, rod stock, rocket motor cases, ordnance
Chopped strands	Strands cut to lengths of $\frac{1}{8}$–2 in.	Premix molding, wet slurry preforming	15–40	Electrical and appliance parts, ordnance components
Reinforcing mats	Continuous or chopped strands in random matting	Matched die molding, hand lay-up, centrifugal casting	20–45	Translucent sheets, truck and auto body panels
Surfacing and overlaying mats	Nonreinforcing random mat	Matched die molding, hand lay-up, and filament winding	5–15	Where smooth surfaces are required—automobile bodies, some housings
Yarns	Twisted strands	Weaving, filament winding	60–80	Aircraft, marine, electrical laminates
Woven fabrics	Woven cloths from glass fiber yarns	Hand lay-up, vacuum bag, autoclave, high-pressure laminating	45–65	Aircraft structures, marine ordnance hardware, electrical flat sheet and tubing
Woven roving	Woven glass fiber strands—coarser and heavier than fabrics		40–70	Marine, large containers
Nonwoven fabrics	Unidirectional and parallel rovings in sheet form	Hand lay-up, filament winding	60–80	Aircraft structures

environmental conditions, then the reinforcement properties of the glass will not be completely realized.

Adhesion is aided by application of coupling agents to the glass, which during processing form a physical and/or chemical bond with the plastic matrix. Coupling agents are applied as the glass fibers are formed (sized) or after the fibers are fabricated into a yarn or fabric (finished).

The mechanical properties of a glass reinforced plastic depend on the combined effect of the amount and arrangement of the glass in the matrix. In general, strength is directly proportional to the amount of glass used in the matrix. The highest strength and modulus are obtained when the fibers (or strands) are parallel to one another. In this type of structure properties are highest in the fiber direction. A random orientation of fibers gives lower properties that are approximately equal in all directions (pseudoisotropic).

Commercial forms of glass fiber reinforcement are described in Table 5 (22).

2.   Glass Reinforcement Processes

Numerous processes are available to produce glass reinforced parts. Design and economics usually dictate which process should be used. The more important processes are summarized in the following paragraphs.

Filament Winding

Commercial applications for filament wound composites include chemical storage tanks, pressure vessels, reinforced pipe, and various electrical devices (insulators). The industry trend is toward cheaper glasses, substituting polyester for more expensive epoxies and higher-speed, more economical winding equipment.

The process consists of winding continuous roving or strand over a rotating mandrel (Fig. 24a). The roving is fed from a creel and then passes through a resin bath (wet winding) or is preimpregnated with resin (dry winding) before being wound on the mandrel. Winding machines lay the glass in a predetermined pattern in one or more layers. After winding, the glass resin matrix is cured in an oven and then the mandrel is removed. Finishing operations are not usually necessary.

Hand Lay-Up

Hand lay-up or contact lay-up is normally used to mold room curing thermosetting polyesters and epoxies. In the first step a catalyzed layer of resin is brushed or sprayed onto a female or male mold surface. After this, a precut glass mat is laid onto the resin layer. Entrapped air is removed with rollers or brushes. This process is repeated until the desired thickness is

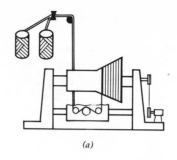

(a)

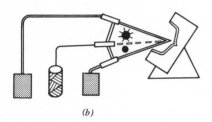

(b)

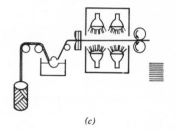

(c)

Figure 24   Glass reinforcement processes (21). (*a*) Filament winding. (*b*) Spray-up. (*c*) Continuous pultrusion.

built up. Curing normally is at room temperature but heat can be applied to accelerate this step. After curing, parts are removed from the mold and are trimmed.

Hand lay-ups are used for low volume production and where other fabrication methods would be too costly. Some typical lay-up applications are boat hulls, custom auto bodies, and radomes.

Variations of hand lay-up techniques use air pressure or vacuum to assist in removing air bubbles (minimize voids) and excess resin.

Spray-Up

The spray-up method is a mechanized version of the hand lay-up method (Fig. 24*b*). It is a more economical method in that glass roving is used instead of the more expensive woven glass (mat). Also, hand labor is reduced using spray-up methods compared to hand lay-up.

Basically, roving is fed through a chopper where it is cut into desired lengths. The chopped roving and resin–catalyst streams are then mixed in the head of a spray gun. The resultant mixture is sprayed over the surface of the mold until the desired thickness is built up. The resin–catalyst streams are combined in a single spray gun or from two guns with intersecting streams.

After the spraying operation the composite mixture is rolled to remove air and smooth the part surface. Curing takes place at room temperature or may be accelerated by heat.

Spray-up applications include shower stalls, bathtubs, boat hulls, truck bodies, furniture, and various types of containers and storage vessels.

Continuous Pultrusion

In this process continuous glass roving is drawn through a resin bath and then through a sizing ring or bushing. The sizing ring is used to remove excess resin and entrapped air and to size the final diameter of the composite. Curing is effected in an oven through which the composite is drawn (Fig. 24c).

Matched Die Molding

This is probably the highest volume method for producing glass reinforced plastics. In this process glass mat, fabric, or preform is placed with resin in the mold cavity. In the molding process matched metal dies form the part under pressure and heat. Cure cycles in the mold range from 1 to 5 min, depending on the resin used and part geometry.

Matched die molding is used for parts of critical contour and dimensions, as in aircraft and space applications. Mold cost is relatively high compared to other glass fabricating processes.

Continuous Laminating

In this process layers of glass fabric pass through a resin bath and are brought together between cellophane sheets. The lay-up passes through a heating zone to cure the resin. Resin content and composite thickness are controlled by squeeze rolls as the plies are brought together.

Injection Molding

Glass is introduced into the plastic via two methods:

1. Glass is compounded into the resin in a separate processing operation. The result is usually a pellet with short glass filaments throughout.

2. Chopped glass strand, milled glass, or glass beads are tumble blended with the plastic granules or pellets just prior to molding.

The injection molding process is similar to molding of unreinforced materials. Care should be taken to minimize shear on the material during molding. Excessive shear reduces the length of fiber in the final part, reducing the strength and modulus. Also excessive shear increases screw wear in the barrel. Parts produced by injection molding are somewhat anisotropic in properties owing to orientation of the glass as it passes through the gates. This can cause shrinkage gradients and part warpage. These effects can be largely eliminated by proper part design and location of gates to provide a random fiber arrangement in the molded part.

In some parts anisotropy might be desirable, for example, maximizing modulus in one direction for maximum bending strength.

I. CLOSING REMARKS

This chapter summarizes a number of important fabrication techniques for both thermoplastic and thermosetting resins. As mentioned earlier, other fabrication techniques exist today but they are of less importance. The future promises to bring more automated and scientific plastic processes into the limelight. Manufacturers are today making blow molding systems capable of producing millions of bottles per year, and injection molding machines capable of molding large appliance and industrial housings. Plastics processing will advance at a rate dictated by the combined knowledge and cooperation of material and machinery manufacturers, processors, designers, and end users of plastic parts.

GENERAL REFERENCES

1. T. H. Clifford, "Predicting Blow-Moldability of High-Density PE, Parts 1–3," *SPE J.*, **25**, (Oct., Nov., Dec. 1969).
2. *Polyethylene Blow Molding—An Operating Manual*, USI Chemical Company, 1963.
3. R. Paci, "Screw Transfer vs. Screw In-Line Thermoset Molding," *SPE J.*, **24**, 56 (March 1968).
4. L. J. Broutman, "Cold Forming of Plastics," *SPE J.*, **25**, 46 (Oct. 1969).
5. R. Roger, "Deep Drawing of ABS Plastic Sheet," *SPE 26th ANTEC Tech. Pap.*, **14**, 231 (May 1968).
6. G. Smoluk and M. Klaus, "Factors Affecting Cold Forming of Thermoplastics," *Mod. Plast.*, **45**, 240 (Nov. 1968).
7. "Metalworking for Plastics, Part I Cold Stamping ABS," *Mod. Plast.*, **45**, 124 (Apr. 1968).
8. "Metalworking for Plastics, Part 2 Warm Forging Thermoplastics," *Mod. Plast.*, **45**, 116 (May 1968).
9. *Modern Plastics Encyclopedia*, McGraw-Hill, New York, 1976–1977. Good introduction to various fabrication methods.

SPECIFIC REFERENCES

10. "The Production of Rotationally Moulded Hollow Articles by the Activated Anionic Polymerization of Lactams," translated from *Kunstoffe*, **58** (1), 6 (June 1968).
11. E. Jones, "Twin-Sheet Forming: A big Future Lies Ahead," *Plast. Technol.*, 47 (April 1969).

12. W. K. McConnel, "Thermoforming," in *Modern Plastics Encyclopedia*, McGraw-Hill, New York, 1969–1970, p. 534.

13. E. Vaill, "Thermoset Injection Molding," SPI-RP/CD Annual Meeting, Feb., 1968.

14. "Injection Molding—A New Day For Thermosets," *Mod. Plast.*, **44**, 96 (Dec. 1966).

15. J. Ferriday, "Injection Moulding of Thermosets," *Plastics*, **33**, 1010 (Sept. 1968).

16. H. R. Lusman, "Foaming Agents," in *Modern Plastics Encyclopedia*, McGraw-Hill, New York, 1969–1970, p. 262.

17. P. M. Coffman, "Forming of Plastics by Metal Working Techniques," *SPE 26th ANTEC Tech. Pap.*, **14**, 225 (May 1968).

18. Editor, "Cold Stamping and Warm Forging," in *Modern Plastics Encyclopedia*, McGraw-Hill, New York, 1969–1970, p. 618.

19. R. R. Kozlowski, "Cold Forming Rigid PVC," *SPE 26th ANTEC Tech. Pap.*, **14**, 236 (May 1968).

20. A. Werner and J. Krimm, "Forging of High Molecular Weight Polyethylene," *SPE J.*, **24**, 76 (Dec. 1968).

21. G. Lubin, (Ed.), *Handbook of Fiberglass and Advanced Plastics Composites*, Van Nostrand-Reinhold, New York, 1969.

22. *Ibid.*, p. 156.

DISCUSSION QUESTIONS AND PROBLEMS

1. Review the various steps in the blow molding process and indicate where orientation of the polymer can occur.

2. Thermoforming usually requires plastic sheets that have the property of "hot melt strength." Explain, and indicate what polymer characteristics might be associated with plastics exhibiting "hot melt strength."

3. Describe the advantages of transfer molding over compression molding of thermosets. Indicate how transfer molding compares with screw injection molding.

4. Compare the processes for making urethane and polystyrene foams.

5. Review the methods of cold forming and especially note the obvious similarity to metalworking processes.

6. Indicate how reinforcing agents modify polymer properties and review the various processes used.

SUBJECT INDEX